INTERNATIONAL UNION OF PURE AND APPLIED CHEMISTRY

ANALYTICAL CHEMISTRY DIVISION
COMMISSION ON SOLUBILITY DATA

SOLUBILITY DATA SERIES

Volume 33

MOLTEN ALKALI METAL ALKANOATES

A complete list of volumes published in the Solubility Data Series will be found on p. 348.

SOLUBILITY DATA SERIES

Editor-in-Chief
A.S. KERTES

Volume 33

MOLTEN ALKALI METAL ALKANOATES

Volume Editor

PAOLO FRANZOSINI *(deceased)*

University of Pavia, Italy

Co-editors

PAOLO FERLONI   ALBERTO SCHIRALDI

GIORGIO SPINOLO

University of Pavia, Italy

Contributors

PRIMO BALDINI   GIULIO D'ANDREA

University of Pavia, Italy

PERGAMON PRESS

OXFORD · NEW YORK · BEIJING · FRANKFURT
SÃO PAULO · SYNDEY · TOKYO · TORONTO

CHEM
sep/ae

U.K.	Pergamon Press plc, Headington Hill Hall, Oxford OX3 0BW, England
U.S.A.	Pergamon Press, Inc., Maxwell House, Fairview Park, Elmsford, New York 10523, U.S.A.
PEOPLE'S REPUBLIC OF CHINA	Pergamon Press, Room 4037, Qianmen Hotel, Beijing, People's Republic of China
FEDERAL REPUBLIC OF GERMANY	Pergamon Press GmbH, Hammerweg 6, D-6242 Kronberg, Federal Republic of Germany
BRAZIL	Pergamon Editora Ltda, Rua Eça de Queiros, 346, CEP 04011, Paraiso, São Paulo, Brazil
AUSTRALIA	Pergamon Press Australia Pty Ltd., P.O. Box 544, Potts Point, N.S.W. 2011, Australia
JAPAN	Pergamon Press, 5th Floor, Matsuoka Central Building, 1-7-1 Nishishinjuku, Shinjuku-ku, Tokyo 160, Japan
CANADA	Pergamon Press Canada Ltd., Suite No. 271, 253 College Street, Toronto, Ontario, Canada M5T 1R5

First edition 1988

The Library of Congress has catalogued this serial title as follows:

Solubility data series.—Vol. 1—Oxford; New York:
Pergamon, c 1979-
v.; 28 cm.
Separately cataloged and classified in LC before no. 18.
ISSN 0191-5622 = Solubility data series.
1. Solubility—Tables—Collected works.
QD543.S6629 541.3′42′05-dc19 85-641351
AACR 2 MARC-S

British Library Cataloguing in Publication Data

Molten alkali metal alkanoates.
1. Molten alkali metal alkanoates.
Solubility
I. Franzosini, Paolo II. Ferloni, Paolo
III. Schiraldi, Alberto IV. Spinolo,
Giorgio V. Baldini, Primo IV. D'Andrea,
Giulio VII. Series
546′.38

ISBN 0-08-032522-X

Printed in Great Britain by A. Wheaton & Co. Ltd., Exeter

CONTENTS

IN MEMORIAM

Dr. Paolo Franzosini, Professor of Physical Chemistry, University of Pavia, Italy, and the Editor of this volume, passed away on January 24, 1986 at the age of 56.

Born in Trecate, near Turin, he received his university education while on a prestigious scholarship at the Collegio Ghislieri of the University of Pavia, graduating with honors in 1952. He then spent two years as a post-doctoral research assistant with Professor Clusius in the Department of Physical Chemistry at the Federal Polytechnic Institute of Zurich. In the years 1955-1960 he served as Assistant Professor at the University of Pavia, and had been promoted to the rank of Associate Professor at the early age of thirty. In 1965 he moved to the University of Camerino as Professor and Chair of Physical Chemistry. In 1968 he was back at his alma mater as Professor of Electrochemistry, and in 1980 as the Chair of Physical Chemistry. In 1979 he served as Visiting Professor at the University of Mogadiscio, Somaliland, and in 1981 at the University of Michigan in Ann Arbor.

In the thirty years of scientific activity, he published over 100 papers, mostly in the field of thermodynamics and thermochemistry of molten salt systems. He made significant contribution to the Atlas of Miscibility Gaps in 1968, and to the Atlas of Phase Diagrams in 1973 on molten salts with organic anions. In 1980 he coedited, with Professor Sanesi, Volume 28 of IUPAC's Chemical Data Series on Thermodynamics and Transport Properties of Organic Salts. In the years 1978-1984 he was member of IUPAC Commission on Thermodynamics. In 1982 he joined the Solubility Data Project and served as the National Representative of Italy to IUPAC's Commission on Solubility Data. At that time he was appointed Editor of the present volume.

He has not seen the completion of this volume. Thanks to the devotion and expertise of his colleagues, Professors Ferloni, Schiraldi and Spinolo of the Department of Physical Chemistry, University of Pavia, this volume, the final major work of Professor Franzosini, was brought to completion.

Paolo was an interesting intellect with a broad European humanitarian culture. He was a Rotarian from an early age, president of the Rotary Club of Pavia, and served in the distinguished position of Rotary District Governor in 1982-3.

Through the generosity of the Franzosini family, it has been possible for the Solubility Data Commission to establish the Paolo Franzosini Endowment Fund, the proceeds of which to assist young scientists in their association with IUPAC's Solubility Data Project.

We are all diminished by his passing.

A.S. Kertes
Editor-in-Chief

Jerusalem, March 1988

FOREWORD

If the knowledge is
undigested or simply wrong,
more is not better

How to communicate and disseminate numerical data effectively in chemical science and technology has been a problem of serious and growing concern to IUPAC, the International Union of Pure and Applied Chemistry, for the last two decades. The steadily expanding volume of numerical information, the formulation of new interdisciplinary areas in which chemistry is a partner, and the links between these and existing traditional subdisciplines in chemistry, along with an increasing number of users, have been considered as urgent aspects of the information problem in general, and of the numerical data problem in particular.

Among the several numerical data projects initiated and operated by various IUPAC commissions, the *Solubility Data Project* is probably one of the most ambitious ones. It is concerned with preparing a comprehensive critical compilation of data on solubilities in all physical systems, of gases, liquids and solids. Both the basic and applied branches of almost all scientific disciplines require a knowledge of solubilities as a function of solvent, temperature and pressure. Solubility data are basic to the fundamental understanding of processes relevant to agronomy, biology, chemistry, geology and oceanography, medicine and pharmacology, and metallurgy and materials science. Knowledge of solubility is very frequently of great importance to such diverse practical applications as drug dosage and drug solubility in biological fluids, anesthesiology, corrosion by dissolution of metals, properties of glasses, ceramics, concretes and coatings, phase relations in the formation of minerals and alloys, the deposits of minerals and radioactive fission products from ocean waters, the composition of ground waters, and the requirements of oxygen and other gases in life support systems.

The widespread relevance of solubility data to many branches and disciplines of science, medicine, technology and engineering, and the difficulty of recovering solubility data from the literature, lead to the proliferation of published data in an ever increasing number of scientific and technical primary sources. The sheer volume of data has overcome the capacity of the classical secondary and tertiary services to respond effectively.

While the proportion of secondary services of the review article type is generally increasing due to the rapid growth of all forms of primary literature, the review articles become more limited in scope, more specialized. The disturbing phenomenon is that in some disciplines, certainly in chemistry, authors are reluctant to treat even those limited-in-scope reviews exhaustively. There is a trend to preselect the literature, sometimes under the pretext of reducing it to manageable size. The crucial problem with such preselection - as far as numerical data are concerned - is that there is no indication as to whether the material was excluded by design or by a less than thorough literature search. We are equally concerned that most current secondary sources, critical in character as they may be, give scant attention to numerical data.

On the other hand, tertiary sources - handbooks, reference books and other tabulated and graphical compilations - as they exist today are comprehensive but, as a rule, uncritical. They usually attempt to cover whole disciplines, and thus obviously are superficial in treatment. Since they command a wide market, we believe that their service to the advancement of science is at least questionable. Additionally, the change which is taking place in the generation of new and diversified numerical data, and the rate at which this is done, is not reflected in an increased third-level service. The emergence of new tertiary literature sources does not parallel the shift that has occurred in the primary literature.

With the status of current secondary and tertiary services being as briefly stated above, the innovative approach of the *Solubility Data Project* is that its compilation and critical evaluation work involve consolidation and reprocessing services when both activities are based on intellectual and scholarly reworking of information from primary sources. It comprises compact compilation, rationalization and simplification, and the fitting of isolated numerical data into a critically evaluated general framework.

The *Solubility Data Project* has developed a mechanism which involves a number of innovations in exploiting the literature fully, and which contains new elements of a more imaginative approach for transfer of reliable information from primary to secondary/tertiary sources. *The fundamental trend of the Solubility Data Project is toward integration of secondary and tertiary services with the objective of producing in-depth critical analysis and evaluation which are characteristic to secondary services, in a scope as broad as conventional tertiary services.*

Fundamental to the philosophy of the project is the recognition that the basic element of strength is the active participation of career scientists in it. Consolidating primary data, producing a truly critically-evaluated set of numerical data, and synthesizing data in a meaningful relationship are demands considered worthy of the efforts of top scientists. Career scientists, who themselves contribute to science by their involvement in active scientific research, are the backbone of the project. The scholarly work is commissioned to recognized authorities, involving a process of careful selection in the best tradition of IUPAC. This selection in turn is the key to the quality of the output. These top experts are expected to view their specific topics dispassionately, paying equal attention to their own contributions and to those of their peers. They digest literature data into a coherent story by weeding out what is wrong from what is believed to be right. To fulfill this task, the evaluator must cover *all* relevant open literature. No reference is excluded by design and every effort is made to detect every bit of relevant primary source. Poor quality or wrong data are mentioned and explicitly disqualified as such. In fact, it is only when the reliable data are presented alongside the unreliable data that proper justice can be done. The user is bound to have incomparably more confidence in a succinct evaluative commentary and a comprehensive review with a complete bibliography to both good and poor data.

It is the standard practice that the treatment of any given solute-solvent system consists of two essential parts: I. Critical Evaluation and Recommended Values, and II. Compiled Data Sheets.

The Critical Evaluation part gives the following information:

(i) a verbal text of evaluation which discusses the numerical solubility information appearing in the primary sources located in the literature. The evaluation text concerns primarily the quality of data after consideration of the purity of the materials and their characterization, the experimental method employed and the uncertainties in control of physical parameters, the reproducibility of the data, the agreement of the worker's results on accepted test systems with standard values, and finally, the fitting of data, with suitable statistical tests, to mathematical functions;

(ii) a set of recommended numerical data. Whenever possible, the set of recommended data includes weighted average and standard deviations, and a set of smoothing equations derived from the experimental data endorsed by the evaluator;

(iii) a graphical plot of recommended data.

The Compilation part consists of data sheets of the best experimental data in the primary literature. Generally speaking, such independent data sheets are given only to the best and endorsed data covering the known range of experimental parameters. Data sheets based on primary sources where the data are of a lower precision are given only when no better data are available. Experimental data with a precision poorer than considered acceptable are reproduced in the form of data sheets when they are the only known data for a particular system. Such data are considered to be still suitable for some applications, and their presence in the compilation should alert researchers to areas that need more work.

The typical data sheet carries the following information:

(i) components - definition of the system - their names, formulas and Chemical Abstracts registry numbers;
(ii) reference to the primary source where the numerical information is reported. In cases when the primary source is a less common periodical or a report document, published though of limited availability, abstract references are also given;
(iii) experimental variables;
(iv) identification of the compiler;
(v) experimental values as they appear in the primary source. Whenever available, the data may be given both in tabular and graphical form. If auxiliary information is available, the experimental data are converted also to SI units by the compiler.

Under the general heading of Auxiliary Information, the essential experimental details are summarized:

(vi) experimental method used for the generation of data;
(vii) type of apparatus and procedure employed;
(viii) source and purity of materials;
(ix) estimated error;
(x) references relevant to the generation of experimental data as cited in the primary source.

This new approach to numerical data presentation, formulated at the initiation of the project and perfected as experience has accumulated, has been strongly influenced by the diversity of background of those whom we are supposed to serve. We thus deemed it right to preface the evaluation/compilation sheets in each volume with a detailed discussion of the principles of the accurate determination of relevant solubility data and related thermodynamic information.

Finally, the role of education is more than corollary to the efforts we are seeking. The scientific standards advocated here are necessary to strengthen science and technology, and should be regarded as a major effort in the training and formation of the next generation of scientists and engineers. Specifically, we believe that there is going to be an impact of our project on scientific-communication practices. The quality of consolidation adopted by this program offers down-to-earth guidelines, concrete examples which are bound to make primary publication services more responsive than ever before to the needs of users. The self-regulatory message to scientists of the early 1970s to refrain from unnecessary publication has not achieved much. A good fraction of the literature is still cluttered with poor-quality articles. The Weinberg report (in 'Reader in Science Information', ed. J. Sherrod and A. Hodina, Microcard Editions Books, Indian Head, Inc., 1973, p. 292) states that 'admonition to authors to restrain themselves from premature, unnecessary publication can have little effect unless the climate of the entire technical and scholarly community encourages restraint...' We think that projects of this kind translate the climate into operational terms by exerting pressure on authors to avoid submitting low-grade material. The type of our output, we hope, will encourage attention to quality as authors will increasingly realize that their work will not be suited for permanent retrievability unless it meets the standards adopted in this project. It should help to dispel confusion in the minds of many authors of what represents a permanently useful bit of information of an archival value, and what does not.

If we succeed in that aim, even partially, we have then done our share in protecting the scientific community from unwanted and irrelevant, wrong numerical information.

A. S. Kertes

PREFACE

1.Phase relationships of alkali alkanoates.

1.1. Solid state transitions of alkali alkanoates.

Most alkali alkanoates (either linear or branched) exhibit polymorphism in the solid state, and the number of phases tends to increase with increasing chain length. However, controversy often exists about the number, nature, and stability range of the polymorphs present in a given salt. Since different hydrates appear as well, "in the literature one may find almost the whole Greek alphabet, primed and unprimed, each notation supposed to define a separate phase. It has been maintained that each such phase is associated with a unique crystal structure, while others have claimed that the different X-ray diffraction patterns do not necessarily represent true crystal structures, but instead are merely associated with different types of disorder of the chains. There is also the additional problem of the descendant phases, i.e., structures that occur at some elevated temperature and remain unaltered at room temperature in a pseudoequilibrium for a long time" (Ref. 1). The present volume discusses solubilities of those linear alkali alkanoates marked with a cross in the scheme below, and, as well, the iso-butanoates and iso-pentanoates of Na and K. No information is available so far on the solubilities of other alkali alkanoates.

Cation	Number of carbon atoms, n_C, 1	2	3	4	5	6	7	8	9	18
Li	x	x	x	x						
Na	x	x	x	x	x	x				x
K	x	x	x	x	x	x	x	x	x	
Rb		x								
Cs		x								

In order to obtain a homogeneous picture of the thermal behavior (in terms of phase transformation temperatures and enthalpy changes) of alkali alkanoates, more than 100 linear and branched homologues belonging to the different alkali families have been submitted to DSC analysis during the last few years in the editor's laboratory. The results obtained on heating are thought to offer an acceptable degree of trustworthiness and internal consistency. Therefore, as useful background material for discussion of the solubility curves, the pertinent superambient solid state transition temperature, T_{trs}, are collected in Tables 1 and 2. These temperatures represent first order, or predominantly first order, phase transitions. For completeness, the temperatures of fusion, T_{fus}, and of clearing, T_{clr}, (when they exist) are also listed in Tables 1 and 2. However, the data on sodium octadecanoate (produced in a different laboratory) are listed separately in Table 4.

The following remarks can be made about the precision and accuracy of the data reported in Tables 1, 2.

Precision is not infrequently better than ± 1 K, although becoming poorer in some cases: in particular, very poor reproducibility was obtained for solid state transitions of sodium methanoate and ethanoate.

Accuracy is thought to be often of the same order of magnitude as precision. However, one must consider that DSC is a dynamic method of investigation, and that some solid state transitions of methanoates and ethanoates are characterized (even on heating at a moderate scanning rate) by a remarkable sluggishness. Consequently, in Tables 1, 2 the T_{trs} data for the shortest homologues can be somewhat too high. This disadvantage tends to decrease in the next higher homologues, and does not involve fusion and clearing. As an example, high accuracy equilibrium adiabatic calorimetric data, taken very recently by Franzosini et al. (Refs. 8, 9) on sodium methanoate and propanoate, are compared in Table 3 with the previous DSC values. The comparison also includes the enthalpy changes involved.

For sodium octadecanoate, reference will be made to the recent DSC data by Forster et al. (Ref. 10), collected in Table 4.

Table 1 - T_{clr}, T_{fus}, and superambient T_{trs} values determined by DSC for 21 linear alkali alkanoates.

n_C	Cation	T_{clr}/K	T_{fus}/K	T'_{trs}/K	T''_{trs}/K	T'''_{trs}/K	T^{IV}_{trs}/K	Ref.
1	Li	-	546±1	496±2	-	-	-	2
	Na	-	530.7±0.5	502±5	-	-	-	3
	K	-	441.9±0.5	418±1	-	-	-	3
2	Li	-	557±2	-	-	-	-	2
	Na	-	601.3±0.5	527±15	465±3	414±10	-	2
	K	-	578.7±0.5	422.2±0.5	-	-	-	2
	Rb	-	514±1	498±1	-	-	-	2
	Cs	-	463±1	-	-	-	-	2
3	Li	-	606.8±0.5*	533±2	-	-	-	2
	Na	-	562.4±0.5	494±1	470.2±0.5	-	-	2
	K	-	638.3±0.5	352.5±0.5	-	-	-	2
4	Li	-	591.7±0.5	-	-	-	-	2
	Na	600.4±0.2	524.5±0.5	508.4±0.5	498.3±0.3	489.8±0.2	450.4±0.5	4
	K	677.3±0.5	626.1±0.7	562.2±0.6	540.8±1.1	467.2±0.5	461.4±1.0	4
5	Na	631±4	498±2	-	-	-	-	5
	K	716±2	586.6±0.7	399.5±0.9	-	-	-	5
6	Na	639.0±0.5	499.6±0.6	473±2	386±2	-	-	5
	K	725.8±0.8	581.7±0.5	-	-	-	-	5
7	K	722±3	571.3±0.9	345.4±0.6	332.0±0.8	-	-	5
8	K	712±2	560.6±0.8	326.6±0.1	-	-	-	6
9	K	707.4±0.8	549.1±0.8	390.5±0.4	367.5±0.5	-	-	6

clr: clearing; fus: fusion; trs: transition

* A metastable fusion point was also detected at $T_{fus(m)}/K$ = 584±1

Table 2 - T_{clr}, T_{fus}, and superambient T_{trs} values determined by DSC for 4 alkali iso-alkanoates.

n_C	Cation	T_{clr}/K	T_{fus}/K	T'_{trs}/K	Ref.
4	Na	-	526.9±0.7	-	7
	K	625.6±0.8	553.9±0.5	424±3	7
5	Na	559±1	461.5±0.6	-	7
	K	679±2	531±3	-	7

clr: clearing; fus: fusion; trs: transition

Table 3 - Comparison between adiabatic calorimetric and DSC data.

n_C	Cation	Quantity	Value	Method	Ref
1	Na	T_{fus}/K	530.46 ± 0.04	ad.cal.	8
			530.7 ± 0.5	DSC	3
		T_{trs}/K	491.5 ± 1	ad.cal.	8
			502 ± 5	DSC	3
		$(\Delta_{fus}H_m/R)/K$	2130 (*)	ad.cal.	8
			$(2.06 \pm 0.05)\ 10^3$	DSC	3
		$(\Delta_{trs}H_m/R)/K$	150 (*)	ad.cal.	8
			$(1.41 \pm 0.05)\ 10^2$	DSC	3
3	Na	T_{fus}/K	561.88 ± 0.03	ad.cal.	9
			562.4 ± 0.5	DSC	2
		T'_{trs}/K	491 ± 1	ad.cal.	9
			494 ± 1	DSC	2
		T''_{trs}/K	467 ± 1	ad.cal.	9
			470.2 ± 0.5	DSC	2
		$(\Delta_{fus}H_m/R)/K$	1597.3 ± 0.6	ad.cal.	9
			$(1.61 \pm 0.05)\ 10^3$	DSC	2
		$(\Delta_{trs}H_m/R)/K$(**)	$(0.91 \pm 0.02)\ 10^3$	ad.cal.	9
			$(0.89 \pm 0.05)\ 10^3$	DSC	2

fus: fusion; trs: transition
(*) Single determination.
(**) Cumulative enthalpy change relevant to both solid state transitions.

Table 4 - T_{clr}, T_{fus}, and superambient T_{trs} values determined by DSC for sodium octadecanoate.

T_{clr}/K	T'_{trs}/K	T_{fus}/K	T''_{trs}/K	T'''_{trs}/K	T^{IV}_{trs}/K	T^{V}_{trs}/K	T^{VI}_{trs}/K
552.7	547.7	527.2	469/476	448	408	390	368

L	NI	NII	SN	SpW	W	SW	CI	CII

clr: clearing; fus: fusion; trs: transition
L: isotropic liquid; N: neat; SN: subneat; SpW: superwaxy; W: waxy; SW: subwaxy; C: crystal.

1.2. Mesomorphism in alkali alkanoates.

Mesomorphic phases (liquid crystalline, or plastic crystalline, or both) can also form in alkali alkanoates, the stability range of the mesomorphic state being intermediate between those of the "true" crystalline and of the "true" liquid phases. In particular, liquid crystals (likely of the smectic type) form in linear alkanoates, starting with butanoate when the cation is either sodium or potassium, from pentanoate when the cation is rubidium, and from hexanoate when the cation is caesium. No liquid crystals form when the cation is lithium. In long chain homologues (which, however, are of little relevance to the present purposes) plastic crystals form for all alkali cations. (See, e.g., the data reported in Table 4 for sodium octadecanoate.)

Unfortunately, the nomenclature employed by different authors is far from homogeneous. In particular, most Russian investigators call "fusion" the transformation of either a crystalline solid or a liquid crystal into an isotropic liquid which is often misleading in the interpretation of phase diagrams. More reasonably, in non - Russian literature a distinction is usually made between clearing temperature, T_{clr} (i.e., the temperature at which a liquid crystal transforms into an isotropic liquid), and fusion temperature, T_{fus} (i.e., the temperature at which a "true" crystal transforms into either an isotropic liquid or a liquid crystal)(*). Further details on these points are given in Section 2.2.

(*) It might be further considered whether in the sequence (met in several long chain alkali alkanoates): crystal --> plastic crystal(s) --> liquid crystal(s) --> isotropic liquid the term "fusion" should be applied to the first or to the second transformation, but such a discussion would be of little relevance here. In Table 4 the term fusion was applied to the transformation from the (plastic crystalline) subneat to the (liquid crystalline) neat I phase.

The literature contains reports on the phase diagrams of 58 binaries (9 with common anion, and 49 with common cation), involving alkali alkanoates which exhibit mesomorphism: they are listed in Table 5 where the component(s) which can exist in the mesomorphic liquid state are underlined.

Unfortunately, information is mostly restricted to the lower boundary of the isotropic liquid field. More details are available only in a limited number of cases among which special interest is to be attached to the following: a) $(C_4H_7O_2)K$ + $(C_4H_7O_2)Na$, b) $(C_4H_7O_2)Li$ + $(C_4H_7O_2)Na$, c) $KC_2H_3O_2$ + $KC_4H_7O_2$, d) $NaC_2H_3O_2$ + $NaC_4H_7O_2$, and e) $NaC_4H_7O_2$ + $NaNO_3$, inasmuch as a comparison is here possible between the results obtained by Prisyazhnyi et al. (Refs. 11, 12), who studied the lower boundaries of both the isotropic liquid and the liquid crystal fields, and those obtained by previous authors, who studied only the lower boundary of the isotropic liquid.

Table 5 - Binaries involving at least one alkali alkanoate able to exist in the liquid crystalline state.

Systems with common anion:

1)	~$(C_4H_7O_2)_2K_2$	$(C_4H_7O_2)_2Mg$	2)	~$(C_4H_7O_2)K$	~$(C_4H_7O_2)Na$
3)	~$(i.C_4H_7O_2)K$	$(i.C_4H_7O_2)Na$	4)	$(C_4H_7O_2)Li$	~$(C_4H_7O_2)Na$
5)	$(C_4H_7O_2)_2Mg$	~$(C_4H_7O_2)_2Na_2$	6)	~$(C_5H_9O_2)K$	~$(C_5H_9O_2)Na$
7)	~$(i.C_5H_9O_2)K$	~$(i.C_5H_9O_2)Na$	8)	$(C_5H_9O_2)_2Mg$	~$(C_5H_9O_2)_2Na_2$
9)	~$(C_6H_{11}O_2)K$	~$(C_6H_{11}O_2)Na$			

Systems with common cation:

10)	$KC_2H_3O_2$	~$KC_4H_7O_2$	11)	$KC_2H_3O_2$	~$Ki.C_4H_7O_2$
12)	$KC_2H_3O_2$	~$KC_5H_9O_2$	13)	$KC_2H_3O_2$	~$Ki.C_5H_9O_2$
14)	$KC_2H_3O_2$	~$KC_6H_{11}O_2$	15)	~$KC_4H_7O_2$	KCNS
16)	~$KC_4H_7O_2$	KNO_2	17)	~$KC_4H_7O_2$	KNO_3
18)	~$Ki.C_4H_7O_2$	KNO_2	19)	~$Ki.C_4H_7O_2$	KNO_3
20)	~$KC_5H_9O_2$	KNO_2	21)	~$KC_5H_9O_2$	KNO_3
22)	~$Ki.C_5H_9O_2$	KNO_2	23)	~$Ki.C_5H_9O_2$	KNO_3
24)	~$KC_6H_{11}O_2$	KNO_2	25)	~$KC_7H_{13}O_2$	KNO_2
26)	~$KC_8H_{15}O_2$	KNO_2	27)	~$KC_9H_{17}O_2$	KNO_2
28)	$NaCHO_2$	~$NaC_4H_7O_2$	29)	$NaCHO_2$	~$Nai.C_5H_9O_2$
30)	$NaC_2H_3O_2$	~$NaC_4H_7O_2$	31)	$NaC_2H_3O_2$	~$NaC_5H_9O_2$
32)	$NaC_2H_3O_2$	~$Nai.C_5H_9O_2$	33)	$NaC_2H_3O_2$	~$NaC_6H_{11}O_2$
34)	~$NaC_4H_7O_2$	$Nai.C_4H_7O_2$	35)	~$NaC_4H_7O_2$	~$Nai.C_5H_9O_2$
36)	~$NaC_4H_7O_2$	~$NaC_6H_{11}O_2$	37)	~$NaC_4H_7O_2$	$NaC_7H_5O_2$
38)	~$NaC_4H_7O_2$	~$NaC_{18}H_{35}O_2$	39)	~$NaC_4H_7O_2$	NaCNS
40)	~$NaC_4H_7O_2$	$NaNO_2$	41)	~$NaC_4H_7O_2$	$NaNO_3$
42)	$Nai.C_4H_7O_2$	~$Nai.C_5H_9O_2$	43)	$Nai.C_4H_7O_2$	~$NaC_6H_{11}O_2$
44)	$Nai.C_4H_7O_2$	~$NaC_{18}H_{35}O_2$	45)	~$NaC_5H_9O_2$	NaCNS
46)	~$NaC_5H_9O_2$	$NaNO_2$	47)	~$NaC_5H_9O_2$	$NaNO_3$
48)	~$Nai.C_5H_9O_2$	~$NaC_6H_{11}O_2$	49)	~$Nai.C_5H_9O_2$	$NaC_7H_5O_2$
50)	~$Nai.C_5H_9O_2$	~$NaC_{18}H_{35}O_2$	51)	~$Nai.C_5H_9O_2$	NaCNS
52)	~$Nai.C_5H_9O_2$	$NaNO_2$	53)	~$Nai.C_5H_9O_2$	$NaNO_3$
54)	~$NaC_6H_{11}O_2$	$NaC_7H_5O_2$	55)	~$NaC_6H_{11}O_2$	~$NaC_{18}H_{35}O_2$
56)	~$NaC_6H_{11}O_2$	NaCNS	57)	~$NaC_6H_{11}O_2$	$NaNO_3$
58)	$NaC_7H_5O_2$	~$NaC_{18}H_{35}O_2$			

~ Compounds which form liquid crystals.

In order to improve homogeneity and succinctness in discussing the systems in Table 5, it seemed convenient to present here a selection of model phase diagrams [see Schemes A, ..., D (*)] to which reference will be made in the subsequent critical evaluations.

(*) Schemes A, ..., D were drawn in the - usually accepted - assumption that any (actually known) transformation

mesomorphic phase <==> isotropic liquid

is first order.

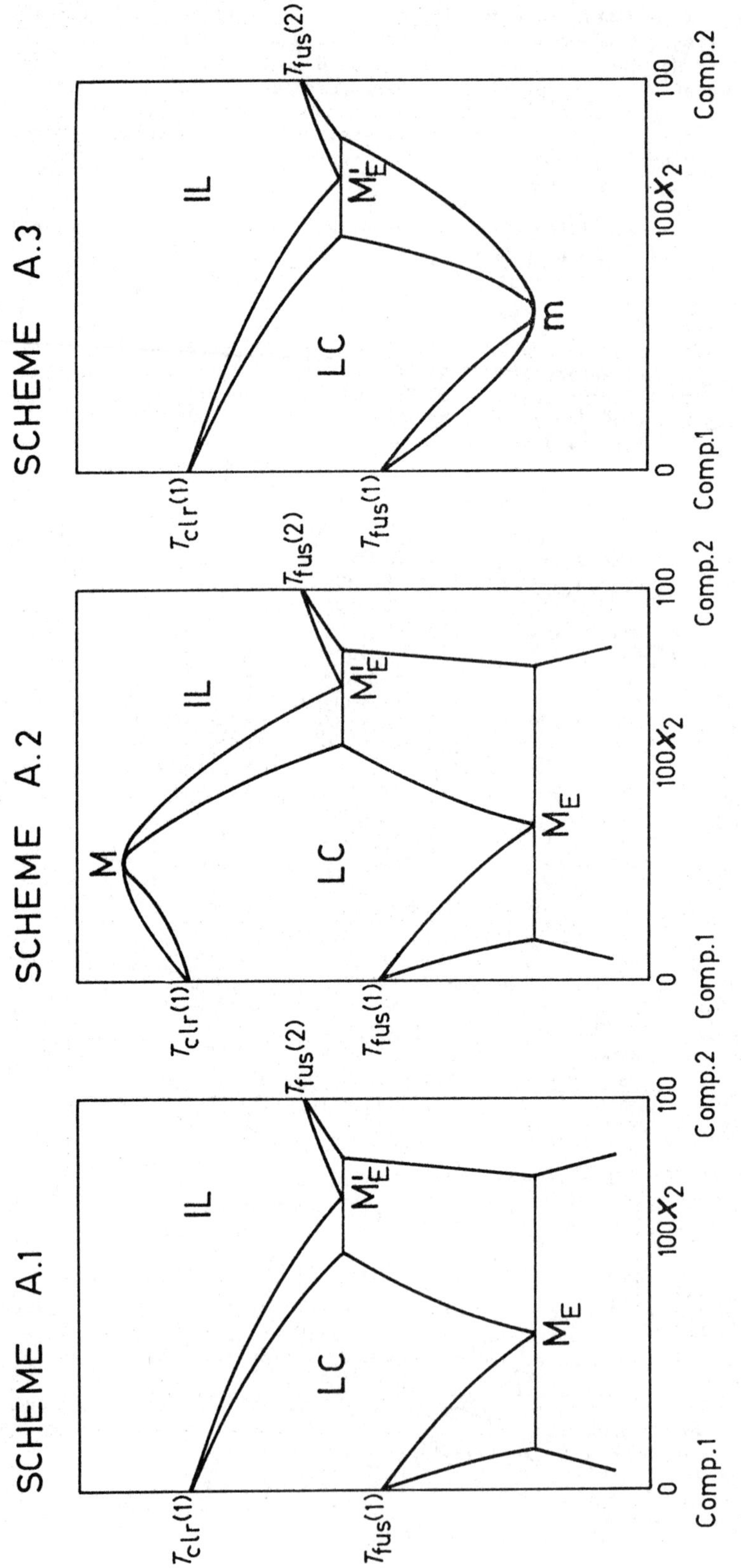
SCHEME A.1
IL
LC
M'E
ME
Tclr(1)
Tfus(1)
Tfus(2)
0
100
Comp.1
100x2
Comp.2
SCHEME A.2
M
IL
LC
M'E
ME
Tclr(1)
Tfus(1)
Tfus(2)
0
100
Comp.1
100x2
Comp.2
SCHEME A.3
IL
LC
M'E
m
Tclr(1)
Tfus(1)
Tfus(2)
0
100
Comp.1
100x2
Comp.2

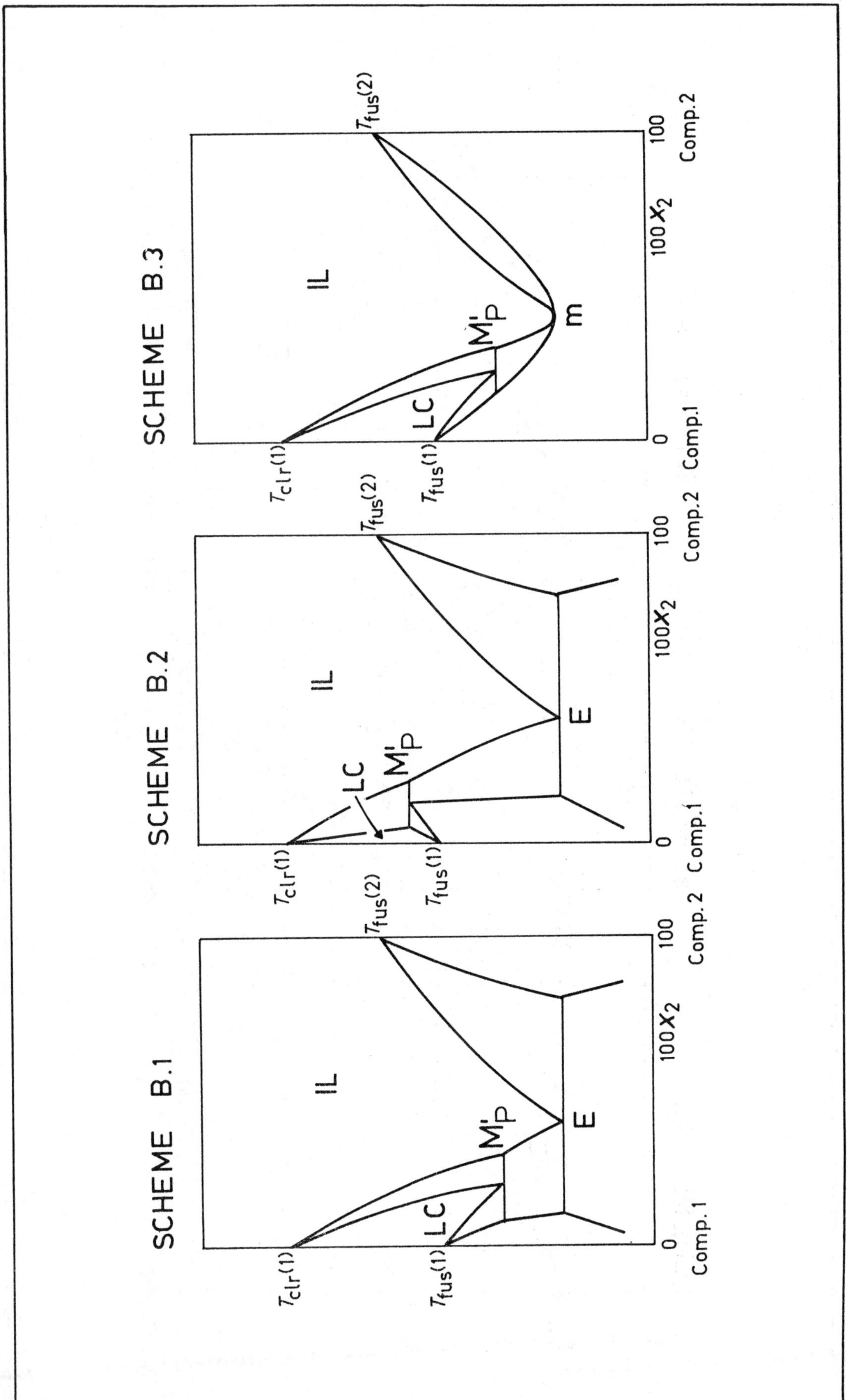
SCHEME B.1
SCHEME B.2
SCHEME B.3
IL
LC
M'P
E
m
$T_{clr}(1)$
$T_{fus}(1)$
$T_{fus}(2)$
0
100
$100x_2$
Comp.1
Comp.2

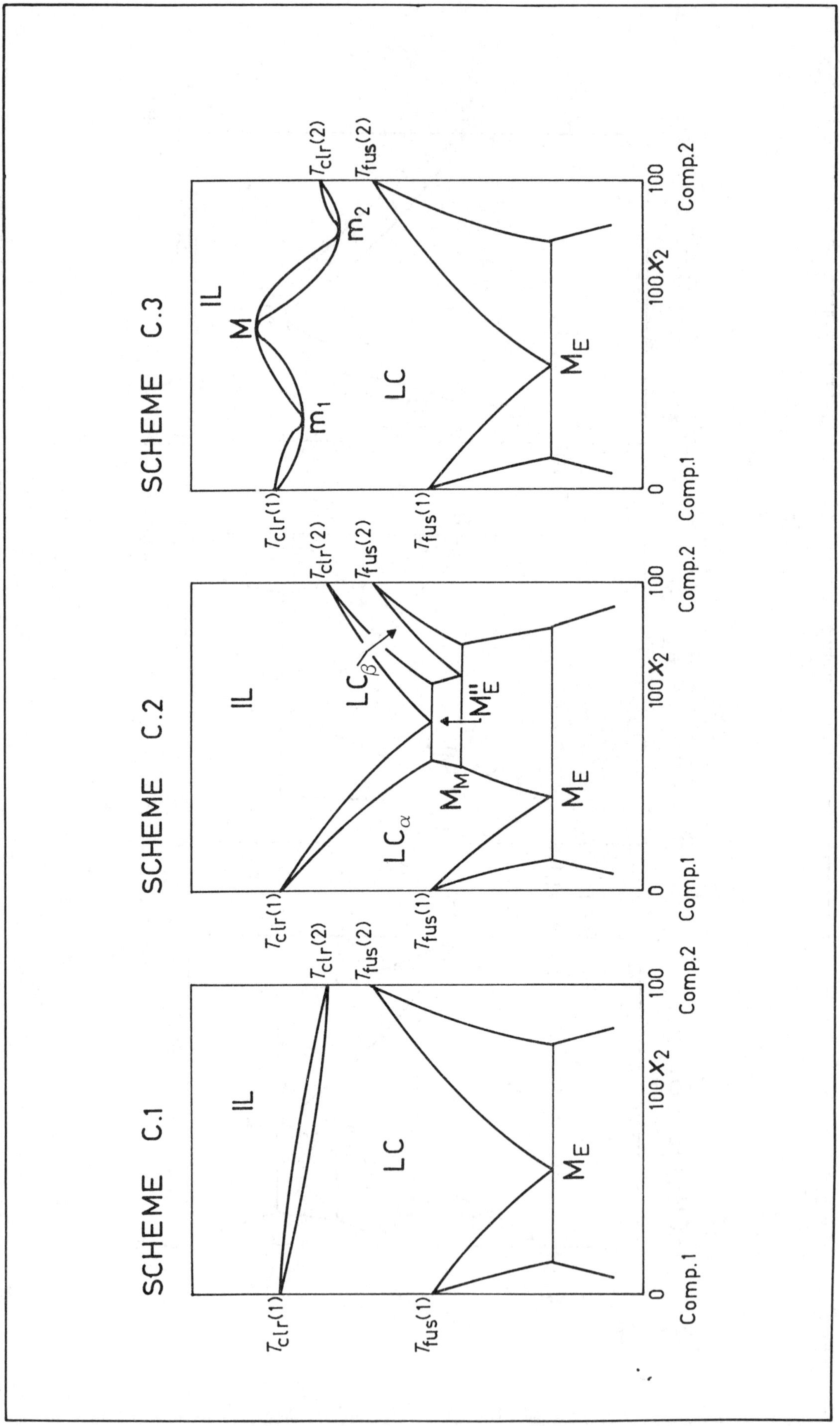

SCHEME C.1
IL
LC
M_E
$T_{clr}(1)$
$T_{fus}(1)$
$T_{clr}(2)$
$T_{fus}(2)$
0 Comp.1
100 Comp.2
$100x_2$
SCHEME C.2
IL
LC_β
LC_α
M_M
M''_E
M_E
SCHEME C.3
IL
M
m_1
m_2
LC

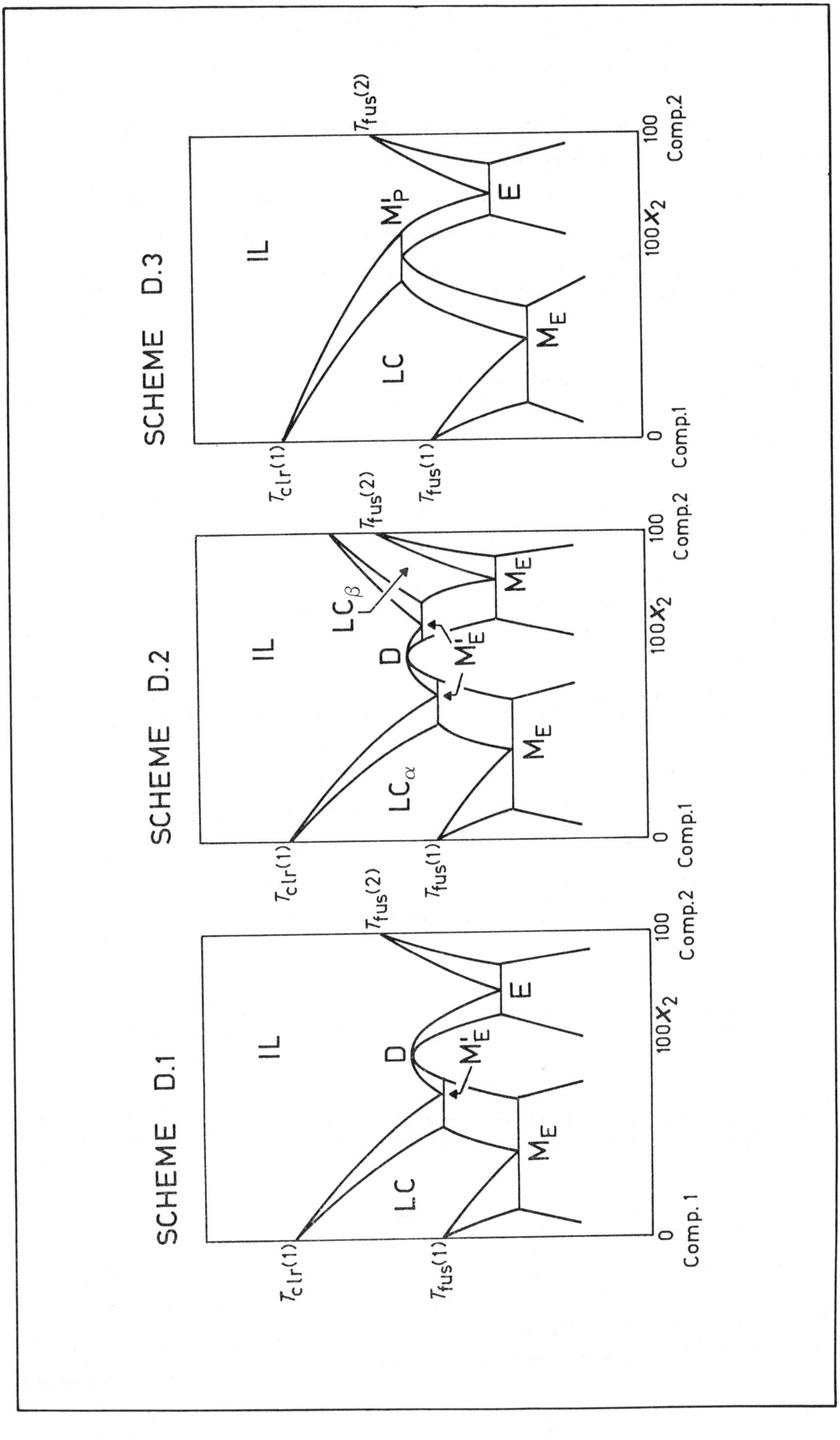
SCHEME D.1
IL
LC
D
M'_E
M_E
E
$T_{clr}(1)$
$T_{fus}(1)$
$T_{fus}(2)$
0
Comp.1
$100x_2$
100
Comp.2
SCHEME D.2
IL
LC_α
LC_β
D
M'_E
M_E
M_E
$T_{clr}(1)$
$T_{fus}(1)$
$T_{fus}(2)$
0
Comp.1
$100x_2$
100
Comp.2
SCHEME D.3
IL
LC
M'_P
E
M_E
$T_{clr}(1)$
$T_{fus}(1)$
$T_{fus}(2)$
0
Comp.1
$100x_2$
100
Comp.2

It must be noted that in these diagrams the presence of liquid crystals causes the occurrence of invariants to which no official designation has been given so far. Therefore, in order to avoid misunderstandings, throughout the present volume we indicate (provisionally) as M_E an invariant at which liquid crystals, LC (and not an isotropic liquid, IL, as in a common eutectic, E), are in equilibrium with two solids; and, respectively, as M'_E an invariant at which an isotropic liquid is in equilibrium with the liquid crystals and one solid phase (and not with two solid phases, as in a common eutectic, E). Here the subscript E is added because the situation of the equilibrium curves involved is similar to that met at a eutectic point.

For situations similar to those met in Schemes B, the subscript P will be used.

By analogy, invariants (met in subsequent Schemes) involving equilibria among two liquid crystalline and one isotropic liquid phase will be indicated as M''_E and M''_P, respectively. Finally, invariants involving two liquid crystalline and one solid phase (i.e., exhibiting a situation similar to that met in a monotectic) will be designated as M_M points.

Scheme A.1.

Component 1 can exist as a liquid crystal between $T_{fus}(1)$ and $T_{clr}(1)$. Component 2 melts at $T_{fus}(2)$, and actually cannot exist as a liquid crystal. Components 1, 2 are assumed to be partially miscible in the solid state (*). The binary invariants are an M_E and an M'_E point.

Scheme A.2.

This differs from Scheme A.1 in that a maximum, M, is present in the liquid crystal - isotropic liquid equilibrium curves.

Scheme A.3.

This differs from Scheme A.1 in that complete mutual solubility (with a minimum, m) is assumed for components 1, 2 in the solid state.

Scheme B.1.

This differs from Scheme A.1 in that the isotropic liquid - liquid crystal equilibrium curves impinge on the liquidus branch richer in component 1. Accordingly the binary invariants are a eutectic, E, and an M'_P point.

Schemes B.2, B.3.

These represent self-explanatory modifications of Scheme B.1.

Scheme C.1.

Both components can exist as liquid crystals between $T_{fus}(1)$ and $T_{clr}(1)$, and between $T_{fus}(2)$ and $T_{clr}(2)$, respectively. Mutual solubility is assumed to be complete in the mesomorphic liquid, and limited in the solid state. The only binary invariant is an M_E point.

Scheme C.2.

This differs from Scheme C.1 in that a limited mutual solubility is assumed for the components in the liquid crystalline state. Besides the M_E point, two more binary invariants exist, i.e., an M_E and an M_M point.

Scheme C.3.

This is a special case (Ref. 13) where four liquid crystal - isotropic liquid diphasic fields ought to be formed, with one maximum, M, and two minima, m_1 and m_2, respectively.

Scheme D.1.

Component 1 can exist as a liquid crystal between $T_{fus}(1)$ and $T_{clr}(1)$. Component 2 melts at $T_{fus}(2)$, and cannot exist as a liquid crystal. An intermediate compound [$(1)_2(2)_3$ in the Scheme; D: dystectic point] is formed, which cannot exist as a liquid crystal. The binary invariants are E, M_E, M'_E.

(*) Indeed, the complete absence of mutual solubility in the solid state has to be considered an exception rather than the rule.

Scheme D.2.

This differs from Scheme D.1 in that liquid crystal formation is assumed for both components.

Scheme D.3.

This differs from Scheme D.1 in that the intermediate compound melts yielding a liquid crystalline and an isotropic liquid phase; consequently, an M'_p point is formed.

Detailed information (updated 1979) on the phase relationships of alkanoates are given in Chapter 1.2 (Ref. 14) of a volume on thermodynamic and transport properties of organic salts published (1980) as a book project of the IUPAC Commission on Thermodynamics.

2. Some ambiguities met in the current literature.

2.1. Transition and peritectic points in binary systems.

Breaks along a branch of the lower boundary of the isotropic liquid region in a binary (where, for simplicity it is assumed that no mesomorphic phases are involved) can arise from the occurrence of

(a) polymorphic (first order) transformations in either component (examples are shown in Schemes E.1 - E.3);
(b) limited mutual solubility of the components in the solid state (an example is shown in Scheme F);
(c) incongruent fusion of an intermediate compound (an example is shown in Scheme G).

(i) In Schemes E.1 - E.3 component 1 undergoes (at constant pressure) the polymorphic (first order) transformation

$$(1)_{II} \underset{}{\overset{T_{trs}(1)}{\rightleftarrows}} (1)_{I}$$

In Scheme E.1, it is further assumed that component 2 (in the solid state) is soluble in neither polymorph of component 1.
At $T_T = T_{trs}(1)$, equilibrium exists among solid polymorphs $(1)_I$ and $(1)_{II}$, and the isotropic liquid of composition x_T:

$$(1)_{II} + (IL)_T \rightleftarrows (1)_I + (IL)_T \quad \text{i.e.} \quad (1)_{II} \rightleftarrows (1)_I$$

Scheme E.2 shows one of the possible situations which can be met when it is assumed that component 2 is soluble only in one polymorph of component 1 [in Scheme E.2: in polymorph $(1)_{II}$ stable at $T \leq T_{trs}(1)$]. At $T_P \neq T_{trs}(1)$ [in Scheme E.2: $T_P > T_{trs}(1)$], equilibrium exists (at constant pressure) among solid polymorph $(1)_I$, solid solutions S [of composition x_S, and formed with polymorph $(1)_{II}$ and component 2], and isotropic liquid of composition x_P:

$$(1)_I + (IL)_P \rightleftarrows S$$

Scheme E.3 shows one of the possible situations which can be met when it is assumed that component 2 is soluble in either polymorph of component 1. At T_P, the phases in invariant equilibrium (at constant pressure) are: the isotropic liquid of composition x_P, and two solid solutions of composition x_{S_β} and x_{S_α}, respectively.

In Schemes E, the isotropic liquid of composition either x_T (Scheme E.1), or x_P (Schemes E.2, E.3) is incongruent with respect to the two solid coexisting phases (i.e., the liquid cannot be synthetized from the solids). When in a binary the incongruent liquid is in equilibrium (at constant pressure) with two solids of the same composition [i.e., in Scheme E.1: polymorphs $(1)_I$ and $(1)_{II}$] the point representing the liquid phase (in Scheme E.1: point T, whose abscissa is x_T) is designated as a transition point.

When in a binary the incongruent liquid is in equilibrium (at constant pressure) with two solids of different composition, Haase and Schoenert (Ref. 15) call the point representing the liquid phase (i.e., in Schemes E.2 and E.3: point P, whose abscissa is x_P) a peritectic point. Haase and Schönert's definition of peritectic

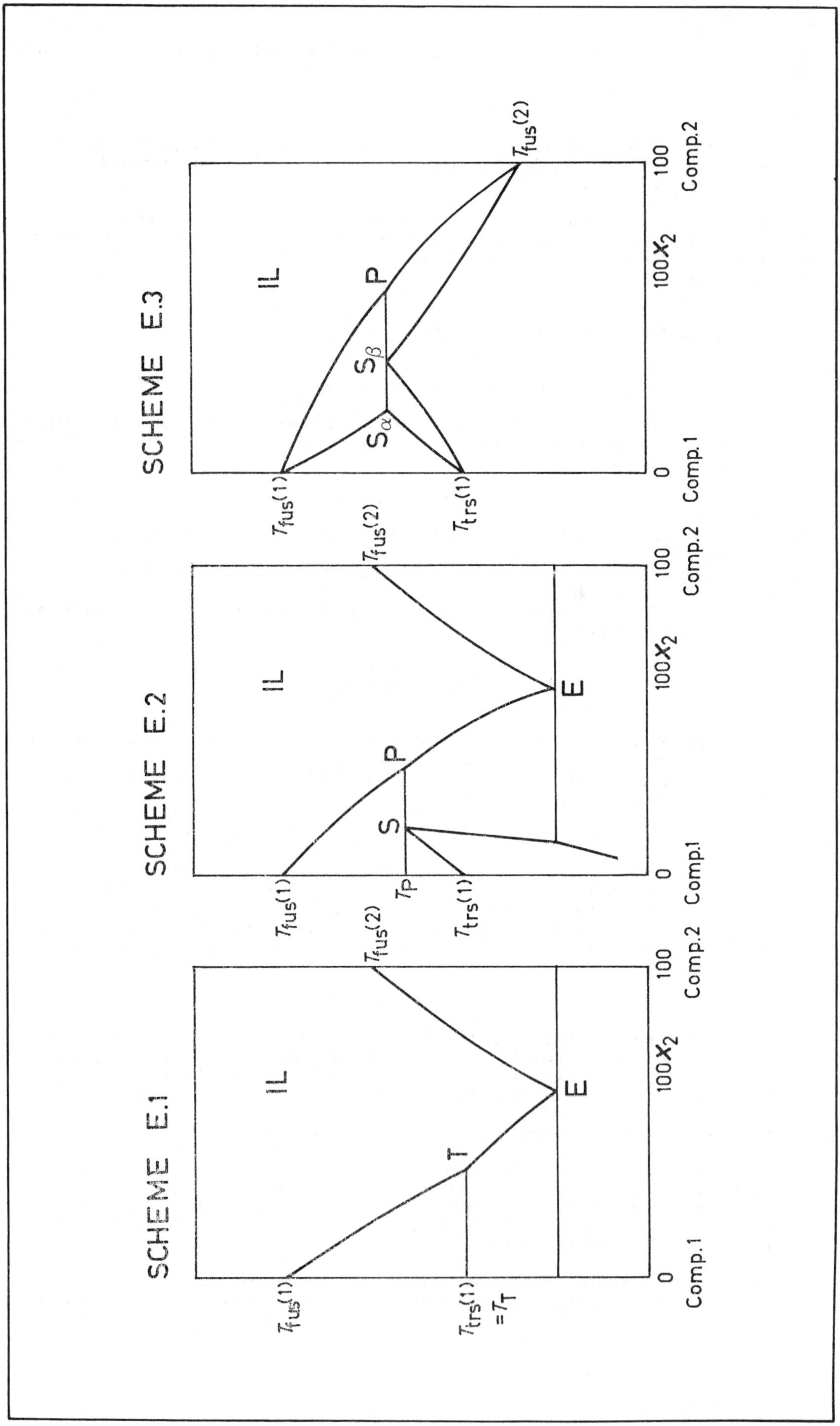
SCHEME E.1
IL
T
E
$T_{fus}(1)$
$T_{trs}(1)$
$=T_T$
$T_{fus}(2)$
0
100
$100x_2$
Comp.1
Comp.2
SCHEME E.2
IL
P
S
E
$T_{fus}(1)$
T_P
$T_{trs}(1)$
$T_{fus}(2)$
0
100
$100x_2$
Comp.1
Comp.2
SCHEME E.3
IL
P
S_β
S_α
$T_{fus}(1)$
$T_{trs}(1)$
$T_{fus}(2)$
0
100
$100x_2$
Comp.1
Comp.2

SCHEME F

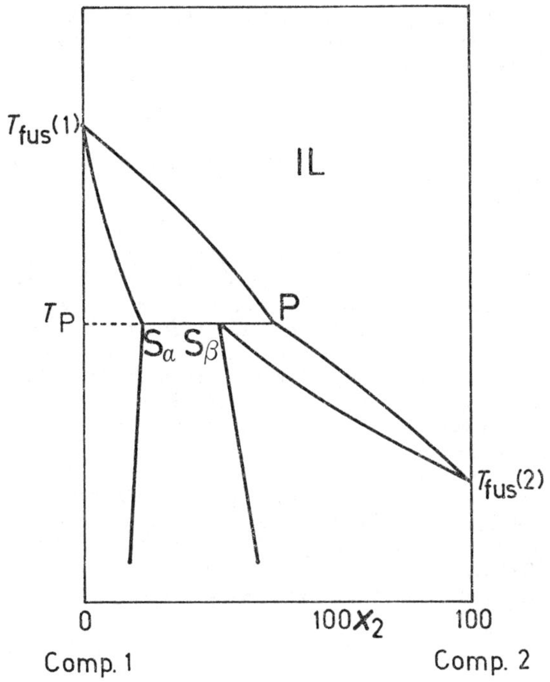

SCHEME G

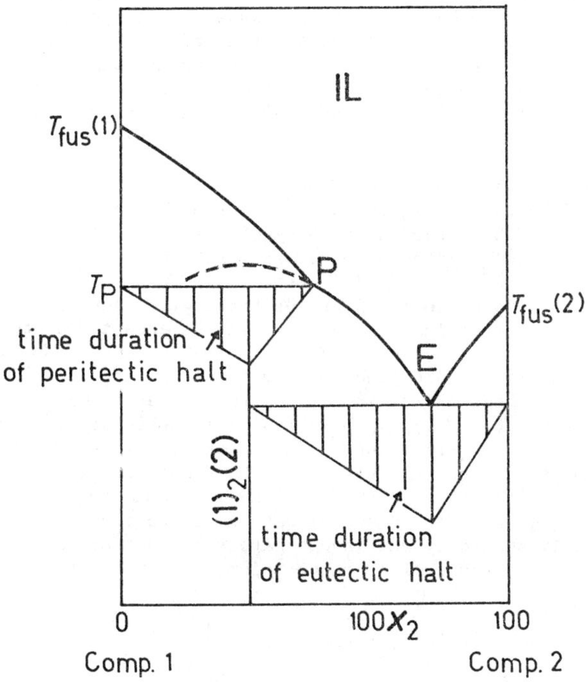

point is convenient for the present purposes, but less restrictive than other widely used formulations. A detailed discussion is unnecessary here, but if any other formulations are accepted, the various situations should be considered accordingly.

(ii) In Scheme F, components 1 and 2 are assumed to exhibit a limited mutual solubility in the solid state. At the peritectic point, P, the equilibrium is:

$$S_\alpha + (IL)_P \rightleftarrows S_\beta$$

(iii) Finally, in Scheme G, the intermediate compound [$(1)_2(2)$, in the specific case] is said to melt incongruently at T_P because (at this temperature) the isotropic liquid, $(IL)_P$, is once more incongruent with respect to both solid phases [i.e., $(1)_2(2)$ and (1)] entering the peritectic equilibrium

$$(1) + (IL)_P \rightleftarrows (1)_2(2)$$

In this case, the invariant, P, is designated either a peritectic point, or transition point of the peritectic type (Ref. 15).

(iv) The situations shown in Schemes E-G, though apparently far different from one another, can be distinguished only with difficulty in the absence of information on the solidus, e.g., on the phase relations of the pure components, the Tammann triangles (examples of with are shown in Scheme G), etc.
Unfortunately, a number of authors, who restricted their investigations to the visual polythermal determination of the liquidus, were frequently inclined either to assume a break on a liquidus branch as an evidence for the existence of an incongruently melting compound (correct only in Scheme G), or to identify the temperature at which a break occurs with that at which a pure component undergoes a phase transition (correct only in Scheme E.1).

As a final remark, it can be added that in the Russian literature an invariant point P is indicated only occasionally as a "peritekticheskaya tochka" (i.e., peritectic point), but most often as a "perekhodnaya tochka" which, however, is translated in English issues of the Russian journals as either transition point (a correct, though rather misleading, literal translation), or peritectic point, etc.

2.2. The case of pure components.

This subject was already mentioned in Section 1.2, but it is now thought useful to comment further on one or two specific papers.

(i) Sokolov, in one of his earliest studies (1954; Ref. 16) on alkanoates, reported that crystalline sodium methanoate, ethanoate, and propanoate melted directly into isotropic liquids, whereas several other sodium alkanoates behaved "as if they had two distinct fusion temperatures: the first one relevant to the transformation from the solid crystalline to the liquid crystalline phase, and the second one to that from the liquid crystalline to the isotropic liquid phase"(*). This statement was in agreement with previous findings by a few former authors known to him; see, e.g., Ref. 17.

Despite the correctness of the above assertions, Sokolov then decided unfortunately (Ref. 16) "to assume as the fusion temperature ... of both the pure salts and mixtures thereof the temperature at which the homogeneity of the melt disappeared"(*) (which, however, is correct only in the absence of mesomorphic phases).

This decision caused a misinterpretation of the topology of several among the many systems studied later on by the Smolensk group, as will be seen throughout the volume.

(ii) In 1956 Sokolov (Ref. 18) provided information, during the 10th Scientific Conference of the Smolensk Medical Institute, on phase transformations occurring in several alkali alkanoates at $T < T_{fus}$. Here fusion is obviously to be intended in Sokolov's sense, i.e., either true fusion or clearing. This information - which concerned only temperatures, and not the nature of the transformations - appeared in the Summaries of Papers presented at the Conference. Sokolov's summary could not be directly consulted but, as far as we know, no numerical data were reported on it. None the less, it has been subsequently quoted by the investigators of the Smolensk group as the pertinent primary source. Comparison of such quotations, however, shows inconsistencies,

------
* Translated from Russian by P. Ferloni, Pavia (Italy).

a few of which are:

(a) There should be either three transitions of sodium propanoate (at 468, 490, and 560 K, respectively) according to Ref. 19, or four (at 350, 468, 490, and 560 K, respectively) according to Ref. 20.

(b) There should be either two transitions of potassium n-butanoate (at 463, and 553-558 K, respectively) according to Ref. 21, or three (at 463, 553-558, and 618 K, respectively) according to Ref. 22.

(c) There should be either two transitions of potassium iso-pentanoate (at 327, and 618 K, respectively) according to Refs. 21, 23, or three (at 473, 493, and 618 K, respectively) according to Ref. 24.

(iii) Because divergent opinions have been expressed rather frequently by different authors about the number, nature, and location of phase transformations occurring in alkanoates, each single situation has been critically evaluated throughout the volume. No discussion, on the contrary, has been made of the phase transformations occurring in the inorganic components of the systems, inasmuch as data from various sources proved usually to be either coincident or negligibly different.

3. Tabulation of the systems.

Binaries with common anion have been listed in Part 1 of the volume following the increasing complexity of the anion, and, for a given anion, following the alphabetical order of the cation.

Example:

Component 1	Component 2
$(CHO_2)_2Ba$	$(CHO_2)_2K_2$
$(CHO_2)_2Ba$	$(CHO_2)_2Na_2$
...	...
$(CHO_2)_2Ca$	$(CHO_2)_2K_2$
...	...
$(CHO_2)K$	$(CHO_2)Li$
...	...
$(C_2H_3O_2)_2Cd$	$(C_2H_3O_2)Cs$
Etc.	Etc.

Binaries with common cation have been listed in Part 2 of the volume following the alphabetical order of the cation, and, for a given cation, following the self-explanatory scheme reported here.

Component 1	Component 2
...	...
KBr	$KCHO_2$
KBr	$KC_2H_3O_2$
...	...
$KCHO_2$	$KC_3H_5O_2$
$KCHO_2$	$KC_4H_7O_2$
$KCHO_2$	$Ki\text{-}C_4H_7O_2$
$KCHO_2$	KCl
Etc.	Etc.

References.

(1) Fontell, K.
Aspects of the Structure of Pure Solid Organic Salts in Thermodynamic and Transport Properties of Organic Salts, IUPAC Chemical Data Series No.28 (Franzosini, P.; Sanesi, M.; Editors), Pergamon Press, Oxford, 1980, 343.

(2) Ferloni, P.; Sanesi, M.; Franzosini, P.
Z. Naturforsch. 1975, A30, 1447.

(3) Braghetti, M.; Berchiesi, G.; Franzosini,P.
Ric.Sci. 1969,39, 576.

(4) Ferloni, P.; Franzosini, P.
Gazz.Chim.Ital. 1975, 105, 391.

(5) Sanesi, M.; Ferloni, P.; Franzosini, P.
Z. Naturforsch. 1977, A32, 1173.

(6) Ferloni, P.; Spinolo, G.; Zangen, M.; Franzosini, P.
Z. Naturforsch. 1977, A32, 329.

(7) Ferloni, P.; Sanesi, M.; Tonelli, P.L.; Franzosini, P.
Z.Naturforsch. 1978, A33, 240.

(8) Franzosini, P.; Plautz, W.A.; Westrum, E.F., Jr.
J. Chem. Thermodyn. 1983, 15, 445.

(9) Franzosini, P.; Westrum, E.F., Jr.; Plautz, W.A.
J. Chem. Thermodyn. 1983, 15, 609.

(10) Forster, G.; Brezesinski, G.; Gerlach, E.; Madicke, A.; Dorfler, H.-D.
Z. Phys. Chem. (Leipzig) 1981, 6, 1009.

(11) Prisyazhnyi, V.D.; Mirnyi, V.N.; Mirnaya, T.A.
Ukr. Khim. Zh. 1983, 49, 659.

(12) Prisyazhnyi, V.D.; Mirnyi, V.N.; Mirnaya, T.A.
Zh. Neorg. Khim. 1983, 28, 253-255; Russ. J. Inorg. Chem. (Engl. Transl.) 1983, 28, 140.

(13) Baum, E.; Demus, D.; Sackmann, H.
Wiss. Z. Univ. Halle 1970, 19, 37.

(14) Sanesi, M.; Cingolani, A.; Tonelli, P.L.; Franzosini, P.
Thermal Properties, in Thermodynamic and Transport Properties of Organic Salts, IUPAC Chemical Data Series No. 28 (Franzosini, P.; Sanesi, M.; Editors), Pergamon Press, Oxford,1980, 29.

(15) Haase, R.; Schonert, H.
Solid-liquid Equilibrium, in International Encyclopedia of Physical Chemistry and Chemical Physics, Topic 13, Vol. 1 (McGlashan, M.L.; Editor), Pergamon Press, Oxford, 1968, 1.

(16) Sokolov, N.M.
Zh. Obshch. Khim. 1954, 24, 1581.

(17) Vorlander. D.
Ber. Dtsch. Chem. Ges. 1910, 43, 3120.

(18) Sokolov, N.M.
Tezisy Dokl. X Nauch. Konf. S.M.I. 1956.

(19) Sokolov, N.M.; Minchenko, S.P.
Zh. Obshch. Khim. 1971, 41, 1656.

(20) Sokolov, N.M.; Pochtakova, E.I.
Zh. Obshch. Khim. 1958, 28, 1397.

(21) Sokolov, N.M.; Minich, M.A.
Zh. Neorg. Khim. 1961, 6, 2558; Russ. J. Inorg. Chem. (Engl. Transl.) 1961, 6, 1293.

(22) Dmitrevskaya, O.I.
Zh. Obshch. Khim. 1958, 28, 2007; Russ. J. Gen. Chem. (Engl. Transl.) 1958, 28, 2046.

(23) Pochtakova, E.I.
Zh. Obshch. Khim. 1963, 33, 342.

(24) Dmitrevskaya, O.I.; Sokolov, N.M.
Zh. Obshch. Khim. 1967, 37, 2160; Russ. J. Gen. Chem. (Engl. Transl.) 1967, 37, 2050.

Thanks are due to Prof. R. Cohen-Adad, Lyon (France), Prof. H. Gasparoux, Bordeaux (France), and Prof. J. W. Lorimer, London (Ontario, Canada) for very valuable comments on the topology of the phase diagrams.

Paolo Franzosini

INTRODUCTION TO THE SOLUBILITY OF SOLIDS IN LIQUIDS

Nature of the Project

The Solubility Data Project (SDP) has as its aim a comprehensive search of the literature for solubilities of gases, liquids, and solids in liquids or solids. Data of suitable precision are compiled on data sheets in a uniform format. The data for each system are evaluated, and where data from different sources agree sufficiently, recommended values are proposed. The evaluation sheets, recommended values, and compiled data sheets are published on consecutive pages.

Definitions

A *mixture* (1, 2) describes a gaseous, liquid, or solid phase containing more than one substance, when the substances are all treated in the same way.

A *solution* (1, 2) describes a liquid or solid phase containing more than one substance, when for convenience one of the substances, which is called the *solvent*, and may itself be a mixture, is treated differently than the other substances, which are called *solutes*. If the sum of the mole fractions of the solutes is small compared to unity, the solution is called a *dilute solution*.

The *solubility* of a substance *B* is the relative proportion of *B* (or a substance related chemically to *B*) in a mixture which is saturated with respect to solid *B* at a specified temperature and pressure. *Saturated* implies the existence of equilibrium with respect to the processes of dissolution and precipitation; the equilibrium may be stable or metastable. The solubility of a substance in metastable equilibrium is usually greater than that of the corresponding substance in stable equliibrium. (Strictly speaking, it is the activity of the substance in metastable equilibrium that is greater.) Care must be taken to distinguish true metastability from supersaturation, where equilibrium does not exist.

Either point of view, mixture or solution, may be taken in describing solubility. The two points of view find their expression in the quantities used as measures of solubility and in the reference states used for definition of activities, activity coefficients and osmotic coefficients.

The qualifying phrase "substance related chemically to *B*" requires comment. The composition of the saturated mixture (or solution) can be described in terms of any suitable set of thermodynamic components. Thus, the solubility of a salt hydrate in water is usually given as the relative proportion of anhydrous salt in solution, rather than the relative proportions of hydrated salt and water.

Quantities Used as Measures of Solubility

1. *Mole fraction* of substance *B*, x_B:

$$x_B = n_B / \sum_{s=1}^{c} n_s \qquad [1]$$

where n_s is the amount of substance of *s*, and *c* is the number of distinct substances present (often the number of thermodynamic components in the system). *Mole per cent* of *B* is 100 x_B.

2. *Mass fraction* of substance *B*, w_B:

$$w_B = m_B' / \sum_{s=1}^{c} m_s' \qquad [2]$$

where m_s' is the mass of substance *s*. *Mass per cent* is 100 w_B. The equivalent terms *weight fraction* and *weight per cent* are not used.

3. *Solute mole (mass) fraction* of solute *B* (3, 4):

$$x_{s,B} = m_B / \sum_{s=1}^{c'} m_s = x_B / \sum_{s=1}^{c'} x_s \qquad [3]$$

$$w_{s,B} = m_B' / \sum_{s=1}^{c'} m_s' = w_B / \sum_{s=1}^{c'} w_s \qquad [3a]$$

where the summation is over the solutes only. For the solvent A, $x_{S,A} = x_A/(1 - x_A)$, $w_{S,A} = w_A/(1 - w_A)$. These quantities are called Jänecke mole (mass) *fractions* in many papers.

4. *Molality* of solute B (1, 2) in a solvent A:

$$m_B = n_B/n_A M_A \quad \text{SI base units: mol kg}^{-1} \quad [4]$$

where M_A is the molar mass of the solvent.

5. *Concentration* of solute B (1, 2) in a solution of volume V:

$$c_B = [B] = n_B/V \quad \text{SI base units: mol m}^{-3} \quad [5]$$

The symbol c_B is preferred to $[B]$, but both are used. The terms *molarity* and *molar* are not used.

Mole and mass fractions are appropriate to either the mixture or the solution point of view. The other quantities are appropriate to the solution point of view only. Conversions among these quantities can be carried out using the equations given in Table I-1 following this Introduction. Other useful quantities will be defined in the prefaces to individual volumes or on specific data sheets.

In addition to the quantities defined above, the following are useful in conversions between concentrations and other quantities.

6. *Density*: $\rho = m/V$ SI base units: kg m^{-3} [6]

7. *Relative density*: d; the ratio of the density of a mixture to the density of a reference substance under conditions which must be specified for both (1). The symbol $d_{t'}^{t}$ will be used for the density of a mixture at t°C, 1 bar divided by the density of water at t'°C, 1 bar. (In some cases 1 atm = 101.325 kPa is used instead of 1 bar = 100 kPa.)

8. *A note on nomenclature*. The above definitions use the nomenclature of the IUPAC *Green Book* (1), in which a solute is called B and a solvent A In compilations and evaluations, the first-named component (component 1) is the solute, and the second (component 2 for a two-component system) is the solvent. The reader should bear these distinctions in nomenclature in mind when comparing nomenclature and theoretical equations given in this Introduction with equations and nomenclature used on the evaluation and compilation sheets.

Thermodynamics of Solubility

The principal aims of the Solubility Data Project are the tabulation and evaluation of: (a) solubilities as defined above; (b) the nature of the saturating phase. Thermodynamic analysis of solubility phenomena has two aims: (a) to provide a rational basis for the construction of functions to represent solubility data; (b) to enable thermodynamic quantities to be extracted from solubility data. Both these are difficult to achieve in many cases because of a lack of experimental or theoretical information concerning activity coefficients. Where thermodynamic quantities can be found, they are not evaluated critically, since this task would involve critical evaluation of a large body of data that is not directly relevant to solubility. The following is an outline of the principal thermodynamic relations encountered in discussions of solubility. For more extensive discussions and references, see books on thermodynamics, e.g., (5-12).

Activity Coefficients (1)

(a) *Mixtures*. The activity coefficient f_B of a substance B is given by

$$RT \ln (f_B x_B) = \mu_B - \mu_B^* \quad [7]$$

where μ_B^* is the chemical potential of pure B at the same temperature and pressure. For any substance B in the mixture,

$$\lim_{x_B \to 1} f_B = 1 \quad [8]$$

(b) *Solutions*.

(i) *Solute B*. The molal activity coefficient γ_B is given by

$$RT \ln(\gamma_B m_B) = \mu_B - (\mu_B - RT \ln m_B)^\infty \quad [9]$$

where the superscript $^\infty$ indicates an infinitely dilute solution. For any solute B,

$$\gamma_B^\infty = 1 \quad [10]$$

Activity coefficients y_B connected with concentrations c_B, and $f_{x,B}$ (called the *rational activity coefficient*) connected with mole fractions x_B are defined in analogous ways. The relations among them are (1, 9), where ρ^* is the density of the pure solvent:

$$f_B = (1 + M_A\sum_s m_s)\gamma_B = [\rho + \sum_s(M_A - M_s)c_s]y_B/\rho^* \quad [11]$$

$$\gamma_B = (1 - \sum_s x_s)f_{x,B} = (\rho - \sum_s M_s c_s)y_B/\rho^* \quad [12]$$

$$y_B = \rho^* f_{x,B}[1 + \sum_s(M_s/M_A - 1)x_B]/\rho = \rho^*(1 + \sum_s M_s m_s)\gamma_B/\rho \quad [13]$$

For an electrolyte solute $B \equiv C_{\nu+}A_{\nu-}$, the activity on the molality scale is replaced by (9)

$$\gamma_B m_B = \gamma_\pm^{\nu} m_B^{\nu} Q^{\nu} \quad [14]$$

where $\nu = \nu_+ + \nu_-$, $Q = (\nu_+^{\nu+}\nu_-^{\nu-})^{1/\nu}$, and $\gamma_\pm$ is the mean ionic activity coefficient on the molality scale. A similar relation holds for the concentration activity, $y_B c_B$. For the mole fractional activity,

$$f_{x,B}x_B = Q^{\nu} f_\pm^{\nu} x_\pm^{\nu} \quad [15]$$

where $x_\pm = (x_+x_-)^{1/\nu}$. The quantities x_+ and x_- are the ionic mole fractions (9), which are

$$x_+ = \nu_+ x_B/[1 + \sum_s(\nu_s - 1)x_s]; \quad x_- = \nu_- x_B[1 + \sum_s(\nu_s - 1)x_s] \quad [16]$$

where ν_s is the sum of the stoichiometric coefficients for the ions in a salt with mole fraction x_s. Note that the mole fraction of solvent is now

$$x_A' = (1 - \sum_s \nu_s x_s)/[1 + \sum_s(\nu_s - 1)x_s] \quad [17]$$

so that

$$x_A' + \sum_s \nu_s x_s = 1 \quad [18]$$

The relations among the various mean ionic activity coefficients are:

$$f_\pm = (1 + M_A\sum_s \nu_s m_s)\gamma_\pm = [\rho + \sum_s(\nu_s M_A - M_s)c_s]y_\pm/\rho^* \quad [19]$$

$$\gamma_\pm = \frac{(1 - \sum_s x_s)f_\pm}{1 + \sum_s(\nu_s - 1)x_s} = (\rho - \sum_s M_s c_s)y_\pm/\rho^* \quad [20]$$

$$y_\pm = \frac{\rho^*[1 + \sum_s(M_s/M_A - 1)x_s]f_\pm}{\rho[1 + \sum_s(\nu_s - 1)x_s]} = \rho^*(1 + \sum_s M_s m_s)\gamma_\pm/\rho \quad [21]$$

(ii) *Solvent, A:*

The *osmotic coefficient*, ϕ, of a solvent *A* is defined as (1):

$$\phi = (\mu_A^* - \mu_A)/RT\, M_A \sum_s m_s \quad [22]$$

where μ_A^* is the chemical potential of the pure solvent.

The *rational osmotic coefficient*, ϕ_x, is defined as (1):

$$\phi_x = (\mu_A - \mu_A^*)/RT\ln x_A = \phi M_A \sum_s m_s/\ln(1 + M_A \sum_s m_s) \quad [23]$$

The activity, a_A, or the activity coefficient, f_A, is sometimes used for the solvent rather than the osmotic coefficient. The activity coefficient is defined relative to pure *A*, just as for a mixture.

For a mixed solvent, the molar mass in the above equations is replaced by the average molar mass; i.e., for a two-component solvent with components *J*, *K*, M_A becomes

$$M_A = M_J + (M_K - M_J)x_{v,K} \quad [24]$$

where $x_{v,K}$ is the solvent mole fraction of component *K*.

The osmotic coefficient is related directly to the vapor pressure, *p*, of a solution in equilibrium with vapor containing *A* only by (12, p.306):

$$\phi M_A\sum_s \nu_s m_s = -\ln(p/p_A^*) + (V_{m,A}^* - B_{AA})(p - p_A^*)/RT \quad [25]$$

where p_A^*, $V_{m,A}^*$ are the vapor pressure and molar volume of pure solvent *A*, and B_{AA} is the second virial coefficient of the vapor.

The Liquid Phase

A general thermodynamic differential equation which gives solubility as a function of temperature, pressure and composition can be derived. The approach is similar to that of Kirkwood and Oppenheim (7); see also (11, 12). Consider a solid mixture containing c thermodynamic components i. The Gibbs-Duhem equation for this mixture is:

$$\sum_{i=1}^{c} x_i'(S_i'dT - V_i'dp + d\mu_i') = 0 \qquad [26]$$

A liquid mixture in equilibrium with this solid phase contains c' thermodynamic components i, where $c' \geqslant c$. The Gibbs-Duhem equation for the liquid mixture is:

$$\sum_{i=1}^{c} x_i(S_i dT - V_i dp + d\mu_i') + \sum_{i=c+1}^{c'} x_i(S_i dT - V_i dp + d\mu_i) = 0 \qquad [27]$$

Subtract [26] from [27] and use the equation

$$d\mu_i = (d\mu_i)_{T,p} - S_i dT + V_i dp \qquad [28]$$

and the Gibbs-Duhem equation at constant temperature and pressure:

$$\sum_{i=1}^{c} x_i(d\mu_i')_{T,p} + \sum_{i=c+1}^{c'} x_i(d\mu_i)_{T,p} = 0 \qquad [29]$$

The resulting equation is:

$$RT\sum_{i=1}^{c} x_i'(d\ln a_i)_{T,p} = \sum_{i=1}^{c} x_i'(H_i - H_i')dT/T - \sum_{i=1}^{c} x_i'(V_i - V_i')dp \qquad [30]$$

where

$$H_i - H_i' = T(S_i - S_i') \qquad [31]$$

is the enthalpy of transfer of component i from the solid to the liquid phase at a given temperature, pressure and composition, with H_i and S_i the partial molar enthalpy and entropy of component i.

Use of the equations

$$H_i - H_i^0 = -RT^2(\partial \ln a_i/\partial T)_{x,p} \qquad [32]$$

and

$$V_i - V_i^0 = RT(\partial \ln a_i/\partial p)_{x,T} \qquad [33]$$

where superscript o indicates an arbitrary reference state gives:

$$RT\sum_{i=1}^{c} x_i' d\ln a_i = \sum_{i=1}^{c} x_i'(H_i^0 - H_i')dT/T - \sum_{i=1}^{c} x_i'(V_i^0 - V_i')dp \qquad [34]$$

where

$$d\ln a_i = (d\ln a_i)_{T,p} + (\partial \ln a_i/\partial T)_{x,p} + (\partial \ln a_i/\partial p)_{x,T} \qquad [35]$$

The terms involving enthalpies and volumes in the solid phase can be written as:

$$\sum_{i=1}^{c} x_i' H_i' = H_s^* \qquad \sum_{i=1}^{c} x_i' V_i' = V_s^* \qquad [36]$$

With eqn [36], the final general solubility equation may then be written:

$$R\sum_{i=1}^{c} x_i' d\ln a_i = (H_s^* - \sum_{i=1}^{c} x_i' H_i^0)d(1/T) - (V_s^* - \sum_{i=1}^{c} x_i' V_i^0)dp/T \qquad [37]$$

Note that those components which are not present in both phases do not appear in the solubility equation. However, they do affect the solubility through their effect on the activities of the solutes.

Several applications of eqn [37] (all with pressure held constant) will be discussed below. Other cases will be discussed in individual evaluations.

(a) *Solubility as a function of temperature.*

Consider a binary solid compound A_nB in a single solvent A. There is no fundamental thermodynamic distinction between a binary compound of A and B which dissociates completely or partially on melting and a solid mixture of A and B; the binary compound can be regarded as a solid mixture of constant composition. Thus, with $c = 2$, $x_A' = n/(n+1)$, $x_B' = 1/(n+1)$, eqn [37] becomes:

$$d\ln(a_A{}^n a_B) = -\Delta H_{AB}{}^0 d(1/RT) \quad [38]$$

where

$$\Delta H_{AB}{}^0 = nH_A + H_B - (n+1)H_s{}^* \quad [39]$$

is the molar enthalpy of melting and dissociation of pure solid A_nB to form A and B in their reference states. Integration between T and T_0, the melting point of the pure binary compound A_nB, gives:

$$\ln(a_A{}^n a_B) = \ln(a_A{}^n a_B)_{T=T_0} - \int_{T_0}^{T} \Delta H_{AB}{}^0 d(1/RT) \quad [40]$$

(i) Non-electrolytes

In eqn [32], introduce the pure liquids as reference states. Then, using a simple first-order dependence of $\Delta H_{AB}{}^*$ on temperature, and assuming that the activitity coefficients conform to those for a simple mixture (6):

$$RT \ln f_A = wx_B{}^2 \qquad RT \ln f_B = wx_A{}^2 \quad [41]$$

then, if w is independent of temperature, eqn [32] and [33] give:

$$\ln\{x_B(1-x_B)^n\} + \ln\left\{\frac{n^n}{(1+n)^{n+1}}\right\} = G(T) \quad [42]$$

where

$$G(T) = -\left\{\frac{\Delta H_{AB}{}^* - T^*\Delta C_p{}^*}{R}\right\}\left\{\frac{1}{T} - \frac{1}{T^*}\right\} + \frac{\Delta C_p{}^*}{R}\ln(T/T^*) - \frac{w}{R}\left\{\frac{x_A{}^2 + nx_B{}^2}{T} - \frac{n}{(n+1)T^*}\right\} \quad [43]$$

where $\Delta C_p{}^*$ is the change in molar heat capacity accompanying fusion plus decomposition of the pure compound to pure liquid A and B at temperature T^*, (assumed here to be independent of temperature and composition), and $\Delta H_{AB}{}^*$ is the corresponding change in enthalpy at $T = T^*$. Equation [42] has the general form:

$$\ln\{x_B(1-x_B)^n\} = A_1 + A_2/(T/K) + A_3\ln(T/K) + A_4(x_A{}^2 + nx_B{}^2)/(T/K) \quad [44]$$

If the solid contains only component B, then $n = 0$ in eqn [42] to [44].

If the infinite dilution reference state is used, then:

$$RT \ln f_{x,B} = w(x_A{}^2 - 1) \quad [45]$$

and [39] becomes

$$\Delta H_{AB}{}^\infty = nH_A{}^* + H_B{}^\infty - (n+1)H_s{}^* \quad [46]$$

where $\Delta H_{AB}{}^\infty$ is the enthalpy of melting and dissociation of solid compound A_nB to the infinitely dilute reference state of solute B in solvent A; $H_A{}^*$ and $H_B{}^\infty$ are the partial molar enthalpies of the solute and solvent at infinite dilution. Clearly, the integral of eqn [32] will have the same form as eqn [35], with $\Delta H_{AB}{}^\infty$ replacing $\Delta H_{AB}{}^*$, $\Delta C_p{}^\infty$ replacing ΔCp^*, and $x_A{}^2 - 1$ replacing $x_A{}^2$ in the last term.

See (5) and (11) for applications of these equations to experimental data.

(ii) Electrolytes

(a) Mole fraction scale

If the liquid phase is an aqueous electrolyte solution, and the solid is a salt hydrate, the above treatment needs slight modification. Using rational mean activity coefficients, eqn [34] becomes:

$$\ln\left\{\frac{x_B^{\nu}(1 - x_B)^n}{[1 + (\nu - 1)x_B]^{n+\nu}}\right\} - \ln\left\{\frac{n^n}{(n + \nu)^{n+\nu}}\right\} + \ln\left\{\left[\frac{f_B}{f_B^*}\right]^{\nu}\left[\frac{f_A}{f_A^*}\right]^n\right\}$$
$$= -\left\{\frac{\Delta H_{AB}^* - T^*\Delta C_p^*}{R}\right\}\left\{\frac{1}{T} - \frac{1}{T^*}\right\} + \frac{\Delta C_p^*}{R}\ln(T/T^*) \qquad [47]$$

where superscript * indicates the pure salt hydrate. If it is assumed that the activity coefficients follow the same temperature dependence as the right-hand side of eqn [47] (13-16), the thermochemical quantities on the right-hand side of eqn [47] are not rigorous thermodynamic enthalpies and heat capacities, but are apparent quantities only. Data on activity coefficients (9) in concentrated solutions indicate that the terms involving these quantities are not negligible, and their dependence on temperature and composition along the solubility-temperature curve is a subject of current research.

A similar equation (with $\nu = 2$ and without the heat capacity terms or activity coefficients) has been used to fit solubility data for some $MOH-H_2O$ systems, where M is an alkali metal (13); enthalpy values obtained agreed well with known values. The full equation has been deduced by another method in (14) and applied to MCl_2-H_2O systems in (14) and (15). For a summary of the use of equation [47] and similar equations, see (14).

(2) Molality scale

Substitution of the mean activities on the molality scale in eqn [40] gives:

$$\nu\ln\left\{\frac{\gamma_{\pm}m_B}{\gamma_{\pm}^* m_B^*}\right\} - \nu(m_B/m_B^* - 1) - \nu\{m_B(\phi - 1)/m_B^* - \phi^* + 1\}$$
$$= G(T) \qquad [48]$$

where $G(T)$ is the same as in eqn [47], $m_B^* = 1/nM_A$ is the molality of the anhydrous salt in the pure salt hydrate and $\gamma_{\pm}$ and ϕ are the mean activity coefficient and the osmotic coefficient, respectively. Use of the osmotic coefficient for the activity of the solvent leads, therefore, to an equation that has a different appearance to [47]; the content is identical. However, while eqn [47] can be used over the whole range of composition ($0 \leqslant x_B \leqslant 1$), the molality in eqn [48] becomes infinite at $x_B = 1$; use of eqn [48] is therefore confined to solutions sufficiently dilute that the molality is a useful measure of composition. The essentials of eqn [48] were deduced by Williamson (17); however, the form used here appears first in the *Solubility Data Series*. For typical applications (where activity and osmotic coefficients are not considered explicitly, so that the enthalpies and heat capacities are apparent values, as explained above), see (18).

The above analysis shows clearly that a rational thermodynamic basis exists for functional representation of solubility-temperature curves in two-component systems, but may be difficult to apply because of lack of experimental or theoretical knowledge of activity coefficients and partial molar enthalpies. Other phenomena which are related ultimately to the stoichiometric activity coefficients and which complicate interpretation include ion pairing, formation of complex ions, and hydrolysis. Similar considerations hold for the variation of solubility with pressure, except that the effects are relatively smaller at the pressures used in many investigations of solubility (5).

(b) *Solubility as a function of composition.*

At constant temperature and pressure, the chemical potential of a saturating solid phase is constant:

$$\mu_{A_nB}^* = \mu_{A_nB}(\text{sln}) = n\mu_A + \mu_B \qquad [49]$$
$$= (n\mu_A^* + \nu_+\mu_+^{\infty} + \nu_-\mu_-^{\infty}) + nRT\ln f_A x_A$$
$$+ \nu RT\ln(\gamma_{\pm}m_{\pm}Q)$$

for a salt hydrate A_nB which dissociates to water (A), and a salt (B), one mole of which ionizes to give ν_+ cations and ν_- anions in a solution in which other substances (ionized or not) may be present. If the saturated solution is sufficiently dilute, $f_A = x_A = 1$, and the quantity K_s in

$$\Delta G^{\infty} = (\nu_+\mu_+^{\infty} + \nu_-\mu_-^{\infty} + n\mu_A^* - \mu_{AB}^*)$$
$$= -RT\ln K_s$$

$$= -\nu RT \ln(Q\gamma_{\pm}m_B) \qquad [50]$$

is called the *solubility product* of the salt. (It should be noted that it is not customary to extend this definition to hydrated salts, but there is no reason why they should be excluded.) Values of the solubility product are often given on mole fraction or concentration scales. In dilute solutions, the theoretical behaviour of the activity coefficients as a function of ionic strength is often sufficiently well known that reliable extrapolations to infinite dilution can be made, and values of K_S can be determined. In more concentrated solutions, the same problems with activity coefficients that were outlined in the section on variation of solubility with temperature still occur. If these complications do not arise, the solubility of a hydrate salt $C_\nu A_\nu \cdot nH_2O$ in the presence of other solutes is given by eqn [50] as

$$\nu \ln\{m_B/m_B(0)\} = -\nu \ln\{\gamma_{\pm}/\gamma_{\pm}(0)\} - n \ln\{a_A/a_A(0)\} \qquad [51]$$

where a_A is the activity of water in the saturated solution, m_B is the molality of the salt in the saturated solution, and (0) indicates absence of other solutes. Similar considerations hold for non-electrolytes.

Consideration of *complex mixed ligand equilibria* in the solution phase are also frequently of importance in the interpretation of solubility equilibria. For nomenclature connected with these equilibria (and solubility equilibria as well), see (19, 20).

The Solid Phase

The definition of solubility permits the occurrence of a single solid phase which may be a pure anhydrous compound, a salt hydrate, a non-stoichiometric compound, or a solid mixture (or solid solution, or "mixed crystals"), and may be stable or metastable. As well, any number of solid phases consistent with the requirements of the phase rule may be present. Metastable solid phases are of widespread occurrence, and may appear as polymorphic (or allotropic) forms or crystal solvates whose rate of transition to more stable forms is very slow. Surface heterogeneity may also give rise to metastability, either when one solid precipitates on the surface of another, or if the size of the solid particles is sufficiently small that surface effects become important. In either case, the solid is not in stable equilibrium with the solution. See (21) for the modern formulation of the effect of particle size on solubility. The stability of a solid may also be affected by the atmosphere in which the system is equilibrated.

Many of these phenomena require very careful, and often prolonged, equilibration for their investigation and elimination. A very general analytical method, the "wet residues" method of Schreinemakers (22), is often used to investigate the composition of solid phases in equilibrium with salt solutions. This method has been reviewed in (23), where [see also (24)] least-squares methods for evaluating the composition of the solid phase from wet residue data (or initial composition data) and solubilities are described. In principle, the same method can be used with systems of other types. Many other techniques for examination of solids, in particular X-ray, optical, and thermal analysis methods, are used in conjunction with chemical analyses (including the wet residues method).

COMPILATIONS AND EVALUATIONS

The formats for the compilations and critical evaluations have been standardized for all volumes. A brief description of the data sheets has been given in the FOREWORD; additional explanation is given below.

Guide to the Compilations

The format used for the compilations is, for the most part, self-explanatory. The details presented below are those which are not found in the FOREWORD or which are not self-evident.

Components. Each component is listed according to IUPAC name, formula, and Chemical Abstracts (CA) Registry Number. The formula is given either in terms of the IUPAC or Hill (25) system and the choice of formula is governed by what is usual for most current users: i.e., IUPAC for inorganic compounds, and Hill system for organic compounds. Components are ordered according to:

(a) saturating components;
(b) non-saturating components in alphanumerical order;
(c) solvents in alphanumerical order.

The saturating components are arranged in order according to a 18-column periodic table with two additional rows:

Columns 1 and 2: H, alkali elements, ammonium, alkaline earth elements
3 to 12: transition elements
13 to 17: boron, carbon, nitrogen groups; chalcogenides, halogens
18: noble gases
Row 1: Ce to Lu
Row 2: Th to the end of the known elements, in order of atomic number.

Salt hydrates are generally not considered to be saturating components since most solubilities are expressed in terms of the anhydrous salt. The existence of hydrates or solvates is carefully noted in the text, and CA Registry Numbers are given where available, usually in the critical evaluation. Mineralogical names are also quoted, along with their CA Registry Numbers, again usually in the critical evaluation.

Original Measurements. References are abbreviated in the forms given by *Chemical Abstracts Service Source Index* (CASSI). Names originally in other than Roman alphabets are given as transliterated by *Chemical Abstracts*.

Experimental Values. Data are reported in the units used in the original publication, with the exception that modern *names* for units and quantities are used; e.g., mass per cent for weight per cent; mol dm^{-3} for molar; etc. Both mass and molar values are given. Usually, only one type of value (e.g., mass per cent) is found in the original paper, and the compiler has added the other type of value (e.g., mole per cent) from computer calculations based on 1983 atomic weights (26).

Errors in calculations and fitting equations in original papers have been noted and corrected, by computer calculations where necessary.

Method. Source and Purity of Materials. Abbreviations used in *Chemical Abstracts* are often used here to save space.

Estimated Error. If these data were omitted by the original authors, and if relevant information is available, the compilers have attempted to estimate errors from the internal consistency of data and type of apparatus used. Methods used by the compilers for estimating and and reporting errors are based on the papers by Ku and Eisenhart (27).

Comments and/or Additional Data. Many compilations include this section which provides short comments relevant to the general nature of the work or additional experimental and thermodynamic data which are judged by the compiler to be of value to the reader.

References. See the above description for Original Measurements.

Guide to the Evaluations

The evaluator's task is to check whether the compiled data are correct, to assess the reliability and quality of the data, to estimate errors where necessary, and to recommend "best" values. The evaluation takes the form of a summary in which all the data supplied by the compiler have been critically reviewed. A brief description of the evaluation sheets is given below.

Components. See the description for the Compilations.

Evaluator. Name and date up to which the literature was checked.

Critical Evaluation

(a) Critical text. The evaluator produces text evaluating *all* the published data for each given system. Thus, in this section the evaluator reviews the merits or shortcomings of the various data. Only published data are considered; even published data can be considered only if the experimental data permit an assessment of reliability.

(b) Fitting equations. If the use of a smoothing equation is justifiable the evaluator may provide an equation representing the solubility as a function of the variables reported on all the compilation sheets.

(c) Graphical summary. In addition to (b) above, graphical summaries are often given.

(d) Recommended values. Data are *recommended* if the results of at least two independent groups are available and they are in good agreement, and if the evaluator has no doubt as to the adequacy and reliability of the applied experimental and computational procedures. Data are considered as *tentative* if only one set of measurements is available, or if the evaluator considers some aspect of the computational or experimental method as mildly undesirable but estimates that it should cause only minor errors. Data are considered as *doubtful* if the evaluator considers some aspect of the computational or experimental method as undesirable but still considers the data to have some value in those instances where the order of magnitude of the solubility is needed. Data determined by an inadequate method or under ill-defined conditions are *rejected*. However references to these data are included in the evaluation together with a comment by the evaluator as to the reason for their rejection.

(e) References. All pertinent references are given here. References to those data which, by virtue of their poor precision, have been rejected and not compiled are also listed in this section.

(f) Units. While the original data may be reported in the units used by the investigators, the final recommended values are reported in S.I. units (1, 28) when the data can be accurately converted.

References

1. Whiffen, D.H., ed., *Manual of Symbols and Terminology for Physicochemical Quantities and Units*. *Pure Applied Chem.* 1979, 51, No. 1.
2. McGlashan, M.L. *Physicochemical Quantities and Units*. 2nd ed. Royal Institute of Chemistry. London. 1971.
3. Jänecke, E. *Z. Anorg. Chem.* 1906, 51, 132.
4. Friedman, H.L. *J. Chem. Phys.* 1960, 32, 1351.
5. Prigogine, I.; Defay, R. *Chemical Thermodynamics*. D.H. Everett, transl. Longmans, Green. London, New York, Toronto. 1954.
6. Guggenheim, E.A. *Thermodynamics*. North-Holland. Amsterdam. 1959. 4th ed.
7. Kirkwood, J.G.; Oppenheim, I. *Chemical Thermodynamics*. McGraw-Hill. New York, Toronto, London. 1961.
8. Lewis, G.N.; Randall, M. (rev. Pitzer, K.S.; Brewer, L.). *Thermodynamics*. McGraw Hill. New York, Toronto, London. 1961. 2nd. ed.
9. Robinson, R.A.; Stokes, R.H. *Electrolyte Solutions*. Butterworths. London. 1959. 2nd ed.
10. Harned, H.S.; Owen, B.B. *The Physical Chemistry of Electrolytic Solutions*. Reinhold. New York. 1958. 3rd ed.
11. Haase, R.; Schönert, H. *Solid-Liquid Equilibrium*. E.S. Halberstadt, trans. Pergamon Press, London, 1969.
12. McGlashan, M.L. *Chemical Thermodynamics*. Academic Press. London. 1979.
13. Cohen-Adad, R.; Saugier, M.T.; Said, J. *Rev. Chim. Miner.* 1973, *10*, 631.
14. Counioux, J.-J.; Tenu, R. *J. Chim. Phys.* 1981, *78*, 815.
15. Tenu, R.; Counioux, J.-J. *J. Chim. Phys.* 1981, *78*, 823.
16. Cohen-Adad, R. *Pure Appl. Chem.* 1985, *57*, 255.
17. Williamson, A.T. *Faraday Soc. Trans.* 1944, 40, 421.
18. Siekierski, S.; Mioduski, T.; Salomon, M. *Solubility Data Series*. Vol. 13. *Scandium, Yttrium, Lanthanum and Lanthanide Nitrates*. Pergamon Press. 1983.
19. Marcus, Y., ed. *Pure Appl. Chem.* 1969, *18*, 459.
20. IUPAC Analytical Division. *Proposed Symbols for Metal Complex Mixed Ligand Equilibria (Provisional)*. *IUPAC Inf. Bull.* 1978, No. 3, 229.
21. Enüstün, B.V.; Turkevich, J. *J. Am. Chem. Soc.* 1960, *82*, 4502.
22. Schreinemakers. F.A.H. *Z. Phys. Chem., Stoechiom. Verwandschaftsl.* 1893, 11, 75.
23. Lorimer, J.W. *Can. J. Chem.* 1981, 59, 3076.
24. Lorimer, J.W. *Can. J. Chem.* 1982, *60*, 1978.
25. Hill, E.A. *J. Am. Chem. Soc.* 1900, *22*, 478.
26. IUPAC Commission on Atomic Weights. *Pure Appl. Chem.* 1984, *56*, 653.
27. Ku, H.H., p. 73; Eisenhart, C., p. 69; in Ku, H.H., ed. *Precision Measurement and Calibration*. NBS Special Publication 300. Vol. 1. Washington. 1969.
28. *The International System of Units*. Engl. transl. approved by the BIPM of *Le Système International d'Unités*. H.M.S.O. London. 1970.

September, 1986

R. Cohen-Adad,
Villeurbanne, France

J. W. Lorimer,
London, Ontario, Canada

M. Salomon,
Fair Haven, New Jersey, U.S.A.

Table I-1

Quantities Used as Measures of Solubility of Solute B
Conversion Table for Multicomponent Systems
Containing Solvent A and Solutes s

	mole fraction $x_B =$	mass fraction $w_B =$	molality $m_B =$	concentration $c_B =$
x_B	x_B	$\dfrac{M_B x_B}{M_A + \sum_s (M_s - M_A) x_s}$	$\dfrac{x_B}{M_A(1 - \sum_s x_s)}$	$\dfrac{\rho x_B}{M_A + \sum_s (M_s - M_A) x_s}$
w_B	$\dfrac{w_B/M_B}{1/M_A + \sum_s (1/M_s - 1/M_A) w_s}$	w_B	$\dfrac{w_B}{M_B(1 - \sum_s w_s)}$	$\rho w_B/M_B$
m_B	$\dfrac{M_A m_B}{1 + M_A \sum_s m_s}$	$\dfrac{M_B m_B}{1 + \sum_s m_s M_s}$	m_B	$\dfrac{\rho m_B}{1 + \sum_s M_s m_s}$
c_B	$\dfrac{M_A c_B}{\rho + \sum_s (M_A - M_s) c_s}$	$M_B c_B/\rho$	$\dfrac{c_B}{\rho - \sum_s M_s c_s}$	c_B

ρ = density of solution
M_A, M_B, M_s = molar masses of solvent, solute B, other solutes s
Formulas are given in forms suitable for rapid computation; all calculations should be made using SI base units.

SYSTEMS WITH COMMON ANION

COMPONENTS:

(1) Barium methanoate (barium formate); $(CHO_2)_2Ba$; [541-43-5]
(2) Potassium methanoate (potassium formate); $(CHO_2)_2K_2$; [590-29-4]

ORIGINAL MEASUREMENTS:

Berchiesi, G.; Cingolani, A.; Leonesi, D.; Piantoni, G.
Can. J. Chem. 1972, **50**, 1972-1975.

VARIABLES:

Temperature.

PREPARED BY:

Baldini, P.

EXPERIMENTAL VALUES:

The experimental values are given only in graphical form (see figure).

Characteristic point(s):

Eutectic, E, at 162.6 ^{o}C and $\mathbf{x}_1$= 0.074 (authors).
Peritectic, P, at 192.2 ^{o}C and $\mathbf{x}_1$= 0.373 (authors).

Note - The investigation was limited to $\mathbf{x}_1 \leq 0.50$ due to thermal instability.

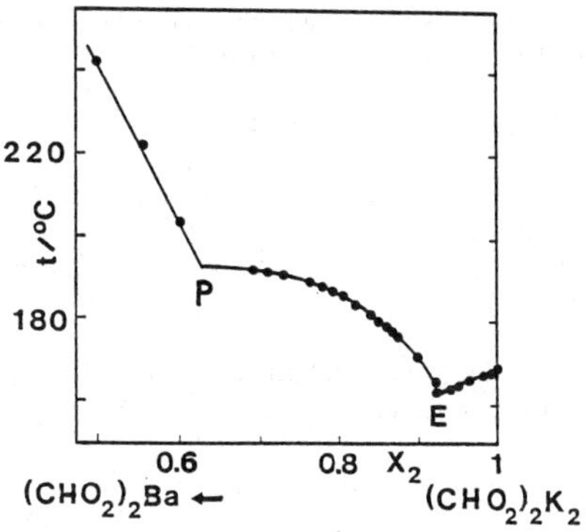

AUXILIARY INFORMATION

METHOD/APPARATUS/PROCEDURE:

A Pyrex device, suitable for work under an inert atmosphere, and allowing one to observe the system visually, was employed (for details, see Ref. 1). The initial crystallization temperatures were measured with a Chromel-Alumel thermocouple checked by comparison with a certified Pt resistance thermometer, and connected with a L&N Type K-3 potentiometer.

NOTE:

The fusion temperature of component 2 read by the compiler on the original plot, i.e., $\mathbf{T}_{fus}(2)\sim$ 169 ^{o}C (442 K) agrees satisfactorily with the value $T_{fus}(2)$= 441.9±0.5 K reported in Table 1 of the Preface. The authors' assertion that the negative deviation with respect to ideality of the liquidus branch richest in component 2 proves poor miscibility of the solid components in this region is reasonable. No assumption is made by the authors about the nature of the peritectic equilibrium.

SOURCE AND PURITY OF MATERIALS:

Component 1: K&K material of stated purity ≥ 99 %.

Component 2: C. Erba RP material of stated purity ≥ 99 %.

ESTIMATED ERROR:

Temperature: accuracy probably ±0.1 K (compiler).

REFERENCES:

(1) Braghetti,M.; Leonesi,D.; Franzosini,P. **Ric. Sci.** 1968, **38**, 116-118.

COMPONENTS:

(1) Barium methanoate (barium formate); $(CHO_2)_2Ba$; [541-43-5]
(2) Sodium methanoate (sodium formate); $(CHO_2)_2Na_2$; [141-53-7]

ORIGINAL MEASUREMENTS:

Berchiesi, G.; Cingolani, A.; Leonesi, D.; Piantoni, G.
Can. J. Chem. 1972, **50**, 1972-1975.

VARIABLES:

Temperature.

PREPARED BY:

Baldini, P.

EXPERIMENTAL VALUES:

The experimental values are given only in graphical form (see figure).

Characteristic point(s):

Eutectic, E, at 224.8 ^oC and x_1= 0.354 (authors).
Peritectic, P, at 242.0 ^oC and x_1= 0.518 (authors).

Note - The investigation was limited to $x_1 \leq 0.55$ due to thermal instability.

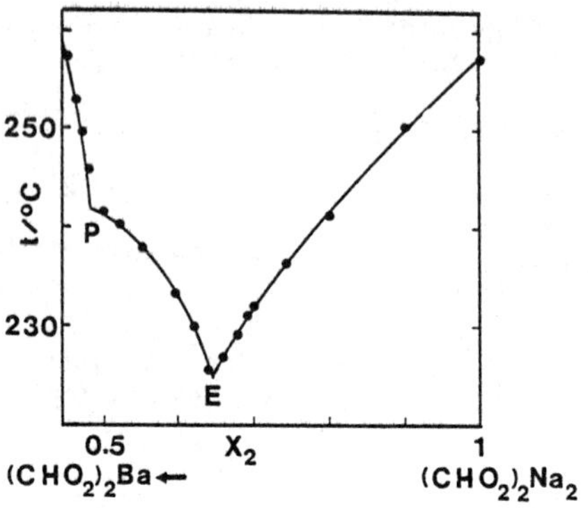

AUXILIARY INFORMATION

METHOD/APPARATUS/PROCEDURE:

A Pyrex device, suitable for work under an inert atmosphere, and allowing one to observe the system visually, was employed (for details, see Ref. 1). The initial crystallization temperatures were measured with a Chromel-Alumel thermocouple checked by comparison with a certified Pt resistance thermometer, and connected with a L&N Type K-3 potentiometer.

NOTE:

The fusion temperature of component 2 read by the compiler on the original plot, i.e., $T_{fus}(2) \sim 258$ ^oC (531 K) agrees satisfactorily with the value $T_{fus}(2) = 530.7 \pm 0.5$ K reported in Table 1 of the Preface. The authors' assertion that the negative deviation with respect to ideality of the liquidus branch richest in component 2 proves poor miscibility of the solid components in this region is reasonable. No assumption is made by the authors about the nature of the peritectic equilibrium.

SOURCE AND PURITY OF MATERIALS:

Component 1: K&K material of stated purity $\geq$ 99 %.

Component 2: C. Erba RP material of stated purity $\geq$ 99 %.

ESTIMATED ERROR:

Temperature: accuracy probably $\pm$0.1 K (compiler).

REFERENCES:

(1) Braghetti,M.; Leonesi,D.; Franzosini,P. **Ric. Sci.** 1968, **38**, 116-118.

COMPONENTS:

(1) Barium methanoate (barium formate); $(CHO_2)_2Ba$; [541-43-5]
(2) Thallium(I) methanoate (thallous formate); $(CHO_2)_2Tl_2$; [992-98-3]

ORIGINAL MEASUREMENTS:

Berchiesi, G.; Cingolani, A.; Leonesi, D.; Piantoni, G.
Can. J. Chem. 1972, **50**, 1972-1975.

VARIABLES:

Temperature.

PREPARED BY:

Baldini, P.

EXPERIMENTAL VALUES:

The experimental values are given only in graphical form (see figure).

Characteristic point(s):

Eutectic, E, at 95.4 °C and x_1= 0.079 (authors).

Note - The investigation was limited to $x_1 \leq 0.09$ due to thermal instability.

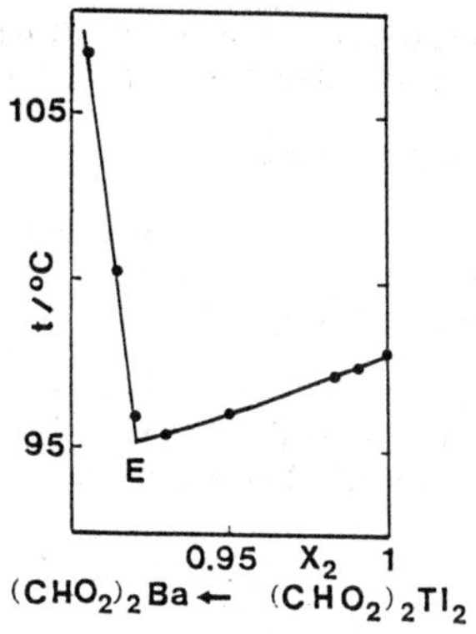

AUXILIARY INFORMATION

METHOD/APPARATUS/PROCEDURE:

A Pyrex device, suitable for work under an inert atmosphere, and allowing one to observe the system visually, was employed (for details, see Ref. 1). The initial crystallization temperatures were measured with a Chromel-Alumel thermocouple checked by comparison with a certified Pt resistance thermometer, and connected with a L&N Type K-3 potentiometer.

NOTE:

The fusion temperature of component 2 read by the compiler on the original plot, i.e., $T_{fus}(2) \sim$ 101 °C (374 K) coincides with the values determined with DSC by Braghetti et al. (Ref. 2), and with DTA by Meisel et al. (Ref. 3), although being 3 K lower than that obtained with hot-stage polarizing microscopy by Baum et al. (Ref. 4).

SOURCE AND PURITY OF MATERIALS:

Component 1: K&K material of stated purity $\geq$ 99 %.
Component 2: BDH material of stated purity $\geq$ 99 %.

ESTIMATED ERROR:

Temperature: accuracy probably ±0.1 K (compiler).

REFERENCES:

(1) Braghetti, M.; Leonesi, D.; Franzosini, P. **Ric. Sci.** 1968, **38**, 116-118.
(2) Braghetti, M.; Berchiesi, G.; Franzosini, P. **Ric. Sci.** 1969, **39**, 576-584.
(3) Meisel, T.; Seybold, K.; Halmos, Z.; Roth, J.; Melykuti, C. **J. Thermal Anal.** 1976, **10**, 419-431.
(4) Baum, E.; Demus, D.; Sackmann, H. **Wiss. Z. Univ. Halle** 1970, **19**, 37-46.

COMPONENTS:

(1) Calcium methanoate (calcium formate); $(CHO_2)_2Ca$; [544-17-2]
(2) Potassium methanoate (potassium formate); $(CHO_2)_2K_2$; [590-29-4]

ORIGINAL MEASUREMENTS:

Berchiesi, G.; Cingolani, A.; Leonesi, D.; Piantoni, G.
Can. J. Chem. 1972, **50**, 1972-1975.

VARIABLES:

Temperature.

PREPARED BY:

Baldini, P.

EXPERIMENTAL VALUES:

The experimental values are given only in graphical form (see figure).

Characteristic point(s):

Eutectic, E, at 163.2 °C and x_1= 0.057 (authors).

Note - The investigation was limited to $x_1 \leq 0.11$ due to thermal instability.

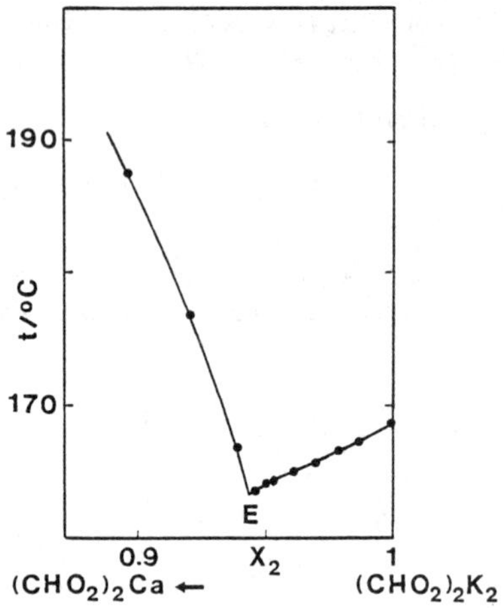

AUXILIARY INFORMATION

METHOD/APPARATUS/PROCEDURE:

A Pyrex device, suitable for work under an inert atmosphere, and allowing one to observe the system visually, was employed (for details, see Ref. 1). The initial crystallization temperatures were measured with a Chromel-Alumel thermocouple checked by comparison with a certified Pt resistance thermometer, and connected with a L&N Type K-3 potentiometer.

NOTE:

The fusion temperature of component 2 read by the compiler on the original plot, i.e., $T_{fus}(2) \sim 169$ °C (442 K) agrees satisfactorily with the value $T_{fus}(2) = 441.9 \pm 0.5$ K reported in Table 1 of the Preface. The authors' assertion that the negative deviation with respect to ideality of the liquidus branch richer in component 2 proves poor miscibility of the solid components in this region is reasonable.

SOURCE AND PURITY OF MATERIALS:

C. Erba RP materials of stated purity ≥ 99 %.

ESTIMATED ERROR:

Temperature: accuracy probably ±0.1 K (compiler).

REFERENCES:

(1) Braghetti, M.; Leonesi, D.; Franzosini, P. **Ric. Sci.** 1968, **38**, 116-118.

COMPONENTS:

(1) Calcium methanoate (calcium formate); $(CHO_2)_2Ca$; [544-17-2]
(2) Sodium methanoate (sodium formate); $(CHO_2)_2Na_2$; [141-53-7]

ORIGINAL MEASUREMENTS:

Berchiesi, G.; Cingolani, A.; Leonesi, D.; Piantoni, G.
Can. J. Chem. 1972, **50**, 1972-1975.

VARIABLES:

Temperature.

PREPARED BY:

Baldini, P.

EXPERIMENTAL VALUES:

The experimental values are given only in graphical form (see figure).

Characteristic point(s):

Eutectic, E, at 233.4 ^{o}C and x_1= 0.243 (authors).

Note - The investigation was limited to $x_1 \leq 0.27$ due to thermal instability.

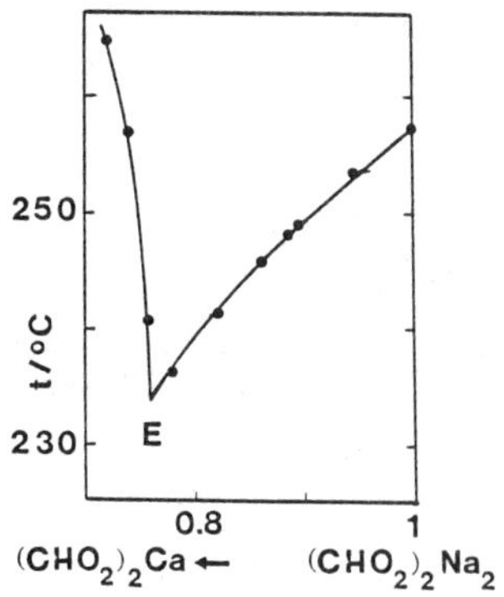

AUXILIARY INFORMATION

METHOD/APPARATUS/PROCEDURE:

A Pyrex device, suitable for work under an inert atmosphere, and allowing one to observe the system visually, was employed (for details, see Ref. 1). The initial crystallization temperatures were measured with a Chromel-Alumel thermocouple checked by comparison with a certified Pt resistance thermometer, and connected with a L&N Type K-3 potentiometer.

NOTE:

The fusion temperature of component 2 read by the compiler on the original plot, i.e., $T_{fus}(2) \sim 258$ ^{o}C (531 K) agrees satisfactorily with the value $T_{fus}(2) = 530.7 \pm 0.5$ K reported in Table 1 of the preface. The authors' assertion that the negative deviation with respect to ideality of the liquidus branch richer in component 2 proves poor miscibility of the solid components in this region is reasonable.

SOURCE AND PURITY OF MATERIALS:

C. Erba RP materials of stated purity $\geq$ 99 %.

ESTIMATED ERROR:

Temperature: accuracy probably $\pm$0.1 K (compiler).

REFERENCES:

(1) Braghetti, M.; Leonesi, D.; Franzosini, P. **Ric. Sci.** 1968, **38**, 116-118.

COMPONENTS:

(1) Calcium methanoate (calcium formate); $(CHO_2)_2Ca$; [544-17-2]
(2) Thallium(I) methanoate (thallous formate); $(CHO_2)_2Tl_2$; [992-98-3]

ORIGINAL MEASUREMENTS:

Berchiesi, G.; Cingolani, A.; Leonesi, D.; Piantoni, G.
Can. J. Chem. 1972, **50**, 1972-1975.

VARIABLES:

Temperature.

PREPARED BY:

Baldini, P.

EXPERIMENTAL VALUES:

The experimental values are given only in graphical form (see figure).

Characteristic point(s):

Eutectic, E, at 94.2 °C and x_1= 0.088 (authors).

Note - The investigation was limited to $x_1 \leq 0.11$ due to thermal instability.

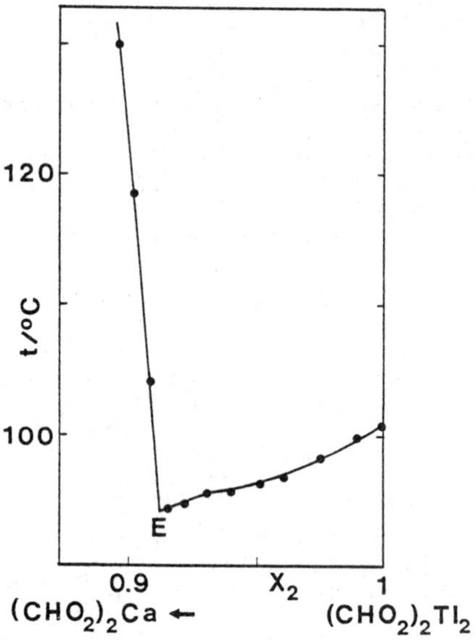

AUXILIARY INFORMATION

METHOD/APPARATUS/PROCEDURE:

A Pyrex device, suitable for work under an inert atmosphere, and allowing one to observe the system visually, was employed (for details, see Ref. 1). The initial crystallization temperatures were measured with a Chromel-Alumel thermocouple checked by comparison with a certified Pt resistance thermometer, and connected with a L&N Type K-3 potentiometer.

NOTE:

The fusion temperature of component 2 read by the compiler on the original plot, i.e., $T_{fus}(2) \sim$ 101 °C (374 K) coincides with the values determined with DSC by Braghetti et al. (Ref. 2), and with DTA by Meisel et al. (Ref. 3), although being 3 K lower than that obtained with hot-stage polarizing microscopy by Baum et al. (Ref. 4).

ESTIMATED ERROR:

Temperature: accuracy probably ±0.1 K (compiler).

SOURCE AND PURITY OF MATERIALS:

Component 1: C. Erba RP material of stated purity ≥ 99 %.
Component 2: BDH material of stated purity ≥ 99 %.

REFERENCES:

(1) Braghetti, M.; Leonesi, D.; Franzosini, P. **Ric. Sci.** 1968, **38**, 116-118.
(2) Braghetti, M.; Berchiesi, G.; Franzosini, P. **Ric. Sci.** 1969, **39**, 576-584.
(3) Meisel, T.; Seybold, K.; Halmos, Z.; Roth, J.; Melykuti, C. **J. Thermal Anal.** 1976, **10**, 419-431.
(4) Baum, E.; Demus, D.; Sackmann, H. **Wiss. Z. Univ. Halle** 1970, **19**, 37-46.

COMPONENTS:	EVALUATOR:
(1) Potassium methanoate (potassium formate); $(CHO_2)K$; [590-29-4] (2) Lithium methanoate (lithium formate); $(CHO_2)Li$; [556-63-8]	Franzosini, P., Dipartimento di Chimica Fisica, Universita´ di Pavia (ITALY).

CRITICAL EVALUATION:

This system was studied by Sokolov and Tsindrik (Ref. 1) as a side of the reciprocal ternary K, Li/CHO_2, NO_3, and by Pochtakova (Ref. 2) as a side of the ternary CHO_2/K, Li, Na. In both cases, the visual polythermal analysis was employed, and the investigation was restricted to the liquidus.

The obtained results, i.e., formation of a 1:1 congruently melting intermediate compound giving a eutectic with either component, are qualitatively similar. It is, however, to be remarked that no explanation is offered by Pochtakova (Ref. 2, where Ref. 1 is quoted) for the considerable difference between the temperature she found (427 K) for the eutectic at $100\mathbf{x}_1$ about 40, and that (413 K) measured previously by Sokolov and Tsindrik (Ref. 1).

The fusion temperatures of the pure components reported in both Ref. 1 and Ref. 2, i.e., $\mathbf{T}_{fus}(1)$= 440 K, $\mathbf{T}_{fus}(2)$= 546 K, are in fair agreement with those listed in Table 1 of the Preface (441.9±0.5 K, 546±1 K). On the contrary, poor correspondence exists between solid state transition temperatures quoted in Ref. 1 from Ref. 3 (i.e., 333, 408, and 430 K for component 1; 360, 388, and 505 for component 2) and those listed in Table 1 of the Preface (418±1 K for component 1, and 496±2 K for component 2).

REFERENCES:

(1) Sokolov, N.M.; Tsindrik, N.M.
Zh. Neorg. Khim. 1969, **14**, 584-590 (*); **Russ. J. Inorg. Chem. (Engl. Transl.)** 1969, **14**, 302-306.

(2) Pochtakova, E.I.
Zh. Neorg. Khim. 1980, 25, 1147-1150; **Russ. J. Inorg. Chem. (Engl. Transl.)** 1980, **25**, 637-639 (*).

(3) Sokolov, N.M.
Tezisy Dokl. X Nauch. Konf. S.M.I. 1956.

COMPONENTS:	ORIGINAL MEASUREMENTS:
(1) Potassium methanoate (potassium formate); $(CHO_2)K$; [590-29-4] (2) Lithium methanoate (lithium formate); $(CHO_2)Li$; [556-63-8]	Sokolov, N.M.; Tsindrik, N.M. **Zh. Neorg. Khim.** 1969, **14**, 584-590 (*); **Russ. J. Inorg. Chem. (Engl. Transl.)** 1969, **14**, 302-306.
VARIABLES:	PREPARED BY:
Temperature.	Baldini, P.

EXPERIMENTAL VALUES:

The results are reported only in graphical form (see figure).

Characteristic point(s):

Eutectic, E_1, at 118 °C and $100x_1$= 75 (authors).
Eutectic, E_2, at 140 °C and $100x_1$= 39.5 (authors).

Intermediate compound(s):

$(CHO_2)_2KLi$ (probable composition), congruently melting (authors).

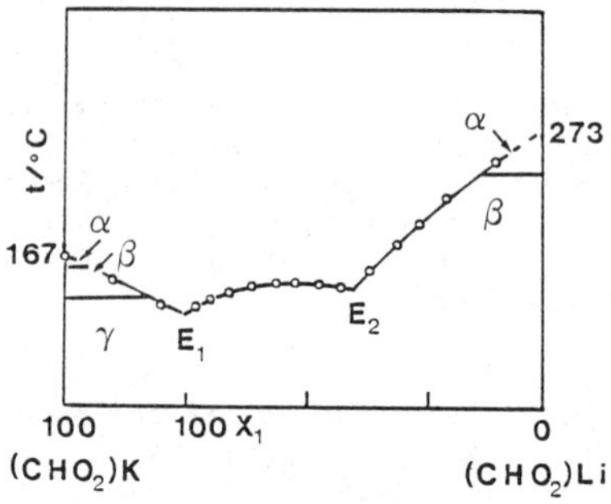

AUXILIARY INFORMATION

METHOD/APPARATUS/PROCEDURE:

Visual polythermal method.

SOURCE AND PURITY OF MATERIALS:

Commercial materials recrystallized.
Component 1 undergoes phase transitions at $t_{trs}(1)/°C$= 60, 135, 157 (Ref. 1) and melts at $t_{fus}(1)/°C$= 167.
Component 2 undergoes phase transitions at $t_{trs}(2)/°C$= 87, 115, 232 and melts at $t_{fus}(2)/°C$= 273.

ESTIMATED ERROR:

Temperature: accuracy probably ±2 K (compiler).

REFERENCES:

(1) Sokolov, N.M.
Tezisy Dokl. X Nauch. Konf. S.M.I. 1956.

COMPONENTS:

(1) Potassium methanoate (potassium formate); $(CHO_2)K$; [590-29-4]
(2) Lithium methanoate (lithium formate); $(CHO_2)Li$; [556-63-8]

ORIGINAL MEASUREMENTS:

Pochtakova, E.I.
Zh. Neorg. Khim. 1980, 25, 1147-1150; **Russ. J. Inorg. Chem. (Engl. Transl.)** 1980, 25, 637-639 (*).

VARIABLES:

Temperature.

PREPARED BY:

Baldini, P.

EXPERIMENTAL VALUES:

The results are reported only in graphical form (see figure).

Characteristic point(s):

Eutectic, E_1, at 121 °C (author) and $100\mathbf{x}_2$= 25 (according to Fig. 1 and Fig. 2 of the original paper, erroneously reported as $100\mathbf{x}_1$ in the text; compiler).
Eutectic, E_2, at 154 °C (author) and $100\mathbf{x}_2$= 60 (according to Fig. 1 and Fig. 2 of the original paper, erroneously reported as $100\mathbf{x}_1$ in the text; compiler).

Intermediate compound(s):

$(CHO_2)_2KLi$, congruently melting at 163 °C (author).

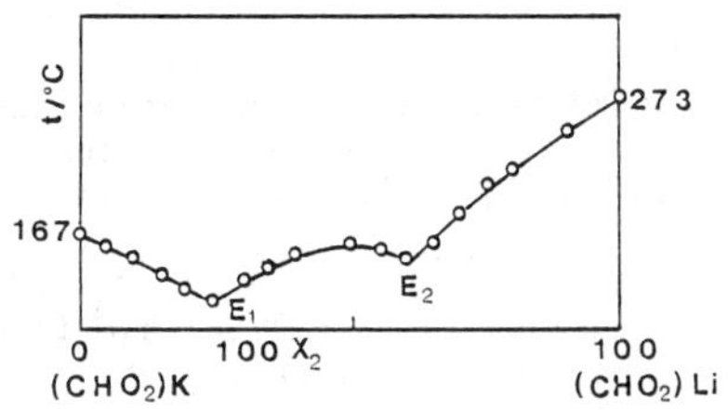

AUXILIARY INFORMATION

METHOD/APPARATUS/PROCEDURE:

Visual polythermal method.

SOURCE AND PURITY OF MATERIALS:

Not stated. Component 1: $t_{fus}(1)/°C$= 167. Component 2: $t_{fus}(2)/°C$= 273.

ESTIMATED ERROR:

Temperature: accuracy probably ±2 K (compiler).

REFERENCES:

COMPONENTS:	EVALUATOR:
(1) Potassium methanoate (potassium formate); $(CHO_2)K$; [590-29-4] (2) Magnesium methanoate (magnesium formate) $(CHO_2)_2Mg$; [557-39-1]	Franzosini, P., Dipartimento di Chimica Fisica, Universita´ di Pavia (ITALY).

CRITICAL EVALUATION:

The system was studied by Berchiesi et al. (Ref. 1), who indicated component 1 as $(CHO_2)_2K_2$, and by Pochtakova (Ref. 2), who indicated component 1 as $(CHO_2)K$ in Table 6 of her paper, and as $(CHO_2)_2K_2$ in Fig. 2. Inspection of text and figures led both the compiler and evaluator to assume the latter formula as the correct one: consequently, a direct comparison is possible between data from either sources.

Comparison makes apparent that Pochtakova (Ref. 2), who seems not to be aware of Ref. 1, could obtain no evidence for the eutectic due to the fact that she performed no measurements at $0 < 100\mathbf{x}_2 \leq 5$, while the eutectic composition (Ref. 1) is $100\mathbf{x}_2 = 1.3$.

It is to be added that: (i) Berchiesi et al.´s fusion temperature of component 1 read by the evaluator on the original plot, i.e., $T_{fus}(1) \simeq 169$ °C (442 K) agrees with the value $T_{fus}(1) = 441.9 \pm 0.5$ K reported in Table 1 of the Preface more satisfactorily than Pochtakova´s figure (440 K); (ii) the solid state transition temperatures quoted for component 1 in Ref. 2 from Ref. 3 (i.e., 333, 408, and 430 K) cannot be identified with the relevant data of Table 1 of the Preface, where a single transition is mentioned which occurs at $T_{trs}(1)/K = 418 \pm 1$; and (iii) Pochtakova´s points are affected by a scattering noticeably larger than Berchiesi et al.´s.

In conclusion, the evaluator recommends the data by Berchiesi et al. (Ref. 1), although regretting that they are presented only in graphical form, and not supported by any investigation of the solidus.

REFERENCES:

(1) Berchiesi, G.; Cingolani, A.; Leonesi, D.; Piantoni, G. **Can. J. Chem.** 1972, 50, 1972-1975.

(2) Pochtakova, E.I. **Zh. Obshch. Khim.** 1974, 44, 241-248.

(3) Sokolov, N.M. **Tezisy Dokl. X Nauch. Konf. S.M.I.** 1956.

COMPONENTS:

(1) Potassium methanoate (potassium formate); $(CHO_2)_2K_2$; [590-29-4]
(2) Magnesium methanoate (magnesium formate); $(CHO_2)_2Mg$; [557-39-1]

ORIGINAL MEASUREMENTS:

Berchiesi, G.; Cingolani, A.; Leonesi, D.; Piantoni, G.
Can. J. Chem. 1972, **50**, 1972-1975.

VARIABLES:

Temperature.

PREPARED BY:

Baldini, P.

EXPERIMENTAL VALUES:

The experimental values are given only in graphical form (see figure).

Characteristic point(s):

Eutectic, E, at 167.4 °C and x_2= 0.013 (authors).

Note - The investigation was limited to $x_1 \geq 0.97$ due to thermal instability.

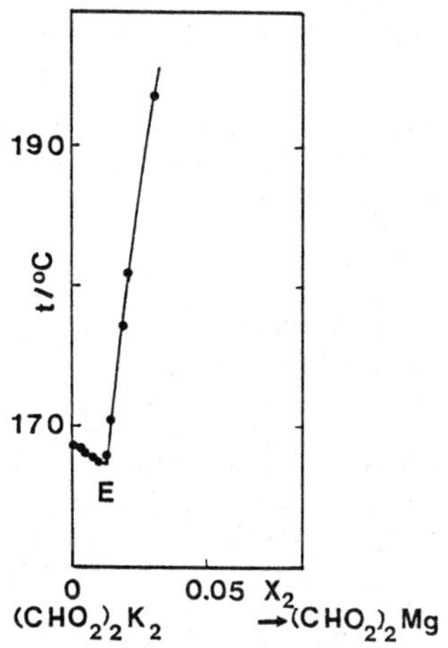

AUXILIARY INFORMATION

METHOD/APPARATUS/PROCEDURE:

A Pyrex device, suitable for work under an inert atmosphere, and allowing one to observe the system visually, was employed (for details, see Ref. 1). The initial crystallization temperatures were measured with a Chromel-Alumel thermocouple checked by comparison with a certified Pt resistance thermometer, and connected with a L&N Type K-3 potentiometer.

SOURCE AND PURITY OF MATERIALS:

Component 1: C. Erba RP material of stated purity ≥ 99 %.
Component 2: K&K material of stated purity ≥ 99 %.

ESTIMATED ERROR:

Temperature: accuracy probably ±0.1 K (compiler).

REFERENCES:

(1) Braghetti,M.; Leonesi,D.; Franzosini,P.
Ric. Sci. 1968, **38**, 116-118.

COMPONENTS:	ORIGINAL MEASUREMENTS:
(1) Potassium methanoate (potassium formate); $(CHO_2)_2K_2$; [590-29-4] (2) Magnesium methanoate (magnesium formate); $(CHO_2)_2Mg$; [557-39-1]	Pochtakova, E.I. **Zh. Obshch. Khim.** 1974, **44**, 241-248.
VARIABLES:	PREPARED BY:
Temperature.	Baldini, P.

EXPERIMENTAL VALUES:

$t/^{\circ}C$	T/K^a	$100x_2$
167	440	0
190	463	5
212	485	7.5
225	498	10
239	512	12.5
243	516	15
257	530	17.5
269	542	20
280	553	22.5
287	560	25

[a] T/K values calculated by the compiler.

Note - The system was investigated at $0 \leq 100x_2 \leq 25$ due to thermal instability of component 2. No characteristic point was observed in the mentioned composition region.

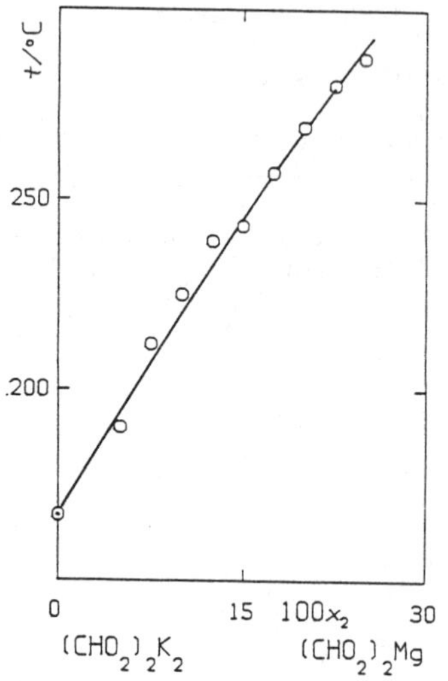

AUXILIARY INFORMATION

METHOD/APPARATUS/PROCEDURE:

Visual polythermal method, supplemented with differential thermal analysis (no numerical DTA data, however, are tabulated by the author).

SOURCE AND PURITY OF MATERIALS:

Materials prepared (Ref. 1) by reacting the proper ("chemically pure") carbonate with a slight excess of methanoic acid of analytical purity.
Component 1 undergoes phase transitions at $t_{trs}(1)/^{\circ}C$= 60, 135, 157 (Ref. 2).
Component 2 undergoes a phase transition at $t_{trs}(2)/^{\circ}C$= 140.

ESTIMATED ERROR:

Temperature: accuracy probably ±2 K (compiler).

REFERENCES:

(1) Sokolov, N.M.
Zh. Obshch. Khim. 1954, 24, 1581-1593.
(2) Sokolov, N.M.
Tezisy Dokl. X Nauch. Konf. S.M.I. 1956.

COMPONENTS:	EVALUATOR:
(1) Potassium methanoate (potassium formate); $(CHO_2)K$; [590-29-4] (2) Sodium methanoate (sodium formate); $(CHO_2)Na$; [141-53-7]	Spinolo, G., Dipartimento di Chimica Fisica, Universita´ di Pavia (ITALY).

CRITICAL EVALUATION:

The binary CHO_2/K, Na was studied by Dmitrevskaya (as a side system of the reciprocal ternary CHO_2, NO_3/K, Na; Ref. 1) and by Leonesi et al. (as a side system of the reciprocal ternary CHO_2, Cl/K, Na; Ref. 2). In both papers, visual observation was employed, and investigation was restricted to the liquidus; moreover, the latter authors listed only the few numerical data which were relevant to their purposes.

The main features of the phase diagrams given in either source exhibit rather close similarities, as shown here:

	Ref. 1	Ref. 2
$T_{fus}(1)$/K:	440	441.9
$T_{fus}(2)$/K:	531	530.7
Intermediate compound (i.c.)	$(CHO_2)_4K_3Na$	$(CHO_2)_4K_3Na$
T_{fus}(i.c.)/K:	-	453.2
Eutectic E_1; T/K:	441	438.2
Eutectic E_1; x_2:	0.505	0.495
Eutectic E_2; T/K:	440	436.7
Eutectic E_2; x_2:	0.04	0.043

It is, however, to be stressed that: (i) Dmitrevskaya´s liquidus branch rich in component 1 exhibits a maximum (unexplained by the author) at 444 K and x_1= 0.98, whereas Leonesi et al. found a monotonically decreasing trend; and (ii) Dmitrevskaya quotes (from Ref. 3) the occurrence of phase transitions in component 1 (at 333, 408, and 430 K), and in component 2 (at 515 K) which have no correspondence in Table 1 of the Preface.

Due to these reasons, and to the higher accuracy to be attributed to the findings by Leonesi et al., the evaluator is inclined to recommend the data listed above under the heading "Ref. 2".

It is finally to be added that previous cryometric work had allowed Leonesi et al. (Ref. 4) to infer, on the basis of the well known equation

$$\lim_{m \to 0} \frac{(\Delta T/m)}{K} = 1 - \rho_0$$

(K: cryometric constant of component 1, used as the solvent; ΔT: experimental freezing point depression; m: molality of component 2, used as the solute), a limiting value ρ_0 about 0.17 for the ratio between the solute concentrations in the solid and liquid phases at equilibrium.

REFERENCES:

(1) Dmitrevskaya, O.I.
Zh. Obshch. Khim. 1958, **28**, 299-304 (*); **Russ. J. Gen. Chem. (Engl. Transl.)** 1958, **28**, 295-300.
(2) Leonesi, D.; Braghetti, M.; Cingolani, A.; Franzosini, P.
Z. Naturforsch. 1970, **25a**, 52-55.
(3) Sokolov, N.M.
Tezisy Dokl. X Nauch. Konf. S.M.I. 1956.
(4) Leonesi, D.; Piantoni, G.; Berchiesi, G.; Franzosini, P.
Ric. Sci. 1968, **38**, 702-705.

COMPONENTS:

(1) Potassium methanoate (potassium formate); $(CHO_2)K$; [590-29-4]
(2) Sodium methanoate (sodium formate); $(CHO_2)Na$; [141-53-7]

ORIGINAL MEASUREMENTS:

Dmitrevskaya, O.I.
Zh. Obshch. Khim. 1958, **28**, 299-304 (*);
Russ. J. Gen. Chem. (Engl. Transl.) 1958, **28**, 295-300.

VARIABLES:

Temperature.

PREPARED BY:

Baldini, P.

EXPERIMENTAL VALUES:

t/°C	T/K[a]	$100x_1$	t/°C	T/K[a]	$100x_1$
258	531	0	176	449	60
252	525	5	179	452	65
246	519	10	180	453	70
240	513	15	182	455	75
232	505	20	181	454	80
222	495	25	179	452	85
209	482	30	173	446	90
198	471	35	167	440	96
186	459	40	168	441	97
176	449	45	171	444	98
168	441	49.5	169	442	99
169	442	50	168	441	99.5
172	445	55	167	440	100

[a] T/K values calculated by the compiler.

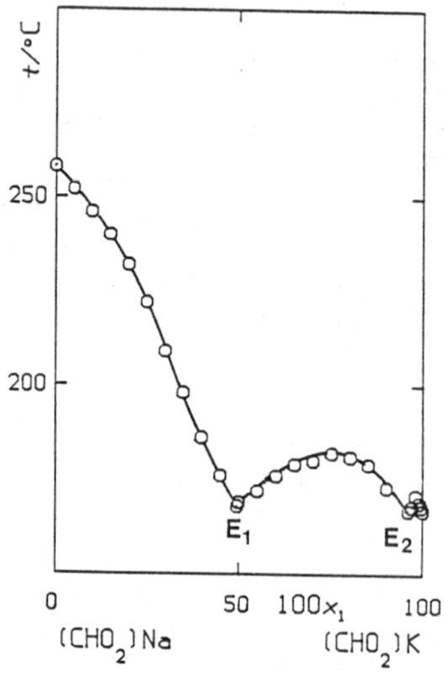

Characteristic point(s):

Eutectic, E_1, at 168 °C and $100x_1$= 49.5 (author).
Eutectic, E_2, at 167 °C and $100x_2$= 4 (author).

Intermediate compound(s):
$(CHO_2)_4K_3Na$ (congruently melting at 182 °C, compiler).

AUXILIARY INFORMATION

METHOD/APPARATUS/PROCEDURE:

Visual polythermal method; temperatures measured with a Nichrome-Constantane thermocouple.

SOURCE AND PURITY OF MATERIALS:

"Chemically pure" materials, recrystallized and dried to constant mass.
Component 1 undergoes phase transitions at $t_{trs}(1)/°C$= 60, 135, 157 (Ref. 1).
Component 2 undergoes a phase transition at $t_{trs}(2)/°C$= 242 (Ref. 1).

ESTIMATED ERROR:

Temperature: accuracy probably ±2 K (compiler).

REFERENCES:

(1) Sokolov, N.M.
Tezisy Dokl. X Nauch. Konf. S.M.I. 1956.

COMPONENTS:

(1) Potassium methanoate (potassium formate); $(CHO_2)K$; [590-29-4]
(2) Sodium methanoate (sodium formate); $(CHO_2)Na$; [141-53-7]

ORIGINAL MEASUREMENTS:

Leonesi, D.; Braghetti, M.; Cingolani, A.; Franzosini, P.
Z. Naturforsch. 1970, **25a**, 52-55.

VARIABLES:

Temperature.

PREPARED BY:

Baldini, P.

EXPERIMENTAL VALUES:

Characteristic point(s):

Eutectic, E_1, at 165.0 °C and $100\mathbf{x}_1$= 50.5 (authors).
Eutectic, E_2, at 163.5 °C and $100\mathbf{x}_1$= 95.7 (authors).

Intermediate compound(s):

$(CHO_2)_4K_3Na$, congruently melting at 180.0 °C (authors).

AUXILIARY INFORMATION

METHOD/APPARATUS/PROCEDURE:

A Pyrex device, suitable for work under an inert atmosphere, and allowing one to observe the system visually, was employed (for details, see Ref. 1). The initial crystallization temperatures were measured with a Chromel-Alumel thermocouple checked by comparison with a certified Pt resistance thermometer, and connected with a L&N Type K-3 potentiometer.

SOURCE AND PURITY OF MATERIALS:

C. Erba RP materials, dried by heating under vacuum.
Component 1: $t_{fus}(1)/^{o}C$= 168.7.
Component 2: $t_{fus}(2)/^{o}C$= 257.5.

ESTIMATED ERROR:

Temperature: accuracy probably ±0.1 K.

REFERENCES:

(1) Braghetti,M.; Leonesi,D.; Franzosini,P. **Ric. Sci.** 1968, **38**, 116-118.

COMPONENTS:

(1) Potassium methanoate (potassium formate); $(CHO_2)_2K_2$; [590-29-4]
(2) Strontium methanoate (strontium formate); $(CHO_2)_2Sr$; [592-89-2]

ORIGINAL MEASUREMENTS:

Berchiesi, G.; Cingolani, A.; Leonesi, D.; Piantoni, G.
Can. J. Chem. 1972, **50**, 1972-1975.

VARIABLES:

Temperature.

PREPARED BY:

Baldini, P.

EXPERIMENTAL VALUES:

The experimental values are given only in graphical form (see figure).

Characteristic point(s):

Eutectic, E, at 153.2 °C and x_2= 0.150 (authors).
Peritectic, P, at 170.8 °C and x_2= 0.327 (authors).

Note - The investigation was limited to $x_1 \geq 0.60$ due to thermal instability.

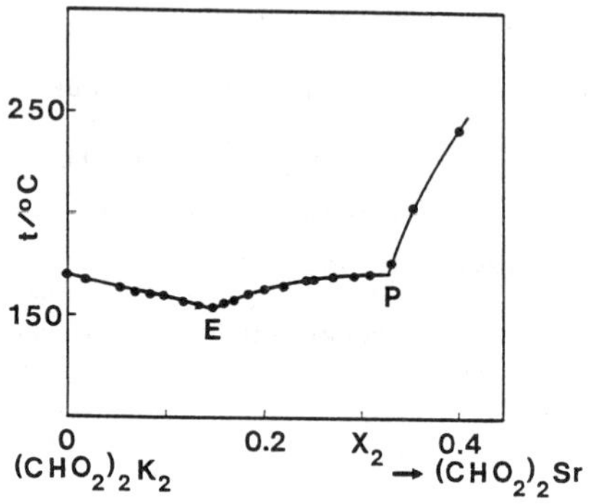

AUXILIARY INFORMATION

METHOD/APPARATUS/PROCEDURE:

A Pyrex device, suitable for work under an inert atmosphere, and allowing one to observe the system visually, was employed (for details, see Ref. 1). The initial crystallization temperatures were measured with a Chromel-Alumel thermocouple checked by comparison with a certified Pt resistance thermometer, and connected with a L&N Type K-3 potentiometer.

NOTE:

The fusion temperature of component 1 read by the compiler on the original plot, i.e., $T_{fus}(1) \sim 169$ °C (442 K) agrees satisfactorily with the value $T_{fus}(1) = 441.9 \pm 0.5$ K reported in Table 1 of the Preface. The authors' assertion that the negative deviation with respect to ideality of the liquidus branch richest in component 2 proves poor miscibility of the solid components in this region is reasonable. No assumption is made by the authors about the nature of the peritectic equilibrium.

SOURCE AND PURITY OF MATERIALS:

Component 1: C. Erba RP material of stated purity $\geq$ 99 %.
Component 2: K&K material of stated purity $\geq$ 99 %.

ESTIMATED ERROR:

Temperature: accuracy probably $\pm$0.1 K (compiler).

REFERENCES:

(1) Braghetti, M.; Leonesi, D.; Franzosini, P. **Ric. Sci.** 1968, **38**, 116-118.

COMPONENTS:	ORIGINAL MEASUREMENTS:
(1) Lithium methanoate (lithium formate); $(CHO_2)Li$; [556-63-8] (2) Sodium methanoate (sodium formate); $(CHO_2)Na$; [141-53-7]	Tsindrik, N.M. **Zh. Obshch. Khim.** 1958, **28**, 830-834.
VARIABLES:	PREPARED BY:
Temperature.	Baldini, P.

EXPERIMENTAL VALUES:

$t/^{o}C$	T/K^{a}	$100\mathbf{x}_1$
258	531	0
250	523	5
242	515	10
232	505	15
224	497	20
214	487	25
204	477	30
194	467	35
182	455	40
176	449	45
170	443	50
175	448	55
184	457	60
196	469	65
208	481	70
220	493	75
232	505	80

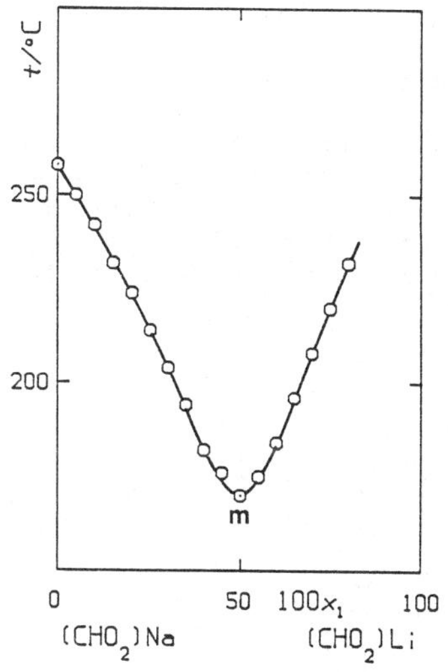

[a] T/K values calculated by the compiler.

Characteristic point(s):

Minimum, m, at 170 ^{o}C and $100\mathbf{x}_2$= 50 (author).

Note - The system was investigated at $0 \leq 100\mathbf{x}_1 \leq 80$.

AUXILIARY INFORMATION

METHOD/APPARATUS/PROCEDURE:

Visual polythermal method; temperatures measured with a Nichrome-Constantane thermocouple.

NOTE:

The fusion temperatures of both components, $\mathbf{T}_{fus}(1)$= 546 K and $\mathbf{T}_{fus}(2)$= 531 K, are in excellent agreement with the corresponding values listed in Table 1 of the Preface. The abscissa of point m, $100\mathbf{x}_2$= 50, coincides with that found by Pochtakova (Ref. 1), whereas its ordinate, 443 K, is somewhat lower than Pochtakova's value, i.e., 449 K.

SOURCE AND PURITY OF MATERIALS:

Materials of analytical purity recrystallized twice (extrapolated $\mathbf{t}_{fus}/^{o}C$ of lithium methanoate: 273; author).

ESTIMATED ERROR:

Temperature: accuracy probably ±2 K (compiler).

REFERENCES:

(1) Pochtakova, E.I.
Zh. Neorg. Khim. 1980, **25**, 1147-1150;
Russ. J. Inorg. Chem. (Engl. Transl.) 1980, 25, 637-639 (*).

COMPONENTS:

(1) Lithium methanoate (lithium formate); $(CHO_2)Li$; [556-63-8]
(2) Sodium methanoate (sodium formate); $(CHO_2)Na$; [141-53-7]

ORIGINAL MEASUREMENTS:

Pochtakova, E.I.
Zh. Neorg. Khim. 1980, 25, 1147-1150; **Russ. J. Inorg. Chem. (Engl. Transl.)** 1980, 25, 637-639 (*).

VARIABLES:

Temperature

PREPARED BY:

Baldini, P.

EXPERIMENTAL VALUES:

The results are reported only in graphical form (see figure).

Characteristic point(s):

Continuous series of solid solutions with a minimum, m, at 176 ^{o}C (according to Fig. 1 and Fig. 2 of the original paper, erroneously reported as 716 in the text; compiler) and $100x_2$= 50 (author).

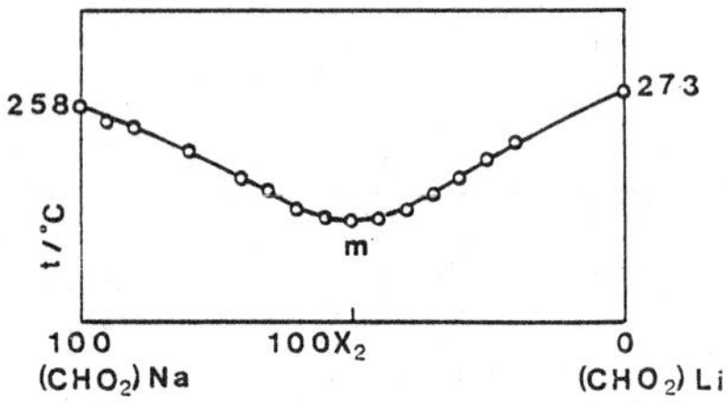

AUXILIARY INFORMATION

METHOD/APPARATUS/PROCEDURE:

Visual polythermal method.

NOTE:

The fusion temperatures of both components, $T_{fus}(1)$= 546 K and $T_{fus}(2)$= 531 K, are in excellent agreement with the corresponding values listed in Table 1 of the Preface. The abscissa of point m, $100x_2$= 50, coincides with that found by Tsindrik (Ref. 1), whereas its ordinate, 449 K, is somewhat higher than Tsindrik's value, i.e., 443 K.

SOURCE AND PURITY OF MATERIALS:

Not stated.
Component 1: $t_{fus}(1)/^{o}C$= 273.
Component 2: $t_{fus}(2)/^{o}C$= 258.

ESTIMATED ERROR:

Temperature: accuracy probably $\pm$2 K (compiler).

REFERENCES:

(1) Tsindrik, N.M.
Zh. Obshch. Khim. 1958, **28**, 830-834.

COMPONENTS:

(1) Magnesium methanoate (magnesium formate); $(CHO_2)_2Mg$; [557-39-1]
(2) Sodium methanoate (sodium formate); $(CHO_2)_2Na_2$; [141-53-7]

ORIGINAL MEASUREMENTS:

Pochtakova, E.I.
Zh. Obshch. Khim. 1974, **44**, 241-248.

VARIABLES:

Temperature.

PREPARED BY:

Baldini, P.

EXPERIMENTAL VALUES:

$t/^{o}C$	T/K^{a}	$100x_1$
258	531	0
257	530	2.5
256	529	5
255	528	7.5
251	524	10
253	526	15
253	526	17.5
252	525	20
257	530	22.5
267	540	25
282	555	27.5
300	573	30

[a] T/K values calculated by the compiler.

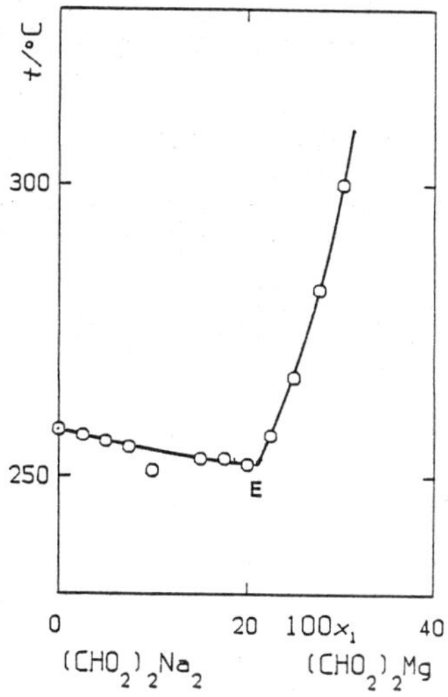

Note - The system was investigated at $0 \leq 100x_1 \leq 30$ due to thermal instability of component 1.

Eutectic, E, at 252 °C and $100x_1$= 21 (author).

AUXILIARY INFORMATION

METHOD/APPARATUS/PROCEDURE:

Visual polythermal method.

NOTE:

Concerning component 2, it can be remarked that the fusion temperature given by the author, $T_{fus}(2)$= 531 K, is in excellent agreement with the value listed in Table 1 of the Preface, i.e., 530.7±0.5 K, whereas the value quoted from Ref. 2 for the solid state transition temperature, $T_{trs}(2)$= 515 K, is noticeably higher than that reported in the Table, i.e., 502±5 K. It can be added that Berchiesi et al. (Ref. 3) asserted they could not investigate this binary due to thermal instability of the mixtures of any composition.

SOURCE AND PURITY OF MATERIALS:

Materials prepared by reacting the proper ("chemically pure") carbonate with a slight excess of methanoic acid of analytical purity (Ref. 1). Component 1 undergoes a phase transition at $t_{trs}(1)/^{o}C$= 140. Component 2 undergoes a phase transition at $t_{trs}(2)/^{o}C$= 242 (Ref. 2).

ESTIMATED ERROR:

Temperature: accuracy probably ±2 K (compiler).

REFERENCES:

(1) Sokolov, N.M.
Zh. Obshch. Khim. 1954, **24**, 1581-1593.
(2) Sokolov, N.M.
Tezisy Dokl. X Nauch. Konf. S.M.I. 1956.
(3) Berchiesi, G.; Cingolani, A.; Leonesi, D.; Piantoni, G.
Can. J. Chem. 1972, **50**, 1972-1975.

COMPONENTS:	ORIGINAL MEASUREMENTS:
(1) Magnesium methanoate (magnesium formate); $(CHO_2)_2Mg$; [557-39-1] (2) Thallium(I) methanoate (thallous formate); $(CHO_2)_2Tl_2$; [992-98-3]	Berchiesi, G.; Cingolani, A.; Leonesi, D.; Piantoni, G. **Can. J. Chem.** 1972, **50**, 1972-1975.
VARIABLES:	PREPARED BY:
Temperature.	Baldini, P.

EXPERIMENTAL VALUES:

The experimental values are given only in graphical form (see figure).

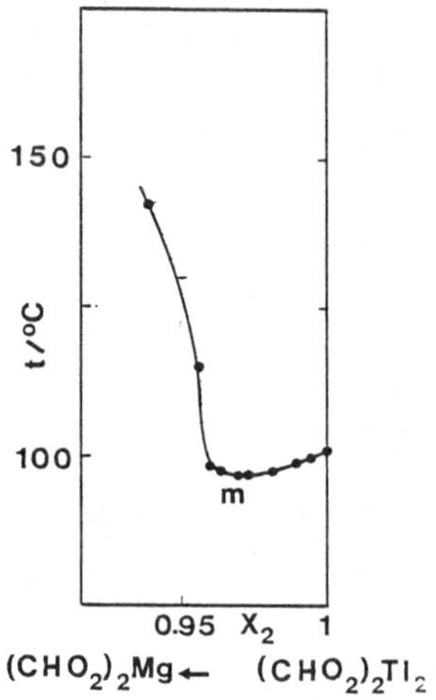

Characteristic point(s):

Minimum, m, at 97.0 °C and x_1= 0.030 (authors).

Note - The investigation was limited to $x_1 \leq 0.06$ due to thermal instability.

AUXILIARY INFORMATION

METHOD/APPARATUS/PROCEDURE:

A Pyrex device, suitable for work under an inert atmosphere, and allowing one to observe the system visually, was employed (for details, see Ref. 1). The initial crystallization temperatures were measured with a Chromel-Alumel thermocouple checked by comparison with a certified Pt resistance thermometer, and connected with a L&N Type K-3 potentiometer.

NOTE:

The fusion temperature of component 2 read by the compiler on the original plot, i.e., $T_{fus}(2) \sim 101$ °C (374 K) coincides with the values determined with DSC by Braghetti et al. (Ref. 2), and with DTA by Meisel et al. (Ref. 3), although being 3 K lower than that obtained with hot-stage polarizing microscopy by Baum et al. (Ref. 4). Solid solutions ought to form.

SOURCE AND PURITY OF MATERIALS:

Component 1: K&K material of stated purity $\geq$ 99 %.
Component 2: BDH material of stated purity $\geq$ 99 %.

ESTIMATED ERROR:

Temperature: accuracy probably $\pm$0.1 K (compiler).

REFERENCES:

(1) Braghetti, M.; Leonesi, D.; Franzosini, P. **Ric. Sci.** 1968, **38**, 116-118.
(2) Braghetti, M.; Berchiesi, G.; Franzosini, P. **Ric. Sci.** 1969, **39**, 576-584.
(3) Meisel, T.; Seybold, K.; Halmos, Z.; Roth, J.; Melykuti, C. **J. Thermal Anal.** 1976, **10**, 419-431.
(4) Baum, E.; Demus, D.; Sackmann, H. **Wiss. Z. Univ. Halle** 1970, **19**, 37-46.

COMPONENTS:

(1) Sodium methanoate (sodium formate); $(CHO_2)_2Na_2$; [141-53-7]
(2) Strontium methanoate (strontium formate); $(CHO_2)_2Sr$; [592-89-2]

ORIGINAL MEASUREMENTS:

Berchiesi, G.; Cingolani, A.; Leonesi, D.; Piantoni, G.
Can. J. Chem. 1972, **50**, 1972-1975.

VARIABLES:

Temperature.

PREPARED BY:

Baldini, P.

EXPERIMENTAL VALUES:

The experimental values are given only in graphical form (see figure).

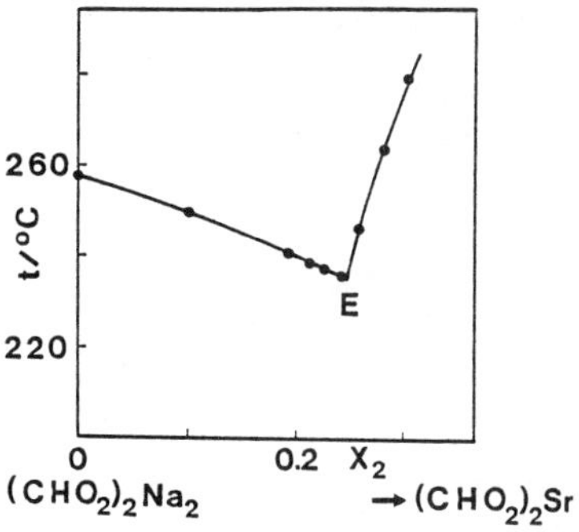

Characteristic point(s):

Eutectic, E, at 235.4 °C and x_2= 0.246 (authors).

Note - The investigation was limited to $x_1 \geq 0.70$ due to thermal instability.

AUXILIARY INFORMATION

METHOD/APPARATUS/PROCEDURE:

A Pyrex device, suitable for work under an inert atmosphere, and allowing one to observe the system visually, was employed (for details, see Ref. 1). The initial crystallization temperatures were measured with a Chromel-Alumel thermocouple checked by comparison with a certified Pt resistance thermometer, and connected with a L&N Type K-3 potentiometer.

NOTE:

The fusion temperature of component 1 read by the compiler on the original plot, i.e., $T_{fus}(1) \sim 258$ °C (531 K) satisfactorily agrees with the value (530.7±0.5 K) reported in Table 1 of the Preface. The authors' assertion that the negative deviation with respect to ideality of the liquidus branch richer in component 2 proves poor miscibility of the solid components in this region is reasonable.

SOURCE AND PURITY OF MATERIALS:

Component 1: C. Erba RP material of stated purity ≥ 99 %.
Component 2: K&K material of stated purity ≥ 99 %.

ESTIMATED ERROR:

Temperature: accuracy probably ±0.1 K (compiler).

REFERENCES:

(1) Braghetti, M.; Leonesi, D.; Franzosini, P. **Ric. Sci.** 1968, **38**, 116-118.

COMPONENTS:	ORIGINAL MEASUREMENTS:
(1) Strontium methanoate (strontium formate); $(CHO_2)_2Sr$; [592-89-2] (2) Thallium(I) methanoate (thallous formate); $(CHO_2)_2Tl_2$; [992-98-3]	Berchiesi, G.; Cingolani, A.; Leonesi, D.; Piantoni, G. **Can. J. Chem.** 1972, **50**, 1972-1975.
VARIABLES:	PREPARED BY:
Temperature.	Baldini, P.

EXPERIMENTAL VALUES:

The experimental values are given only in graphical form (see figure).

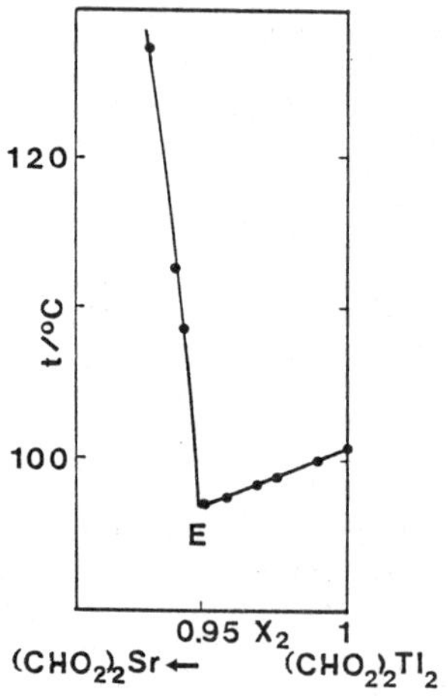

Characteristic point(s):

Eutectic, E, at 96.8 °C and x_1= 0.051 (authors).

Note - The investigation was limited to $x_1 \leq 0.07$ due to thermal instability.

AUXILIARY INFORMATION

METHOD/APPARATUS/PROCEDURE:

A Pyrex device, suitable for work under an inert atmosphere, and allowing one to observe the system visually, was employed (for details, see Ref. 1). The initial crystallization temperatures were measured with a Chromel-Alumel thermocouple checked by comparison with a certified Pt resistance thermometer, and connected with a L&N Type K-3 potentiometer.

NOTE:

The fusion temperature of component 2 read by the compiler on the original plot, i.e., $T_{fus}(2) \sim 101$ °C (374 K) coincides with the values determined with DSC by Braghetti et al. (Ref. 2), and with DTA by Meisel et al. (Ref. 3), although being 3 K lower than that obtained with hot-stage polarizing microscopy by Baum et al. (Ref. 4).

SOURCE AND PURITY OF MATERIALS:

Component 1: K&K material of stated purity $\geq$ 99 %.
Component 2: BDH material of stated purity $\geq$ 99 %.

ESTIMATED ERROR:

Temperature: accuracy probably $\pm$0.1 K (compiler).

REFERENCES:

(1) Braghetti, M.; Leonesi, D.; Franzosini, P. **Ric. Sci.** 1968, **38**, 116-118.
(2) Braghetti, M.; Berchiesi, G.; Franzosini, P. **Ric. Sci.** 1969, **39**, 576-584.
(3) Meisel, T.; Seybold, K.; Halmos, Z.; Roth, J.; Melykuti, C. **J. Thermal Anal.** 1976, **10**, 419-431.
(4) Baum, E.; Demus, D.; Sackmann, H. **Wiss. Z. Univ. Halle** 1970, **19**, 37-46.

COMPONENTS:

(1) Cadmium ethanoate (cadmium acetate); $(C_2H_3O_2)_2Cd$; [543-90-8]
(2) Cesium ethanoate (cesium acetate); $(C_2H_3O_2)Cs$; [3396-11-0]

ORIGINAL MEASUREMENTS:

Nadirov, E.G.; Bakeev, M.I.
Tr. Khim.-Metall. Inst. Akad. Nauk Kaz. SSR 1974, **25**, 129-141.

VARIABLES:

Temperature.

PREPARED BY:

Baldini, P.

EXPERIMENTAL VALUES:

t/°C	T/K[a]	$100x_2$
228	501	40
242	515	45
249	522	50
256[b]	529	58
257	530	60
247	520	65
235	508	70
221	494	75
196	469	80
167	440	85
174	447	90
178	451	95
181	454	100

[a] T/K values calculated by the compiler.
[b] 456 °C in the original table (compiler).

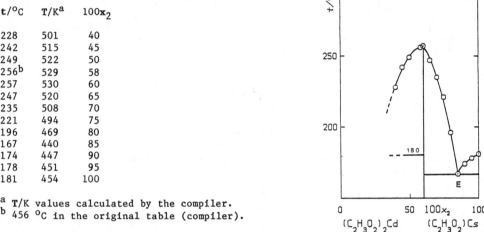

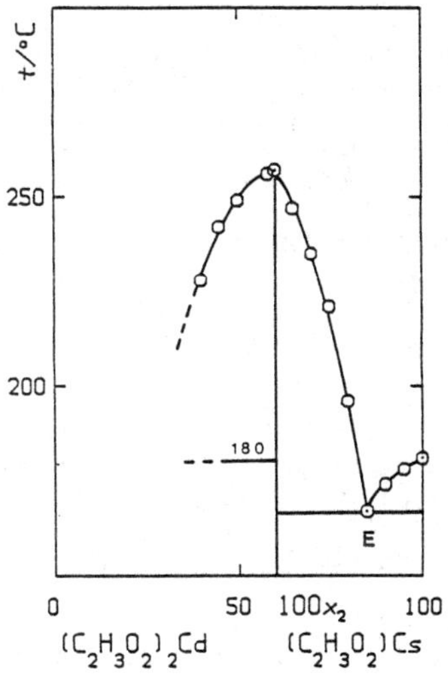

Characteristic point(s):
Eutectic, E, at 167 °C (164 °C according to Fig. 9 of the original paper; compiler) and $100x_2$= 85 (authors).

Intermediate compound(s):
$(C_2H_3O_2)_7Cd_2Cs_3$, congruently melting at 257 °C (255 °C, thermographic analysis), and exhibiting a polymorphic transition (at 130 °C, thermographic analysis; 133 °C, conductometry).

Note - The system was investigated at $40 \leq 100x_2 \leq 100$.

AUXILIARY INFORMATION

METHOD/APPARATUS/PROCEDURE:

Visual polythermal method; temperatures measured with a Chromel-Alumel thermocouple and a PP potentiometer. Additional investigations were performed by means of thermographical analysis, electrical conductometry, and X-ray diffractometry.

NOTE:

The occurrence of the intermediate compound is supported by X-ray diffractometry, and seems reliable. According to the authors, this compound has a density of 2.472 g cm^{-3}. Although the $T_{fus}(2)$ value (454 K) given in this paper is lower than the corresponding one from Table 1 of the Preface, i.e., 463 K, the general trend of the phase diagram should be considered as substantially correct.

SOURCE AND PURITY OF MATERIALS:

Not stated.

ESTIMATED ERROR:

Temperature: accuracy probably ±2 K (compiler).

REFERENCES:

COMPONENTS:	EVALUATOR:
(1) Cadmium ethanoate (cadmium acetate); $(C_2H_3O_2)_2Cd$; [543-90-8] (2) Potassium ethanoate (potassium acetate); $(C_2H_3O_2)K$; [127-08-2]	Schiraldi, A., Dipartimento di Chimica Fisica, Universita' di Pavia (ITALY).

CRITICAL EVALUATION:

This system was studied by Lehrman and Schweitzer (Ref. 1), Il'yasov (Ref. 2), Pavlov and Golubkova (Ref. 3), and Nadirov and Bakeev (Ref. 4), with significantly discrepant results.

Lehrman and Schweitzer (Ref. 1), and Pavlov and Goblubkova (Ref. 3) claim the existence of three congruently melting intermediate compounds, and four eutectics; however, both the coordinates of the eutectics, and the compositions and the fusion temperatures of the intermediate compounds given in either paper do not allow one to reconcile the phase diagram proposed in Ref. 1 with that reported in Ref. 3.

According to Il'yasov (Ref. 2), a single eutectic should exist [at 505 K (232 °C) and $100\mathbf{x}_2$= 75] within the composition range he investigated, viz., $0 \leq 100\mathbf{x}_1 \leq 43$ (the corresponding compositions given in the original paper refer to equivalent fractions of potassium ethanoate).

Finally, according to Nadirov and Bakeev (Ref. 4), a eutectic at either 461, or 469, or 476 K (188, 196, 203 °C, respectively) dependently on the method employed for the determination, and $100\mathbf{x}_2$= 54, and an intermediate compound, $(C_2H_3O_2)_8CdK_6$, incongruently melting at either 518, or 524, or 526 K (245C, 251C, 253 °C, respectively) dependently on the method employed for the determination, are the characteristic features of the system.

The general disagreement existing among the above mentioned authors seems not to be attributed to differences in the purity of the alkanoates they employed, although this factor might play some role in the case of Lehrman and Schweitzer (Ref. 1), inasmuch as they report a fusion temperature of component 2, $T_{fus}(2)$= 565 K (292 °C), which is significantly lower than the generally accepted value of about 579 K (578.7 ± 0.5 K, in Table 1 of the Preface).

Indeed, it seems more likely that the formation of complex ions in the melt (Ref. 4) might affect the results obtained with techniques (e.g., the visual polythermal method) implying the observation of the system during cooling. Should these complex ions be sufficiently stable, the actual liquidus might be different as a consequence of largely different cooling rates.

Taking into account this possibility, the evaluator is inclined to consider as more reliable the phase diagram suggested by Nadirov and Bakeev (Ref. 4), as it is supported by results obtained with several investigation methods, including X-ray diffractometry which was employed to confirm the existence of the intermediate compound $(C_2H_3O_2)_8CdK_6$.

Some doubt, however, might subsist about the interpretation of the slope variation Nadirov and Bakeev (Ref. 4) observed in the plot electric conductivity vs. T, as due to an allotropic transition of potassium ethanoate at 467 K (194 °C). According to Table 1 of the Preface, inter alia, a solid state transition in this salt is to be expected only at $T_{trs}(2)$= 422.2 ± 0.5 K.

REFERENCES:

(1) Lehrman, A.; Schweitzer, D.
J. Phys. Chem. 1954, **58**, 383-384.
(2) Il'yasov, I.I.
Zh. Obshch. Khim, 1962, 32, 347-349.
(3) Pavlov, V.L.; Golubkova, V.V.
Vestn. Kiev. Politekh. Inst. Ser. Khim. Mashinostr. Tekhnol. 1969, No. 6, 76-79.
(4) Nadirov, E.G.; Bakeev, M.I.
Tr. Khim.-Metall. Inst. Akad. Nauk. Kaz. SSR 1974, 25, 129-141.

COMPONENTS:	ORIGINAL MEASUREMENTS:
(1) Cadmium ethanoate (cadmium acetate); $(C_2H_3O_2)_2Cd$; [543-90-8] (2) Potassium ethanoate (potassium acetate); $(C_2H_3O_2)K$; [127-08-2]	Lehrman, A.; Schweitzer, D. **J. Phys. Chem.** 1954, **58**, 383-384.
VARIABLES:	PREPARED BY:
Temperature.	Baldini, P.

EXPERIMENTAL VALUES:

t/°C	T/K[a]	$100x_1$
292	565	0.0
289	562	10.0
246	519	20.0
183[b]	456	20.0
195	468	30.0
202	475	33.3
196	469	35.0
188[b]	461	35.0
203	476	38.0
213	486	40.0
217	490	41.0
221	494	42.86
216	489	44.44
201[b]	474	44.44
206	479	48.0
210	483	50.0
205	478	52.0
202	475	55.0
187[b]	460	55.0
190	463	60.0
220	493	70.0

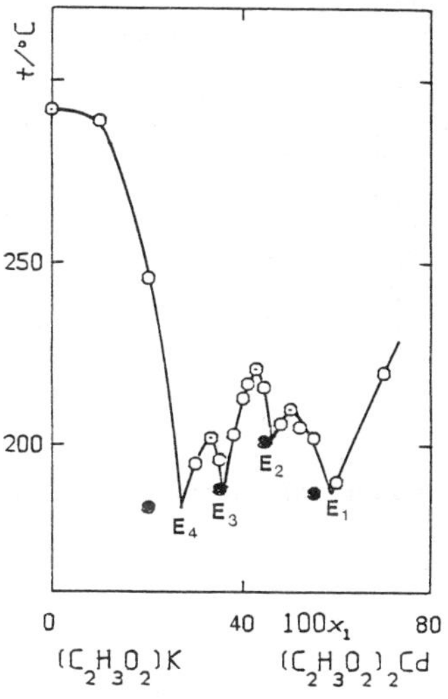

[a] T/K values calculated by the compiler.
[b] Eutectic temperatures (filled circles in the figure).

Characteristic point(s):

Eutectic, E_1, at 187 °C (authors) and $100x_2$= 41 (compiler).
Eutectic, E_2, at 201 °C (authors) and $100x_2$= 54 (compiler).
Eutectic, E_3, at 188 °C (authors) and $100x_2$= 64 (compiler).
Eutectic, E_4, at 183 °C (authors) and $100x_2$= 73 (compiler).

Intermediate compound(s):

$(C_2H_3O_2)_3CdK$, congruently melting at 210 °C (authors).
$(C_2H_3O_2)_{10}Cd_3K_4$, congruently melting at 221 °C (authors).
$(C_2H_3O_2)_4CdK_2$, congruently melting at 202 °C (authors).

AUXILIARY INFORMATION

METHOD/APPARATUS/PROCEDURE:	SOURCE AND PURITY OF MATERIALS:
A molten salt bath was employed to melt the mixtures placed in a 2.5x20 cm Pyrex tube. The beginning of crystallization (under stirring and by seeding) was observed visually and the corresponding temperature was measured with a potentiometer (16 mV full scale) and a Copper-Constantane thermocouple (whose emf could be read to ±0.02 mV), calibrated at the boiling points of water and benzophenone, and at the fusion points of tin and potassium nitrate.	Component 1: "C.P." material added with a few drops of glacial ethanoic acid and dried in an oven at 140 °C. Component 2: "Analytical Reagent" material dried at 140 °C for one week. ESTIMATED ERROR: Temperature: accuracy probably ±0.5 K (compiler).

COMPONENTS:

(1) Cadmium ethanoate (cadmium acetate); $(C_2H_3O_2)_2Cd$; [543-90-8]
(2) Potassium ethanoate (potassium acetate); $(C_2H_3O_2)_2K_2$; [127-08-2]

ORIGINAL MEASUREMENTS:

Il´yasov, I.I.
Zh. Obshch. Khim. 1962, **32**, 347-349.

VARIABLES:

Temperature.

PREPARED BY:

Baldini, P.

EXPERIMENTAL VALUES:

$t/^{\circ}C$	T/K^a	$100x_1$
306	579	0
303	576	5
292	565	15
285	558	20
277	550	25
263	536	30
248	521	35
232	505	40
235	508	45
237	510	50
239	512	55
242	515	60

[a] T/K values calculated by the compiler.

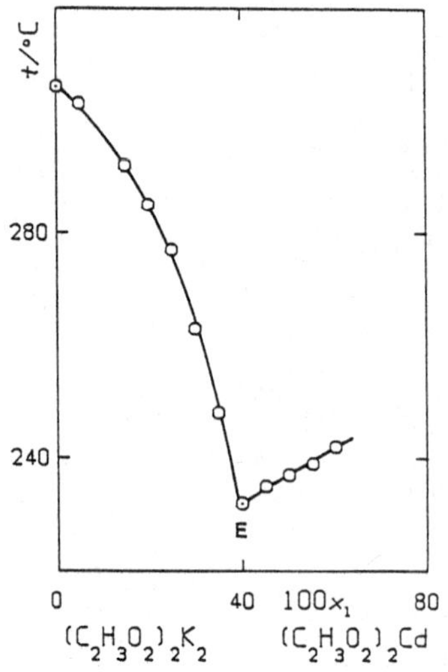

Characteristic point(s):

Eutectic, E, at 232 °C and $100x_2$= 60 (author).

Note - The system was investigated at $0 \leq 100x_1 \leq 60$.

AUXILIARY INFORMATION

METHOD/APPARATUS/PROCEDURE:

Visual polythermal method.

SOURCE AND PURITY OF MATERIALS:

Not stated.

ESTIMATED ERROR:

Temperature: accuracy probably ±2 K (compiler).

REFERENCES:

COMPONENTS:	ORIGINAL MEASUREMENTS:
(1) Cadmium ethanoate (cadmium acetate); $(C_2H_3O_2)_2Cd$; [543-90-8] (2) Potassium ethanoate (potassium acetate); $(C_2H_3O_2)K$; [127-08-2]	Pavlov, V.L.; Golubkova, V.V. **Vestn. Kiev. Politekh. Inst. Ser. Khim. Mashinostr. Tekhnol.** 1969, No. 6, 76-79.
VARIABLES:	PREPARED BY:
Temperature.	Baldini, P.

EXPERIMENTAL VALUES:

$t/°C$	T/K[a]	$100x_1$	$t/°C$	T/K[a]	$100x_1$
300	573	5.0	152	425	49.9
298	571	7.1	176	449	55.0
290	563	9.9	178	451	55.2
278	551	11.9	150[b]	423	55.2
268	541	15.0	160	433	60.1
264	537	17.0	148[b]	421	60.1
166[b]	439	17.0	160	433	60.2
232	505	20.1	148[b]	421	60.2
166[b]	439	20.1	192	465	65.1
220	493	22.1	150	423	65.1
166[b]	439	22.1	198	471	66.9
164	437	25.0	220	493	69.7
166	439	25.1	220	493	70.1
178	451	30.1	150	423	70.1
200	473	34.9	242	515	75.5
148[b]	421	34.9	238	511	77.0
188	461	37.5	232	505	80.0
178	451	39.9	230	503	82.0
148[b]	421	39.9	240	513	85.0
164	437	45.1	242	515	85.3
172	445	46.9	248	521	95.1
188	461	49.9			

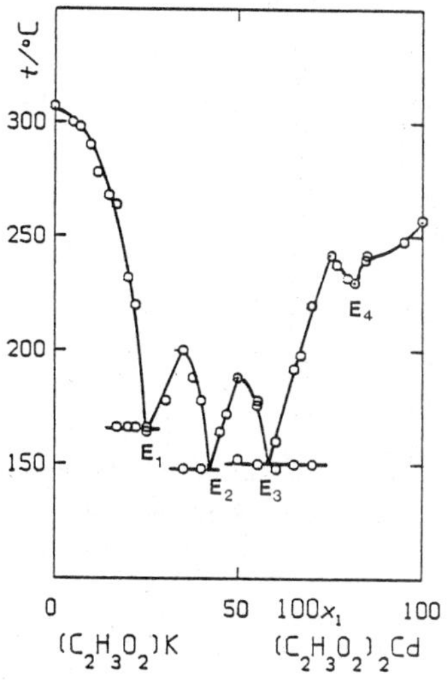

[a] T/K values calculated by the compiler.
[b] Eutectic temperatures.

Characteristic point(s):

Eutectic, E_1, at 166 °C and $100x_1$= 24 (authors).
Eutectic, E_2, at 148 °C and $100x_1$= 42 (authors).
Eutectic, E_3, at 150 °C and $100x_1$= 58 (authors).
Eutectic, E_4, at 230 °C and $100x_1$= 82 (authors).

Intermediate compound(s):

$(C_2H_3O_2)_4CdK_2$, congruently melting at 200 °C (authors).
$(C_2H_3O_2)_3CdK$, congruently melting at 188 °C (authors).
$(C_2H_3O_2)_7Cd_3K$, congruently melting at 242 °C (authors).

AUXILIARY INFORMATION

METHOD/APPARATUS/PROCEDURE:	SOURCE AND PURITY OF MATERIALS:
Visual polythermal method and time-temperature curves. Mixtures prepared in a glove-box.	Component 1 of analytical purity, dehydrated ($T_{fus}(1)$= 257-258°C, 530-531 K). Component 2 of analytical purity, heated at 110-140 °C to constant mass ($T_{fus}(2)$= 306-308°C, 579-581 K).
ESTIMATED ERROR:	
Temperature: accuracy probably ±2 K (compiler).	

COMPONENTS:	ORIGINAL MEASUREMENTS:
(1) Cadmium ethanoate (cadmium acetate); $(C_2H_3O_2)_2Cd$; [543-90-8] (2) Potassium ethanoate (potassium acetate); $(C_2H_3O_2)K$; [127-08-2]	Nadirov, E.G.; Bakeev, M.I. **Tr. Khim.-Metall. Inst. Akad. Nauk Kaz. SSR** 1974, 25, 129-141.
VARIABLES:	PREPARED BY:
Temperature.	Baldini, P.

EXPERIMENTAL VALUES:

$t/^{o}C$	T/K^{a}	$100x_2$
239	512	25
222	495	40
213	486	45
205	478	50
203	476	54
231	504	60
245	518	65
248	521	70
250	523	75
252	525	80
257	530	85
282	555	90
306	579	100

[a] T/K values calculated by the compiler.

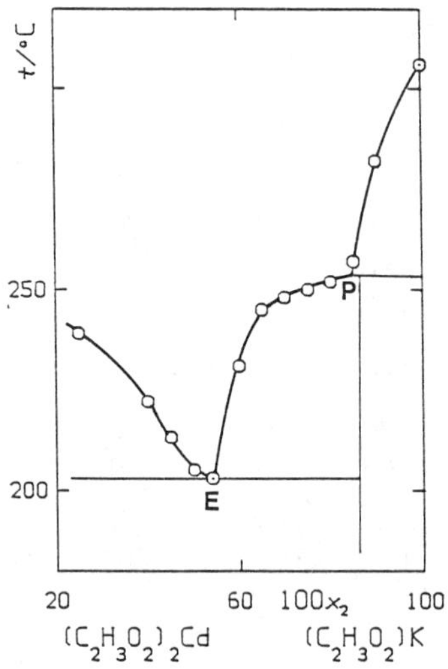

Characteristic point(s):

Eutectic, E, at 203 °C (visual polythermal method, initial crystallization), or 196 °C (thermographical analysis, fusion temperature), or 188 °C (conductometry, fusion temperature), and $100x_2$= 54 (authors).

Peritectic, P, at 253 °C (visual polythermal method), or 245 °C (thermographical analysis), or 251°C (conductometry, Fig.3 of the original paper), erroneously reported as 215 °C in the text (compiler), and $100x_2 \sim$ 84 (compiler).

Intermediate compound: $(C_2H_3O_2)_8CdK_6$, incongruently melting.

Note 1 - The system has been investigated at $25 \leq 100x_2 \leq 100$.

Note 2 - At about 194 °C abrupt changes (to be related to a polymorphic transition; authors) occur in the electrical conductivity of the mixtures with $100x_2$= 85, 90, 95.

AUXILIARY INFORMATION

METHOD/APPARATUS/PROCEDURE:	SOURCE AND PURITY OF MATERIALS:
Visual polythermal method; temperatures measured with a Chromel-Alumel thermocouple and a PP potentiometer. Additional investigations have been performed by means of thermographical analysis, electrical conductometry, and X-ray diffractometry.	Not stated. ESTIMATED ERROR: Temperature: accuracy probably ±2 K (compiler).

COMPONENTS:

(1) Cadmium ethanoate (cadmium acetate); $(C_2H_3O_2)_2Cd$; [543-90-8]
(2) Sodium ethanoate (sodium acetate); $(C_2H_3O_2)Na$; [127-09-3]

EVALUATOR:

Schiraldi, A.,
Dipartimento di Chimica Fisica,
Universita´ di Pavia (ITALY).

CRITICAL EVALUATION:

This system was studied by Il´yasov (Ref. 1), and Pavlov and Golubkova (Ref. 2). The former author claims the diagram to be of the simple eutectic type, with the invariant at 528 K (255 °C) and $100\mathbf{x}_2$= 68 (the eutectic composition is given in Ref. 1 as $100\mathbf{x}_2$= 52 since it refers to the equivalent fraction of component 2), whereas Pavlov and Golubkova suggest the existence of the intermediate compound $(C_2H_3O_2)_4CdNa_2$, congruently melting at 527 K (254 °C), and, accordingly, of two eutectics, E_1, E_2, occurring at 496 K (223 °C) and $100\mathbf{x}_2$= 75, and at 507 K (234 °C) and $100\mathbf{x}_2$= 58, respectively.

Although the experimental data by Pavlov and Golubkova seem more detailed than those by Il´yasov, the evaluator has no arguments to definitely prefer the diagram shown in Ref. 2, ruling out that of Ref. 1.

As a comment, one may notice that the fusion temperature of the intermediate compound given in Ref. 2 is close to that of the eutectic reported in Ref. 1. This might suggest undercooling of Pavlov and Golubkova´s samples. In any case, the existence of the intermediate compound suggested by the latter authors should be confirmed with X-ray diffractometry.

It is finally to be added that the fusion temperature of component 2 by Il´yasov (601 K) meets that listed in Table 1 of the Preface (601.3$\pm$0.5 K), whereas the value by Pavlov and Golubkova (595 K) is significantly lower.

REFERENCES:

(1) Il´yasov, I.I.
Zh. Obshch. Khim. 1962, 32, 347-349.

(2) Pavlov, V.L.; Golubkova, V.V.
Vestn. Kiev. Politekh. Inst. Ser. Khim. Mashinostr. Tekhnol. 1969, No. 6, 76-79.

COMPONENTS:

(1) Cadmium ethanoate (cadmium acetate); $(C_2H_3O_2)_2Cd$; [543-90-8]
(2) Sodium ethanoate (sodium acetate); $(C_2H_3O_2)_2Na_2$; [127-09-3]

ORIGINAL MEASUREMENTS:

Il´yasov, I.I.
Zh. Obshch. Khim. 1962, 32, 347-349.

VARIABLES:

Temperature.

PREPARED BY:

Baldini, P.

EXPERIMENTAL VALUES:

$t/^oC$	T/K^a	$100x_1$
328	601	0
318	591	10
315	588	15
309	582	20
297	570	30
287	560	35
277	550	40
261	534	45
259	532	50
264	537	55
267	540	60
267	540	65

[a] T/K values calculated by the compiler.

Characteristic point(s):

Eutectic, E, at 255 ^oC and $100x_2$= 52 (author).

Note - The system was investigated at $0 \leq 100x_1 \leq 65$.

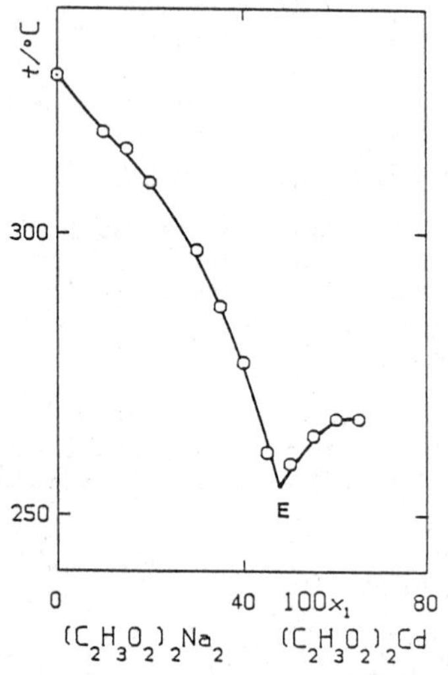

AUXILIARY INFORMATION

METHOD/APPARATUS/PROCEDURE:

Visual polythermal method.

SOURCE AND PURITY OF MATERIALS:

Not stated.

ESTIMATED ERROR:

Temperature: accuracy probably ±2 K (compiler).

REFERENCES:

COMPONENTS:	ORIGINAL MEASUREMENTS:
(1) Cadmium ethanoate (cadmium acetate); $(C_2H_3O_2)_2Cd$; [543-90-8] (2) Sodium ethanoate (sodium acetate); $(C_2H_3O_2)Na$; [127-09-3]	Pavlov, V.L.; Golubkova, V.V. **Vestn. Kiev. Politekh. Inst. Ser. Khim. Mashinostr. Tekhnol.** 1969, No. 6, 76-79.
VARIABLES:	PREPARED BY:
Temperature.	Baldini, P.

EXPERIMENTAL VALUES:

$t/^{o}C$	T/K^{a}	$100\mathbf{x}_1$
322	595	0
314	587	6.2
300	573	12.3
274	547	17.9
244	517	23.6
228	501	27.1
254	527	34.8
236	509	40.2
313	586	8.2
291	564	13.2
259	532	19.2
223[b]	496	19.2
233	506	26.3
223[b]	496	26.3
251	524	34.8
232	505	45.4
234[b]	507	45.4
249	522	51.6
234[b]	507	51.6
253	526	58.7
257-258	530-531	100

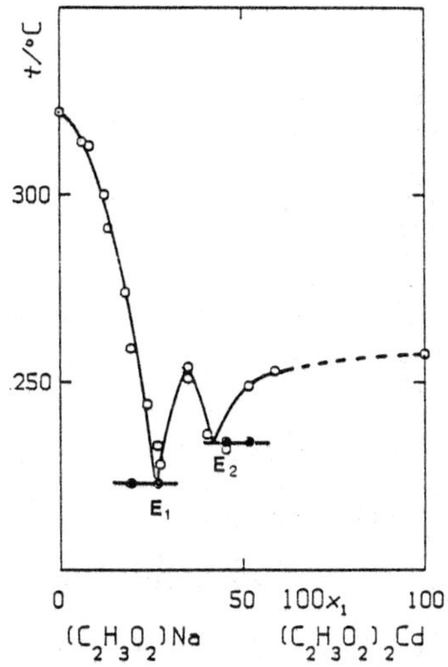

[a] T/K values calculated by the compiler.
[b] Eutectic temperatures (filled circles in the figure).

Characteristic point(s):
Eutectic, E_1, at 223 ^{o}C and $100\mathbf{x}_1$= 25 (authors).
Eutectic, E_2, at 234 ^{o}C and $100\mathbf{x}_1$= 42 (authors).

Intermediate compound(s):
$(C_2H_3O_2)_4CdNa_2$, congruently melting at 254 ^{o}C (authors).

AUXILIARY INFORMATION

METHOD/APPARATUS/PROCEDURE:	SOURCE AND PURITY OF MATERIALS:
Visual polythermal method and time - temperature curves; temperatures measured with a Copper-Constantane thermocouple. Mixtures prepared in a glove-box and added with 1-3 drops anhydrous ethanoic acid to prevent thermal decomposition of component 1.	Component 1 of analytical purity, in part dehydrated, and in part recrystallized from aqueous (2%) ethanoic acid and then dehydrated. Component 2 of analytical purity, recrystallized and then heated at 110-140 ^{o}C to constant mass.
	ESTIMATED ERROR:
	Temperature: accuracy probably ± 2 K (compiler).

COMPONENTS:

(1) Cadmium ethanoate (cadmium acetate); $(C_2H_3O_2)_2Cd$; [543-90-8]
(2) Rubidium ethanoate (rubidium acetate); $(C_2H_3O_2)Rb$; [563-67-7]

ORIGINAL MEASUREMENTS:

Nadirov, E.G.; Bakeev, M.I.
Tr. Khim.-Metall. Inst. Akad. Nauk Kaz. SSR 1974, **25**, 129-141.

VARIABLES:

Temperature.

PREPARED BY:

Baldini, P.

EXPERIMENTAL VALUES:

$t/^{\circ}C$	T/K^a	$100x_2$
236	509	40
233	506	50
231	504	60
228	501	65
217	490	70
215	488	75
206	479	80
192	465	84.1
179	452	86
198	471	87
214	487	90
231	504	95
237	510	100

[a] T/K values calculated by the compiler.

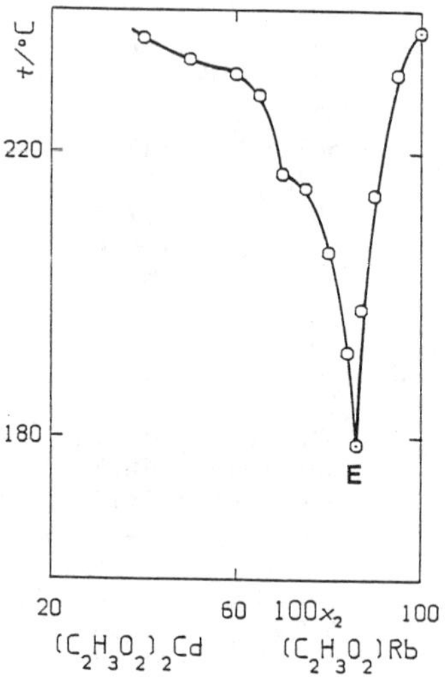

Characteristic point: Eutectic, E, at 179 °C (visual polythermal method, initial crystallization), or 145 °C (fusion temperature by thermographical analysis), or 169 °C (fusion temperature by conductometry), and $100x_2$= 86 (authors).

Intermediate compound: $(C_2H_3O_2)_4CdRb_2$, incongruently melting at 219 °C (visual polythermal method), 192 °C (thermographical analysis), or 206 °C (conductometry).

Note - The system has been investigated at $40 \leq 100x_2 \leq 100$.

AUXILIARY INFORMATION

METHOD/APPARATUS/PROCEDURE:

Visual polythermal method; temperatures measured with a Chromel-Alumel thermocouple and a PP potentiometer. Additional investigations were performed by means of thermographical analysis and electrical conductometry.

NOTE:

The occurrence of intermediate compounds in the binaries $C_2H_3O_2$/Cd, K and $C_2H_3O_2$/Cd, Cs was claimed by the same authors in the same paper, and supported with X-ray diffraction patterns: for the present system, on the contrary, no analogous evidence was given. Moreover, the exceedingly large differences among the eutectic temperatures obtained with different techniques is to be stressed.

SOURCE AND PURITY OF MATERIALS:

Not stated.

ESTIMATED ERROR:

Temperature: accuracy probably ±2 K (compiler).

REFERENCES:

COMPONENTS:	EVALUATOR:
(1) Cesium ethanoate (cesium acetate) $(C_2H_3O_2)Cs$; [3396-11-0] (2) Potassium ethanoate (potassium acetate) $(C_2H_3O_2)K$; [127-08-2]	Schiraldi, A., Dipartimento di Chimica Fisica, Universita' di Pavia (ITALY).

CRITICAL EVALUATION:

Results on this binary have been repeatedly reported by Diogenov et al. (Refs. 1-3) as a part of their investigations on ternary and reciprocal ternary systems. These authors, who carried out visual polythermal observations on the liquidus, define the system as of the eutectic type with the invariant at either 405 K (132 °C; Ref. 1), or 403 K (130 °C; Ref. 2), or 413 K (140 °C; Ref. 3), and $100\mathbf{x}_2$= 28.5. It is not clear whether the different eutectic temperatures given in Refs. 1-3 come from different sets of measurements or depend on adjustments suggested by the general topology of the particular ternary studied in each paper. A knee in the liquidus branch richer in component 1 (Ref. 1) has been interpreted by these authors as due to a phase transition occurring in this salt at 447 K (174 °C). Diogenov et al. also claimed in a previous paper (Ref. 4) the occurrence in component 2 of a phase transition at 565-566 K (292-293 °C).

The DTA investigations by Storonkin et al. (Ref. 5) give further support to the fact that the system is of the eutectic type although the temperature (412 K) and composition ($100\mathbf{x}_2$= 32) of the invariant have been singled out by extrapolation of the two liquidus branches. According to Fig. 3 of the original paper (Ref. 5), the authors assume that the eutectic equilibrium covers the composition range from $100\mathbf{x}_2$= 0 to $100\mathbf{x}_2$= 100. They do not mention, however, the occurrence of any allotropic transition in either component: according to Table 1 this ought to be correct for what concerns component 1, whereas component 2 ought to undergo a phase transition at 422.2 ± 0.5 K.

Storonkin et al. (Ref. 5) ascribe the differences between their and Diogenov et al.'s diagram to the higher purity of the salts they employed: indeed, the fusion temperature they report for component 1 [$\mathbf{T}_{fus}(1)/K$ = 467] is much closer to that listed in Table 1 of the Preface (463 ± 1) than that given by Diogenov et al. (453).

As a conclusion, the following remarks should be taken into account.

(i) The phase transition temperature reported for cesium ethanoate by Diogenov et al. seems to be unreliable.

(ii) The phase transition temperature reported for potassium ethanoate in Ref. 4 (565-566 K) seems also to be unreliable, as it cannot be identified with any transition temperature found by other investigators (Ref. 6).

(iii) The eutectic temperature reported by Storonkin et al., viz., 412 K, seems satisfactorily supported by their DTA results, as well as the trend of the liquidus branch richer in cesium ethanoate. On the contrary, there is some doubt about the reliability of the other liquidus branch which, according to these authors, does not show any "knee" to be possibly matched with the expected (see above) phase transition of potassium ethanoate. Consequently, the eutectic composition (attained by extrapolation of the liquidus branches) cannot be considered more reliable than that reported by Diogenov.

(iv) Finally, the complete immiscibility in the solid state should be more carefully verified, e.g., by further DTA or DSC investigations extended to extreme compositions.

REFERENCES:

(1) Nurminskii, N.N. and Diogenov, G.G.; **Zh. Neorg. Khim.** 1960, **5**, 2084-2087; **Russ. J. Inorg. Chem. (Engl. Transl.) 1960, 5,** 1011-1013 (*).
(2) Diogenov, G.G. and Sergeeva, G.S.; **Zh. Neorg. Khim.** 1965, **10**, 292-294; **Russ. J. Inorg. Chem. (Engl. Transl.)** 1965, **10**, 153-154 (*).
(3) Diogenov, G.G. and Morgen, L.T.; **Fiz.-Khim. Issled. Rasplavov Solei, Irkutsk** 1975, 59-61.
(4) Diogenov, G.G.; Nurminskii, N.N. and Gimel'shtein, V.G.; **Zh. Neorg. Khim.** 1957, **2**, 1596-1600 ; **Russ. J. Inorg. Chem. (Engl. Transl.)** 1957, **2(7)**, 237-245.
(5) Storonkin, A.V.; Vasil'kova, I.V. and Tarasov, A.A.; **Vestn. Leningr. Univ., Fiz., Khim.** 1977, **(4)**, 80-85.
(6) Sanesi, M.; Cingolani, A.; Tonelli, P.L.; Franzosini, P.; **Thermal Properties**, in **Thermodynamic and Transport Properties of Organic Salts**, IUPAC Chemical Data Series No. 28 (Franzosini, P.; Sanesi, M.; Editors), Pergamon Press, Oxford 1980, 29-115.

COMPONENTS:

(1) Cesium ethanoate (cesium acetate); $(C_2H_3O_2)Cs$; [3396-11-0]
(2) Potassium ethanoate (potassium acetate); $(C_2H_3O_2)K$; [127-08-2]

ORIGINAL MEASUREMENTS:

Nurminskii, N.N.; Diogenov, G.G.
Zh. Neorg. Khim. 1960, **5**, 2084-2087; **Russ. J. Inorg. Chem. (Engl. Transl.)** 1960, **5**, 1011-1013 (*).

VARIABLES:

Temperature.

PREPARED BY:

Baldini, P.

EXPERIMENTAL VALUES:

t/°C	T/K[a]	$100x_2$
180	453	0
176	449	2
172	445	4.5
167	440	7.5
163	436	13.0
160	433	15.0
154	427	18.0
148	421	21.5
138	411	25.0
135	408	27.0
135	408	30.5
148	421	34.0
153	426	35.0
164	437	38.5
180	453	43.0
192	465	47.5
204	477	51.5
212	485	55.5
220	493	59.0
228	501	62.0
248	521	71.5
272	545	82.5

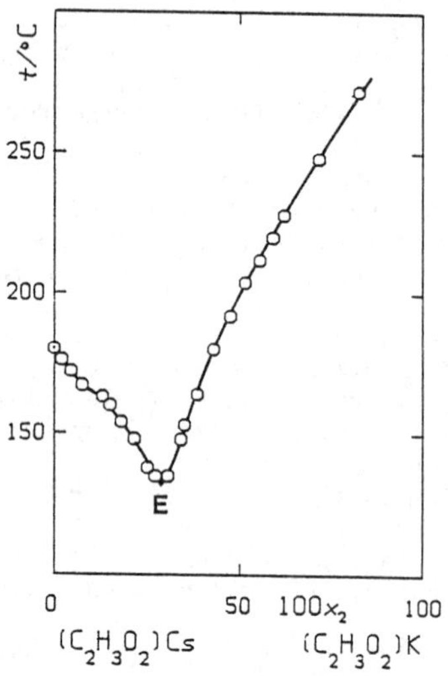

[a] **T**/K values calculated by the compiler.

Characteristic point(s):
Eutectic, E, at 132 °C and $100x_2$= 28.5 (authors).

AUXILIARY INFORMATION

METHOD/APPARATUS/PROCEDURE:

Visual polythermal method. Temperatures measured with a Chromel-Alumel thermocouple and a 17 mV millivoltmeter.

SOURCE AND PURITY OF MATERIALS:

Not stated.
Component 1 undergoes a phase transition at $t_{trs}(1)$/°C= 174 and melts at $t_{fus}(1)$/°C= 182 (Fig. 1 of the original paper), or 180 (table).
Component 2 melts at $t_{fus}(2)$/°C= 310 (Fig. 1).

ESTIMATED ERROR:

Temperature: accuracy probably ±2 K (compiler).

REFERENCES:

COMPONENTS:

(1) Cesium ethanoate (cesium acetate); $(C_2H_3O_2)Cs$; [3396-11-0]
(2) Potassium ethanoate (potassium acetate); $(C_2H_3O_2)K$; [127-08-2]

ORIGINAL MEASUREMENTS:

Diogenov, G.G.; Sergeeva, G.S.
Zh. Neorg. Khim. 1965, **10**, 292-294; **Russ. J. Inorg. Chem. (Engl. Transl.)** 1965, **10**, 153-154 (*).

VARIABLES:

Temperature.

PREPARED BY:

Baldini, P.

EXPERIMENTAL VALUES:

The authors refer to Ref. 1 for the experimental values, although giving a different eutectic temperature.

Characteristic point(s):

Eutectic, E, at 130 °C and $100\mathbf{x}_2$= 28.5 (authors).

AUXILIARY INFORMATION

METHOD/APPARATUS/PROCEDURE:

Visual polythermal method. Temperatures measured with a Chromel-Alumel thermocouple.

SOURCE AND PURITY OF MATERIALS:

Not stated.
Component 1: $\mathbf{t}_{fus}(1)/°C$= 180 (Fig. 1 of the original paper).
Component 2: $\mathbf{t}_{fus}(2)/°C$= 310 (Fig. 1).

ESTIMATED ERROR:

Temperature: accuracy probably ±2 K (compiler).

REFERENCES:

(1) Nurminskii, N.N.; Diogenov, G.G. **Zh. Neorg. Khim.** 1960, 5, 2084-2087; **Russ. J. Inorg. Chem., (Engl. Transl.)** 1960, **5**, 1011-1013.

COMPONENTS:

(1) Cesium ethanoate (cesium acetate); $(C_2H_3O_2)Cs$; [3396-11-0]
(2) Potassium ethanoate (potassium acetate); $(C_2H_3O_2)K$; [127-08-2]

ORIGINAL MEASUREMENTS:

Diogenov, G.G.; Morgen, L.T.
Fiz.-Khim. Issled. Rasplavov Solei, Irkutsk, 1975, 59-61.

VARIABLES:

Temperature.

PREPARED BY:

Baldini, P.

EXPERIMENTAL VALUES:

The authors refer to Ref. 1 for the experimental values, although giving a different eutectic temperature.

Characteristic point(s):

Eutectic, E, at 140 °C and $100\mathbf{x}_1$= 71.5 (authors).

AUXILIARY INFORMATION

METHOD/APPARATUS/PROCEDURE:

Visual polythermal method. Temperatures measured with a Chromel-Alumel thermocouple and a millivoltmeter.

SOURCE AND PURITY OF MATERIALS:

Not stated.
Component 1: $t_{fus}(1)/^{o}C$= 187 (Fig. 1 of the original paper).
Component 2: $t_{fus}(2)/^{o}C$= 308 (Fig. 1).

ESTIMATED ERROR:

Temperature: accuracy probably ± 2 K (compiler).

REFERENCES:

(1) Nurminskii, N.N.; Diogenov, G.G.
Zh. Neorg. Khim. 1960, 5, 2084-2087; **Russ. J. Inorg. Chem., (Engl. Transl.) 1960, 5,** 1011-1013.

COMPONENTS:

(1) Cesium ethanoate (cesium acetate); $(C_2H_3O_2)Cs$; [3396-11-0]
(2) Potassium ethanoate (potassium acetate); $(C_2H_3O_2)K$; [127-08-2]

ORIGINAL MEASUREMENTS:

Storonkin, A.V.; Vasil'kova, I.V.; Tarasov, A.A.
Vestn. Leningr. Univ., Fiz., Khim. 1977, (4), 80-85.

VARIABLES:

Temperature.

PREPARED BY:

Baldini, P.

EXPERIMENTAL VALUES:

Data reported only in graphical form (see figure).

Characteristic point(s):

Eutectic, E, at 412 K and $100x_1$= 68 (authors).

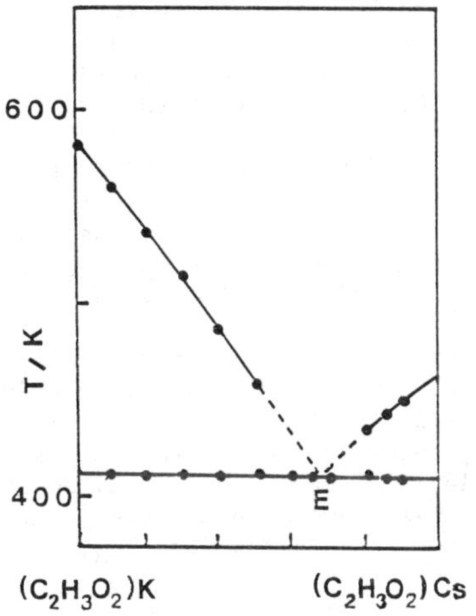

AUXILIARY INFORMATION

METHOD/APPARATUS/PROCEDURE:

DTA and "contact polythermal method" under polarized light. IR spectra were also used to state the existence of intermediate compound(s).

SOURCE AND PURITY OF MATERIALS:

Component 1 synthetized from Cs_2CO_3 and ethanoic acid ($T_{fus}(1)/K$= 467; authors). Component 2 of analytical purity recrystallized twice from water and dried under vacuum ($T_{fus}(2)/K$= 584; authors). The purity of both components was checked by thermographical analysis. The mixtures were prepared in a glove box.

ESTIMATED ERROR:

Temperature: accuracy probably ±2 K (compiler).

REFERENCES:

COMPONENTS:	EVALUATOR:
(1) Cesium ethanoate (cesium acetate); $(C_2H_3O_2)Cs$; [3396-11-0] (2) Lithium ethanoate (lithium acetate); $(C_2H_3O_2)Li$; [546-89-4]	Franzosini, P., Dipartimento di Chimica Fisica, Universita' di Pavia (ITALY).

CRITICAL EVALUATION:

This binary was first studied as a side of the ternary $C_2H_3O_2$/Cs, Li, Rb (Ref. 1), and re-determined by the same group ten years later (Ref. 2). Due to more accurate experimental methods (DTA and X-ray diffractometry) employed in the latter paper (Ref. 2), the phase diagram therein shown seems much more reliable than the previous one (Ref. 1).

Accordingly, the system is to be considered as characterized (Ref. 2) by the occurrence of a single intermediate compound, $(C_2H_3O_2)_3CsLi_2$, congruently melting at 563 K (290 °C), and by two eutectics, at 420 K (147 °C) and $100\mathbf{x}_1 = 77$, and at 520 K (247 °C) and $100\mathbf{x}_1 = 12$, respectively.

The main difference of this phase diagram with respect to that presented in the previous work (Ref. 1) is the lack of a further intermediate compound, $(C_2H_3O_2)_2CsLi$ (incongruently melting). Consequently to this lack, however, a large part of the phase diagram of the ternary $C_2H_3O_2$/Cs, Li, Rb (Ref. 1) ought to be redrawn, which, at the present time has not been done, at least as far as the evaluator knows.

The fusion temperatures of component 1 and component 2 as given in Refs. 1, 2 (458-459 K, and 561-563 K, respectively) are not far from those listed in Table 1 of the Preface (463±1 K, and 557±2 K, respectively). Moreover, no mention is made of the occurrence of phase transitions in either component, which is again in agreement with Table 1 of the Preface, although in disagreement with the fact that in other papers by the same group (see, e.g., Ref. 3) component 1 is described as undergoing a phase transition at 477 K (174 °C).

REFERENCES:

(1) Diogenov, G.G.; Sarapulova, I.F.
Zh. Neorg. Khim. 1964, **9**, 482-487; **Russ. J. Inorg. Chem. (Engl. Transl.)** 1964, **9(2)**, 265-267.

(2) Sarapulova, I.F.; Kashcheev, G.N.; Diogenov, G.G.
Nekotorye Vopr. Khimii Rasplavlen. Solei i Produktov Destrusktii Sapropelitov, Irkutsk 1974, 3-10.

(3) Nurminskii, N.N.; Diogenov, G.G.
Zh. Neorg. Khim. 1960, **5**, 2084-2087; **Russ. J. Inorg. Chem. (Engl. Transl.)** 1960, **5**, 1011-1013 (*).

COMPONENTS:

(1) Cesium ethanoate (cesium acetate); $(C_2H_3O_2)Cs$; [3396-11-0]
(2) Lithium ethanoate (lithium acetate); $(C_2H_3O_2)Li$; [546-89-4]

ORIGINAL MEASUREMENTS:

Diogenov, G.G.; Sarapulova, I.F.
Zh. Neorg. Khim. 1964, **9**, 482-487; **Russ. J. Inorg. Chem. (Engl. Transl.)** 1964, 9(2), 265-267 (*).

VARIABLES:

Temperature.

PREPARED BY:

Baldini, P.

EXPERIMENTAL VALUES:

$t/^{o}C$	T/K^{a}	$100x_2$	$t/^{o}C$	T/K^{a}	$100x_2$
185	458	0	247	520	44.0
182	455	2.5	261	534	47.0
175	448	7.0	271	544	50.0
163	436	12.5	282	555	55.0
147	420	19.5	288	561	59.5
140	413	24.5	292	565	64.5
160	433	26.5	292	565	70.0
170	443	27.5	283	556	75.5
184	457	29.5	270	543	81.5
196	469	32.0	250	523	86.5
208	481	34.0	250	523	90.0
219	492	36.5	267	540	93.0
230	503	40.0	279	552	96.5
232	505	42.0	290	563	100.0
238	511	43.5			

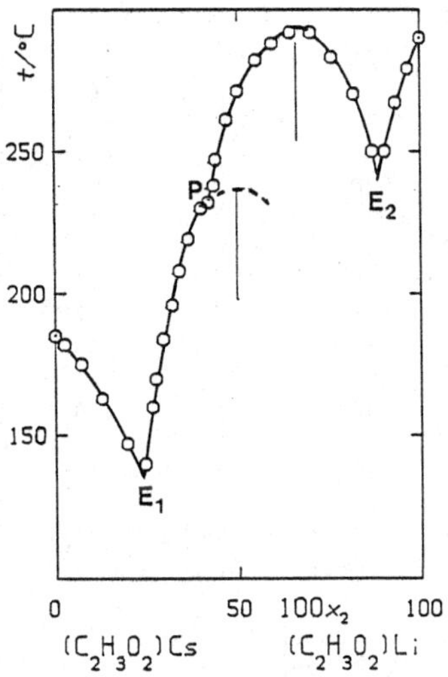

[a] T/K values calculated by the compiler.

Characteristic point(s):
Eutectic, E_1, at 135 °C and $100x_1$= 76 (authors).
Peritectic, P, at 233 °C and $100x_2$= 42.5 (authors).
Eutectic, E_2, at 240 °C and $100x_1$= 12 (authors).

Intermediate compound(s):
$(C_2H_3O_2)_2CsLi$, incongruently melting.
$(C_2H_3O_2)_3CsLi_2$, congruently melting at 293 °C (according to the text and Fig. 2 of the original paper); at 295 °C (according to Fig. 1 of the original paper).

AUXILIARY INFORMATION

METHOD/APPARATUS/PROCEDURE:

Visual polythermal method. Temperatures measured by means of a Chromel-Alumel thermocouple.

SOURCE AND PURITY OF MATERIALS:

Not stated.

ESTIMATED ERROR:

Temperature: accuracy probably ±2 K (compiler).

REFERENCES:

COMPONENTS:	ORIGINAL MEASUREMENTS:
(1) Cesium ethanoate (cesium acetate); $(C_2H_3O_2)Cs$; [3396-11-0] (2) Lithium ethanoate (lithium acetate); $(C_2H_3O_2)Li$; [546-89-4]	Sarapulova, I.F.; Kashcheev, G.N.; Diogenov, G.G. **Nekotorye Vopr. Khimii Rasplavlen. Solei i Produktov Destruktsii Sapropelitov, Irkutsk,** 1974, 3-10.
VARIABLES:	PREPARED BY:
Temperature.	Baldini, P.

EXPERIMENTAL VALUES:

$t/^{o}C$	T/K^{a}	$100\mathbf{x}_1$	$t/^{o}C$	T/K^{a}	$100\mathbf{x}_1$
288	561	0	275^{bc}	548	50
284^{bc}	557	0	147^{bd}	420	50
253	526	10	247	520	55
252^{bc}	525	10	147^{bd}	420	55
247^{be}	520	10	182	455	70
268	541	20	172^{bc}	445	70
267^{bc}	540	20	148^{bd}	421	70
247^{be}	520	20	147	420	80
283	556	25	147^{bc}	420	80
283^{bc}	556	25	156^{bc}	429	85
246^{be}	519	25	147^{bd}	420	85
293	566	33	186	459	100
290^{bc}	563	33	185^{bc}	458	100
273	546	50	35^{f}	308	100

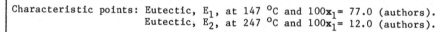

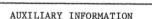

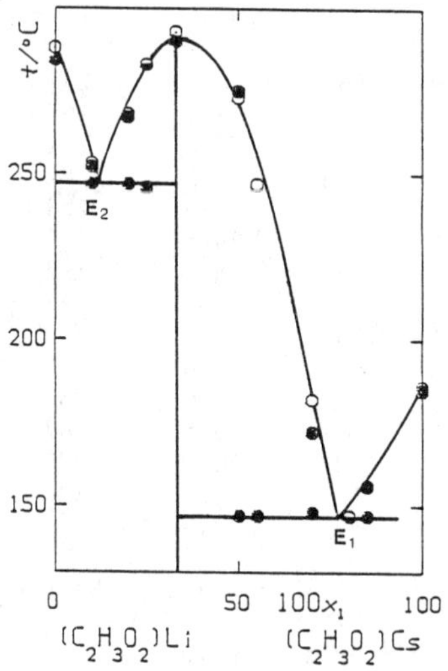

[a] T/K values calculated by the compiler.
[b] Differential thermal analysis (filled circles
[c] Initial fusion.
[d] Eutectic stop (E_1).
[e] Eutectic stop (E_2).
[f] Solid state transition.

Characteristic points: Eutectic, E_1, at 147 ^{o}C and $100\mathbf{x}_1$= 77.0 (authors).
Eutectic, E_2, at 247 ^{o}C and $100\mathbf{x}_1$= 12.0 (authors).

Intermediate compound: $(C_2H_3O_2)_3CsLi_2$, congruently melting at 293 ^{o}C (290 ^{o}C by DTA).

AUXILIARY INFORMATION

METHOD/APPARATUS/PROCEDURE:

A thermographical analysis was performed with a Kurnakov pyrometer mod. 1959 (reference material: Al_2O_3). Only heating traces (at the heating rate of 5-6 $^{o}C/min$) were recorded due to the tendency of the melts to undercool. Supplementary visual polythermal observations are also tabulated. X-ray diffraction patterns were used to obtain information on the intermediate compound.

SOURCE AND PURITY OF MATERIALS:

Not stated. Component 1 undergoes a phase transition at $t_{trs}(1)/^{o}C$ = 35.

ESTIMATED ERROR:

Temperature: accuracy probably ± 2 K (compiler).

REFERENCES:

COMPONENTS:

(1) Cesium ethanoate (cesium acetate); $(C_2H_3O_2)Cs$; [3396-11-0]
(2) Sodium ethanoate (sodium acetate); $(C_2H_3O_2)Na$; [127-09-3]

EVALUATOR:

Schiraldi, A.,
Dipartimento di Chimica Fisica,
Universita´ di Pavia (ITALY).

CRITICAL EVALUATION:

This binary was first investigated as a side of the ternary $C_2H_3O_2$/Cs, Na, Rb by Diogenov and Sarapulova (Ref. 1), who reported a eutectic at 388 K (115 °C) and $100x_1$= 68, on the basis of visual polythermal observations.

The liquidus by these authors shows a knee at about 585 K and $100x_1$ about 5, which might be identified with the phase transition of $(C_2H_3O_2)Na$ reported by Diogenov at 596 K (323 °C; Ref. 2), and by Gimel´shtein and Diogenov at 583-584 K (310-311 °C; Ref. 3). However, such figures do not meet any of the high temperature T_{trs} values by other authors (Ref. 4), which range between 511-513 and 527±15 K.

Substantially analogous results, including the knee (for which no explanation is offered), have been reported also by Storonkin et al. (Ref. 5) for the liquidus branch richer in component 2. The other branch by these authors, however, lies significantly above the corresponding curve by Diogenov and Sarapulova: the difference has been attributed by Storonkin et al. to the higher purity of the cesium ethanoate they employed.
According to the latter authors (Ref. 5), who carried out DTA determinations through most of the composition range, the eutectic temperature is 392 K, and the eutectic composition (which was obtained by extrapolation, due to the tendency to undercool of the melts of composition close to x_E) is $100x_1$= 64.

In the opinion of the evaluator, the following points should be remarked.

(i) Neither Ref. 1 nor Ref. 5 report the phase transition of sodium ethanoate observed by other authors (Ref. 4) at 510-530 K, i.e., well above the eutectic temperature of the binary.

(ii) No comment is explicitely made in either work on the apparent knee of the liquidus branch richer in component 2.

(iii) No experimental support is given to rule out the occurrence of solid solutions in the regions of the phase diagram close to the pure components.

(iv) The phase transition of cesium ethanoate observed by Nurminskii and Diogenov (Ref. 6) at 447 K is neither confirmed nor mentioned in the present investigation (Ref. 1) by the same group.

Accordingly, it seems justified to cast some doubts about the reliability of the upper part of the liquidus branch richer in component 2, whereas the eutectic temperature (390±2 K) and composition ($100x_2$= 66±2) seem satisfactorily supported by the data available.

REFERENCES:

(1) Diogenov, G.G.; Sarapulova, I.F.
Zh. Neorg. Khim. 1964, **9**, 1499-1502; **Russ. J. Inorg. Chem. (Engl. Transl.)** 1964, **9**,814-816.
(2) Diogenov, G.G.
Zh. Neorg. Khim. 1956, **1(4)**, 799-805; **Russ. J. Inorg. Chem. (Engl. Transl.)** 1956, **1(4), 199-205.**
(3) Gimel´shtein, V.G.; Diogenov, G.G.
Zh. Neorg. Khim. 1958, 3, 1644-49 ; **Russ. J. Inorg. Chem. (Engl. Transl.)** 1958, **3(7)**, 230-236.
(4) Sanesi, M.; Cingolani, A.; Tonelli, P.L.; Franzosini, P.
Thermal Properties, in **Thermodynamic and Transport Properties of Organic Salts,** IUPAC Chemical Data Series No. 28 (Franzosini, P.; Sanesi, M.; Editors), Pergamon Press, Oxford, 1980, 29-115.
(5) Storonkin, A.V.; Vasil´kova, I.V. and Tarasov, A.A.
Vestn. Leningr. Univ., Fiz., Khim. 1977, **(4)**, 80-85.
(6) Nurminskii, N.N. and Diogenov, G.G.
Zh. Neorg. Khim. 1960, **5**, 2084-2087; **Russ. J. Inorg. Chem. (Engl. Transl.)** 1960, **5**, 1011-1013.

COMPONENTS:

(1) Cesium ethanoate (cesium acetate); $(C_2H_3O_2)Cs$; [3396-11-0]
(2) Sodium ethanoate (sodium acetate); $(C_2H_3O_2)Na$; [127-09-3]

ORIGINAL MEASUREMENTS:

Diogenov, G.G.; Sarapulova, I.F.
Zh. Neorg. Khim. 1964, **9**, 1499-1502 (*);
Russ. J. Inorg. Chem. (Engl. Transl.) 1964, **9**, 814-816.

VARIABLES:

Temperature.

PREPARED BY:

Baldini, P.

EXPERIMENTAL VALUES:

$t/^{o}C$	T/K^{a}	$100x_2$	$t/^{o}C$	T/K^{a}	$100x_2$
180	453	0	153	426	42.0
178	451	2.7	175	448	47.5
175	448	5.0	200	473	54.0
167	440	10.0	224	497	60.3
156	429	16.0	243	516	66.2
145	418	21.0	258	531	72.0
134	407	26.0	275	548	79.0
125	398	29.5	287	560	84.5
120	393	31.0	296	569	89.0
120	393	33.0	304	577	93.5
126	399	35.5	310	583	95.5
140	413	38.5	335	608	100

[a] T/K values calculated by the compiler.

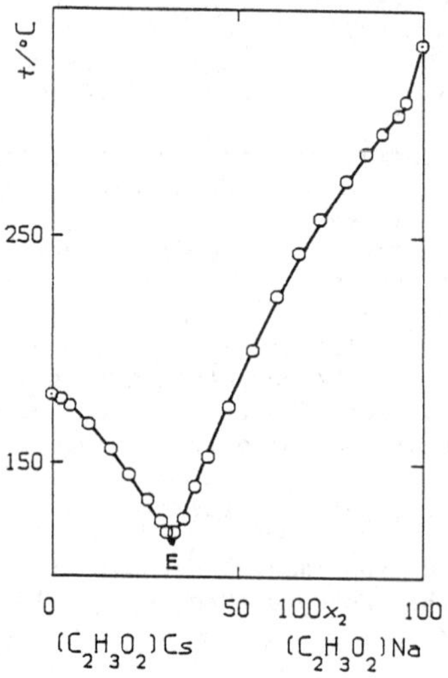

Characteristic point(s):

Eutectic, E, at 115 °C and $100x_1 = 68$ (authors).

AUXILIARY INFORMATION

METHOD/APPARATUS/PROCEDURE:

Visual polythermal method; temperatures measured with a Chromel-Alumel thermocouple.

SOURCE AND PURITY OF MATERIALS:

"Chemically pure" materials, recrystallized twice and dehydrated by prolonged heating (Ref. 1). Component 2 undergoes a phase transition at $t_{trs}(2)/^{o}C = 335$.

ESTIMATED ERROR:

Temperature: accuracy probably ±2 K (compiler).

REFERENCES:

(1) Diogenov, G.G.; Sarapulova, I.F.
Zh. Neorg. Khim. 1964, **9**, 1292-1294;
Russ. J. Inorg. Chem. (Engl. Transl.) 1964, **9**, 704-706.

COMPONENTS:	ORIGINAL MEASUREMENTS:
(1) Cesium ethanoate (cesium acetate); $(C_2H_3O_2)Cs$; [3396-11-0] (2) Sodium ethanoate (sodium acetate); $(C_2H_3O_2)Na$; [127-09-3]	Storonkin, A.V.; Vasil´kova, I.V.; Tarasov, A.A. **Vestn. Leningr. Univ., Fiz., Khim.** 1977, (4), 80-85.
VARIABLES:	PREPARED BY:
Temperature.	Baldini, P.

EXPERIMENTAL VALUES:

Data presented only in graphical form (see figure).

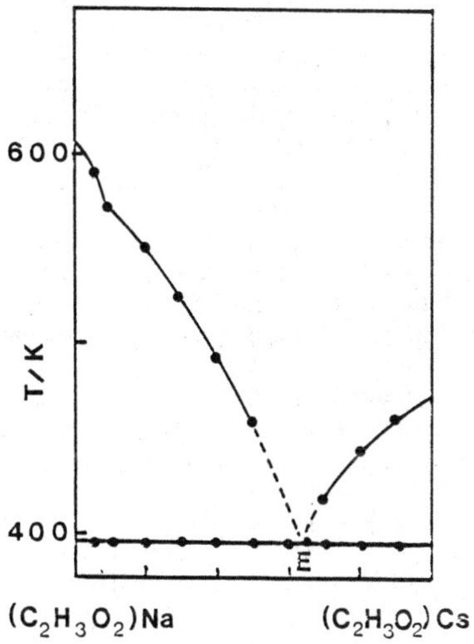

Characteristic point(s):

Eutectic, E, at 392 K and $100x_1$= 64 (authors).

Note - Undercooling does not allow one to draw the liquidus with accuracy at compositions close to the eutectic.

AUXILIARY INFORMATION

METHOD/APPARATUS/PROCEDURE:	SOURCE AND PURITY OF MATERIALS:
DTA and "contact polythermal method" under polarized light. IR spectra were also used to state the existence of intermediate compound(s).	Component 1 synthetized from Cs_2CO_3 and ethanoic acid ($T_{fus}(1)/K$= 467; authors). Component 2 of analytical purity recrystallized twice from water and dried under vacuum ($T_{fus}(2)/K$= 607; authors). The purity of both components was checked by thermographical analysis. The mixtures were prepared in a glove box.
	ESTIMATED ERROR:
	Temperature: accuracy probably ± 2 K (compiler).
	REFERENCES:

COMPONENTS:	EVALUATOR:
(1) Cesium ethanoate (cesium acetate); $(C_2H_3O_2)Cs$; [3396-11-0] (2) Rubidium ethanoate (rubidium acetate); $(C_2H_3O_2)Rb$; [563-67-7]	Schiraldi, A., Dipartimento di Chimica Fisica, Universita´ di Pavia (ITALY).

CRITICAL EVALUATION:

This binary was studied as a side of the ternary $C_2H_3O_2$/Cs, Na, Rb (Ref. 1), and of the reciprocal ternary Cs, Rb/$C_2H_3O_2$, NO_2 (Ref. 2), respectively.

Both papers give substantially analogous results, i.e., a liquidus with a minimum at 446 K (173 °C) and $100\mathbf{x}_1$= 72 (Ref. 1), and at 445 K (172 °C) and $100\mathbf{x}_1$= 71 (Ref. 2), respectively. It is, however, not clear whether the slight differences in the coordinates of the minimum as given in Ref. 1 and Ref. 2, respectively, come from different sets of determinations, or from a suitable adjustment improving the overall presentation of the ternary involved. It is also to be remarked that, although coming from the same group, a significant difference exists between the $\mathbf{T}_{fus}(2)$ values given in Ref. 1 (453.2 K) and Ref. 2 (460 K), the corresponding value given in Table 1 being 463±1 K.

Moreover, in neither paper the phase transition of rubidium ethanoate, occurring at either 489-493 K (Ref. 3), or 498±1 (Preface, Table 1) is explicitely mentioned, although, e.g., it might reasonably justify the knee observed at about 498 K (Ref. 1) in the liquidus branch richer in component 2.

The inspection of the liquidus of both ternaries mentioned above strongly supports the occurrence of solid solutions in the $C_2H_3O_2$/Cs, Rb side binary. However, the limits of the T, $\mathbf{x}_2$ field covered in the binary by these solutions seem poorly defined, in particular for what concerns the compositions close to pure component 2, and for temperatures close to the transition temperature of this salt. Thence, in the evaluator´s opinion, an investigation of the solidus would be desirable, in order to attain more satisfactory information about these points.

REFERENCES:

(1) Diogenov, G.G.; Sarapulova, I.F.
Zh. Neorg. Khim. 1964, **9**, 1499-1502; **Russ. J. Inorg. Chem. (Engl. Transl.)** 1964, **9**, 814-816.

(2) Diogenov, G.G.; Morgen, L.T.
Fiz.-Khim. issled. Rasplavov Solei, Irkutsk, 1975, 62-64.

(3) Gimel´shtein, V.G.; Diogenov, G.G.
Zh. Neorg. Khim. 1958, 3, 1644-1649; **Russ. J. Inorg. Chem. (Engl. Transl.)** 1958, **3(7)**, 230-236.

COMPONENTS:

(1) Cesium ethanoate (cesium acetate); $(C_2H_3O_2)Cs$; [3396-11-0]
(2) Rubidium ethanoate (rubidium acetate); $(C_2H_3O_2)Rb$; [563-67-7]

ORIGINAL MEASUREMENTS:

Diogenov, G.G.; Sarapulova, I.F.
Zh. Neorg. Khim. 1964, **9**, 1499-1502 (*);
Russ. J. Inorg. Chem. (Engl. Transl.) 1964, **9**, 814-816.

VARIABLES:

Temperature.

PREPARED BY:

Baldini, P.

EXPERIMENTAL VALUES:

$t/^oC$	T/K^a	$100x_1$
240.0[b]	513.2	0
232.5	505.7	3.5
226.5	499.7	5.3
224.4	497.6	8.5
223.0	496.2	14.0
216.5	489.7	21.0
208.0	481.2	27.5
202.0	475.2	33.5
197.0	470.2	40.0
190.0	463.2	47.2
185.0	458.2	53.5
180.0	453.2	60.0
175.0	448.2	66.5
172.5	445.7	71.7
173.7	446.9	77.0
175.0	448.2	82.5
177.5	450.7	87.5
179.0	452.2	93.0
180.0	453.2	100.0

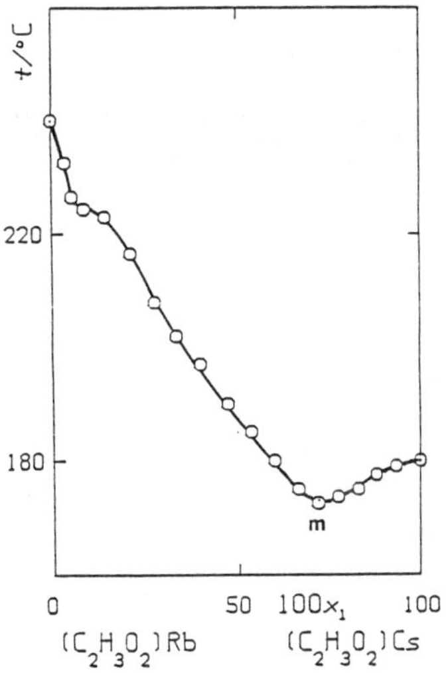

[a] T/K values calculated by the compiler.
[b] 238 in Fig. 1 of the original paper (compiler).

Characteristic point(s):

Continuous series of solid solutions with a minimum, m, at 173 ^oC and $100x_1$ about 72 (authors).

AUXILIARY INFORMATION

METHOD/APPARATUS/PROCEDURE:

Visual polythermal method; temperatures measured with a Chromel-Alumel thermocouple.

SOURCE AND PURITY OF MATERIALS:

"Chemically pure" materials, recrystallized twice and dehydrated by prolonged heating (Ref. 1).

ESTIMATED ERROR:

Temperature: accuracy probably ± 2 K (compiler).

REFERENCES:

(1) Diogenov, G.G.; Sarapulova, I.F. **Zh. Neorg. Khim.** 1964, 9, 1292-1294; **Russ. J. Inorg. Chem. (Engl. Transl.)** 1964, **9**, 704-706.

COMPONENTS:	ORIGINAL MEASUREMENTS:
(1) Cesium ethanoate (cesium acetate); $(C_2H_3O_2)Cs$; [3396-11-0] (2) Rubidium ethanoate (rubidium acetate); $(C_2H_3O_2)Rb$; [563-67-7]	Diogenov, G.G.; Morgen, L.T. **Fiz.-Khim. Issled. Rasplavov Solei, Irkutsk,** 1975, 62-64.
VARIABLES:	PREPARED BY:
Temperature.	Baldini, P.

EXPERIMENTAL VALUES:

Characteristic point(s):

Continuous series of solid solutions with a minimum, m, at 172 °C (authors) and $100\mathbf{x}_1$ about 71 (compiler).

AUXILIARY INFORMATION

METHOD/APPARATUS/PROCEDURE:	SOURCE AND PURITY OF MATERIALS:
Visual polythermal method; temperatures measured with a Chromel-Alumel thermo couple.	Not stated. Component 1: $t_{fus}(1)/°C = 187.$ Component 2: $t_{fus}(2)/°C = 238.$
	ESTIMATED ERROR:
	Temperature: accuracy probably ± 2 K (compiler).
	REFERENCES:

COMPONENTS:

(1) Cesium ethanoate (cesium acetate); $(C_2H_3O_2)Cs$; [3396-11-0]
(2) Zinc ethanoate (zinc acetate); $(C_2H_3O_2)_2Zn$; [557-34-6]

ORIGINAL MEASUREMENTS:

Pavlov, V.L.; Golubkova, V.V.
Visn. Kiiv. Univ., Ser. Khim., Kiev, 1972, **No.** 13, 28-30.

VARIABLES:

Temperature.

PREPARED BY:

Baldini, P.

EXPERIMENTAL VALUES:

The results are reported only in graphical form (see figure).

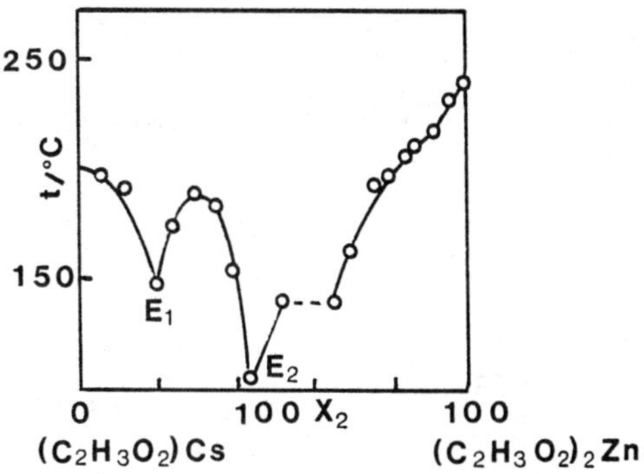

Characteristic point(s):

Eutectic, E_1, at 140 °C and $100\mathbf{x}_2$= 20 (authors).
Eutectic, E_2, at 104 °C and $100\mathbf{x}_2$= 45 (authors).

Note - Glasses form at $50 \leq 100\mathbf{x}_2 \leq 60$.

Intermediate compound(s):

$(C_2H_3O_2)_4Cs_2Zn$, congruently melting at 190 °C (authors).

AUXILIARY INFORMATION

METHOD/APPARATUS/PROCEDURE:

The visual polythermal method as well as time-temperature curves were employed. The temperatures were measured with a Chromel-Alumel thermocouple checked at the freezing temperatures of Zn, $K_2Cr_2O_7$, Cd, Sn, and benzoic acid.

NOTE:

The formation of glasses in this system seems likely. Accordingly, one should expect marked undercooling over a large composition range which would make the results of visual polythermal observations less reliable than usual. The lack of any further experimental evidence (e.g., from X-ray diffractometry) justifies casting doubts about the actual existence of the intermediate compound(s).

SOURCE AND PURITY OF MATERIALS:

Component 1: obtained by reacting Cs_2CO_3 and ethanoic acid, and kept in a dessiccator in the presence of P_2O_5 until constant mass.
Component 2: $(C_2H_3O_2)_2Zn.2H_2O$ of analytical purity dried to constant mass at 110 °C.

ESTIMATED ERROR:

Temperature: accuracy probably ± 2 K (compiler).

REFERENCES:

COMPONENTS:	EVALUATOR:
(1) Potassium ethanoate (potassium acetate); $(C_2H_3O_2)K$; [127-08-2] (2) Lithium ethanoate (lithium acetate); $(C_2H_3O_2)Li$; [546-89-4]	Spinolo, G., Dipartimento di Chimica Fisica, Universita' di Pavia (ITALY).

CRITICAL EVALUATION:

The system potassium ethanoate - lithium ethanoate was investigated by Diogenov (visual polythermal analysis, 1956; Ref. 1), Pochtakova (visual polythermal analysis, 1965; Ref. 2), Sokolov and Tsindrik (visual polythermal analysis, supplemented with DTA, 1969; Ref. 3), and Gimel'shtein (DTA, supplemented with X-ray patterns, 1970, 1971; Refs. 4, 5, respectively).

Phase transitions are reported at 571 K (298 °C) by Diogenov (Ref. 1), at 331 and 428 K (58 and 155 °C, respectively) by Sokolov (Ref. 6, quoted in Refs. 2, 3), and at 428 K (155 °C) by Gimel'shtein (Ref. 5) for component 1; at 540 K (267 °C) by Diogenov (Ref. 1), and at 405 K (132 °C) by Gimel'shtein (Ref. 5) for component 2. In Table 1 of the Preface mention is made of a transition at 422.2±0.5 K for component 1, whereas no transition is reported for component 2.

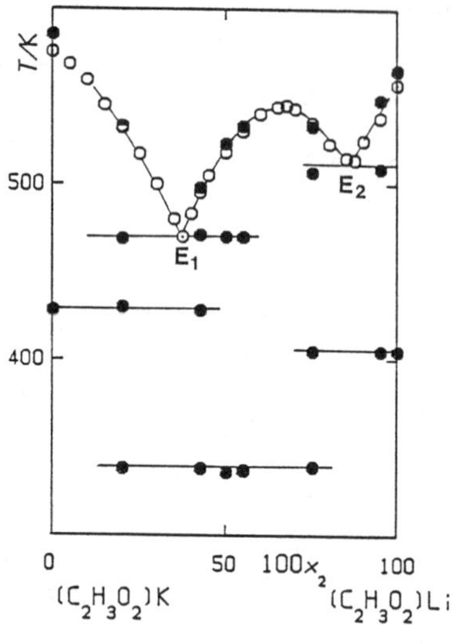

Diogenov (Ref. 1) investigated the binary concerned here as a side system of the ternary $C_2H_3O_2$/K, Li, Na, and claimed the existence of two congruently melting intermediate compounds, i.e., $(C_2H_3O_2)_2KLi$ and $(C_2H_3O_2)_3KLi_2$, respectively. The existence of the former, inferred by Diogenov from discontinuities observed in the liquidus of the binary itself and of two internal cuts of the ternary, was denied by all subsequent authors. In particular, no evidence of the existence of a crystallization field attributable to a 1:1 compound was found either by Pochtakova (Ref. 2) in her re-investigation of the ternary $C_2H_3O_2$/K, Li, Na, or by Sokolov and Tsindrik (Ref. 3), and Gimel'shtein (Ref. 4) in their studies of the topology of the reciprocal ternary K, Li/$C_2H_3O_2$, NO_3. The thermographical traces recorded by Gimel'shtein (and detailed in Ref. 5) support satisfactorily the assertion that in the mixtures of potassium and lithium ethanoates only the intermediate compound $(C_2H_3O_2)_3KLi_2$ does form, which melts congruently at 547±2 K (Refs. 2, 4, 5), and gives eutectics with each of the component salts.

In the figure, the visual data by Pochtakova (Ref. 2) are plotted, along with the thermographical ones obtained by Gimel'shtein (Ref. 5) to give a comprehensive and reasonably reliable representation of the liquidus, solidus, and subsolidus.
The main discrepancies between the two authors occur in the fusion temperatures of the pure components:

$T_{fus}(1)$/K= 575, 585 (Refs. 2, 5, respectively);
$T_{fus}(2)$/K= 557, 565 (Refs. 2, 5, respectively).

The more correct probably are those reported in Ref. 2, which are closer to $T_{fus}(1)$/K= 578.7±0.5, and $T_{fus}(2)$/K= 557±2, reported in Table 1 of the Preface. These discrepancies, however, do not affect substantially the overall features of the phase diagram.

REFERENCES:

(1) Diogenov, G.G.; **Zh. Neorg. Khim.** 1956, 1, 2551-2555 (*); **Russ. J. Inorg. Chem. (Engl. Transl.)** 1956, 1(11), 122-126.
(2) Pochtakova, E.I.; **Zh. Neorg. Khim.** 1965, **10**, 2333-2338 (*); **Russ. J. Inorg. Chem. (Engl. Transl.)** 1965, **10**, 1268-1271.
(3) Sokolov, N.M.; Tsindrik, N.M.; **Zh. Neorg. Khim.** 1969, **14**, 584-590 (*); **Russ. J. Inorg. Chem., (Engl. Transl.)** 1969, **14**, 302-306.
(4) Gimel'shtein, V.G. **Symposium, "Fiziko-Khimicheskii Analiz Solevykh Sistem", Irkutsk,** 1970, 39-45.
(5) Gimel'shtein, V.G. **Tr. Irkutsk. Politekh. Inst.** 1971, No. 66, 80-100.
(6) Sokolov, N.M. **Tezisy Dokl. X Nauch. Konf. S.M.I.** 1956.

COMPONENTS:	ORIGINAL MEASUREMENTS:
(1) Potassium ethanoate (potassium acetate); $(C_2H_3O_2)K$; [127-08-2] 2) Lithium ethanoate (lithium acetate); $(C_2H_3O_2)Li$; [546-89-4]	Diogenov, G.G. **Zh. Neorg. Khim.** 1956, 1, 2551-2555 (*); **Russ. J. Inorg. Chem. (Engl. Transl.)** 1956, 1(11), 122-126.
VARIABLES:	PREPARED BY:
Temperature.	Baldini, P.

EXPERIMENTAL VALUES:

$t/^oC$	T/K^a	$100x_2$	$t/^oC$	T/K^a	$100x_2$
310.5	583.5	0	247	520	52.5
308	581	2.5	258	531	55.2
305	578	3.7	268	541	58.5
295	568	7.5	273	546	62.5
285	558	11.5	274	547	67.8
274	547	15.5	267	540	72.5
262	535	19.3	257	530	76.3
251	524	22.5	245	518	80.0
234	507	26.7	235	508	82.3
221	494	29.5	225	498	84.0
209	482	31.7	229	502	85.0
197	470	34.0	240	513	86.7
187	460	35.5	252	525	89.4
191	464	37.5	262	535	93.0
208	481	40.0	270	543	94.5
221	494	42.5	278	551	96.0
232	505	46.0	291	564	100.0
236	509	50.0			

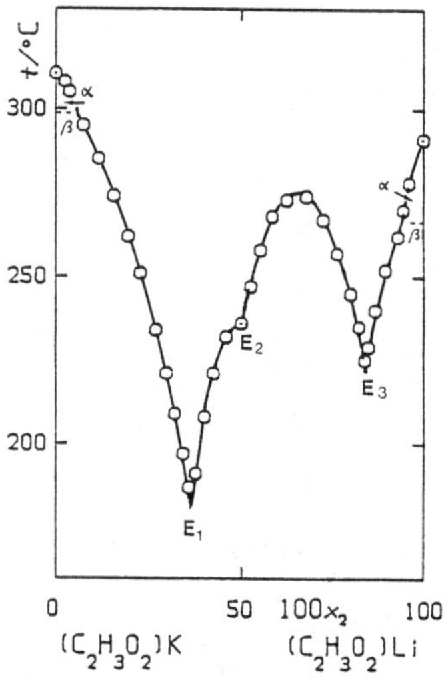

[a] T/K values calculated by the compiler.

Characteristic point(s): Eutectic, E_1, at 181 ^oC and $100x_2$= 36 (author).
Eutectic, E_2, at 236 ^oC and $100x_2$= 51.5 (author).
Eutectic, E_3, at 222 ^oC and $100x_2$= 84.5 (author).

Intermediate compound(s): $(C_2H_3O_2)_2KLi$, congruently melting at 236 ^oC.
$(C_2H_3O_2)_3KLi_2$, congruently melting at 275 ^oC.

AUXILIARY INFORMATION

METHOD/APPARATUS/PROCEDURE:	SOURCE AND PURITY OF MATERIALS:
Visual polythermal method; temperatures measured with a Chromel-Alumel thermocouple.	"Chemically pure" materials, recrystallized twice and dehydrated by prolonged heating. Components 1 and 2 undergo phase transitions at $t_{trs}(1)/^oC$= 298 and $t_{trs}(2)/^oC$= 267, respectively, according to Fig. 1 of the original paper (compiler).
	ESTIMATED ERROR:
	Temperature: accuracy probably ± 2 K (compiler).
	REFERENCES:

COMPONENTS:

(1) Potassium ethanoate (potassium acetate); $(C_2H_3O_2)K$; [127-08-2]
(2) Lithium ethanoate (lithium acetate); $(C_2H_3O_2)Li$; [546-89-4]

ORIGINAL MEASUREMENTS:

Pochtakova, E.I.
Zh. Neorg. Khim. 1965, 10, 2333-2338 (*); **Russ. J. Inorg. Chem. (Engl. Transl.)** 1965, 10, 1268-1271.

VARIABLES:

Temperature.

PREPARED BY:

Baldini, P.

EXPERIMENTAL VALUES:

$t/^{o}C$	T/K^{a}	$100\mathbf{x}_2$	$t/^{o}C$	T/K^{a}	$100\mathbf{x}_2$
302	575	0	257	530	55
295	568	5	267	540	60
286	559	10	271	544	65
272	545	15	272	545	67.5
259	532	20	270	543	70
244	517	25	262	535	75
227	500	30	250	523	80
207	480	35	242	515	85
197	470	37.5	241	514	87.5
210	483	40	252	525	90
222	495	42.5	265	538	95
232	505	45	284	557	100
245	518	50			

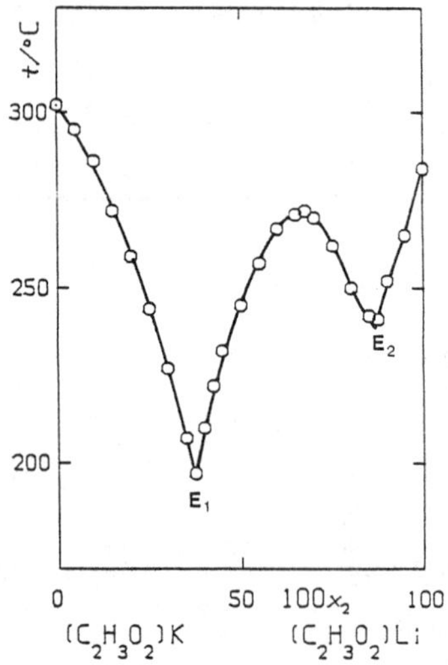

[a] T/K values calculated by the compiler.

Characteristic point(s):

Eutectic, E_1, at 197 ^{o}C and $100\mathbf{x}_2$= 37.6 (authors).
Eutectic, E_2, at 238 ^{o}C and $100\mathbf{x}_2$= 87 (authors).

Intermediate compound(s):

$(C_2H_3O_2)_3KLi_2$, congruently melting at 272 ^{o}C (authors).

AUXILIARY INFORMATION

METHOD/APPARATUS/PROCEDURE:

Visual polythermal method.

SOURCE AND PURITY OF MATERIALS:

Not stated.
Component 1 undergoes phase transitions at $t_{trs}(1)/^{o}C$= 58, 155 (Ref. 1).

ESTIMATED ERROR:

Temperature: accuracy probably ± 2 K (compiler).

REFERENCES:

(1) Sokolov, N.M.
Tezisy Dokl. X Nauch. Konf. S.M.I. 1956.

COMPONENTS:

(1) Potassium ethanoate (potassium acetate); $(C_2H_3O_2)K$; [127-08-2]
(2) Lithium ethanoate (lithium acetate); $(C_2H_3O_2)Li$; [546-89-4]

ORIGINAL MEASUREMENTS:

Sokolov, N.M.; Tsindrik, N.M.
Zh. Neorg. Khim. 1969, 14, 584-590 (*);
Russ. J. Inorg. Chem. (Engl. Transl.) 1969, **14**, 302-306.

VARIABLES:

Temperature.

PREPARED BY:

Baldini, P.

EXPERIMENTAL VALUES:

The results are reported only in graphical form (see figure).

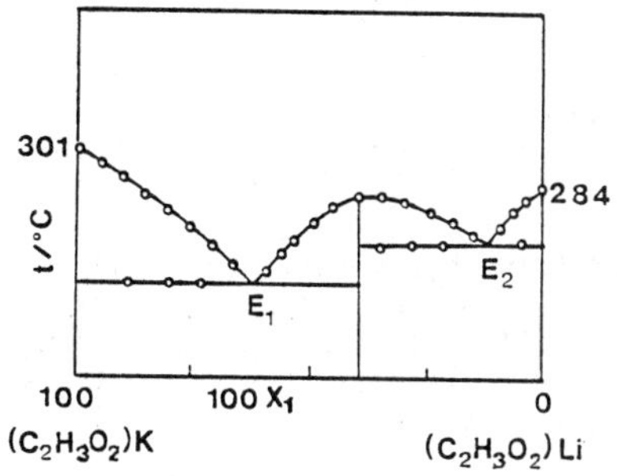

Characteristic point(s):

Eutectic, E_1, at 197 °C and $100\mathbf{x}_1$= 62 (authors).
Eutectic, E_2, at 234 °C and $100\mathbf{x}_1$= 13 (authors).

Intermediate compound(s):

$(C_2H_3O_2)_3KLi_2$, congruently melting (authors).

AUXILIARY INFORMATION

METHOD/APPARATUS/PROCEDURE:

Visual polythermal method, supplemented with differential thermal analysis.

SOURCE AND PURITY OF MATERIALS:

Commercial materials recrystallized. Component 1 undergoes phase transitions at $t_{trs}(1)/^{o}C$= 58, 155 (Ref. 1) and melts at $t_{fus}(1)/^{o}C$= 301.
Component 2 melts at $t_{fus}(2)/^{o}C$= 284.

ESTIMATED ERROR:

Temperature: accuracy probably ±2 K (compiler).

REFERENCES:

(1) Sokolov, N.M.
Tezisy Dokl. X Nauch. Konf. S.M.I. 1956.

COMPONENTS:	ORIGINAL MEASUREMENTS:
(1) Potassium ethanoate (potassium acetate); $(C_2H_3O_2)K$; [127-08-2] (2) Lithium ethanoate (lithium acetate); $(C_2H_3O_2)Li$; [546-89-4]	Gimel'shtein, V.G. **Symposium, "Fiziko-Khimicheskii Analiz Solevykh Sistem", Irkutsk**, 1970, 39-45.
VARIABLES:	PREPARED BY:
Temperature.	Baldini, P.

EXPERIMENTAL VALUES:

Characteristic point(s):

Eutectic, E_1, at 197 °C and $100\mathbf{x}_2$= 37.5 (author).
Eutectic, E_2, at 234 °C and $100\mathbf{x}_2$= 87 (author).

Intermediate compound(s):

$(C_2H_3O_2)_3KLi_2$, congruently melting at 275 °C, and undergoing a phase transition at 65 °C (author).

AUXILIARY INFORMATION

METHOD/APPARATUS/PROCEDURE:	SOURCE AND PURITY OF MATERIALS:
Thermographical analysis.	Not stated. Component 2 undergoes a phase transition at $t_{trs}(2)/^{o}C$= 132.
	ESTIMATED ERROR:
	Temperature: accuracy probably ±2 K (compiler).
	REFERENCES:

COMPONENTS:

(1) Potassium ethanoate (potassium acetate); $(C_2H_3O_2)K$; [127-08-2]
(2) Lithium ethanoate (lithium acetate); $(C_2H_3O_2)Li$; [546-89-4]

ORIGINAL MEASUREMENTS:

Gimel'shtein, V.G.
Tr. Irkutsk. Politekh. Inst. 1971, No. 66, 80-100.

VARIABLES:

Temperature.

PREPARED BY:

Baldini, P.

EXPERIMENTAL VALUES:

$t/^{\circ}C$	T/K^{a}	$100x_2$	$t/^{\circ}C$	T/K^{a}	$100x_2$
312	585	0	260	533	55.0
155	428	0	197	470	55.0
260	533	20.0	64	337	55.0
196	469	20.0	260	533	75.0
157	430	20.0	234	507	75.0
65	338	20.0	132	405	75.0
225	498	42.5	66	339	75.0
198	471	42.5	275	548	95.0
155	428	42.5	236	509	95.0
65	338	42.5	132	405	95.0
250	523	50.0	292	565	100
197	470	50.0	132	405	100
63	336	50.0			

[a] T/K values calculated by the compiler.

The meaning of the data listed in the table becomes apparent by observing the figure reported in the critical evaluation.

Characteristic point(s):
Eutectic, E_1, at 197 $^{\circ}C$ and $100x_2$= 37.5 (author).
Eutectic, E_2, at 237 $^{\circ}C$ (234 $^{\circ}C$ according to Fig. 4) and $100x_2$= 87.0 (author).

Intermediate compound:
$(C_2H_3O_2)_3KLi_2$, congruently melting at 275 $^{\circ}C$ (author), and undergoing a phase transition at 65 $^{\circ}C$ (author).

AUXILIARY INFORMATION

METHOD/APPARATUS/PROCEDURE:

Differential thermal analysis (using a derivatograph with automatic recording of the heating curves) and room temperature X-ray diffractometry (using a URS-501M apparatus) were employed.

NOTE:

The coordinates of the characteristic points were stated by the author on the basis of his own DTA measurements, and of previous literature data (Refs. 1, 2). X-ray patterns were taken at $100x_2$= 45, 70.

SOURCE AND PURITY OF MATERIALS:

Not stated.
Component 1 melts at $t_{fus}(1)/^{\circ}C$= 312 (310 according to Fig. 4 of the original paper; compiler), and undergoes a phase transition at $t_{trs}(1)/^{\circ}C$= 155.
Component 2 melts at $t_{fus}(2)/^{\circ}C$= 292 (291 according to Fig. 4 of the original paper; compiler), and undergoes a phase transition at $t_{trs}(2)/^{\circ}C$= 132.

ESTIMATED ERROR:

Temperature: accuracy probably ± 2 K (compiler).

REFERENCES:

(1) Pochtakova, E.I.
Zh. Neorg. Khim. 1965, **10**, 2333-2338.
(2) Sokolov, N.M.; Tsindrik, N.M.
Zh. Neorg. Khim. 1969, **14**, 584-590.

COMPONENTS:

(1) Potassium ethanoate (potassium acetate);
$(C_2H_3O_2)_2K_2$; [127-08-2]
(2) Magnesium ethanoate (magnesium acetate);
$(C_2H_3O_2)_2Mg$; [142-72-3]

ORIGINAL MEASUREMENTS:

Pochtakova, E.I.
Zh. Obshch. Khim. 1974, **44**, 241-248.

VARIABLES:

Temperature.

PREPARED BY:

Baldini, P.

EXPERIMENTAL VALUES:

t/°C	T/K[a]	$100x_2$	t/°C	T/K[a]	$100x_2$
302	575	0	251	524	32.5
301	574	5	242	515	35
297	570	7.5	233[bc]	506	36.5
293	566	10	233[bd]	506	36.5
297[bc]	570	10	175[bf]	448	36.5
236[bd]	509	10	244	517	37.5
155[be]	428	10	254	527	40
290	563	12.5	264	537	42.5
286	559	15	273	546	45
282	555	17.5	282	555	47.5
278	551	20	292	565	50
273	546	22.5	296[bc]	569	50
265	538	25	233[bd]	506	50
276[bc]	549	25	166[bf]	439	50
235[bd]	508	25	310	583	55
155[be]	428	25	310[bc]	583	55
263	536	27.5	232[bd]	505	55
257	530	30	153[be]	426	55

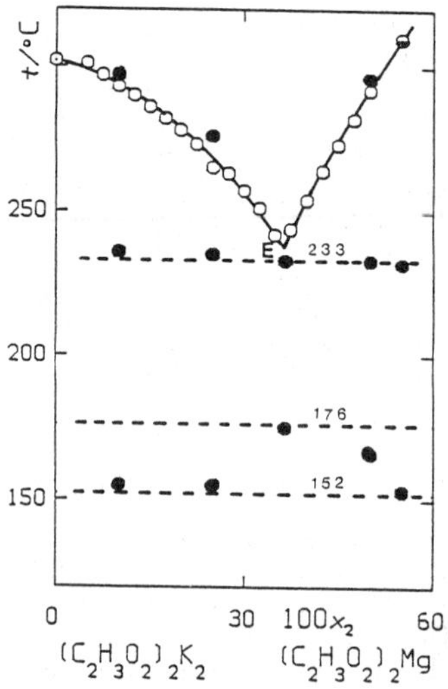

[a] T/K values calculated by the compiler.
[b] Differential thermal analysis (filled circles in the figure).
[c] Initial crystallization.
[d] Eutectic stop.
[e] First transition in the system.
[f] Second transition in the system.

Characteristic point: Eutectic, E, at 238 °C (extrapolated, visual polythermal method), or 233 °C (differential thermal analysis), and $100x_2$= 36.5 (author).

AUXILIARY INFORMATION

METHOD/APPARATUS/PROCEDURE:

Visual polythermal method, supplemented with differential thermal analysis.

NOTE:

The system was investigated only at $0 \leq 100x_2 \leq 55$ due to thermal instability of component 2.
The fusion temperature of component 1 (575 K) is not far below that reported in Table 1 of the Preface (578.7$\pm$0.5 K), where, however, only one solid state transition (at 422.2$\pm$0.5 K) is mentioned, instead of the two ones (at 428 and 331 K, respectively) quoted by Pochtakova from Ref. 1.

SOURCE AND PURITY OF MATERIALS:

Component 1: "chemically pure" material recrystallized and dried at 200 °C to constant mass (phase transitions at $t_{trs}(1)$/°C = 58, 155; Ref. 1). Component 2: prepared (Ref. 2) by reacting the ("chemically pure") carbonate with a slight excess of ethanoic acid of analytical purity (phase transitions at $t_{trs}(2)$/°C = 152, 176).

ESTIMATED ERROR:

Temperature: accuracy probably $\pm$2 K (compiler).

REFERENCES:

(1) Sokolov, N.M.
Tezisy Dokl. X Nauch. Konf. S.M.I. 1956.
(2) Sokolov, N.M.
Zh. Obshch. Khim. 1954, **24**, 1581-1593.

COMPONENTS:	EVALUATOR:
(1) Potassium ethanoate (potassium acetate); $(C_2H_3O_2)K$; [127-08-2] (2) Sodium ethanoate (sodium acetate); $(C_2H_3O_2)Na$; [127-09-3]	Franzosini, P., Dipartimento di Chimica Fisica, Universita' di Pavia (ITALY).

CRITICAL EVALUATION:

This system has been the most widely studied during the last 70 years. The opinions by the different authors are summarized hereafter.

(1) Baskov (1915; Ref. 1).
$T_{fus}(1)$= 568.2 K (295.0 °C); $T_{fus}(2)$= 593.2 K (320.0 °C); continuous series of solid solutions with a minimum, m, at 496.2 K (223.0 °C) and $100x_2$= 46 (method: thermal analysis; liquidus and solidus investigated).

(2) Bergman; Evdokimova (1956; Ref. 2).
$T_{fus}(1)$= 575 K (302 °C); $T_{fus}(2)$= 599 K (326 °C); T_{trs} (2)= 527 K (254 °C); eutectic, E, at 497 K (224 °C) and $100x_2$= 45 (method: visual polythermal analysis, supplemented with three DTA records; liquidus and solidus investigated).

(3) Diogenov; Erlykov (1958; Ref. 3).
$T_{fus}(1)$= 583.5 K (310.5 °C); $T_{trs}(1)$= 569 K (296 °C); $T_{fus}(2)$= 610 K (337 °C); $T_{trs}(2)$= 599 K (326 °C); continuous series of solid solutions with a minimum, m, at 501 K (228 °C) and $100x_2$= 45 (method: visual polythermal analysis; liquidus only investigated).

(4) Golubeva; Bergman; Grigor'eva (1958; Ref. 4).
Intermediate compound $(C_2H_3O_2)_3K_2Na$, incongruently melting at 513 K (240 °C) (method: visual polythermal analysis).

(5) Sokolov; Pochtakova (1958; Ref. 5).
$T_{fus}(1)$= 574 K (301 °C); $T_{fus}(2)$= 604 K (331 °C); [$T_{trs}(2)$= 527 K (254 °C); quoted by the authors from Ref. 2]; eutectic, E_1, at 513 K (240 °C) and $100x_2$= 38.5; eutectic, E_2, at 508 K (235 °C) and $100x_2$= 46.5; intermediate compound, $(C_2H_3O_2)_5K_3Na_2$, congruently melting at 514 K (241 °C) (method: visual polythermal analysis; liquidus only investigated).

(6) Nesterova; Bergman (1960; Ref. 6).
$T_{fus}(1)$= 579 K (306 °C); $T_{fus}(2)$= 601 K (328 °C); peritectic, P, at 511 K (238 °C) and $100x_2$= 36.5; eutectic, E, at 505 K (232 °C) and $100x_2$= 50; intermediate compound, $(C_2H_3O_2)_3K_2Na$, incongruently melting (method: visual polythermal analysis; liquidus only investigated).

(7) Il'yasov; Bergman (1960; Ref. 7).
$T_{fus}(1)$= 579 K (306 °C); $T_{fus}(2)$= 601 K (328 °C); peritectic, P, at 523-529 K (250-256 °C) and $100x_2$= 35; eutectic, E, at 513 K (240 °C) and $100x_2$= 50; intermediate compound, $(C_2H_3O_2)_3K_2Na$, incongruently melting (method: visual polythermal analysis; liquidus only investigated).

(8) Diogenov; Sarapulova (1964; Ref. 8).
$T_{fus}(1)$= 583 K (310 °C); $T_{fus}(2)$= 608 K (335 °C); eutectic, E_1, at 513 K (240 °C) (composition not reported); eutectic, E_2, at 508 K (235 °C) (composition not reported); intermediate compound, $(C_2H_3O_2)_5K_3Na_2$, congruently melting (method: visual polythermal analysis).

(9) Sokolov; Pochtakova (1967; Ref. 9).
$T_{fus}(1)$= 575 K (302 °C); $T_{fus}(2)$= 604 K (331 °C); solid state transitions at 428 and 331 K (155 and 58 °C) for component 1, at 511, 403, 391, and 331 K (238, 130, 118 , and 58 °C) for component 2; eutectic, E_1, at 513 K (240 °C) and $100x_2$= 38.5; eutectic, E_2, at 506 K (233 °C) and $100x_2$= 46.5; intermediate compound, $(C_2H_3O_2)_5K_3Na_2$, congruently melting at 513-514 K (240-241 °C) (method: thermographical analysis, supplemented with visual polythermal measurements and microscopic observations in polarized light).

(10) Diogenov; Chumakova (1975; Ref. 10).
$T_{fus}(1)$= 575 K (302 °C); $T_{fus}(2)$= 599 K (326 °C); peritectic, P, at 513 K (240 °C) (composition not reported); eutectic, E, at 510 K (237 °C) (composition not reported); intermediate compound, $(C_2H_3O_2)_5K_3Na_2$, incongruently melting (method: visual polythermal analysis).

COMPONENTS:	EVALUATOR:
(1) Potassium ethanoate (potassium acetate); $(C_2H_3O_2)K$; [127-08-2] (2) Sodium ethanoate (sodium acetate); $(C_2H_3O_2)Na$; [127-09-3]	Franzosini, P., Dipartimento di Chimica Fisica, Universita' di Pavia (ITALY).

CRITICAL EVALUATION (cont.d):

(11) Storonkin; Vasil'kova; Tarasov (1977; Ref. 11).
$T_{fus}(1)$= 584 K (311 °C); $T_{fus}(2)$= 607 K (334 °C); eutectic, E, at 511 K (238 °C) and $100x_2$= 46 (method: differential thermal analysis and "contact polythermal method" under polarized light, supplemented with IR spectroscopy).

Information from different sources on the thermophysics of both components is conflicting, possibly due - inter alia - to hygroscopicity, and to the fact that solid state transitions are characterized by a remarkable sluggishness.

$T_{fus}(1)$ values ranging between 565 and 584 K, and $T_{fus}(2)$ values ranging between 592 and 610 K can be found in the literature (Ref. 12). The DSC data from Preface Table 1, i.e., $T_{fus}(1)$= 578.7±0.5 K and $T_{fus}(2)$= 601.3±0.5 K, are thought to be reasonably trustworthy, being supported by independent cryometric measurements by the same group (Ref. 12). Concerning in particular the T_{fus} data given in Refs. 1-11, the following remarks can be made. Poor reliability seems to be attached to the fusion temperatures from Refs. 1, 3, 8, 10, 11. Indeed: (i) Baskov (Ref. 1), who studied the system in 1915, might have not had at disposal high purity samples, thus obtaining too low T_{fus} values [$T_{fus}(1)$= 568.2 K; $T_{fus}(2)$= 593.2 K]; (ii) Diogenov et al.'s figures [$T_{fus}(1)$= 583.5 K (1958; Ref. 3), 583 K (1964; Ref. 8), and 575 K (1975; Ref. 10); $T_{fus}(2)$= 610 K (1958; Ref. 3), 608 K (1964; Ref. 8), and 599 K (1975; Ref. 10)] look as doubtful, due to excessive fluctuation; (iii) Storonkin et al.'s figures [$T_{fus}(1)$= 584 K; $T_{fus}(2)$= 607 K (1977; Ref. 11)] seem also to be doubtful and for the same reason, inasmuch as in previous papers Storonkin, Vasil'kova, and Potemin (1974; Ref. 13) gave $T_{fus}(2)$= 601 K, while Potemin, Tarasov, and Panin (1973; Ref. 14) gave $T_{fus}(1)$= 581 K, $T_{fus}(2)$= 604 K. Instead, the agreement with T_{fus} data from Preface Table 1 is satisfactory for the most recent figures by Bergman et al. (Refs. 6, 7), and still acceptable for those by Sokolov and Pochtakova (Refs. 5, 9).

As for the solid state transitions, the situation is rather puzzling, as shown in the following table.

Salt	T_{trs}/K	Method	Year	Ref.
$C_2H_3O_2K$	428, 331	Vis. pol.	1956	15
	565-566	Vis. pol.	1957	16
	569	Vis. pol.	1958	3
	423	Dilat.,DTA	1966	17
	(503, 433, 353)	-	1966	18
	428, about 348	X-ray	1972	19
	422.2±0.5	DSC	1975	Preface, Table 1
	413-423	DTA	1976	20
$C_2H_3O_2Na$	527	Vis. pol.	1956	2
	596	Vis. pol.	1956	21
	511-513, 403, 391, 331	Vis. pol.	1956	15
	599	Vis. pol.	1958	3
	583-584	Vis. pol.	1958	22
	527±15, 465±3, 414±10	DSC	1975	Preface, Table 1
	337	DTA	1976	23

Vis. pol.: visual polythermal analysis; Dilat.: dilatometry;
(...): provisional data.

Potassium ethanoate was submitted to X-ray investigation by Hatibarua and Parry (Ref. 19), who obtained evidence for a monoclinic -> monoclinic transformation at about 348 K, and for a monoclinic -> orthorhombic transformation at 428 K. Allowance being made for some fluctuations in the T_{trs} values, it can be asserted that the occurrence of the former transition is supported by Sokolov's (Ref. 15), and Hazlewood et al.'s (Ref. 18) findings, while on the occurrence of the latter transition all the authors concerned agree, but for Diogenov et al. (Refs. 3, 16). These, in turn, are alone in claiming that component 1 undergoes a transformation at a temperature as high as 560-570 K: the evaluator, however, is inclined to think that the existence of the latter

COMPONENTS:	EVALUATOR:
(1) Potassium ethanoate (potassium acetate); $(C_2H_3O_2)K$; [127-08-2] (2) Sodium ethanoate (sodium acetate); $(C_2H_3O_2)Na$; [127-09-3]	Franzosini, P., Dipartimento di Chimica Fisica, Universita' di Pavia (ITALY).

CRITICAL EVALUATION (cont.d):

transformation is quite doubtful.

The number and location of solid state transitions in sodium ethanoate is still an open question, and the pertinent data are the most uncertain among those listed in Preface Table 1. It can only be said that the occurrence of a transition at 510-530 K seems to be reasonably supported (Refs. 2, 15, and Preface Table 1), whereas insufficient experimental evidence has been provided so far for the remaining transitions, including that reported by Diogenov et al. (Refs. 21, 3, 22) at 580-600 K.

Concerning the topology of the phase diagram, the evaluator is inclined not to take into account the findings by: (i) Baskov (Ref. 1), because reasonable doubts exist - as said above - about the purity of the salts he could have at disposal in 1915; (ii) Diogenov et al. (Refs. 3, 8, 10), for both the above made remarks on the phase transformation temperatures they report, and their conflicting assertions on the phase relations (continuous series of solid solutions in Ref. 3; congruently melting intermediate compound in Ref. 8; incongruently melting intermediate compound in Ref. 10).

Storonkin et al. (Ref. 11) quoted in their paper Refs. 1-5, 7, 8, and - inter alia - asserted correctly that it is hard to state the composition of an incongruently melting intermediate compound on the only basis of visual observations carried out on the liquidus. They asserted also that: (i) due to undercooling of the molten mixtures of composition $50 \leq 100\mathbf{x}_1 \leq 60$, no reliable information could be drawn from their liquidus on the formation of any intermediate compound; and (ii) their supplementary IR measurements gave no evidence of the existence of such compounds. Accordingly, they claimed the occurrence of a eutectic as the only invariant, and singled out its composition ($100\mathbf{x}_2 = 46$) by extrapolation of the part of the liquidus branches they were able to investigate. Storonkin et al. (Ref. 11), however, employed salts on the purity of which doubts - as said above - are not unreasonable, and were not aware of the more recent paper by Sokolov and Pochtakova (Ref. 9).

Bergman et al. in their oldest paper (Ref. 2) claimed the existence of a eutectic, but subsequently changed their mind (Refs. 4, 6, 7), and asserted that the incongruently melting compound $(C_2H_3O_2)_3K_2Na$ was formed. It can be observed that the fusion temperatures of the pure components given in their most recent paper (Ref. 7), i.e., $\mathbf{T}_{fus}(1)/K = 579$ and $\mathbf{T}_{fus}(2)/K = 601$, are in excellent agreement with the corresponding values listed in Table 1 of the Preface (578.7±0.5 K, and 601.3±0.5 K, respectively), and that they make no mention of difficulties in measuring the liquidus. The composition they stated for the intermediate compound, however, was not supported by any investigation of the solidus, and poor reliability is to be attached to the peritectic temperature they suggested (511 K in Ref. 6; 523-529 K in Ref. 7).

Finally, Sokolov and Pochtakova (Refs. 5, 9) in their more recent paper (Ref. 9) employed thermographical analysis to support the assertion already made in Ref. 5 that the intermediate compound $(C_2H_3O_2)_5K_3Na_2$ is formed in the binary. They too seem not to have met special difficulties in measuring the liquidus.

In conclusion, the evaluator is inclined to think that:

- in the composition range $40 \leq 100\mathbf{x}_2 \leq 100$ a eutectic exists at 508±3 K and $100\mathbf{x}_2 = 48 \pm 2$;

- an intermediate compound is likely formed: it ought to have composition $(C_2H_3O_2)_5K_3Na_2$, and melt congruently (thus giving origin to a second eutectic in the composition range $0 \leq 100\mathbf{x}_2 \leq 40$);

- limited mutual solubility exists on both sides of the diagram;

The second conclusion is based on Sokolov and Pochtakova's (Refs. 5, 9) information, which seems the most reliable at disposal so far, although being not fully free from criticisms (see, e.g., the above made remarks on the solid state transformations occurring in pure components).

The last assertion is supported by the findings of Sokolov and Pochtakova (Ref. 9), and Storonkin et al. (Ref. 11). Moreover, Braghetti et al. (Ref. 24) found for sodium

COMPONENTS:	EVALUATOR:
(1) Potassium ethanoate (potassium acetate); $(C_2H_3O_2)K$; [127-08-2] (2) Sodium ethanoate (sodium acetate); $(C_2H_3O_2)Na$; [127-09-3]	Franzosini, P., Dipartimento di Chimica Fisica, Universita' di Pavia (ITALY).

CRITICAL EVALUATION (cont.d):

ethanoate dissolved in potassium ethanoate a limiting value

$$\lim_{m \to 0} (\Delta T/m) = 14.6 \text{ K molality}^{-1}$$

(ΔT: experimental freezing point depression; m: molality of the solute), whereas the cryometric constant of potassium ethanoate is 18.0±0.3 K molality^{-1} (Ref. 24).

REFERENCES:

(1) Baskov, A.; **Zh. Russk. Fiz.-Khim. Obshch.** 1915, **47**, 1533-1535.
(2) Bergman, A.G.; Evdokimova, K.A. **Izv. Sektora Fiz.-Khim. Anal., Inst. Obshchei i Neorg. Khim. Akad. Nauk** SSSR 1956, **27**, 296-314.
(3) Diogenov, G.G.; Erlykov, A.M. **Nauch. Dokl. Vysshei Shkoly, Khim. i Khim. Tekhnol.** 1958, **No. 3**, 413-416.
(4) Golubeva, M.S.; Bergman, A.G.; Grigor'eva, E.A. **Uch. Zap. Rostovsk.-na-Donu Gos. Univ.** 1958, **41**, 145-154.
(5) Sokolov, N.M.; Pochtakova, E.I. **Zh. Obshch. Khim.** 1958, **28**, 1397-1404.
(6) Nesterova, A.K.; Bergman, A.G. **Zh. Obshch. Khim.** 1960, **30**, 317-320; **Russ. J. Gen. Chem., Engl. Transl.**, 1960, **30**, 339-342 (*).
(7) Il'yasov, I.I.; Bergman, A.G. **Zh. Obshch. Khim.** 1960, **30**, 355-358.
(8) Diogenov, G.G.; Sarapulova, I.F. **Zh. Neorg. Khim.** 1964, **9**, 1292-1294 (*); **Russ. J. Inorg. Chem., Engl. Transl.**, 1964, **9**, 704-706.
(9) Sokolov, N.M.; Pochtakova, E.I. **Zh. Obshch. Khim.** 1967, **37**, 1420-1422.
(10) Diogenov, G.G.; Chumakova, V.P. **Fiz.-Khim. Issled. Rasplavov Solei, Irkutsk,** 1975, 7-12.
(11) Storonkin, A.V.; Vasil'kova, I.V.; Tarasov, A.A. **Vestn. Leningr. Univ., Fiz., Khim.** 1977, (4), 80-85.
(12) Sanesi, M.; Cingolani, A.; Tonelli, P.L.; Franzosini, P. **Thermal Properties**, in **Thermodynamic and Transport Properties of Organic Salts**, IUPAC Chemical Data Series No. 28 (Franzosini, P.; Sanesi, M.; Editors), Pergamon Press, Oxford, 1980, 29-115.
(13) Storonkin, A.V.; Vasil'kova, I.V.; Potemin, S.S. **Vestn. Leningr. Univ., Fiz., Khim.** 1974(16), 73-76.
(14) Potemin, S.S.; Tarasov, A.A.; Panin, O.B. Vestn. Leningr. Univ., Fiz., Khim. 1973(1), 86-89.
(15) Sokolov, N.M. **Tezisy Dokl. Nauch. Konf. S.M.I.** 1956, as quoted in Ref. 9.
(16) Diogenov, G.G.; Nurminskii, N.N.; Gimel'shtein, V.G. **Zh. Neorg. Khim.** 1957, 2, 1596-1600; **Russ. J. Inorg. Chem. (Engl. Transl.)** 1957, **2**(7), 237-245.
(17) Bouaziz, R.; Basset, J.Y. **Compt. Rend.** 1966, **263**, 581-584.
(18) Hazlewood, F.J.; Rhodes, E.; Ubbelohde, A.R. **Trans. Faraday Soc.** 1966, **62**, 3101-3113.
(19) Hatibarua, J.R.; Parry, G.S. **Acta Cryst.** 1972, **B28**, 3099-3100.
(20) Poppl, L. **Proc. Eur. Symp. Thermal Anal., 1st**, 1976, 237-240.
(21) Diogenov, G.G. **Zh. Neorg. Khim.** 1956, 1, 799-805; **Russ. J. Inorg. Chem. (Engl. Transl.)** 1956, **1**(4), 199-205.
(22) Gimel'shtein, V.G.; Diogenov, G.G. **Zh. Neorg. Khim.** 1958, 3, 1644-1649; **Russ. J. Inorg. Chem. (Engl. Transl.)** 1958, **3**(7), 230-236.
(23) Roth, J.; Meisel, T.; Seybold, K.; Halmos, Z. **J. Thermal Anal.** 1976, **10**, 223-232.
(24) Braghetti, M.; Leonesi, D.; Franzosini, P. **Ric. Sci.** 1968, **38**, 116-118.

COMPONENTS:

(1) Potassium ethanoate (potassium acetate); $(C_2H_3O_2)K$; [127-08-2]
(2) Sodium ethanoate (sodium acetate); $(C_2H_3O_2)Na$; [127-09-3]

ORIGINAL MEASUREMENTS:

Baskov, A.
Zh. Russk. Fiz.-Khim. Obshch. 1915, 47, 1533-1535.

VARIABLES:

Temperature.

PREPARED BY:

Baldini, P.

EXPERIMENTAL VALUES:

$t/°C$[a]	T/K[b]	$t/°C$[c]	T/K[b]	$100x_2$
295.0	568.2	295.0	568.2	0.0
288.0[d]	561.2	263.5	536.7	12.0
260.0	533.2	250.0	523.2	24.0
237.0	510.2	228.5	501.7	37.5
231.0	504.2	223.5	496.7	40.5
223.0	496.2	223.0	496.2	46.5
		232.0	505.2	52.0
253.0[e]	526.2	240.5	513.7	58.5
271.5	544.7	256.5	529.7	66.5
293.0	566.2	277.2	550.4	78.0
307.5	580.7	295.0	568.2	87.5
320.0	593.2	320.0	593.2	100.0

[a] Starting of crystallization.
[b] T/K values calculated by the compiler.
[c] End of crystallization.
[d] 238.0 in the original text (correction compatible with Fig. 1 of the text; compiler).
[e] 233.0 in the original text (correction compatible with Fig. 1 of the text; compiler).

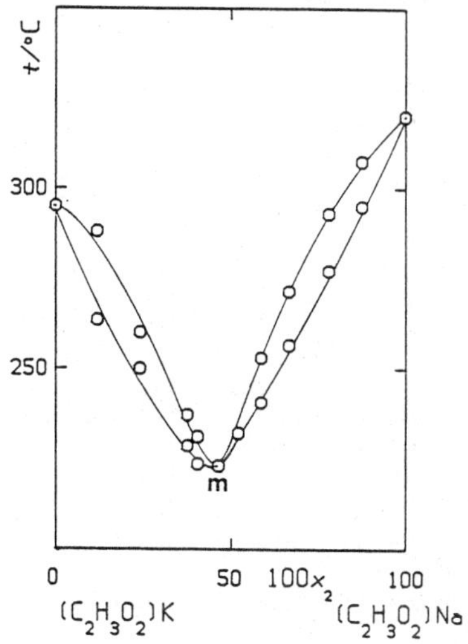

Characteristic point(s):

Minimum, m, at 233 °C (author), or 223 °C (compiler), and $100x_2$= 46; none of the cooling curves shows a eutectic stop (author).

AUXILIARY INFORMATION

METHOD/APPARATUS/PROCEDURE:

Thermal analysis.

SOURCE AND PURITY OF MATERIALS:

Materials dehydrated by heating, then cooled in a dessiccator before use.

ESTIMATED ERROR:

Temperature: accuracy not evaluable (compiler).

REFERENCES:

COMPONENTS:

(1) Potassium ethanoate (potassium acetate); $(C_2H_3O_2)K$; [127-08-2]
(2) Sodium ethanoate (sodium acetate); $(C_2H_3O_2)Na$; [127-09-3]

ORIGINAL MEASUREMENTS:

Bergman, A.G.; Evdokimova, K.A.
Izv. Sektora Fiz.-Khim. Anal., Inst. Obshchei i Neorg. Khim. Akad. Nauk SSSR 1956, **27**, 296-314.

VARIABLES:

Temperature.

PREPARED BY:

Baldini, P.

EXPERIMENTAL VALUES:

t/°C	T/K[a]	$100\mathbf{x}_2$	t/°C	T/K[a]	$100\mathbf{x}_2$
302	575	0.0	238	511	52.3
298	571	2.4	249	522	58.4
296	569	4.6	256	529	61.3
293	566	7.9	265	538	64.0
288	561	11.8	268	541	66.1
280	553	15.5	274	547	68.0
274	547	18.8	277	550	69.9
267	540	23.0	281	554	71.9
256	529	28.1	285	558	73.7
245	518	33.0	290	563	75.7
242	515	34.9	294	567	77.3
238	511	37.2	295	568	79.4
233	506	39.3	298	571	81.0
231	504	41.8	302	575	83.0
227	500	44.2	309	582	86.0
227	500	46.9	314	587	88.0
231	504	49.4	326	599	100

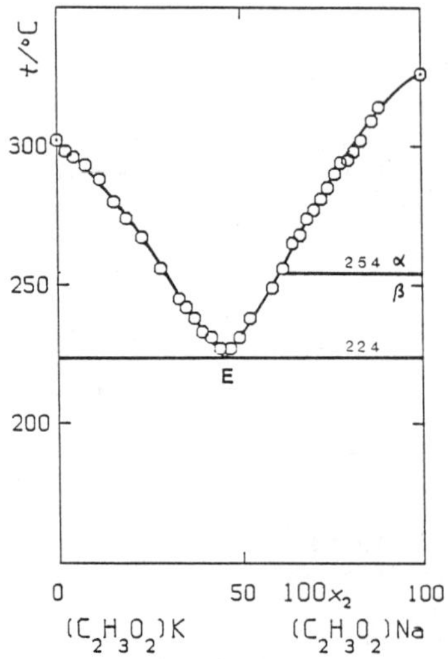

[a] T/K values calculated by the compiler.

Characteristic point(s):

Eutectic, E, at 224 °C and $100\mathbf{x}_2$= 45 (authors).

Note - DTA heating traces were recorded on three mixtures ($100\mathbf{x}_2$= 5, 20, 80, respectively) previously melted and cooled quickly, in order to confirm the eutectic temperature.

AUXILIARY INFORMATION

METHOD/APPARATUS/PROCEDURE:

Visual polythermal method: the temperatures of starting crystallization were measured with a Nichrome-Constantane thermocouple and a millivoltmeter (17 mV full-scale).

SOURCE AND PURITY OF MATERIALS:

"Chemically pure" $(C_2H_3O_2)K$ and $(C_2H_3O_2)Na.3H_2O$ were dried to constant mass. Component 2 undergoes a phase transition at $\mathbf{t}_{trs}(2)$/°C= 254.

ESTIMATED ERROR:

Temperature: accuracy probably ±2 K (compiler).

REFERENCES:

COMPONENTS:	ORIGINAL MEASUREMENTS:
(1) Potassium ethanoate (potassium acetate); $(C_2H_3O_2)K$; [127-08-2] (2) Sodium ethanoate (sodium acetate); $(C_2H_3O_2)Na$; [127-09-3]	Diogenov, G.G.; Erlykov, A.M. **Nauch. Dokl. Vysshei Shkoly, Khim. i Khim. Tekhnol.** 1958, **No. 3**, 413-416.
VARIABLES:	PREPARED BY:
Temperature.	Baldini, P.

EXPERIMENTAL VALUES:

t/°C	T/K[a]	$100x_1$	t/°C	T/K[a]	$100x_1$
337	610	0	229	502	52.0
336	609	1.2	228	501	52.7
332	605	3.3	229.5	502.5	55.5
326	599	4.5	231.5	504.5	57.5
326	599	7.5	235	508	60.0
321	594	10.5	237.5	510.5	62.0
316	589	13.7	242	515	64.0
310	583	17.0	250	523	68.5
305	578	19.9	254	527	70.0
296	569	23.2	264	537	75.0
289	562	26.0	274	547	80.0
282	555	28.5	282	555	84.5
274	547	31.0	284	557	85.5
266	539	34.0	288	561	87.8
260	533	37.3	292	565	89.5
250	523	40.6	295	568	91.7
243	516	43.3	297	570	94.0
240	513	44.5	303	576	96.5
234	507	47.5	308	581	98.0
231	504	49.5	310.5	583.5	100.0
230	503	51.3			

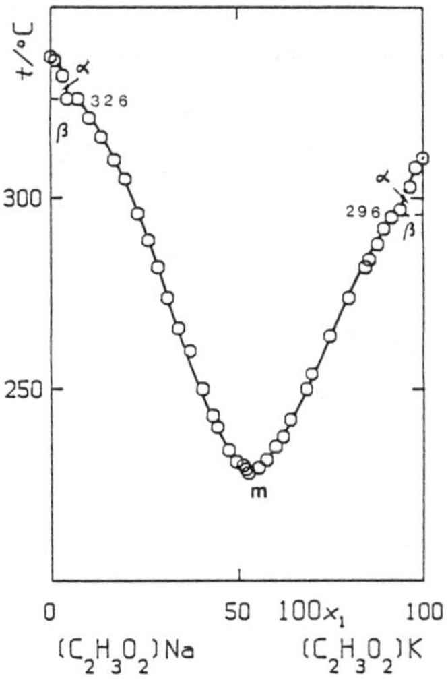

[a] T/K values calculated by the compiler.

Characteristic point(s):

Minimum, m, at 228 °C and $100x_2$= 45 (authors).

AUXILIARY INFORMATION

METHOD/APPARATUS/PROCEDURE:	SOURCE AND PURITY OF MATERIALS:
Visual polythermal method.	Not stated. Component 1 undergoes a phase transition at $t_{trs}(1)$/°C= 296. Component 2 undergoes a phase transition at $t_{trs}(2)$/°C= 326.
	ESTIMATED ERROR:
	Temperature: accuracy probably ±2 K (compiler).
	REFERENCES:

COMPONENTS:	ORIGINAL MEASUREMENTS:
(1) Potassium ethanoate (potassium acetate); $(C_2H_3O_2)K$; [127-08-2] (2) Sodium ethanoate (sodium acetate); $(C_2H_3O_2)Na$; [127-09-3]	Golubeva, M.S.; Bergman, A.G.; Grigor´eva, E.A. **Uch. Zap. Rostovsk.-na-Donu Gos. Univ.** 1958, 41, 145-154.
VARIABLES:	PREPARED BY:
Temperature.	Baldini, P.

EXPERIMENTAL VALUES:

Intermediate compound(s):

$(C_2H_3O_2)_3K_2Na$, melting with decomposition at 240 °C (authors).

AUXILIARY INFORMATION

METHOD/APPARATUS/PROCEDURE:

Visual polythermal method; temperatures measured with a Chromel-Alumel thermocouple.

SOURCE AND PURITY OF MATERIALS:

Materials of analytical purity recrystallized twice, and dehydrated before use.

ESTIMATED ERROR:

Temperature: accuracy probably ±2 K (compiler).

REFERENCES:

COMPONENTS:	ORIGINAL MEASUREMENTS:
(1) Potassium ethanoate (potassium acetate); $(C_2H_3O_2)K$; [127-08-2] (2) Sodium ethanoate (sodium acetate); $(C_2H_3O_2)Na$; [127-09-3]	Sokolov, N.M.; Pochtakova, E.I. **Zh. Obshch. Khim.** 1958, **28**, 1397-1404.
VARIABLES:	PREPARED BY:
Temperature.	Baldini, P.

EXPERIMENTAL VALUES:

t/°C	T/K[a]	$100x_1$	t/°C	T/K[a]	$100x_1$
331	604	0	38	511	55
322	595	5	238	511	57.5
315	588	10	241	514	60
307	580	15	240	513	61.5
299	572	20	242	515	62.5
291	564	25	246	519	65
282	555	30	251	524	67.5
273	546	35	255	528	70
264	537	40	263	536	75
258	531	42.5	272	545	80
254	527	45	282	555	85
247	520	47.5	290	563	90
244	517	50	298	571	95
238	511	52.5	301	574	100
235	508	53.5			

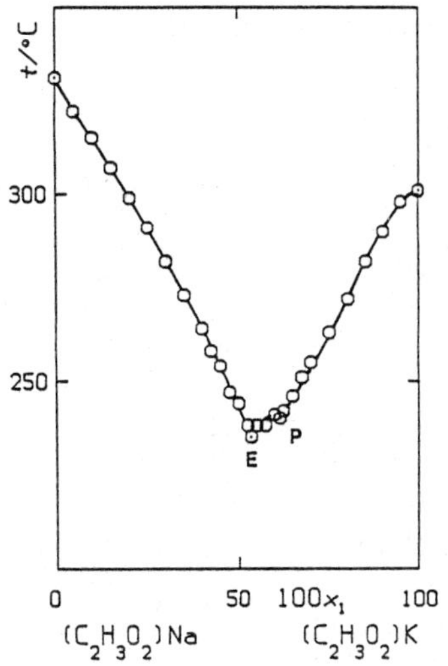

[a] T/K values calculated by the compiler.

Characteristic point(s):

Peritectic, P, (eutectic in the compiler's opinion) at 240 °C and $100x_1$= 61.5 (authors).
Eutectic, E, at 235 °C and $100x_1$= 53.5 (authors).

Intermediate compound(s):

$(C_2H_3O_2)_5K_3Na_2$ (congruently melting at 241 °C, compiler).

AUXILIARY INFORMATION

METHOD/APPARATUS/PROCEDURE:

Visual polythermal method.

SOURCE AND PURITY OF MATERIALS:

"Chemically pure" materials were employed. Component 2 undergoes a phase transition at $t_{trs}(2)$/°C= 254 (Ref. 1).

ESTIMATED ERROR:

Temperature: accuracy probably ±2 K (compiler).

REFERENCES:

(1) Bergman, A.G.; Evdokimova, K.A. **Izv. Sektora Fiz.-Khim. Anal.** 1956, **27**, 296-314.

COMPONENTS:	ORIGINAL MEASUREMENTS:
(1) Potassium ethanoate (potassium acetate); $(C_2H_3O_2)K$; [127-08-2] (2) Sodium ethanoate (sodium acetate); $(C_2H_3O_2)Na$; [127-09-3]	Nesterova, A.K.; Bergman, A.G. **Zh. Obshch. Khim.** 1960, **30**, 317-320; **Russ. J. Gen. Chem. (Engl. Transl.)**, 1960, **30**, 339-342 (*).
VARIABLES:	PREPARED BY:
Temperature.	Baldini, P.

EXPERIMENTAL VALUES:

$t/^{o}C$	T/K^{a}	$100\mathbf{x}_2$
306	579	0
290	563	5
283	556	10
276	549	15
267	540	20
259	532	25
250	523	30
241	514	35
237	510	40
235	508	45
232	505	50
243	516	55
253	526	60
263	536	65
273	546	70

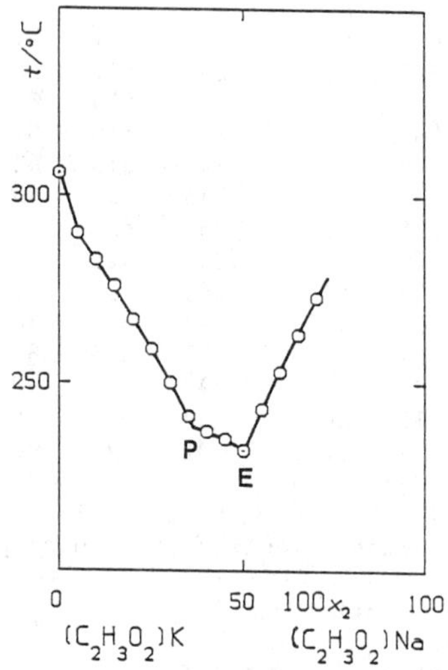

[a] T/K values calculated by the compiler.

Characteristic point(s):

Peritectic, P, at 238 ^{o}C and $100\mathbf{x}_2$= 36.5 (authors).
Eutectic, E, at 232 ^{o}C and $100\mathbf{x}_2$= 50 (authors).

Intermediate compound(s):

$(C_2H_3O_2)_3K_2Na$, melting with decomposition (authors).

AUXILIARY INFORMATION

METHOD/APPARATUS/PROCEDURE:	SOURCE AND PURITY OF MATERIALS:
Visual polythermal method; temperatures measured with a thermometer (accuracy: $\pm$0.5 ^{o}C). A glycerol bath was employed.	"Chemically pure", recrystallized materials were used. Component 2: $\mathbf{t}_{fus}(2)/^{o}C$= 328 (Fig. 2 of the original paper).
	ESTIMATED ERROR:
	Temperature: accuracy $\pm$0.5 K (authors).
	REFERENCES:

COMPONENTS:

(1) Potassium ethanoate (potassium acetate); $(C_2H_3O_2)K$; [127-08-2]
(2) Sodium ethanoate (sodium acetate); $(C_2H_3O_2)Na$; [127-09-3]

ORIGINAL MEASUREMENTS:

Il'yasov, I.I.; Bergman, A.G.
Zh. Obshch. Khim. 1960, **30**, 355-358.

VARIABLES:

Temperature.

PREPARED BY:

Baldini, P.

EXPERIMENTAL VALUES:

$t/^oC$	T/K[a]	$100x_1$[b]
328	601	0.0
300	573	20.0
280	553	30.0
258	531	40.0
250	523	45.0
240	513	50.0
247	520	55.0
250	523	60.0
256	529	65.0
266	539	70.0
271	544	75.0
279	552	80.0
292	565	90.0
306	579	100.0

[a] T/K values calculated by the compiler.
[b] Erroneously reported as x_2 in Table 1 of the original paper (compiler).

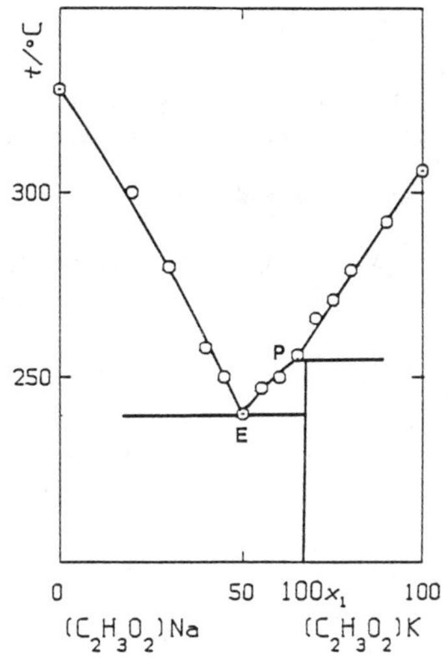

Characteristic point(s):
Peritectic, P, at 255 ^oC (as reported in the text and in Fig. 2 of the original paper, or at 256 ^oC as reported in Table 1 of the original paper, or at 250 ^oC as reported in Fig. 1 of the original paper; compiler) and $100x_1$= 65 (authors).
Eutectic, E, at 240 ^oC and $100x_1$= 50 (authors).

Intermediate compound(s):
$(C_2H_3O_2)_3K_2Na$, incongruently melting (authors).

AUXILIARY INFORMATION

METHOD/APPARATUS/PROCEDURE:

Visual polythermal method; temperatures measured with a Nichrome-Constantane thermocouple and a millivoltmeter.

SOURCE AND PURITY OF MATERIALS:

Not stated.

ESTIMATED ERROR:

Temperature: accuracy probably ± 2 K (compiler).

REFERENCES:

COMPONENTS:	ORIGINAL MEASUREMENTS:
(1) Potassium ethanoate (potassium acetate); $(C_2H_3O_2)K$; [127-08-2] (2) Sodium ethanoate (sodium acetate); $(C_2H_3O_2)Na$; [127-09-3]	Diogenov, G.G.; Sarapulova, I.F. **Zh. Neorg. Khim.** 1964, 9, 1292-1294 (*); **Russ. J. Inorg. Chem. (Engl. Transl.)**, 1964, 9, 704-706.
VARIABLES:	PREPARED BY:
Temperature.	Baldini, P.

EXPERIMENTAL VALUES:

Characteristic point(s):
Eutectic, E_1, at 240 °C; composition not stated (authors).
Eutectic, E_2, at 235 °C; composition not stated (authors).

Intermediate compound(s):

$(C_2H_3O_2)_5K_3Na_2$ (congruently melting, compiler).

AUXILIARY INFORMATION

METHOD/APPARATUS/PROCEDURE:	SOURCE AND PURITY OF MATERIALS:
Visual polythermal method; temperature measured with a Chromel-Alumel thermocouple.	"Chemically pure" materials, recrystallized twice and dehydrated by prolonged heating at about 300 °C were employed. Component 1: $t_{fus}(1)/^{o}C$= 310. Component 2: $t_{fus}(2)/^{o}C$= 335 (authors).
	ESTIMATED ERROR:
	Not evaluable (compiler).
	REFERENCES:

COMPONENTS:	ORIGINAL MEASUREMENTS:
(1) Potassium ethanoate (potassium acetate); $(C_2H_3O_2)K$; [127-08-2] (2) Sodium ethanoate (sodium acetate); $(C_2H_3O_2)Na$; [127-09-3]	Sokolov, N.M.; Pochtakova, E.I. **Zh. Obshch. Khim.** 1967, **37**, 1420-1422.
VARIABLES:	PREPARED BY:
Temperature.	Baldini, P.

EXPERIMENTAL VALUES:

t/°C	T/K[a]	$100x_1$	t/°C	T/K[a]	$100x_1$
318[b]	591	10	120[f]	393	60
310[c]	583	10	60[g]	333	60
60[g]	333	10	233[b]	506	61.5
95[h]	368	10	233[d]	506	61.5
308[b]	581	15	118[f]	391	61.5
233[d]	506	15	60[g]	333	61.5
60[g]	333	15	246[b]	519	65
278[b]	551	30	240[d]	513	65
233[d]	506	30	198[e]	471	65
120[f]	393	30	122[f]	395	65
58[g]	331	30	268[b]	541	75
248[b]	521	50	240[c]	513	75
238[d]	511	50	240[d]	513	75
190[e]	463	50	120[f]	393	75
120[f]	393	50	60[g]	333	75
233[b]	506	53.5	286[b]	559	85
233[d]	506	53.5	270[c]	543	85
115[f]	388	53.5	120[f]	393	85
239[b]	512	58	60[g]	333	85
233[d]	506	58	204[h]	477	85
196[e]	469	58	300[b]	573	95
117[f]	390	58	300[c]	573	95
60[g]	333	58	142[f]	415	95
240[b]	513	60	60[g]	333	95
240[d]	513	60	187[h]	460	95
198[e]	471	60			

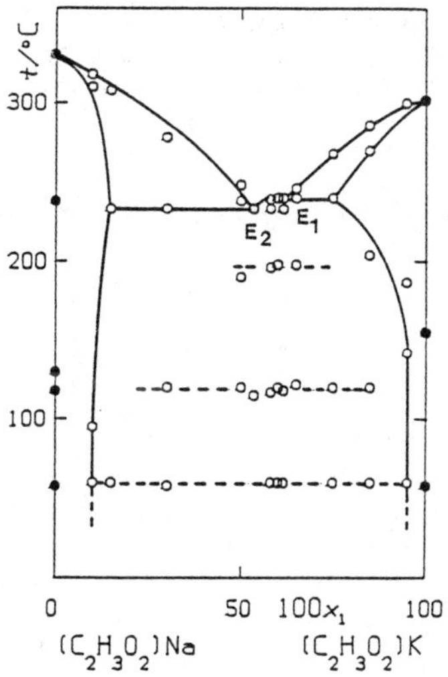

[a] **T**/K values calculated by the compiler.
[b] Temperatures of starting crystallization (authors).
[c] Temperatures of ending crystallization (authors).
[d] Eutectic temperatures (authors).
[e] Solid-solid transition of the intermediate compound (authors).
[f] Interaction of the intermediate compound with the solid solution rich in component 1 (authors).
[g] Reaction $2[(C_2H_3O_2)_3K_2Na] = (C_2H_3O_2)_5K_3Na + (C_2H_3O_2)K$ (authors).
[h] Limits of the solid solution regions (authors).

Characteristic point(s): Eutectic, E_1, at 240 °C and $100x_1$= 61.5 (compiler).
Eutectic, E_2, at 233 °C and $100x_1$= 53.5 (compiler).
Intermediate compound: $(C_2H_3O_2)_5K_3Na_2$ congruently melting at 240 °C (compiler), or 241 °C according to the figure of the original paper.

AUXILIARY INFORMATION

METHOD/APPARATUS/PROCEDURE:	SOURCE AND PURITY OF MATERIALS:
Thermographical analysis (with recording of the heating traces), supplemented with (not detailed) visual polythermal measurements, and microscopic observations on solid (previously melted) samples in polarized light.	"Chemically pure" materials employed. Component 1 melts at 302 °C and undergoes phase transitions at $t_{trs}(1)$/°C= 58, 155 (Ref. 1). Component 2 melts at 331 °C and undergoes phase transitions at $t_{trs}(2)$/°C= 58, 118, 130, 238 (Ref. 1).
ESTIMATED ERROR:	REFERENCES:
Temperature: accuracy probably ±2 K (compiler).	(1) Sokolov, N.M. **Tezisy Dokl. X Nauch. Konf. S.M.I.** 1956

COMPONENTS:

(1) Potassium ethanoate (potassium acetate); $(C_2H_3O_2)K$; [127-08-2]
(2) Sodium ethanoate (sodium acetate); $(C_2H_3O_2)Na$; [127-09-3]

ORIGINAL MEASUREMENTS:

Diogenov, G.G.; Chumakova, V.P.
Fiz.-Khim. Issled. Rasplavov Solei, Irkutsk, 1975, 7-12.

VARIABLES:

Temperature.

PREPARED BY:

Baldini, P.

EXPERIMENTAL VALUES:

Eutectic, E, at 238 °C; composition not stated (authors).
Peritectic, P, at 240 °C; composition not stated (authors).

Intermediate compound(s):

$(C_2H_3O_2)_5K_3Na_2$, incongruently melting (authors).

AUXILIARY INFORMATION

METHOD/APPARATUS/PROCEDURE:

Visual polythermal analysis.

SOURCE AND PURITY OF MATERIALS:

Not stated.
Component 1: $t_{fus}(1)/^{o}C = 302$.
Component 2: $t_{fus}(2)/^{o}C = 326$ (Fig. 1 of the original paper).

ESTIMATED ERROR:

Temperature: accuracy probably ± 2 K (compiler).

REFERENCES:

COMPONENTS:	ORIGINAL MEASUREMENTS:
(1) Potassium ethanoate (potassium acetate); $(C_2H_3O_2)K$; [127-08-2] (2) Sodium ethanoate (sodium acetate); $(C_2H_3O_2)Na$; [127-09-3]	Storonkin, A.V.; Vasil´kova, I.V.; Tarasov, A.A. **Vestn. Leningr. Univ., Fiz., Khim.** 1977, (4), 80-85.
VARIABLES:	PREPARED BY:
Temperature.	Baldini, P.

EXPERIMENTAL VALUES:

Data presented in graphical form (see figure).

Characteristic point(s):

Eutectic, E, at 511 K and $100\mathbf{x}_1 = 54$ (authors), singled out by extrapolation.

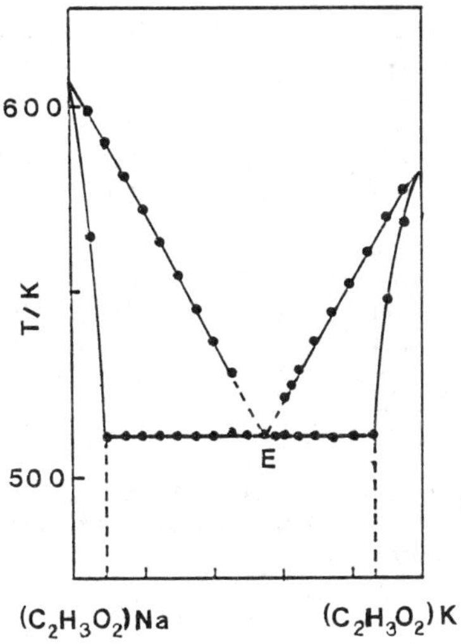

AUXILIARY INFORMATION

METHOD/APPARATUS/PROCEDURE:	SOURCE AND PURITY OF MATERIALS:
DTA and "contact polythermal method" under polarized light. IR spectra were used to deny the existence of any intermediate compound.	Both components of analytical purity recrystallized twice from water and dried under vacuum (T_{fus}/K= 584 and 607, respectively, authors). The purity of both components was checked with thermographical analysis. The mixtures were prepared in a glove box.
	ESTIMATED ERROR:
	Temperature: accuracy probably ±2 K (compiler).
	REFERENCES:

COMPONENTS:

(1) Potassium ethanoate (potassium acetate); $(C_2H_3O_2)K$; [127-08-2]
(2) Lead ethanoate (lead acetate); $(C_2H_3O_2)_2Pb$; [15347-57-6]

ORIGINAL MEASUREMENTS:

Lehrman, A.; Leifer, E.
J. Amer. Chem. Soc. 1938, **60**, 142-144.

VARIABLES:

Temperature.

PREPARED BY:

Baldini, P.

EXPERIMENTAL VALUES:

t/°C	T/K[a]	$100x_2$	t/°C	T/K[a]	$100x_2$
292	565	0	193	466	50.8
278	551	13.6	180	453	55.2
259	532	20.1	177	450	58.0
227	500	25.9	162	435	60.6
221	494	28.3	159	432	62.5
174.9[b]	448.1	28.3	169	442	65.3
181	454	30.6	148[c]	421	65.3
183	456	33.7	169	442	66.8
182	455	39.7	164	437	71.2
180	453	44.3	134	407	78.1
180	453	44.6	132.2[b]	405.4	78.1
190	463	47.4	168	441	83.1
169.5[b]	442.7	47.4	188	461	90.8
194	467	50.6	204	477	100

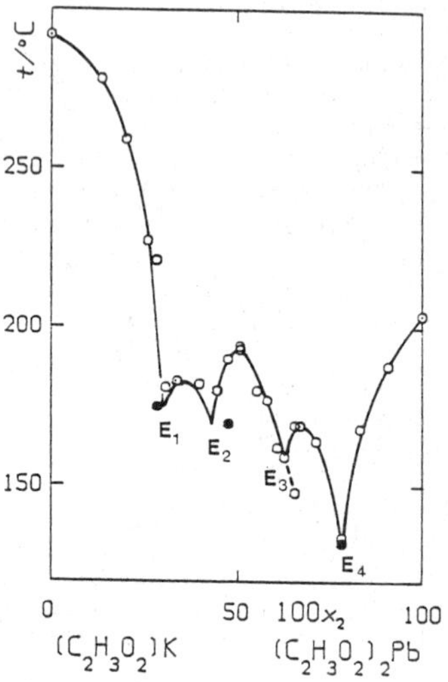

[a] T/K values calculated by the compiler;
[b] Eutectic temperatures (filled circles);
[c] Metastable.

Characteristic point(s):

Eutectic, E_1, at 174.9 °C; composition not stated (about $100x_2$= 30, compiler).
Eutectic, E_2, at 169.5 °C; composition not stated (about $100x_2$= 43, compiler).
Eutectic, E_3, at 159.9 °C; and $100x_2$= 62.5 (authors).
Eutectic, E_4, at 132.2 °C; composition not stated (about $100x_2$= 79, compiler).

Intermediate compounds: $(C_2H_3O_2)_4K_2Pb$, congruently melting at 183 °C (compiler).
$(C_2H_3O_2)_3KPb$, congruently melting at 194 °C (compiler).
$(C_2H_3O_2)_5KPb_2$, congruently melting at 169 °C (compiler).

AUXILIARY INFORMATION

METHOD/APPARATUS/PROCEDURE:

The mixtures (20-35 g) were weighed into 2.5x20 cm Pyrex tubes, then suspended in a bath of the molten eutectic of Ca, K, Li nitrates. When necessary to prevent decomposition, two drops of glacial ethanoic acid were added. Due to the tendency to supercool, it was preferred to take the temperatures of complete melting. Cooling curves were used to obtain a few eutectic temperatures. Temperatures were measured mainly with a Copper-Constantane thermocouple (checked at the boiling point of water, and at the melting points of Sn, KNO_3, and of the Sn-Pb eutectic mixture). In a few cases a mercury thermometer was employed.

NOTE:

It can be remarked that the fusion temperature of component 1 found by Lehrman and Leifer does not agree with recent literature data which range mostly between 574 and 584 K (Ref. 1).

SOURCE AND PURITY OF MATERIALS:

Component 1: material of "chemically pure" grade, recrystallized from distilled water, then dried in an oven at 100 °C for one week, and at 140 °C for six hours before weighing.
Component 2: material of "chemically pure" grade, recrystallized from distilled water acidified with ethanoic acid, then dried at 100 °C.

ESTIMATED ERROR:

Temperature: accuracy ±0.5 K (authors).

REFERENCES:

(1) Sanesi, M.; Cingolani, A.; Tonelli, P.L.; Franzosini, P. **Thermal Properties**, in **Thermodynamic and Transport Properties of Organic Salts,** IUPAC Chemical Data Series No. 28 (Franzosini, P.; Sanesi, M.; Editors), Pergamon Press, Oxford, 1980, 29-115.

COMPONENTS:

(1) Potassium ethanoate (potassium acetate); $(C_2H_3O_2)K$; [127-08-2]
(2) Rubidium ethanoate (rubidium acetate); $(C_2H_3O_2)Rb$; [563-67-7]

EVALUATOR:

Spinolo, G.,
Dipartimento di Chimica Fisica,
Universita´ di Pavia (ITALY).

CRITICAL EVALUATION:

This binary was first studied with the visual polythermal method by Diogenov and Sarapulova (Ref. 1). Subsequently, Sarapulova et al. (Ref. 2) carried out a thermographical analysis of the system, supplemented with a few visual observations, and X-ray diffractograms recorded on the pure components and five (previously melted) intermediate mixtures.

Only minor differences occur between the liquidus curves by either source. The fusion temperatures of the pure components, i.e., $T_{fus}(1)$= 583 K (Refs. 1, 2), and $T_{fus}(2)$= 509 K (visual; Refs. 1, 2) or 511 K (thermographical; Ref. 2) are acceptable, although somewhat lower than the corresponding values listed in Table 1 of the Preface, i.e., $T_{fus}(1)$= 578.7±0.5 K, and $T_{fus}(2)$= 514±1 K. Poorer agreement, on the contrary, exists between the solid state transition temperatures reported in Ref. 2 (i.e., 327 K and 428 K for component 1, and 488 K for component 2), and those listed in Table 1 of the Preface (i.e., 422.2±0.5 K for component 1, and 498±1 K for component 2).

On the basis of the X-ray patterns mentioned above, Sarapulova et al. (Ref. 2) assert that complete miscibility exists even at room temperature, although giving no information about the phase of component 1 they assume to be involved in these solid solutions.

In the evaluator´s opinion, doubts are to be cast about the solid state transition at 327 K in component 1. Should it actually exist, the lower part of the diagram shown in Ref. 2 would require completion, whereas, in its absence, the picture of the phase relations would be substantially correct.

REFERENCES:

(1) Diogenov, G.G.; Sarapulova, I.F.
Zh. Neorg. Khim. 1964, **9**, 1292-1294 (*); **Russ. J. Inorg. Chem. (Engl. Transl.)**, 1964, **9**, 704-706.

(2) Sarapulova, I.F.; Kashcheev, G.N.; Diogenov, G.G.
Nekotorye Vopr. Khimii Rasplavlen. Solei i Produktov Destruktsii Sapropelitov, Irkutsk, 1974, 3-10.

COMPONENTS:

(1) Potassium ethanoate (potassium acetate); $(C_2H_3O_2)K$; [127-08-2]
(2) Rubidium ethanoate (rubidium acetate); $(C_2H_3O_2)Rb$; [563-67-7]

ORIGINAL MEASUREMENTS:

Diogenov, G.G.; Sarapulova, I.F.
Zh. Neorg. Khim. 1964, **9**, 1292-1294 (*); **Russ. J. Inorg. Chem. (Engl. Transl.)**, 1964, **9**, 704-706.

VARIABLES:

Temperature.

PREPARED BY:

Baldini, P.

EXPERIMENTAL VALUES:

$t/^{\circ}C$	T/K^a	$100x_1$	$t/^{\circ}C$	T/K^a	$100x_1$
236	509	0	259	532	50.0
236	509	4.0	263	536	55.0
235	508	8.5	267	540	60.0
235	508	11.3	271	544	64.5
235	508	14.5	274	547	68.0
237	510	18.0	279	552	74.0
239	512	22.0	287	560	82.0
242	515	25.5	292	565	88.0
246	519	30.0	294	567	90.5
249	522	34.5	300	573	95.0
253	526	40.0	310	583	100
256	529	45.5			

$^a T/K$ values calculated by the compiler.

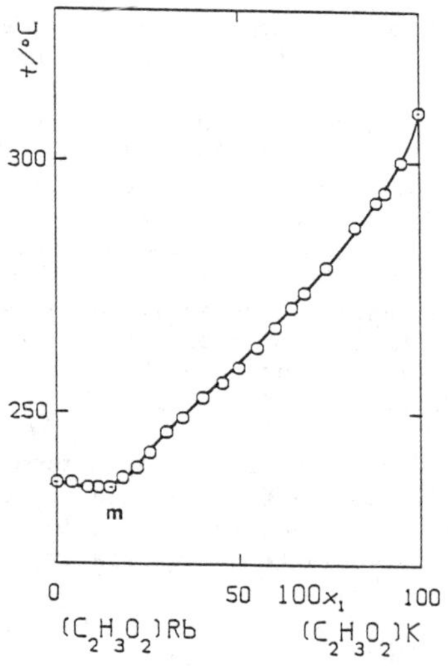

Characteristic point(s):

Continuous series of solid solutions with a minimum (erroneously indicated as a maximum in the text; compiler), m, at 235 °C and $100x_2$ about 85.

AUXILIARY INFORMATION

METHOD/APPARATUS/PROCEDURE:

Visual polythermal analysis; temperature measured with a Chromel-Alumel thermocouple.

SOURCE AND PURITY OF MATERIALS:

Component 1: "chemically pure" material, recrystallized twice and dehydrated by prolonged heating at about 300 °C.
Component 2: prepared from rubidium carbonate.

ESTIMATED ERROR:

Temperature: accuracy probably ±2 K (compiler).

REFERENCES:

COMPONENTS:

(1) Potassium ethanoate (potassium acetate); $(C_2H_3O_2)K$; [127-08-2]
(2) Rubidium ethanoate (rubidium acetate); $(C_2H_3O_2)Rb$; [563-67-7]

ORIGINAL MEASUREMENTS:

Sarapulova, I.F.; Kashcheev, G.N.; Diogenov, G.G.
Nekotorye Vopr. Khimii Rasplavlen. Solei i Produktov Destruktsii Sapropelitov, Irkutsk, 1974, 3-10.

VARIABLES:

Temperature.

PREPARED BY:

Baldini, P.

EXPERIMENTAL VALUES:

t/°C	T/K[a]	$100x_1$	t/°C	T/K[a]	$100x_1$
236	509	0	180[be]	453	40
238[bc]	511	0	266	539	60
238[bd]	511	0	267[bc]	540	60
215[be]	488	0	258[bd]	531	60
236	509	10	160[be]	433	60
240[bc]	513	10	289[bc]	562	80
204[be]	477	10	283[bd]	556	80
243[bc]	516	20	153[be]	426	80
238[bd]	511	20	310	583	100
197[be]	470	20	308[bc]	581	100
255	528	40	308[bd]	581	100
257[bc]	530	40	155[be]	428	100
249[bd]	522	40			

[a] T/K values calculated by the compiler.
[b] Differential thermal analysis (filled circles in the figure).
[c] Liquidus.
[d] Solidus.
[e] Solid state transition.

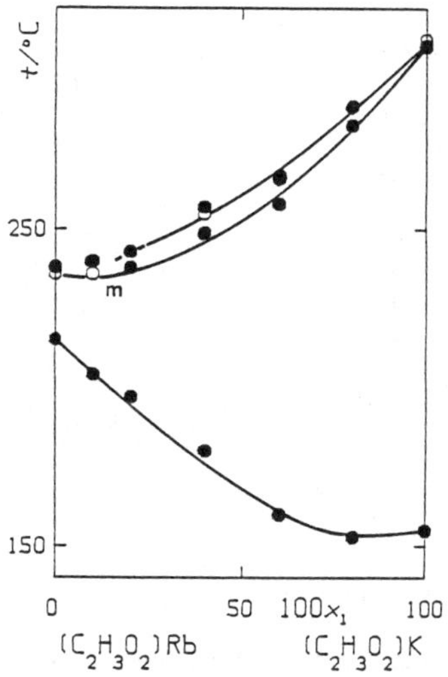

Characteristic point(s):

Continuous series of solid solutions with a minimum, m, at 235 °C and $100x_2$ about 85 (authors).

AUXILIARY INFORMATION

METHOD/APPARATUS/PROCEDURE:

A thermographical analysis was performed with a Kurnakov pyrometer Mod. 1959 (reference material: Al_2O_3). Only heating traces (at the heating rate of 5-6 °C/min) were recorded due to the tendency of the melts to undercool. Supplementary visual polythermal observations are also tabulated. X-ray diffraction patterns were used to obtain information on the solid solutions.

SOURCE AND PURITY OF MATERIALS:

Not stated. Component 1 undergoes phase transitions at $t_{trs}(1)$/°C= 54, 155. Component 2 undergoes a phase transition at $t_{trs}(2)$/°C= 215.

ESTIMATED ERROR:

Temperature: accuracy probably ±2 K (compiler).

REFERENCES:

COMPONENTS:

(1) Potassium ethanoate (potassium acetate); $(C_2H_3O_2)K$; [127-08-2]
(2) Zinc ethanoate (zinc acetate); $(C_2H_3O_2)_2Zn$; [557-34-6]

ORIGINAL MEASUREMENTS:

Nadirov, E.G.; Bakeev, M.I.
Tr. Khim.-Metall. Inst. Akad. Nauk Kaz. SSR 1974, **25**, 115-128.

VARIABLES:

Temperature

PREPARED BY:

Baldini, P.

EXPERIMENTAL VALUES:

t/°C	T/K[a]	$100x_1$	t/°C	T/K[a]	$100x_1$
236	509	0	229	502	60
222	495	10	236	509	65.5
201	474	20	238	511	70
169	442	30	241	514	75
188	461	35	245	518	80
209	482	40	247	520	85
213	486	45	263	536	90
217	490	50	292	565	94.6
219	492	52	306	579	100
222	495	55			

[a] T/K values calculated by the compiler.

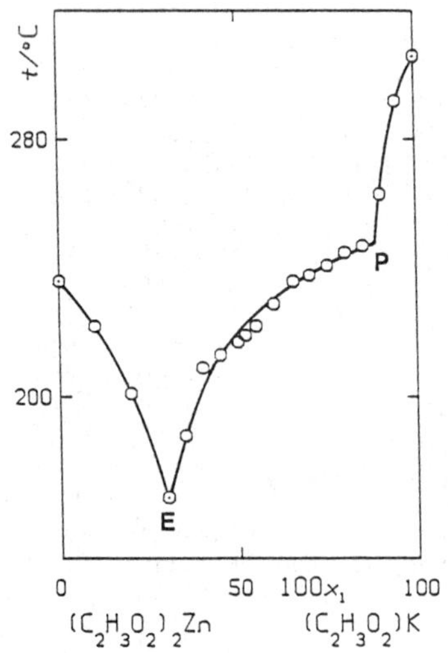

Characteristic point(s):

Eutectic, E, at 169 °C and $100x_1$= 30.
Peritectic, P, at 248 °C (visual polythermal analysis) or at 242 °C (conductometry) and $100x_1$= 88 (according to Fig. 6 of the original paper; compiler).

Intermediate compound(s):

$(C_2H_3O_2)_{10}K_8Zn$, incongruently melting (it undergoes a phase transition at 161 °C, conductometry).

AUXILIARY INFORMATION

METHOD/APPARATUS/PROCEDURE:

Visual polythermal analysis supplemented with conductometry and occasionally with X-ray investigations. Temperatures of initial crystallization measured with a thermocouple.

NOTE:

It can be observed that the fusion temperature of component 1 reported by Nadirov and Bakeev (579 K) is in fair agreement with the corresponding value listed in Table 1 of the Preaface (578.7±0.5 K), whereas the fusion temperature of component 2 (509 K) is noticeably lower than other recent data by different investigators (Ref. 1).

SOURCE AND PURITY OF MATERIALS:

Component 1: material recrystallized three times and dried at 110-120 °C. Component 2: $(C_2H_3O_2)_2Zn.2H_2O$ of analytical purity, recrystallized twice and dried at 140 °C.

ESTIMATED ERROR:

Temperature: accuracy probably ± 2 K (compiler).

REFERENCES:

(1) Sanesi, M.; Cingolani, A.; Tonelli, P.L.; Franzosini, P. **Thermal Properties**, in **Thermodynamic and Transport Properties of Organic Salts**, IUPAC Chemical Data Series No. 28 (Franzosini, P.; Sanesi, M.; Editors), Pergamon Press, Oxford 1980, 29-115.

COMPONENTS:	EVALUATOR:
(1) Lithium ethanoate (lithium acetate); $(C_2H_3O_2)Li$; [546-89-4] (2) Sodium ethanoate (sodium acetate); $(C_2H_3O_2)Na$; [127-09-3]	Franzosini, P., Dipartimento di Chimica Fisica Universita´ di Pavia (ITALY).

CRITICAL EVALUATION:

This system was investigated by Diogenov (Ref. 1), by Pochtakova (Ref. 2), and again by Diogenov and Chumakova (Ref. 3) with substantially discrepant conclusions.

Diogenov, in his earlier paper (Ref. 1), claimed the existence of: (i) eutectic, E_1, occurring at 499-500 K (226-227 ^oC),and (likely) $100\mathbf{x}_1$= 81.5 (the latter figure being quoted in Ref. 4, which is a later paper by the same author); (ii) eutectic, E_2, occurring at 433 K (160 ^oC) and $100\mathbf{x}_1$= 57; and (iii) the intermediate compound $(C_2H_3O_2)_5Li_4Na$, congruently melting at 500 K (227 ^oC).

These results, however, were not confirmed in Ref. 3, where Diogenov and Chumakova reported approximately the same coordinates for E_1, viz., 492-494 K (219-221 ^oC) and $100\mathbf{x}_1$ about 78, but completely different fusion temperature for E_2, viz., either 486 K (213 ^oC; Fig. 2 of the original paper), or 449 K (176 ^oC; Fig. 4 of the original paper). Moreover they suggested for the intermediate compound a new formula, i.e., $(C_2H_3O_2)_4Li_3Na$.

Finally, it is to be noted that the fusion temperatures given in Refs. 1, 3 for component 2 differ by 11 K, and the phase transitions reported in Ref. 1, i.e., $\mathbf{T}_{trs}(1)$= 530 K (257 ^oC), and $\mathbf{T}_{trs}(2)$= 596 K (323 ^oC), do not meet any value of Table 1 of the Preface.

In conclusion, the poor reproducibility of the results by Diogenov´s group does not allow one to take them into consideration for assessing the actual diagram of this system.

Conversely, Pochtakova´s data (Ref. 2) seem more reliable, although among the phase transition temperatures of component 2 quoted by the author from Ref. 5, i.e., 331, 391, 403, and 511 K (58, 118, 130, and 238 ^oC, respectively), only two can be identified with those listed in Preface, Table 1. This disagreement, however, does not seem, in the evaluator´s opinion, to involve heavily the reliability of the liquidus, due also to the fact that the fusion temperatures of both pure components (604 K for component 2, and 557 K for component 1, respectively) are close to those reported in Preface, Table 1 (601.3$\pm$0.5 and 557$\pm$2 K, respectively).

Accordingly, the phase diagram by Pochtakov can be accepted with some confidence: in particular, the composition of the congruently melting intermediate compound, i.e., $(C_2H_3O_2)_5Li_3Na_2$, seems satisfactorily defined by the dome exhibited by the liquidus.

REFERENCES:

(1) Diogenov, G.G.
Zh. Neorg. Khim. 1956, 1, 799-805(*); **Russ. J. Inorg. Chem. (Engl. Transl.)** 1956, **1** (4), 199-205.

(2) Pochtakova, E.I.
Zh. Neorg. Khim. 1965, 10, 1333-2338 (*); **Russ. J. Inorg. Chem. (Engl. Transl.)** 1965, **10**, 1268-1271.

(3) Diogenov, G.G.; Chumakova, V.P.
Fiz.-Khim. Issled. Rasplavov Solei. Irkutsk. 1975, 7-12.

(4) Diogenov, G.G.
Zh. Neorg. Khim. 1956, 1, 2551-2555; **Russ. J. Inorg. Chem. (Engl.Transl.) 1956, 1** (11), 122-126 (*).

(5) Sokolov, N.M.
Tezisy Dokl. X Nauch. Konf. S.M.I. 1956.

COMPONENTS:	ORIGINAL MEASUREMENTS:
(1) Lithium ethanoate (lithium acetate); $(C_2H_3O_2)Li$; [546-89-4] (2) Sodium ethanoate (sodium acetate); $(C_2H_3O_2)Na$; [127-09-3]	Diogenov, G.G. **Zh. Neorg. Khim.** 1956, 1, 799-805 (*); **Russ. J. Inorg. Chem. (Engl. Transl.)** 1956, 1 (4), 199-205.
VARIABLES:	PREPARED BY:
Temperature.	Baldini, P.

EXPERIMENTAL VALUES:

$t/^{o}C$	T/K^{a}	$100x_1$	$t/^{o}C$	T/K^{a}	$100x_1$
337	610	0	165	438	56.5
333	606	2	169	442	58.5
326	599	4	185	458	61.5
320	593	7	206	479	66
316	589	10	216	489	70.3
310	583	14	225	498	75.5
303	576	18	227	500	80
295	568	22.3	232	505	82
285	558	26	248	521	86
277	550	29.3	253	526	90.3
266	539	33	266	539	94.5
237	510	41.5	277	550	96
222	495	45.5	284	557	97
208	481	49.2	291	564	100
188	461	53			

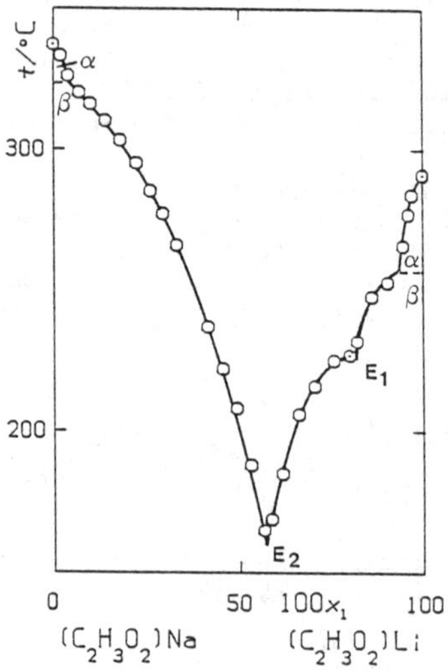

[a] **T**/K values calculated by the compiler.

Characteristic point(s):
Eutectic, E_1, at 227 ^{o}C (226 ^{o}C according to Fig. 1 of the original paper, compiler) and $100x_1$= 81.5 (according to Ref.1 where the author, quoting the present paper, reports: "The eutectic of the compound $4CH_3COOLi.CH_3COON$ and lithium acetate melts at 226° and corresponds to 81.5 % lithium acetate"; compiler).
Eutectic, E_2, at 160 ^{o}C and $100x_1$= 57 (author).

Intermediate compound(s):
$(C_2H_3O_2)_5Li_4Na$, congruently melting at 227 ^{o}C (author).

AUXILIARY INFORMATION

METHOD/APPARATUS/PROCEDURE:

Visual polythermal analysis.

SOURCE AND PURITY OF MATERIALS:

Not stated. Component 1 undergoes phase transition at $t_{trs}(1)/^{o}C$= 257. Component 2 undergoes phase transition at $t_{trs}(2)/^{o}C$= 323.

ESTIMATED ERROR:

Temperature: accuracy probably ± 2 K (compiler).

REFERENCES:

(1) Diogenov, G.G.
Zh. Neorg. Khim. 1956, 1, 2551-2555;
Russ. J. Inorg. Chem. (Engl. Transl.) 1956, 1 (11), 122-126 (*).

COMPONENTS:

(1) Lithium ethanoate (lithium acetate); $(C_2H_3O_2)Li$; [546-89-4]
(2) Sodium ethanoate (sodium acetate); $(C_2H_3O_2)Na$; [127-09-3]

ORIGINAL MEASUREMENTS:

Pochtakova, E.I.
Zh. Neorg. Khim. 1965, **10**, 2333-2338 (*);
Russ. J. Inorg. Chem., Engl. Transl., 1965, **10**, 1268-1271.

VARIABLES:

Temperature

PREPARED BY:

Baldini, P.

EXPERIMENTAL VALUES:

t/°C	T/K[a]	$100x_1$	t/°C	T/K[a]	$100x_1$
331	604	0	224	497	57.5
322	595	5	227	500	60
314	587	10	224	497	62.5
301	574	15	223	496	65
289	562	20	222	495	67.5
277	550	25	222	495	70
265	538	30	229	502	72.5
251	524	35	234	507	75
236	509	40	241	514	80
219	492	45	259	532	90
213	486	47.5	273	546	95
219	492	50	284	557	100
222	495	52.5			

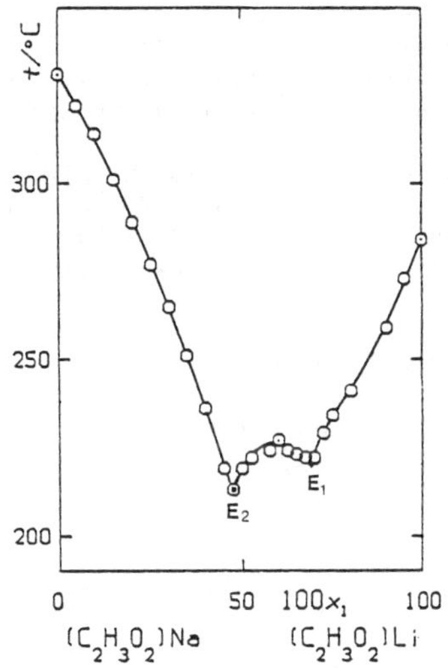

[a] T/K values calculated by the compiler.

Characteristic point(s):

Eutectic, E_1, at 219 °C and $100x_1$= 69 (author).
Eutectic, E_2, at 213 °C and $100x_1$= 47.5 (author).

Intermediate compound(s):

$(C_2H_3O_2)_5Li_3Na_2$, congruently melting at 227 °C (author).

AUXILIARY INFORMATION

METHOD/APPARATUS/PROCEDURE:

Visual polythermal method.

SOURCE AND PURITY OF MATERIALS:

Not stated.
Component 2 undergoes phase transitions at $t_{trs}(2)$/°C= 58, 118, 130, 238 (Ref. 1).

ESTIMATED ERROR:

Temperature: accuracy probably ±2 K (compiler).

REFERENCES:

(1) Sokolov, N.M.
Tezisy Dokl. X Nauch. Konf. S.M.I. 1956.

COMPONENTS:

(1) Lithium ethanoate (lithium acetate); $(C_2H_3O_2)Li$; [546-89-4]
(2) Sodium ethanoate (sodium acetate); $(C_2H_3O_2)Na$; [127-09-3]

ORIGINAL MEASUREMENTS:

Diogenov, G.G.; Chumakova, V.P.
Fiz.-Khim. Issled. Rasplavov Solei, Irkutsk, 1975, 7-12.

VARIABLES:

Temperature.

PREPARED BY:

Baldini, P.

EXPERIMENTAL VALUES:

Eutectic, E_1, at 219 °C (Fig. 2 of the original paper) or 221 °C (Fig. 4); composition not stated ($100\mathbf{x}_1$ about 78 in compiler's graphical estimation from Fig. 4).
Eutectic, E_2, at 213 °C (Fig. 2 of the original paper) or 176 °C (Fig. 4); composition not stated ($100\mathbf{x}_1$ about 54 in compiler's graphical estimation from Fig. 4).

Intermediate compound(s):

$(C_2H_3O_2)_4Li_3Na$, congruently melting at 226 °C (authors).

AUXILIARY INFORMATION

METHOD/APPARATUS/PROCEDURE:

Visual polythermal method.

SOURCE AND PURITY OF MATERIALS:

Not stated.
Component 1: $t_{fus}(1)/°C = 291$ (Fig. 3 of the original paper).
Component 2: $t_{fus}(2)/°C = 326$ (Fig. 1 of the original paper).

ESTIMATED ERROR:

Temperature: accuracy probably ± 2 K (compiler).

REFERENCES:

COMPONENTS:

(1) Lithium ethanoate (lithium acetate); $(C_2H_3O_2)Li$; [546-89-4]
(2) Rubidium ethanoate (rubidium acetate); $(C_2H_3O_2)Rb$; [563-67-7]

EVALUATOR:

Schiraldi, A.,
Dipartimento di Chimica Fisica,
Universita´ di Pavia (ITALY).

CRITICAL EVALUATION:

This system was studied twice by Diogenov´s group, as a side of the ternary $C_2H_3O_2$/Cs, Li, Rb (Ref. 1), and as a side of the reciprocal ternary $C_2H_3O_2$, NO_3/Li, Rb (Ref. 2), respectively.

In both papers two eutectics are reported, viz., E_1 at 509 K (236 °C), and either $100\mathbf{x}_1$= 88.5 (Ref. 1), or $100\mathbf{x}_1$= 88 (Ref. 2), and E_2 at either 449 K (176 °C; Ref. 1), or 460 K (187 °C; Ref. 2), and $100\mathbf{x}_1$= 26.

In Ref. 1, however, Diogenov and Sarapulova report two intermediate compounds, i.e., $(C_2H_3O_2)_5Li_2Rb_3$ and $(C_2H_3O_2)_5Li_3Rb_2$ [congruently melting at 518 K (245 °C) and 582 K (309 °C), respectively], and consequently a third invariant, whilst Diogenov et al. report in Ref. 2 a single intermediate compound, $(C_2H_3O_2)_3Li_2Rb$ [congruently melting at 573 K (300 °C)].

Due to the detailed experimental evidence (obtained, inter alia, with X-ray diffractometry) given in Ref. 2, the evaluator thinks that the existence of the latter compound should be considered as reasonably assessed. On the contrary, the existence of both $(C_2H_3O_2)_5Li_2Rb_3$ and $(C_2H_3O_2)_5Li_3Rb_2$ does not seem adequately supported.

It is to be noticed that some discrepancies exist between the phase transition temperatures reported in Ref. 2 and those given in Table 1 of the Preface, viz., $\mathbf{T}_{fus}(1)$= 564 K (291 °C), to be identified with 557±2 K, $\mathbf{T}_{trs}(1)$= 405 K (132 °C), with no correspondence, $\mathbf{T}_{fus}(2)$= 509 K (236 °C), to be identified with 514±1 K, and $\mathbf{T}_{trs}(2)$= 479 K (206 °C), to be identified with 498±1 K. These discrepancies, however, do not imply significant changes in the liquidus by Diogenov et al. (Ref. 2): the evaluator is consequently inclined to consider the presentation by these authors as sufficiently reliable.

REFERENCES:

(1) Diogenov, G.G.; Sarapulova, I.F.
Zh. Neorg. Khim. 1964, **9(2)**, 482-487; **Russ. J. Inorg. Chem. (Engl. Transl.)** 1964, **9**, 265-267 (*).

(2) Diogenov, G.G.; Erlykov, A.M.; Gimel´shtein, V.G.
Zh. Neorg. Khim. 1974, **19**, 1955-1960; **Russ. J. Inorg. Chem. (Engl. Transl.)** 1974, **19**, 1069-1073 (*).

COMPONENTS:

(1) Lithium ethanoate (lithium acetate); $(C_2H_3O_2)Li$; [546-89-4]
(2) Rubidium ethanoate (rubidium acetate); $(C_2H_3O_2)Rb$; [563-67-7]

ORIGINAL MEASUREMENTS:

Diogenov, G.G.; Sarapulova, I.F. **Zh. Neorg. Khim.** 1964, **9(2)**, 482-487; **Russ. J. Inorg. Chem., Engl. Transl.**, 1964, **9**, 265-267 (*).

VARIABLES:

Temperature.

PREPARED BY:

Baldini, P.

EXPERIMENTAL VALUES:

t/oC	T/K[a]	$100x_1$	t/oC	T/K[a]	$100x_1$
240	513	0	283	556	47.0
234	507	3.5	288	561	48.5
225	498	8.3	299	572	52.0
216	489	12.0	304	577	55.0
213	486	14.0	309	582	60.0
208	481	16.5	309	582	64.0
203	476	18.5	307	580	67.5
195	468	21.0	298	571	74.0
187	460	23.0	289	562	78.0
181	454	24.5	267	540	83.5
185	458	26.5	257	530	85.0
203	476	29.0	242	515	88.0
207	480	29.5	241	514	89.5
213	486	30.5	246	519	90.5
224	497	32.0	258	531	92.0
236	509	34.0	265	538	93.0
242	515	37.5	272	545	94.5
260	533	42.0	282	555	97.0
273	546	44.5	290	563	100.0

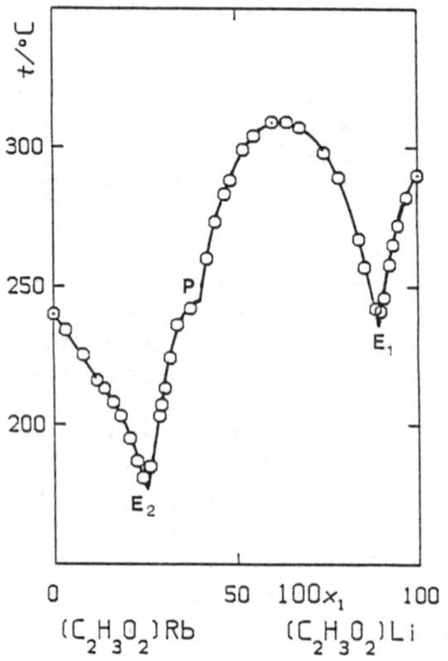

[a] T/K values calculated by the compiler.

Characteristic point(s): Eutectic, E_1, at 236 °C and $100x_1$= 88.5 (authors).
Peritectic, P, at 245 °C and $100x_1$= 40 (compiler).
Eutectic, E_2, at 176 °C and $100x_1$= 26 (authors).
Intermediate compound(s): $(C_2H_3O_2)_5Li_2Rb_3$, melting at 245 °C (authors).
$(C_2H_3O_2)_5Li_3Rb_2$, congruently melting at 309 °C (authors).

AUXILIARY INFORMATION

METHOD/APPARATUS/PROCEDURE:

Visual polythermal analysis. Temperatures measured with a Chromel-Alumel thermocouple.

SOURCE AND PURITY OF MATERIALS:

Not stated.

ESTIMATED ERROR:

Temperature: accuracy probably ±2 K (compiler).

REFERENCES:

COMPONENTS:	ORIGINAL MEASUREMENTS:
(1) Lithium ethanoate (lithium acetate); $(C_2H_3O_2)Li$; [546-89-4] (2) Rubidium ethanoate (rubidium acetate); $(C_2H_3O_2)Rb$; [563-67-7]	Diogenov, G.G.; Erlykov, A.M.; Gimel´shtein, V.G. **Zh. Neorg. Khim.** 1974, **19**, 1955-1960; **Russ. J. Inorg. Chem., Engl. Transl.**, 1974, **19**, 1069-1073 (*).
VARIABLES:	PREPARED BY:
Temperature.	Baldini, P.

EXPERIMENTAL VALUES:

The results are reported only in graphical form (see figure).

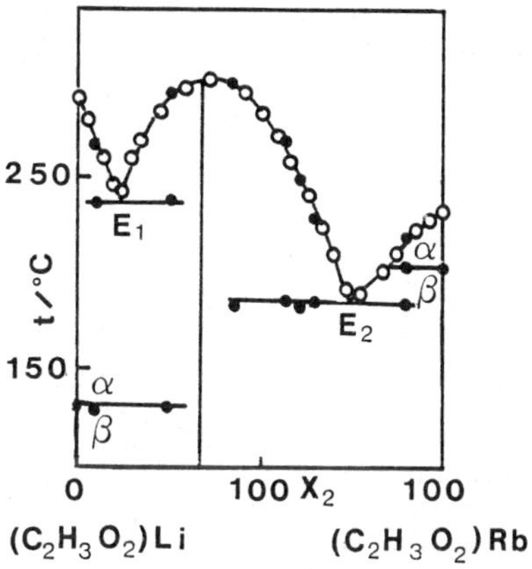

Characteristic point(s):

Eutectic, E_1, at 236 ^oC and $100x_2$= 12 (authors).
Eutectic, E_2, at 187 ^oC and $100x_2$= 74 (authors).

Intermediate compound(s):

$(C_2H_3O_2)_3Li_2Rb$, congruently melting at 300 ^oC (authors).

AUXILIARY INFORMATION

METHOD/APPARATUS/PROCEDURE:	SOURCE AND PURITY OF MATERIALS:
The data were obtained by visual polythermal and thermographical analysis (empty and filled circles in the figure, respectively), supplemented with a few X-ray diffraction patterns.	Not stated. Component 1 melts at 291 ^oC and undergoes a phase transition at 132 ^oC. Component 2 melts at 236 ^oC and undergoes a phase transition at 206 ^oC.
	ESTIMATED ERROR:
	Temperature: precision probably ± 2 K (compiler).
	REFERENCES:

COMPONENTS:

(1) Lithium ethanoate (lithium acetate); $(C_2H_3O_2)Li$; [546-89-4]
(2) Zinc ethanoate (zinc acetate); $(C_2H_3O_2)_2Zn$; [557-34-6]

ORIGINAL MEASUREMENTS:

Pavlov, V.L.; Golubkova, V.V.
Visn. Kiiv. Univ., Ser. Khim., Kiev, 1972, **No. 13,** 28-30.

VARIABLES:

Temperature.

PREPARED BY:

Baldini, P.

EXPERIMENTAL VALUES:

The results are reported only in graphical form (see figure).

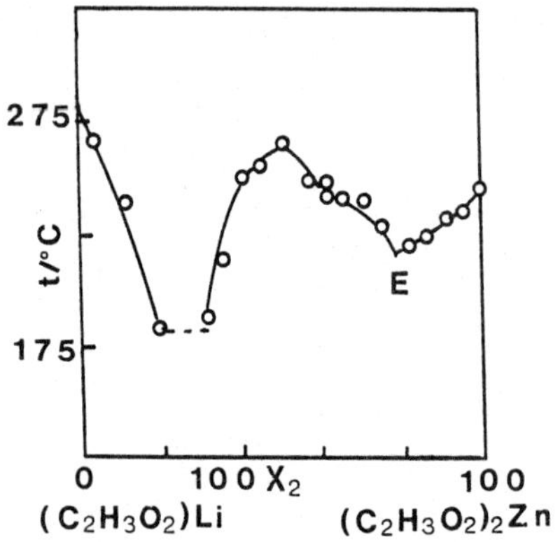

Characteristic point(s):

Eutectic, E, at 220 °C and $100x_2$= 75 (authors).

Note - Glasses form at 15 $\leq$ $100x_2$ $\leq$ 30.

Intermediate compound(s):

$(C_2H_3O_2)_3LiZn$, congruently melting at 265 °C (authors).
$(C_2H_3O_2)_5LiZn_2$, incongruently melting at 240 °C (authors).

AUXILIARY INFORMATION

METHOD/APPARATUS/PROCEDURE:

Visual polythermal analysis as well as time-temperature curves were employed. The temperatures were measured with a Chromel-Alumel thermocouple checked at the freezing temperatures of Zn, $K_2Cr_2O_7$, Cd, Sn, and benzoic acid.

NOTE:

The formation of glasses in this system is reasonable. Accordingly, one should expect a marked tendency of the molten mixtures to supercool, which might cause the polythermal observations to be less reliable than usual.

SOURCE AND PURITY OF MATERIALS:

Component 1: either $(C_2H_3O_2)Li.2H_2O$ of analytical purity, or material obtained by reacting Li_2CO_3 and ethanoic acid; both materials dehydrated in an oven at 105-110 °C.
Component 2: $(C_2H_3O_2)_2Zn.2H_2O$ of analytical purity dried to constant mass at 110 °C.

ESTIMATED ERROR:

Temperature: accuracy probably ±2 K (compiler).

REFERENCES:

COMPONENTS:	EVALUATOR:
(1) Magnesium ethanoate (magnesium acetate); $(C_2H_3O_2)_2Mg$; [142-72-3] (2) Sodium ethanoate (sodium acetate); $(C_2H_3O_2)_2Na_2$; [127-09-3]	Schiraldi, A., Dipartimento di Chimica Fisica, Universita' di Pavia (ITALY).

CRITICAL EVALUATION:

This system has been investigated only by Pochtakova (Ref. 1) who reports the results of visual polythermal observations supplemented with DTA investigations, both in numerical and graphical form.

The trend of the accessible part of the liquidus ($0 \leq 100\mathbf{x}_1 \leq 70$) has been interpreted by the author as follows: the occurrence of the intermediate compound $(C_2H_3O_2)_4MgNa_2$, congruently melting at 533 K (260 °C), splits the diagram into two eutectic subsystems whose invariants are E_1, at 529 K (256 °C) and $100\mathbf{x}_2$= 40.0, and E_2, at 528 K (255 °C) and $100\mathbf{x}_2$= 57.5. The author suggests also that the intermediate compound undergoes an alpha-beta transition at 493 K (220 °C), and a lattice readjustment of the beta form at 373 K (100 °C).

For an evaluation of the reliability of the above conclusions, the following discrepancies between the text or tables and the original plot must be mentioned.

(i) In the experimental section of the paper two solid-solid transitions are reported for component 1 at 425 K (152 °C) and 449 K (176 °C), respectively, whilst the corresponding figures on the plot are 425 K (152 °C) and 445 K (172 °C).

(ii) The table summarizing the visual polythermal data reports two temperature values at $100\mathbf{x}_1$= 50, the first of which - possibly due to a misprint - probably corresponds to $100\mathbf{x}_1$= 30.

(iii) The table collecting the DTA results reports, at $100\mathbf{x}_1$= 60, five temperature values, one of which (236 °C) is neither included in the phase diagram nor otherwise discussed in the text.

(iv) No DTA evidence for the lattice readjustment at 373 K is provided at the composition of the intermediate compound.

(v) DTA measurements carried out at $100\mathbf{x}_2 > 50$ did not allow the author to obtain evidence for either the transition of the intermediate compound at 493 K, or the lattice readjustment at 373 K.

(vi) DTA measurements carried out on the mixtures did not allow the author to obtain evidence for the solid state transitions of the pure components. It is however to be stressed that the transition temperatures of sodium ethanoate are quoted by the author from Ref. 2.

In conclusion the upper part of the phase diagram given in the paper seems to be supported adequately by the experimental results, whereas the system is still to be considered as largely unexplored below the eutectic lines.

REFERENCES:

(1) Pochtakova, E.I.
Zh. Obshch. Khim. 1974, **44**, 241-248.

(2) Sokolov, N.M.
Tezisy Dokl. X Nauchn. Konf. S.M.I. 1956.

COMPONENTS:	ORIGINAL MEASUREMENTS:
(1) Magnesium ethanoate (magnesium acetate); $(C_2H_3O_2)_2Mg$; [142-72-3] (2) Sodium ethanoate (sodium acetate); $(C_2H_3O_2)_2Na_2$; [127-09-3]	Pochtakova, E.I. **Zh. Obshch. Khim.** 1974, **44**, 241-248.
VARIABLES:	PREPARED BY:
Temperature.	Baldini, P.

EXPERIMENTAL VALUES:

$t/^{o}C$	T/K^{a}	$100x_1$
331	604	0
329	602	2.5
326	599	5
324	597	7.5
320	593	10
321^{bc}	594	10
255^{bd}	528	10
313	586	15
310	583	17.5
306	579	20
297	570	25
290	563	27.5
288	561	30^{i}
284	557	32.5
275	548	35
269	542	37.5
261	534	40
255	528	42.5
255^{bc}	528	42.5
255^{bd}	528	42.5
256	529	45
257	530	47.5
260	533	50
260^{bc}	533	50
220^{bg}	593	50
259	532	52.5
258	531	55
260^{bc}	533	56.5
258^{be}	531	56.5
100^{bf}	373	56.5
220^{bg}	493	56.5
257	530	57.5
256	529	60
258^{bc}	531	60
258^{be}	531	60
100^{df}	373	60
220^{bg}	493	60
236^{bh}	509	60
268	541	65
272	545	67.5

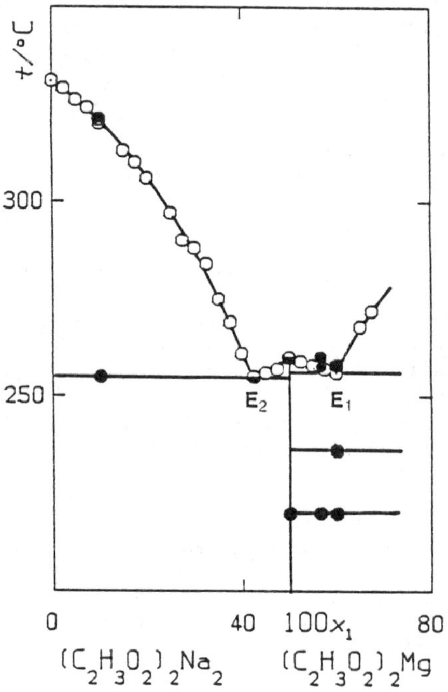

a T/K values calculated by the compiler.
b Differential thermal analysis (filled circles in the figure).
c Initial crystallization.
d First eutectic stop.
e Second eutectic stop.
f First transition of the system.
g Second transition of the system.
h Third transition of the system.
i 50 in the original text (corrected by the compiler).

Note - The system was investigated at $0 < 100x_1 < 67.5$ due to thermal instability of component 1.

COMPONENTS:	ORIGINAL MEASUREMENTS:
(1) Magnesium ethanoate (magnesium acetate); $(C_2H_3O_2)_2Mg$; [142-72-3] (2) Sodium ethanoate (sodium acetate); $(C_2H_3O_2)_2Na_2$; [127-09-3]	Pochtakova, E.I. **Zh. Obshch. Khim.** 1974, **44**, 241-248.
VARIABLES:	PREPARED BY:
Temperature.	Baldini, P.

EXPERIMENTAL VALUES: (continued)

Characteristic point(s):

Eutectic, E_1, at 256 °C (extrapolated, visual polythermal analysis), or 258 °C (differential thermal analysis), and $100\mathbf{x}_1$= 60 (author).

Eutectic, E_2, at 255 °C and $100\mathbf{x}_1$= 42.5 (author).

Intermediate compound(s):

$(C_2H_3O_2)_4MgNa_2$, congruently melting at 260 °C (author).

AUXILIARY INFORMATION

METHOD/APPARATUS/PROCEDURE:

Visual polythermal analysis, supplemented with differential thermal analysis.

SOURCE AND PURITY OF MATERIALS:

Component 1: prepared (Ref. 1) by reacting the ("chemically pure") carbonate with a slight excess of ethanoic acid of analytical purity [phase transitions at $t_{trs}(1)$/°C= 152, 176].
Component 2: "chemically pure" material recrystallized and dried at 200 °C to constant mass [phase transitions at $t_{trs}(2)$/°C= 238-240, 130, 118, 58, Ref. 2].

ESTIMATED ERROR:

Temperature: accuracy probably ±2 K (compiler).

REFERENCES:

(1) Sokolov, N.M.
Zh. Obshch. Khim. 1954, **24**, 1581-1593
(2) Sokolov, N.M.
Tezisy Dokl. X Nauch. Konf. S.M.I. 1956.

COMPONENTS:	EVALUATOR:
(1) Sodium ethanoate (sodium acetate); $(C_2H_3O_2)Na$; [127-09-3] (2) Rubidium ethanoate (rubidium acetate); $(C_2H_3O_2)Rb$; [563-67-7]	Schiraldi, A., Dipartimento di Chimica Fisica Universita´ di Pavia (ITALY).

CRITICAL EVALUATION:

This system was studied twice in Gimel´shtein´s laboratory [Ref. 1: visual polythermal analysis (empty circles in the figure); Ref. 2: DTA (filled circles in the figure)] with substantially analogous results for the liquidus: an intermediate compound, $(C_2H_3O_2)_4NaRb_3$, congruently melting at 452-453 K (179 °C, Ref. 1; 180 °C, Ref. 2), forms eutectics with both pure components, at 418-419 K (145-146 °C) and $100\mathbf{x}_1$= 38-38.5, and at 451-453 K (178-180 °C) and $100\mathbf{x}_1$= 23.5, respectively.

Discrepancies, however, exist between Ref.s 1 and 2 about the phase transition temperatures of the pure components.

As for component 1, Gimel´shtein and Diogenov (Ref. 1) report $\mathbf{T}_{trs}(1)$= 583-584 K (310-311 °C), while Gimel´shtein (Ref. 2) gives $T_{trs}(1)$= 543 K (270 °C). The former figure exceeds largely the highest $\mathbf{T}_{trs}(1)$ value listed in Table 1 of the Preface, viz., 527±15 K, while the latter one lies just above the upper uncertainty limit of Table 1 value.

As for component 2, 493 K (220 °C) and 479 K (206 °C) are reported in Ref. 1 and Ref. 2, respectively, as the transition temperature: the former value is close to, while the latter one is significantly lower than that listed in Table 1 of the Preface, viz., 498±1 K.

X-ray diffractometric results were claimed (Ref. 2) to support the existence of the intermediate compound, and to suggest that this should decompose into a solid solution just below 383 K (110 °C). The second assertion, however, does not seem convincing, inasmuch as it would imply a change in the solid from a state of miscibility at lower temperatures into a state of immiscibility at higher temperatures.

Finally, the assumption of the congruent fusion of the intermediate compound does not seem adequately supported: the shape of the liquidus could as well suggest the occurrence of a peritectic equilibrium, e.g., in connection with the incongruent fusion of the compound.

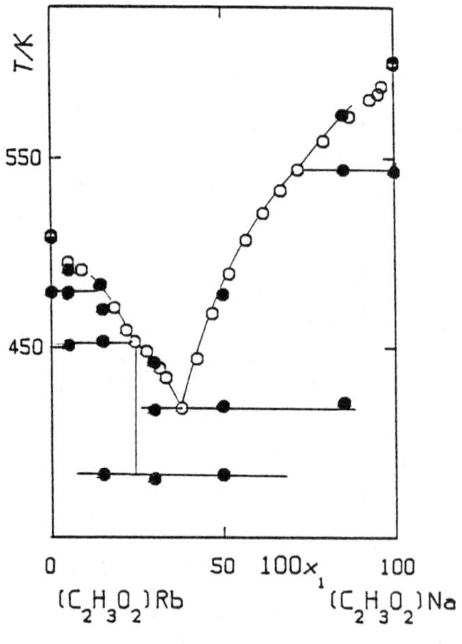

REFERENCES:

(1) Gimel´shtein, V.G.; Diogenov, G.G. **Zh. Neorg. Khim.** 1958, **3**, 1644-1649 (*); **Russ. J. Inorg. Chem. (Engl. Transl.)** 1958, **3 (7)**, 230-237.

(2) Gimel´shtein, G.G.; **Tr. Irkutsk. Politech. Inst.** 1971, No. 66, 80-100.

COMPONENTS:

(1) Sodium ethanoate (sodium acetate); $(C_2H_3O_2)Na$; [127-09-3]
(2) Rubidium ethanoate (rubidium acetate); $(C_2H_3O_2)Rb$; [563-67-7]

ORIGINAL MEASUREMENTS:

Gimel'shtein, V.G.; Diogenov, G.G.
Zh. Neorg. Khim. 1958, 3, 1644-1649 (*);
Russ. J. Inorg. Chem. (Engl. Transl.) 1958, 3 (7), 230-237.

VARIABLES:

Temperature

PREPARED BY:

Baldini, P.

EXPERIMENTAL VALUES:

t/°C	T/K[a]	$100x_1$	t/°C	T/K[a]	$100x_1$
236	509	0	195	468	47
222	495	5	216	489	52
218	491	9	234	507	57
210	483	14.5	248	521	62
198	471	18.5	260	533	67
186	459	22	271	544	72
180	453	24.5	286	559	79.5
175	448	28	299	572	87
166	439	31.7	308	581	93
161	434	33.5	311	584	95.5
145	418	38	315	588	96.5
171	444	42.5	327	600	100

[a] T/K values calculated by the compiler.

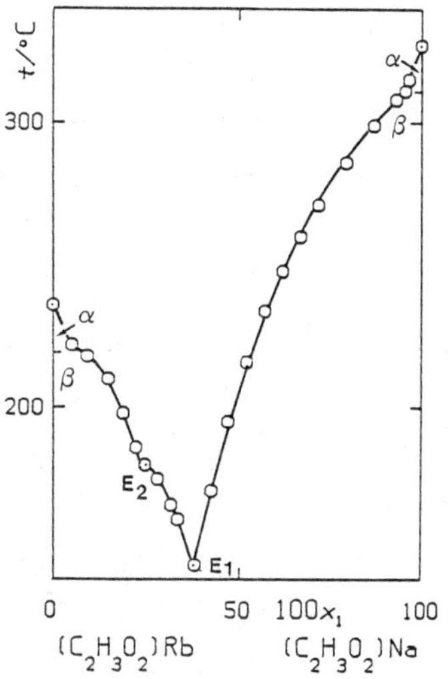

Characteristic point(s):
Eutectic, E_1, at 145 °C (according to Fig. 2 of the original paper, or at 146 °C according to Fig. 1 of the original paper, and not at 179 °C as reported in the text; compiler) and $100x_1$= 38 (authors).
Eutectic, E_2, at 180 °C (according to Fig. 2 of the original paper, or at 179 °C according to Fig. 1 of the original paper; compiler) and $100x_1$ about 23.5 (compiler).

Intermediate compound(s):
$(C_2H_3O_2)_4NaRb_3$, congruently melting at 180 °C (authors).

AUXILIARY INFORMATION

METHOD/APPARATUS/PROCEDURE:

Visual polythermal analysis. Temperatures measured with a Chromel-Alumel thermocouple and a 17-mV-range millivoltmeter. Mixtures being hygroscopic, the method of additions with determination of the sample mass by difference was employed in order to avoid hydration.

SOURCE AND PURITY OF MATERIALS:

Not stated.
Component 1 undergoes a phase transition at $t_{trs}(1)$/°C= 311 (310 °C according to Fig. 2 of the original paper; compiler).
Component 2 undergoes a phase transition at $t_{trs}(2)$/°C= 220.

ESTIMATED ERROR:

Temperature: accuracy probably ±2 K (compiler).

REFERENCES:

COMPONENTS:	ORIGINAL MEASUREMENTS:
(1) Sodium ethanoate (sodium acetate); $(C_2H_3O_2)Na$; [127-09-3] (2) Rubidium ethanoate (rubidium acetate); $(C_2H_3O_2)Rb$; [563-67-7]	Gimel´shtein, V.G. **Tr. Irkutsk. Politekh. Inst.** 1971, No. 66, 80-100.
VARIABLES:	PREPARED BY:
Temperature	Baldini, P.

EXPERIMENTAL VALUES:

t/°C	T/K[a]	$100x_2$	t/°C	T/K[a]	$100x_2$
328	601	0	108	381	70.0
270	543	0	197	470	85.0
300	573	15.0	180	453	85.0
271	544	15.0	110	383	85.0
148	421	15.0	218	491	95.0
205	478	50.0	206	479	95.0
146	419	50.0	178	451	95.0
110	383	50.0	235	508	100
169	442	70.0	206	479	100
144	417	70.0			

[a] **T**/K values calculated by the compiler.

Characteristic point(s):

Eutectic, E_1, at 178 °C and $100x_1$= 23.5 (author).
Eutectic, E_2, at 146 °C and $100x_1$= 38.5 (author).

Intermediate compound(s):

$(C_2H_3O_2)_4NaRb_3$, congruently (compiler) melting at 179 °C (author), and undergoing a transformation at 110 °C (author).

AUXILIARY INFORMATION

METHOD/APPARATUS/PROCEDURE:

Differential thermal analysis (using a derivatograph with automatic recording of the heating curves) and room temperature X-ray diffractometry (using a URS-501M apparatus) were employed.

NOTE - 1

The meaning of the data listed in the table becomes apparent by observing the figure reported in the critical evaluation.

NOTE - 2

The coordinates of the characteristic points were stated by the author on the basis of his own DTA measurements, and of previous literature data (Ref. 1). X-ray patterns were taken at $100x_1$= 27.5.

SOURCE AND PURITY OF MATERIALS:

Not stated.
Component 1 melts at $t_{fus}(1)$/°C= 328 (327 according to Fig. 7 of the original paper; compiler), and undergoes a phase transition at $t_{trs}(1)$/°C= 270.
Component 2 melts at $t_{fus}(2)$/°C= 235 (236 according to Fig. 7 of the original paper; compiler), and undergoes a phase transition at $t_{trs}(2)$/°C= 206.

ESTIMATED ERROR:

Temperature: accuracy probably ±2 K (compiler).

REFERENCES:

(1) Gimel´shtein, V.G.; Diogenov, G.G.
Zh. Neorg. Khim. 1958, **3**, 1644-1649.

COMPONENTS:	EVALUATOR:
(1) Sodium ethanoate (sodium acetate); $(C_2H_3O_2)Na$; [127-09-3] (2) Zinc ethanoate (zinc acetate); $(C_2H_3O_2)_2Zn$; [557-34-6]	Schiraldi, A., Dipartimento di Chimica Fisica, Universita´ di Pavia (ITALY).

CRITICAL EVALUATION:

This system was studied by Lehrman and Skell (Ref. 1), Pavlov and Golubkova (Ref. 2), and Nadirov and Bakeev (Ref. 3).

A qualitative agreement exists between Refs. 1 and 2, as both of them report a phase diagram characterized by two eutectics, E_1 and E_2, and the congruently melting intermediate compound $(C_2H_3O_2)_4Na_2Zn$. Differences between these papers concern the coordinates of the eutectics: according to Ref. 1, E_1 should occur at 491-493 K (218-220 °C) and $100\mathbf{x}_2$ about 28, and E_2 at 548.5-551.8 K (175.3-178.6 °C) and $100\mathbf{x}_2$ about 54, whereas, according to Ref. 2, the invariants should be at 473 K (200 °C) and $100\mathbf{x}_2$= 25, and at 413 K (140 °C) and $100\mathbf{x}_2$= 50, respectively.

The phase diagram suggested in Ref. 3 shows in turn: (i) a single eutectic at either 415, or 421 K (either 142, or 148 °C, according to visual polythermal and conductometric investigations, respectively) and $100\mathbf{x}_2$= 57; (ii) a peritectic at either 480, or 477, or 484 K (either 207, or 204, or 211 °C, according to visual polythermal, conductometric, and thermographical results, respectively), and, possibly, $100\mathbf{x}_2$= 33.3; and (iii) the intermediate compound $(C_2H_3O_2)_4Na_2Zn$ reported here as incongruently melting.

In the evaluator´s opinion, the discrepancies among the diagrams suggested by the different authors should be attributed mainly to different degrees of accuracy in the determination of the actual liquidus temperatures. In this connection, it is important to stress that Lehrman and Skell observed a tendency of the melts to supercool and, in particular, found at temperatures below 483 K extremely viscous melts "so that great difficulty was experienced in obtaining crystallization and reproducible melting points" (Ref. 1). Consequently, in the case of the present binary, poorly reliable results can be reasonably expected both by techniques implying observations performed on cooling (as visual polythermal analysis), and by techniques (as conductometry) implying observations performed on heating at constant rate. Accordingly, the diagrams by Pavlov and Golubkova (based only on visual polythermal observations), and by Nadirov and Bakeev (based mainly on visual polythermal and conductometric investigations) probably suffer from limited accuracy.

In conclusion, the evaluator is inclined to consider as more reliable the findings by Lehrman and Skell (who employed very small heating rates), viz.: (i) the presence of the intermediate compound $(C_2H_3O_2)_4Na_2Zn$, congruently melting at about 500 K; and (ii) the occurrence of two eutectics, E_1 at about 490 K and $100\mathbf{x}_2$ about 28, and E_2 at about 550 K and 100 $\mathbf{x}_2$ about 54.

REFERENCES:

(1) Lehrman, A.; Skell, P.
J. Am. Chem. Soc. 1939, **61**, 3340-3342.

(2) Pavlov, V.L.; Golubkova, V.V.
Visn. Kiiv. Univ., Ser. Khim., Kiev, 1972, No. 13, 28-30.

(3) Nadirov, E.G.; Bakeev, M.I.
Tr. Khim.-Metall. Inst. Akad. Nauk Kaz. SSR 1974, **25**, 115-128.

COMPONENTS:

(1) Sodium ethanoate (sodium acetate); $(C_2H_3O_2)Na$; [127-09-3]
(2) Zinc ethanoate (zinc acetate); $(C_2H_3O_2)_2Zn$; [557-34-6]

ORIGINAL MEASUREMENTS:

Lehrman, A.; Skell, P.
J. Amer. Chem. Soc. 1939, **61**, 3340-3342.

VARIABLES:

Temperature

PREPARED BY:

Baldini, P.

EXPERIMENTAL VALUES:

t/°C	T/K[a]	$100x_2$	t/°C	T/K[a]	$100x_2$
328.3	601.5	0	203.9	477.1	46.0
313.4	586.6	10.1	198.0	471.2	48.0
277.5	550.7	20.0	194.6	467.8	49.0
261.4	534.6	22.6	192.5	465.7	50.0
233.2	506.4	26.5	189.1	462.3	51.0
218.0[b]	491.2	26.5	183.0	456.2	52.0
223.3	496.5	28.0	178.6[c]	451.8	52.0
220.0[b]	493.2	28.0	180.9	454.1	55.0
225.7	498.9	30.0	175.3[c]	448.5	55.0
227.2	500.4	33.3	197.8	471.0	60.0
227.1	500.3	33.3	212.5	485.7	66.7
227.1	500.3	33.3	221.8	495.0	70.0
226.4	499.6	35.0	236.5	509.7	80.0
220.2	493.4	40.0	242.4	515.6	100.0

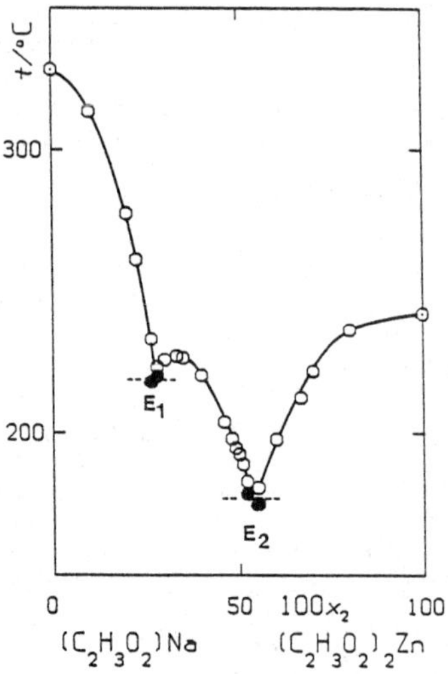

[a] T/K values calculated by the compiler.
[b] Eutectic stop (E_1); filled circles in the figure.
[c] Eutectic stop (E_2); filled circles in the figure.

Characteristic point(s):
Eutectic, E_1, at 218-220 °C and $100x_2$ about 28 (compiler).
Eutectic, E_2, at 175.3-178.6 °C and $100x_2$ about 54 (compiler).

Intermediate compound(s):
$(C_2H_3O_2)_4Na_2Zn$, congruently melting at 227.1±0.1 °C (compiler).

AUXILIARY INFORMATION

METHOD/APPARATUS/PROCEDURE:

The salts, contained into 2.5x20 cm Pyrex tube and added with 5 drops of glacial ethanoic acid, were heated in bath formed with the eutectic mixture of calcium, potassium, and lithium nitrates.
The temperature of disappearance of the last crystal as the mixture was heated under stirring was measured with Copper-Constantane thermocouple and potentiometer. The fusion temperatures tabulated come from three or more determinations ranging within 1 K. The eutectic stops relevant to E_1 were measured by means of time - temperature cooling curves.

SOURCE AND PURITY OF MATERIALS:

Materials of not stated source, recrystallized from dilute ethanoic acid, and dehydrated according to Ref. 1.

ESTIMATED ERROR:

Temperature: accuracy ± 0.5 K (compiler).

REFERENCES:

(1) Davidson, A.W.; McAllister
J. Amer. Chem. Soc. 1930, 52, 519-527.

COMPONENTS:

(1) Sodium ethanoate (sodium acetate); $(C_2H_3O_2)Na$; [127-09-3]
(2) Zinc ethanoate (zinc acetate); $(C_2H_3O_2)_2Zn$; [557-34-6]

ORIGINAL MEASUREMENTS:

Pavlov, V.L.; Golubkova, V.V.
Visn. Kiiv. Univ., Ser. Khim., Kiev, 1972, **No. 13,** 28-30.

VARIABLES:

Temperature.

PREPARED BY:

Baldini, P.

EXPERIMENTAL VALUES:

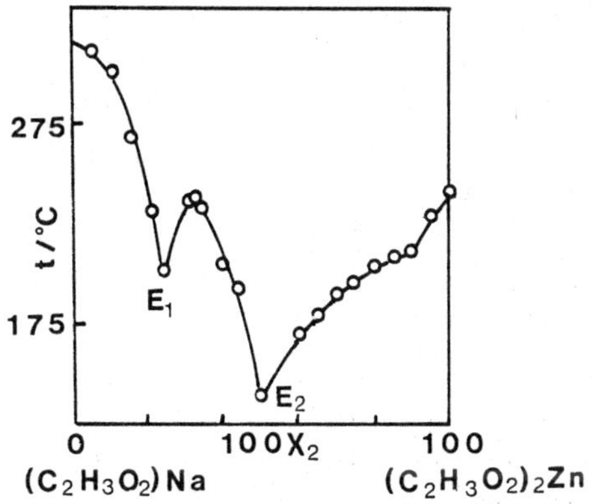

The results are reported only in graphical form (see figure).

Characteristic point(s):

Eutectic, E_1, at 200 °C and $100x_2$= 25 (authors).
Eutectic, E_2, at 140 °C and $100x_2$= 50 (authors).

Intermediate compound(s):

$(C_2H_3O_2)_4Na_2Zn$, congruently melting at 240 °C (authors).

AUXILIARY INFORMATION

METHOD/APPARATUS/PROCEDURE:

Visual polythermal analysis as well as time-temperature curves were employed. The temperatures were measured with a Chromel-Alumel thermocouple checked at the freezing temperatures of Zn, $K_2Cr_2O_7$, Cd, Sn, and benzoic acid.

SOURCE AND PURITY OF MATERIALS:

Component 1: $(C_2H_3O_2)Na.3H_2O$ of analytical purity recrystallized from water and dried in an oven at 110-120 °C to constant mass.
Component 2: $(C_2H_3O_2)_2Zn.2H_2O$ of analytical purity dried to constant mass at 110 °C.

ESTIMATED ERROR:

Temperature: accuracy probably ±2 K (compiler).

REFERENCES:

COMPONENTS:	ORIGINAL MEASUREMENTS:
(1) Sodium ethanoate (sodium acetate); $(C_2H_3O_2)Na$; [127-09-3] (2) Zinc ethanoate (zinc acetate); $(C_2H_3O_2)_2Zn$; [557-34-6]	Nadirov, E.G.; Bakeev, M.I. **Tr. Khim.-Metall. Inst. Akad. Nauk Kaz. SSR** 1974, **25**, 115-128.
VARIABLES:	PREPARED BY:
Temperature	Baldini, P.

EXPERIMENTAL VALUES:

$t/^oC$	T/K^a	$100x_1$
236	509	0
229	502	10
217	490	20
191	464	30
174	447	35
165	438	40
142	415	43
177	450	50
183	456	52
189	462	55
202	475	60
206	479	66.7
214	487	70
234	507	75
261	534	80
279	552	85
298	571	90
313	586	95
332	605	100

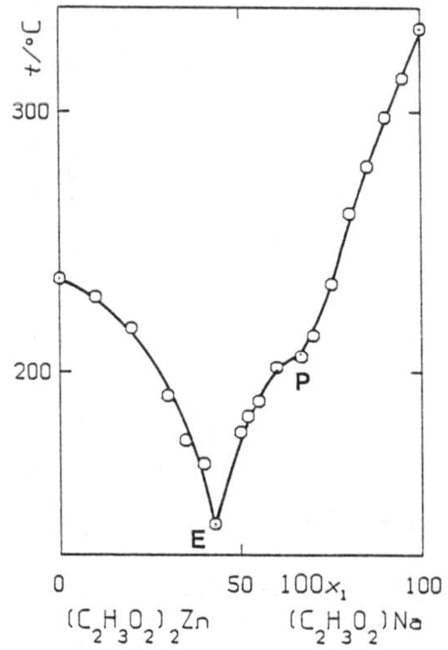

[a] T/K values calculated by the compiler.

Characteristic point(s):

Eutectic, E, at either 142 ^oC (visual polythermal analysis), or 148 ^oC (conductometry), and $100x_1$= 43.

Peritectic, P, at either 207 ^oC (visual polythermal analysis), or 204 ^oC (conductometry), or 211 ^oC (thermographical analysis), and at a not clearly specified composition [in compiler's opinion, the coordinates of the peritectic might be 206 ^oC (visual polythermal analysis; tabulated value) and $100x_1$= 66.7].

Intermediate compound(s):

$(C_2H_3O_2)_4Na_2Zn$, incongruently melting.

AUXILIARY INFORMATION

METHOD/APPARATUS/PROCEDURE:	SOURCE AND PURITY OF MATERIALS:
Visual polythermal analysis supplemented with conductometry and occasionally with thermographical investigations. Temperatures of initial crystallization measured with a thermocouple.	Component 1: "chemically pure" hydrated $C_2H_3O_2Na$ recrystallized twice and dried at 130 ^oC. Component 2: $(C_2H_3O_2)_2Zn.2H_2O$ of analytical purity, recrystallized twice and dried at 140 ^oC.
	ESTIMATED ERROR:
	Temperature: accuracy probably ± 2 K (compiler).

COMPONENTS:	ORIGINAL MEASUREMENTS:
(1) Lead(II) ethanoate (lead acetate); $(C_2H_3O_2)_2Pb$; [15347-57-6] (2) Zinc ethanoate (zinc acetate); $(C_2H_3O_2)_2Zn$; [557-34-6]	Petersen, J. **Z. Elektrochem.** 1914, **20**, 328-332.
VARIABLES:	PREPARED BY:
Temperature.	Baldini, P.

EXPERIMENTAL VALUES:

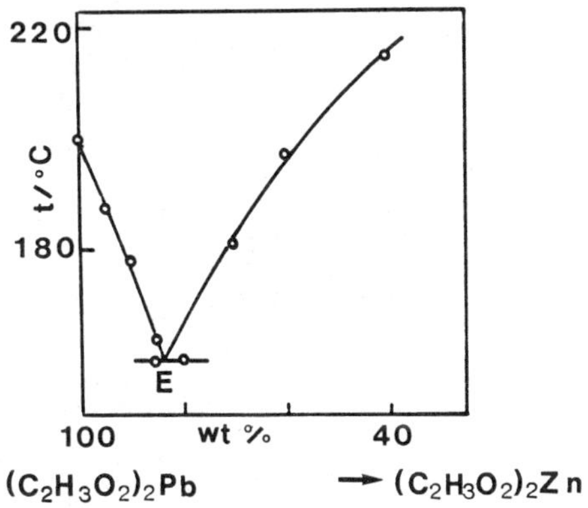

The results are reported only in graphical form (see figure).

Characteristic point(s):

Eutectic, E, at 160 °C and $100x_2$ about 25 (author).

AUXILIARY INFORMATION

METHOD/APPARATUS/PROCEDURE:	SOURCE AND PURITY OF MATERIALS:
Mixtures contained in a glass tube and heated in a sulfuric acid bath. NOTE: $T_{fus}(1)$ and $T_{fus}(2)$ are in reasonable agreement with the data by other authors (Ref. 1). The general features of the diagram seem to be reliable.	Not stated. Component 1: $t_{fus}(1)/°C= 204$. Component 2: $t_{fus}(2)/°C= 244$. ESTIMATED ERROR: Temperature: accuracy not evaluable (compiler). REFERENCES: (1) Sanesi, M.; Cingolani, A.; Tonelli, P.L.; Franzosini, P.; **Thermal Properties**, in **Thermodynamic and Transport Properties of Organic Salts**, IUPAC Chemical Data Series No. 28 (Franzosini, P.; Sanesi, M.; Editors), Pergamon Press, Oxford, 1980, 29-115.

COMPONENTS:

(1) Rubidium ethanoate (rubidium acetate); $(C_2H_3O_2)Rb$; [563-67-7]
(2) Zinc ethanoate (zinc acetate); $(C_2H_3O_2)_2Zn$; [557-34-6]

ORIGINAL MEASUREMENTS:

Nadirov, E.G.; Bakeev, M.I.
Tr. Khim.-Metall. Inst. Akad. Nauk Kaz. SSR 1974, **25**, 115-128.

VARIABLES:

Temperature.

PREPARED BY:

Baldini, P.

EXPERIMENTAL VALUES:

$t/^{\circ}C$	T/K^a	$100\mathbf{x}_1$
236	509	0
223	496	10
219	492	15
212	485	20
198	471	30
182	455	35
159	432	40
173	446	45
187	460	50
196	469	55
204	477	60
209	482	65
217	490	70
223	496	75
230	503	80
232	505	85
235	508	90
236	509	93.7
237	510	100

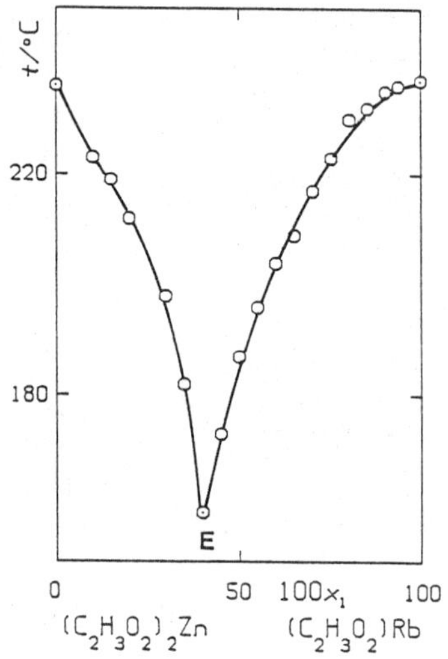

[a] T/K values calculated by the compiler.

Characteristic point(s):

Eutectic, E, at either 159 °C (visual polythermal analysis), or 163 °C (conductometry), and $100\mathbf{x}_1$= 40

AUXILIARY INFORMATION

METHOD/APPARATUS/PROCEDURE:

Visual polythermal analysis supplemented with conductometry, and occasionally with thermographical and X-ray investigations. Temperatures of initial crystallization measured with a thermocouple.

NOTE 1:

The mixtures at $55 \leq 100\mathbf{x}_1 \leq 80$ tend to form glasses.

NOTE 2:

The $T_{fus}(1)$ and $T_{fus}(2)$ values given here are lower than the corresponding values from Preface 1 [$T_{fus}(1)$= 514 K] and from Ref. 1 [$T_{fus}(2)$= 514-533 K], respectively. In Fig. 8 of the original paper the authors report an isothermal line at 404 K (131 °C) which is not discussed in the text. The ability to form glasses might imply poor reliability of the eutectic coordinates; however, the classification of the diagram as of the simple eutectic type might be accepted with some confidence.

SOURCE AND PURITY OF MATERIALS:

Component 1: material recrystallized three times and dried at 110-120 °C. Component 2: $(C_2H_3O_2)_2Zn.2H_2O$ of analytical purity, recrystallized twice and dried at 140 °C.

ESTIMATED ERROR:

Temperature: accuracy probably ± 2 K (compiler).

REFERENCES:

(1) Sanesi, M.; Cingolani, A.; Tonelli, P.L.; Franzosini, P.; **Thermal Properties**, in **Thermodynamic and Transport Properties of Organic Salts**, IUPAC Chemical Data Series No. 28 (Franzosini, P.; Sanesi, M.; Editors), Pergamon Press, Oxford, 1980, 29-115.

COMPONENTS:

(1) Potassium propanoate (potassium propionate); $(C_3H_5O_2)K$; [327-62-8]

(2) Lithium propanoate (lithium propionate); $(C_3H_5O_2)Li$; [6531-45-9]

ORIGINAL MEASUREMENTS:

Sokolov, N.M.; Tsindrik, N.M.
Zh. Neorg. Khim. 1969, 14, 584-590 (*);
Russ. J. Inorg. Chem. (Engl. Transl.) 1969, 14, 302-306.

VARIABLES:

Temperature.

PREPARED BY:

Baldini, P.

EXPERIMENTAL VALUES:

The results are reported only in graphical form (see figure).

Characteristic point(s):

Eutectic, E_1, at 291 °C (authors) and $100x_1$= 67.5 (according to Fig. 1 of the original paper; erroneously reported as 19 in the text; compiler).
Eutectic, E_2, at 279 °C (authors) and $100x_1$= 19 (according to Fig. 1 of the original paper; erroneously reported as 67.5 in the text; compiler).

Intermediate compound(s):

$(C_3H_5O_2)_2KLi$ (probable composition), congruently melting (authors).

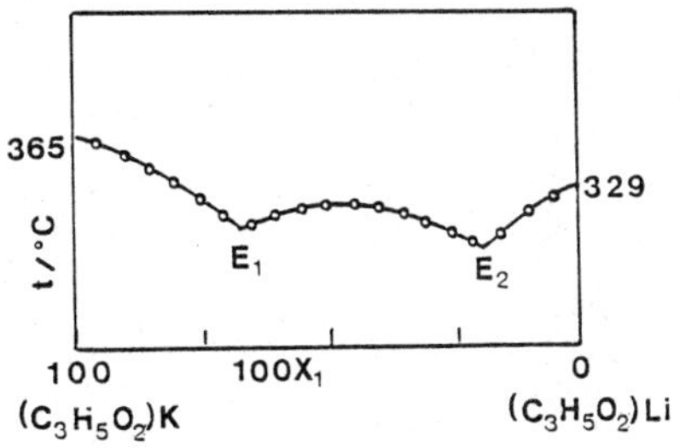

AUXILIARY INFORMATION

METHOD/APPARATUS/PROCEDURE:

Visual polythermal analysis.

NOTE:

The fusion temperature of component 2 is about 5 K lower than that listed in Preface, Table 1 (606.8±0.5 K), whereas $T_{fus}(1)$ meets satisfactorily the value (638.3±0.5 K) given in the table. The general features of the diagram should be considered with some confidence.

SOURCE AND PURITY OF MATERIALS:

Materials prepared by reacting "chemically pure" carbonates with propanoic acid of analytical purity (Ref. 1).
Component 1 undergoes a phase transition at $t_{trs}(1)/°C$= 68 (Ref. 2) and melts at $t_{fus}(1)/°C$= 365.
Component 2 undergoes a phase transition at $t_{trs}(2)/°C$= 265 and melts at $t_{fus}(2)/°C$= 329.

ESTIMATED ERROR:

Temperature: accuracy probably ±2 K (compiler).

REFERENCES:

(1) Sokolov, N.M.
Zh. Obshch. Khim. 1954, **24**, 1581-1593 (this is Ref. 2 in the original paper, not Ref.3 as quoted by the authors).
(2) Sokolov, N.M.
Tezisy Dokl. X **Nauch. Konf.** S.M.I. 1956.

COMPONENTS:

(1) Potassium propanoate (potassium propionate); $(C_3H_5O_2)_2K_2$; [327-62-8]
(2) Magnesium propanoate (magnesium propionate); $(C_3H_5O_2)_2Mg$; [557-27-7]

ORIGINAL MEASUREMENTS:

Pochtakova, E.I.
Zh. Obshch. Khim. 1974, **44**, 241-248.

VARIABLES:

Temperature.

PREPARED BY:

Baldini, P.

EXPERIMENTAL VALUES:

t/°C	T/K[a]	$100x_2$	t/°C	T/K[a]	$100x_2$
365	638	0	312	585	32.5
358	631	2.5	305	578	35
354	627	5	306[bc]	579	36.5
352	625	7.5	306[bd]	579	36.5
347	620	10	236[be]	509	36.5
345	618	12.5	306	579	37.5
343	616	15	315	588	40
341	614	17.5	320	593	42.5
335	608	20	322	595	45
332	605	22.5	322[bc]	595	45
328	601	25	306[bd]	579	45
330[bc]	603	25	235[be]	508	45
304[bd]	577	25	324	597	47.5
236[be]	509	25	324	597	50
324	597	27.5	324	597	52.5
318	591	30	324	597	55

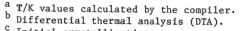

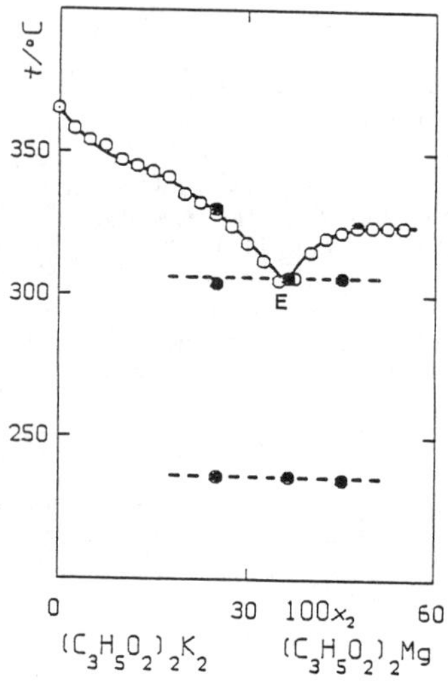

[a] T/K values calculated by the compiler.
[b] Differential thermal analysis (DTA).
[c] Initial crystallization.
[d] Eutectic stop.
[e] First transition of the system.

Characteristic point(s): Eutectic, E, at 306 °C (DTA), and $100x_2$= 36.5 (author).

AUXILIARY INFORMATION

METHOD/APPARATUS/PROCEDURE:

Visual polythermal analysis, supplemented with differential thermal analysis.

SOURCE AND PURITY OF MATERIALS:

Materials prepared (Ref. 1) by reacting the proper ("chemically pure") carbonate with a slight excess of propanoic acid of analytical purity. Component 1 undergoes phase transitions at $t_{trs}(1)/°C$= 68, 330 (Ref. 2). Component 2 undergoes phase transitions at $t_{trs}(2)/°C$= 185, 200, 217, 246.

REFERENCES:

(1) Sokolov, N.M. **Zh. Obshch. Khim.** 1954, **24**, 1581-1593.
(2) Sokolov, N.M. **Tezisy Dokl. X Nauch. Konf. S.M.I.** 1956.
(3) Sanesi, M.; Cingolani, A.; Tonelli, P.L.; Franzosini, P.; **Thermal Properties,** in Thermodynamic and Transport Properties of Organic Salts, IUPAC Chemical Data Series No.28 (Franzosini, P.; Sanesi, M.; Editors), Pergamon Press, Oxford, 1980, 29-115.

NOTES:

The system was investigated at $0 \leq 100x_2 \leq 55$ due to thermal instability of component 2.
The fusion temperature of component 1 is in fair agreement with that listed in Preface, Table 1, whereas discrepancies exist for the solid state transition temperatures of the same component. Moreover, it is worth mentioning that Pochtakova´s paper is the only source of information (see Ref. 3) for what concerns the solid state transitions of magnesium propanoate.

ESTIMATED ERROR:

Temperature: accuracy probably ±2 K (compiler).

COMPONENTS:	EVALUATOR:
(1) Potassium propanoate (potassium propionate); $(C_3H_5O_2)K$; [327-62-8] (2) Sodium propanoate (sodium propionate); $(C_3H_5O_2)Na$; [137-40-6]	Franzosini, P., Dipartimento di Chimica Fisica, Universita' di Pavia (ITALY).

CRITICAL EVALUATION:

This binary was studied by visual polythermal analysis in Sokolov's laboratory as a side system of two reciprocal ternaries [i.e., K, Na/$C_2H_3O_2$, $C_3H_5O_2$ (Ref. 1), and K, Na/$C_3H_5O_2$, NO_3 (Ref. 2)] with almost identical results.

The occurrence of eutectics at 583-585 K (310-312 °C) and $100\mathbf{x}_1$= 66, and at 560-561 K (287-288 °C) and $100\mathbf{x}_1$= 8 is to be held for certain, as well as the existence of a congruently melting intermediate compound. However, the composition of the latter as claimed by the authors [i.e., $(C_3H_5O_2)_5K_3Na_2$], although possible, does not seem fully proved due to the fluctuation of the experimental points, and the lack of data other than the visual polythermal ones.

The fusion temperature of component 1 (638 K) is in fair agreement with that (638.3±0.5 K) listed in Table 1 of the Preface, whereas the fusion temperature of component 2 (571 K) has to be considered as too high, inasmuch as the DSC value given in Table 1 of the Preface, (562.4±0.2 K) was subsequently confirmed by that obtained with adiabatic calorimetry (561.88±0.03 K; Table 3).

Rather puzzlingly, for the solid state transition temperature of component 1 far different values are quoted [from the same source (Ref. 3)] in Ref. 1 and Ref. 2, i.e., 603 and 341 K, respectively. Both figures are in turn different from that reported in Table 1 of the Preface (352.5±0.5 K).

Again from Ref. 3, solid state transitions are quoted in both Ref. 1 and Ref. 2 as occurring in component 2 at $\mathbf{T}_{trs}(2)$/K= 350, 468, 490, and 560. Doubts, however, are to be cast about the existence of the lowest transition as well as of the highest one, inasmuch as DSC provided evidence for the occurrence of only two solid state transformations (at 470.2±0.5 and 494±1 K, respectively; Preface, Table 1) which was subsequently confirmed with adiabatic calorimetry (Preface, Table 3).

REFERENCES:

(1) Sokolov, N.M.; Pochtakova, E.I.
Zh. Obshch. Khim. 1958, **28**, 1397-1404.

(2) Dmitrevskaya, O.I.; Sokolov, N.M.
Zh. Obshch. Khim. 1958, **28**, 2920-2926 (*); **Russ. J. Gen. Chem.(Engl. Transl.)** 1958, **28**, 2949-2954.

(3) Sokolov, N.M.; **Tezisy Dokl. X Nauch. Konf. S.M.I.** 1956.

COMPONENTS:

(1) Potassium propanoate (potassium propionate); $(C_3H_5O_2)K$; [327-62-8]
(2) Sodium propanoate (sodium propionate); $(C_3H_5O_2)Na$; [137-40-6]

ORIGINAL MEASUREMENTS:

Sokolov, N.M.; Pochtakova, E.I, **Zh. Obshch. Khim.** 1958, **28**, 1397-1404.

VARIABLES:

Temperature.

PREPARED BY:

Baldini, P.

EXPERIMENTAL VALUES:

t/°C	T/K[a]	$100x_1$	t/°C	T/K[a]	$100x_1$
298	571	0	317	590	50
292	565	5	318	591	55
290	563	7.5	319	592	60
288	561	8	312	585	66
294	567	10	313	586	67.5
303	576	15	317	590	70
307	580	20	324	597	75
310	583	25	340	613	85
312	585	30	348	621	90
315	588	35	358	631	95
316	589	40	365	638	100
317	590	45			

[a] **T**/K values calculated by the compiler.

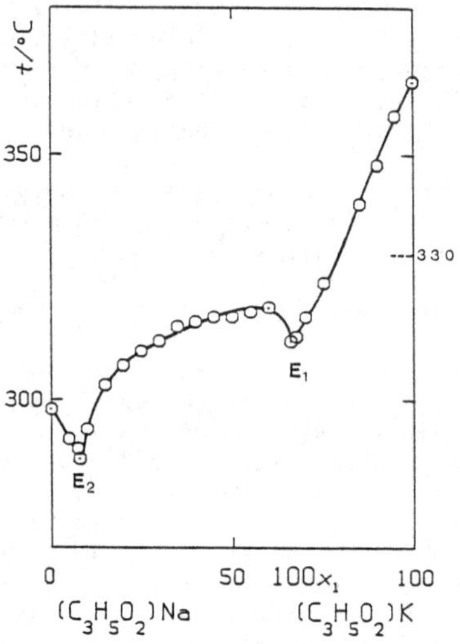

Characteristic point(s):

Eutectic, E_1, at 312 °C and $100x_1$= 66 (authors).
Eutectic, E_2, at 288 °C and $100x_1$= 8 (authors).

Intermediate compound(s):

$(C_3H_5O_2)_5K_3Na_2$, congruently melting at 319 °C.

AUXILIARY INFORMATION

METHOD/APPARATUS/PROCEDURE:

Visual polythermal analysis.

SOURCE AND PURITY OF MATERIALS:

Both components were prepared from commercial propanoic acid (distilled before use) and the proper "chemically pure" carbonate; the solids recovered were recrystallized from butanol. Component 1 undergoes a phase transition at $t_{trs}(1)$/°C= 330 (Ref. 1). Component 2 undergoes phase transitions at $t_{trs}(2)$/°C= 77, 195, 217, 287 (Ref. 1).

ESTIMATED ERROR:

Temperature: accuracy probably ±2 K (compiler).

REFERENCES:

(1) Sokolov, N.M.; **Tezisy Dokl. X Nauch. Konf. S.M.I.** 1956.

COMPONENTS:	ORIGINAL MEASUREMENTS:
(1) Potassium propanoate (potassium propionate); $(C_3H_5O_2)K$; [327-62-8] (2) Sodium propanoate (sodium propionate); $(C_3H_5O_2)Na$; [137-40-6]	Dmitrevskaya, O.I.; Sokolov, N.M. **Zh. Obshch. Khim.** 1958, **28**, 2920-2926 (*); **Russ. J. Gen. Chem. (Engl. Transl.)** 1958, **28**, 2949-2954.
VARIABLES:	PREPARED BY:
Temperature.	Baldini, P.

EXPERIMENTAL VALUES:

$t/^oC$	T/K^a	$100x_1$	$t/^oC$	T/K^a	$100x_1$
298	571	0	318	591	55
292	565	5	319	592	60
287	560	8	314	587	65
294	567	10	310	583	66
303	576	15	316	589	70
307	580	20	322	595	75
310	583	30	340	613	85
315	588	35	351	624	90
316	589	40	358	631	95
317	590	45	365	638	100
317	590	50			

[a] T/K values calculated by the compiler.

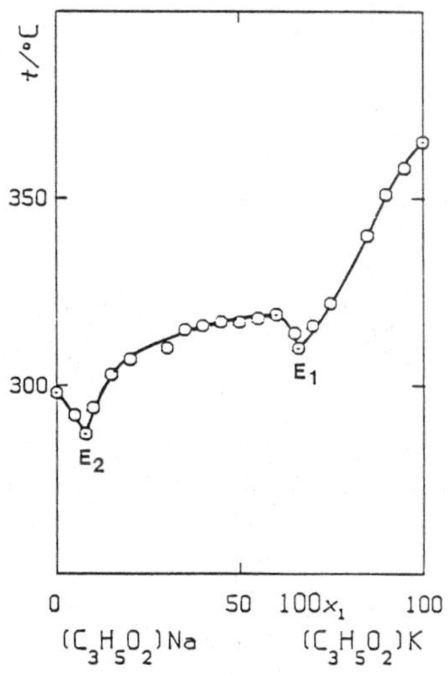

Characteristic point(s):

Eutectic, E_1, at 310 ^oC and $100x_1$= 66 (authors).
Eutectic, E_2, at 287 ^oC and $100x_1$= 8 (authors).

Intermediate compound(s):

$(C_3H_5O_2)_5K_3Na_2$ (probable composition), congruently melting at 319 ^oC (authors).

AUXILIARY INFORMATION

METHOD/APPARATUS/PROCEDURE:

Visual polythermal analysis. Temperature of initial crystallization measured with a Nichrome-Constantane thermocouple checked at the boiling point of water, and at the fusion points of benzoic acid, mannitol, succinic acid, silver nitrate, tin, potassium nitrate, and potassium dichromate. Mixtures melted in a glass tube inserted into a wider tube to ensure uniform heating. Glass fiber stirrer used.

SOURCE AND PURITY OF MATERIALS:

Components prepared by adding a small excess of distilled commercial propanoic acid to a solution of the proper "chemically pure" hydrogen carbonate; the solids recovered after evaporation of the solvent were recrystallized from butanol. Component 1 undergoes a phase transition at $t_{trs}(1)/^oC$= 68 (Ref. 1).
Component 2 undergoes phase transitions at $t_{trs}(2)/^oC$= 77, 195, 217, 287 (Ref. 1).

ESTIMATED ERROR:

Temperature: accuracy probably ± 2 K (compiler).

REFERENCES:

(1) Sokolov, N.M.; **Tezisy Dokl. X Nauch. Konf. S.M.I.** 1956.

COMPONENTS:	ORIGINAL MEASUREMENTS:
(1) Lithium propanoate (lithium propionate); $(C_3H_5O_2)Li$; [6531-45-9] (2) Sodium propanoate (sodium propionate); $(C_3H_5O_2)Na$; [137-40-6]	Tsindrik, N.M.; Sokolov, N.M. **Zh. Obshch. Khim.**, 1958, **28**, 1404-1410 (*); **Russ. J. Gen. Chem. (Engl. Transl.)** 1958, **28**, 1462-1467.
VARIABLES:	PREPARED BY:
Temperature.	Baldini, P.

EXPERIMENTAL VALUES:

$t/^oC$	T/K^a	$100\mathbf{x}_2$
329	602	0
322	595	5
316	589	10
304	577	15
292	565	20
278	551	25
262	535	30
246	519	35
226	499	40
218	491	42.5
206	479	45
196	469	47.5
194	467	48
198	471	50
204	477	52.5
212	485	55
216	489	57.5
224	497	60
248	521	70
266	539	80
282	555	90
298	571	100

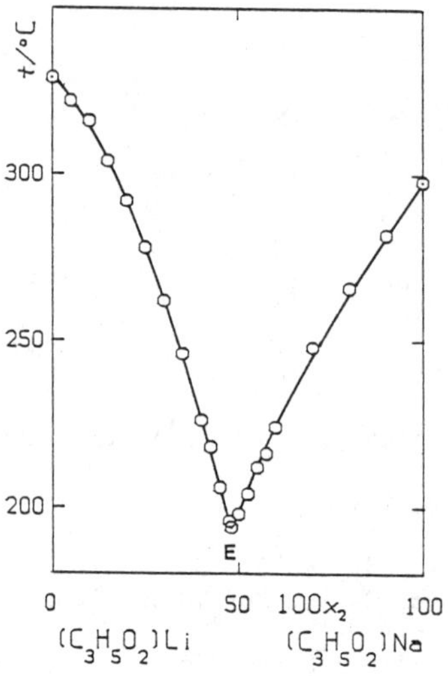

[a] **T**/K values calculated by the compiler.

Characteristic point(s): Eutectic, E, at 194 ^oC and $100\mathbf{x}_2$= 48 (authors).

AUXILIARY INFORMATION

METHOD/APPARATUS/PROCEDURE:

Visual polythermal analysis.

NOTE:

The fusion temperature of component 2 (571 K) is to be considered as too high, inasmuch as the DSC value (562.4$\pm$0.2 K) given in Preface, Table 1 was subsequently confirmed by that obtained with adiabatic calorimetry (561.88$\pm$0.03 K; Preface, Table 3). For the same component, both DSC and adiabatic calorimetry proved (Preface, Table 1 and Table 3, respectively) the occurrence of only two (instead of four, as quoted by the authors from Ref. 2) solid state transitions. Nevertheless, the main features of the diagram are to be looked at with sufficient confidence.

SOURCE AND PURITY OF MATERIALS:

Components prepared from propanoic acid and the proper hydrogen carbonate (Ref. 1), and recrystallized from **n**-butanol. Component 2 undergoes phase transitions at $t_{trs}(2)/^oC$= 77, 195, 217, 287 (Ref. 2).

ESTIMATED ERROR:

Temperature: accuracy probably $\pm$2 K (compiler).

REFERENCES:

(1) Sokolov, N.M.
Zh. Obshch. Khim. 1954, **24**, 1150-1156.
(2) Sokolov, N.M.
Tezisy Dokl. X Nauch. Konf. S.M.I. 1956.

COMPONENTS:	EVALUATOR:
(1) Magnesium propanoate (magnesium propionate); $(C_3H_5O_2)_2Mg$; [557-27-7] (2) Sodium propanoate (sodium propionate); $(C_3H_5O_2)_2Na_2$; [137-40-6]	Franzosini, P., Dipartimento di Chimica Fisica, Universita´ di Pavia (ITALY).

CRITICAL EVALUATION:

This binary was studied by Pochtakova (Ref. 1) both with visual polythermal and DTA investigation. In order to evaluate the trustworthiness of her results, the following points have to be considered.

(i) The fusion temperature of component 1 (577 K) coincides with the DSC value by Ferloni et al. (Ref. 2).

(ii) Pochtakova´s solid state transition temperatures of the same component (i.e., 458, 473, 490, and 519 K) represent the only source of information on this subject.

(iii) The fusion temperature of component 2 (571 K) has to be considered as too high, inasmuch as the DSC value (562.4±0.2 K) given in Preface, Table 1 was subsequently confirmed by that obtained with adiabatic calorimetry (561.88±0.03 K; Preface, Table 3).

(iv) As for the solid state transitions of the same component quoted by Pochtakova from Ref. 3 as occurring at $T_{trs}(2)/K$= 350, 468, 490, and 560, heavy doubts are to be cast about the existence of the lowest and highest ones inasmuch as DSC provided evidence for only two solid state transformations (at 470.2±0.5, and 494±1 K, respectively; Preface, Table 1) which was subsequently confirmed with adiabatic calorimetry (Preface, Table 3).

(v) Indeed, the DTA traces recorded at $100x_1$= 2.5, 4, 25, and 42.5 seem to be consistent with the existence of only two solid state transitions of component 2; moreover, they support the occurrence of eutectic E_2, and tend to prove the absence of solid solutions between component 2 and the intermediate compound.

(vi) The DTA traces recorded at $100x_1$= 60, 65, and 75 are somewhat embarrassing because all of them support the occurrence of eutectic E_1, but evidence for solid state transitions of component 1 is offered only by the trace taken at $100x_1$= 60 for what concerns the transition at 473 K, and by that taken at $100x_1$= 65 for what concerns the transition at 458 K.

(vii) No explanation is given by the author for the discontinuities exhibited at temperatures far above the liquidus by the DTA traces taken at $100x_1$= 60, and 65.

In conclusion, the evaluator is inclined to consider as satisfactorily supported by the experimental evidence:

(i) the occurrence of the congruently melting intermediate compound $(C_3H_5O_2)_4MgNa_2$;

(ii) the occurrence of eutectics E_1 and E_2, located as suggested by Pochtakova; and

(iii) the phase relations relevant to solidus and subsolidus at $0 \leq 100x_1 \leq 50$ as suggested by Pochtakova.

On the contrary, the knees occurring in the liquidus branch richest in component 1 as well as in that richest in component 2, the nature of possible transformations occurring in the melt, and the phase relations relevant to solidus and subsolidus at $50 \leq 100x_1 \leq 100$ seem to need further investigation.

REFERENCES:

(1) Pochtakova, E.I.
Zh. Obshch. Khim. 1974, **44**, 241-248.

(2) Ferloni, P.; Sanesi, M.; Franzosini, P.
Z. Naturforsch. 1976, **31a**, 679-682.

(3) Sokolov, N.M.
Tezisy Dokl. X Nauch. Konf. S.M.I. 1956.

COMPONENTS:

(1) Magnesium propanoate (magnesium propionate); $(C_3H_5O_2)_2Mg$; [557-27-7]
(2) Sodium propanoate (sodium propionate); $(C_3H_5O_2)_2Na_2$; [137-40-6]

ORIGINAL MEASUREMENTS:

Pochtakova, E.I.
Zh. Obshch. Khim. 1974, **44**, 241-248.

VARIABLES:

Temperature.

PREPARED BY:

Baldini, P.

EXPERIMENTAL VALUES:

t/°C	T/K[a]	$100x_1$	t/°C	T/K[a]	$100x_1$
298	571	0	247	520	47.5
288	561	2.5	248	521	50
288[bc]	561	2.5	244[bc]	517	50
196[bh]	469	2.5	245	518	52.5
214[bj]	487	2.5	243	516	55
288[bk]	561	2.5	241	514	60
288[bc]	561	4	240[bc]	513	60
216[bj]	489	4	235[be]	508	60
287	560	5	200[bi]	473	60
284	557	10	286[bl]	559	60
281	554	15	246	519	62.5
274	547	20	239	512	65
270	543	25	235[bc]	508	65
270[bc]	543	25	235[be]	508	65
240[bd]	513	25	185[bg]	458	65
196[bh]	469	25	284[bl]	557	65
264	537	30	244	517	67.5
258	531	35	245	518	70
252	525	37.5	249	522	75
247	520	40	250[bc]	523	75
244	517	42.5	236[be]	509	75
242[bc]	515	42.5	118[bf]	391	75
242[bd]	515	42.5	253	526	80
193[bh]	466	42.5	285	558	90
246	519	45	304	577	100

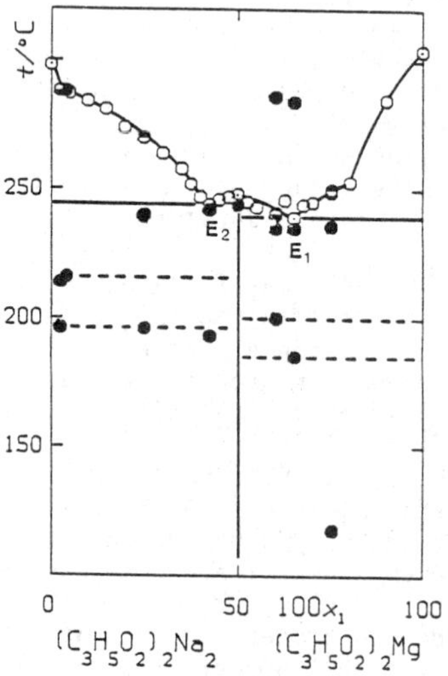

[a] T/K values calculated by the compiler.
[b] Differential thermal analysis.
[c] Initial crystallization.
[d] First eutectic stop.
[e] Second eutectic stop.
[f] First transition of the system.
[g] Second transition of the system.
[h] Third transition of the system.
[i] Fourth transition of the system.
[j] Fifth transition of the system.
[k] Sixth transition of the system.
[l] Seventh transition of the system (no explanation is offered by the author for the occurrence of this point above the liquidus, compiler).

Characteristic point(s):

Eutectic, E_1, at 239 °C (235 °C by D.T.A.), and $100x_1$ = 65 (author).
Eutectic, E_2, at 244 °C (242 °C by D.T.A.), and $100x_1$ = 42.5 (author).

Intermediate compound: $(C_3H_5O_2)_4MgNa_2$, congruently melting at 248 °C (244 °C by D.T.A.).

METHOD/APPARATUS/PROCEDURE:

Visual polythermal analysis, supplemented with differential thermal analysis.

SOURCE AND PURITY OF MATERIALS:

Materials prepared (Ref. 1) by reacting the proper ("chemically pure") carbonate with a slight excess of propanoic acid of analytical purity.
Component 1 undergoes phase transitions at $t_{trs}(1)$/°C = 185, 200, 217, 246.
Component 2 undergoes phase transitions at $t_{trs}(2)$/°C = 77, 195, 217, 287 (Ref. 2).

ESTIMATED ERROR:

Temperature: precision probably ±2 K (compiler).

REFERENCES:

(1) Sokolov, N.M.
Zh. Obshch. Khim. 1954, **24**, 1581-1593.
(2) Sokolov, N.M.
Tezisy Dokl. X Nauch. Konf.S.M.I. 1956.

COMPONENTS:	EVALUATOR:
(1) Potassium butanoate (potassium butyrate); $(C_4H_7O_2)_2K_2$; [589-39-9] (2) Magnesium butanoate (magnesium butyrate); $(C_4H_7O_2)_2Mg$; [556-45-6]	Franzosini, P., Dipartimento di Chimica Fisica, Universita' di Pavia (ITALY).

CRITICAL EVALUATION:

This binary was studied only by Pochtakova (Ref. 1) who, on the basis of her visual polythermal and DTA results, asserted the occurrence of a congruently melting intermediate compound, i.e., $(C_4H_7O_2)_4K_2Mg$, forming (possibly simple) eutectics with either component.

Component 1, however, goes through the liquid crystalline state before transformation into a clear melt. Therefore, the topology of the phase diagram at $0 \leq 100x_2 \leq 50$ should be described more correctly with reference to Scheme B.1 of the Preface, and an invariant type M'_p (undetected by Pochtakova) should also exist.

The following points are still worth mentioning.

(i) Pochtakova's fusion temperature of component 1 (677 K) coincides with the clearing temperature (677.3$\pm$0.5 K) listed in Preface, Table 1 for the same component, whereas her $T_{fus}(2)$ value (575 K) is noticeably higher than data by other authors reported in Ref. 2.

(ii) Among the phase transformation temperatures of component 1 quoted in Ref. 1 from Ref. 3 (i.e., 618, 553-558, and 463 K) the first one can be reasonably identified with the fusion temperature (626.1$\pm$0.7 K) listed in Preface, Table 1, whereas the second and third ones lie each halfway between the two pairs of solid state transition temperatures (i.e., 562.2$\pm$0.6 and 540.8$\pm$1.1, and 467.2$\pm$0.5 and 461.4$\pm$1.0, respectively) also reported in Table 1 of the Preface.

(iii) No explanation is given by the author for the discontinuities observed at temperatures (643 and 624 K, respectively) far above the liquidus in the DTA traces taken at $100x_2$= 25 and 50.

(iv) The author's explanation, that the discontinuities observed at temperatures corresponding to the lowest section of the subsolidus might be due to transformation (at about 445 K) of the intermediate compound into a metastable phase turning to stable at 370-400 K, should be more detailed and better supported.

In conclusion it seems to the evaluator that the composition of the intermediate compound, the location of both eutectics, the liquidus dome, and the liquidus branch richest in component 2 are sufficiently well assessed, whereas the remaining part of the diagram needs several refinements to become satisfactory.

REFERENCES:

(1) Pochtakova, E.I.
Zh. Obshch. Khim. 1974, **44**, 241-248.

(2) Sanesi, M.; Cingolani, A.; Tonelli, P.L.; Franzosini, P.
Thermal Properties, in **Thermodynamic and Transport Properties of Organic Salts,** IUPAC Chemical Data Series No. 28 (Franzosini, P.; Sanesi, M.; Editors), Pergamon Press, Oxford, 1980, 29-115.

(3) Sokolov, N.M.
Tezisy Dokl. X Nauch. Konf. S.M.I. 1956.

COMPONENTS:	ORIGINAL MEASUREMENTS:
(1) Potassium butanoate (potassium butyrate); $(C_4H_7O_2)_2K_2$; [589-39-9] (2) Magnesium butanoate (magnesium butyrate); $(C_4H_7O_2)_2Mg$; [556-45-6]	Pochtakova, E.I. **Zh. Obshch. Khim.** 1974, **44**, 241-248.
VARIABLES:	PREPARED BY:
Temperature.	Baldini, P.

EXPERIMENTAL VALUES:

t/°C	T/K[a]	$100x_2$	t/°C	T/K[a]	$100x_2$
404	677	0	318[bc]	591	50
389	662	5	170[bh]	443	50
392[bc]	665	5	354[bl]	627	50
294[bd]	567	5	316	589	55
196[bi]	469	5	309	582	60
272[bj]	545	5	297	570	65
352[bk]	625	5	306[bc]	579	65
390[bc]	663	9	238[be]	511	65
305[bd]	578	9	164[bh]	445	65
172[bh]	445	9	283	556	70
196[bi]	469	9	272	545	72.5
277[bj]	550	9	262	535	75
349[bk]	622	9	252	525	77.5
376	649	10	248	521	80
368	641	15	232[bc]	505	81.5
359	632	17.5	232[be]	505	81.5
361	634	20	122[bf]	395	81.5
348	621	25	172[bh]	445	81.5
342[bc]	575	25	245	518	82.5
294[bd]	567	25	252	525	85
172[bh]	445	25	268	541	90
194[bi]	467	25	273[bc]	546	90
370[bl]	643	25	232[be]	505	90
331	604	30	100[bf]	373	90
322	595	32.5	136[bg]	409	90
307	580	35	174[bh]	447	90
302[bc]	575	36	288	561	95
302[bd]	575	36	284[bc]	557	95
173[bh]	446	36	230[be]	503	95
304	577	37.5	105[bf]	378	95
305	578	40	168[bh]	441	95
313	586	45	302	575	100
317	590	50			

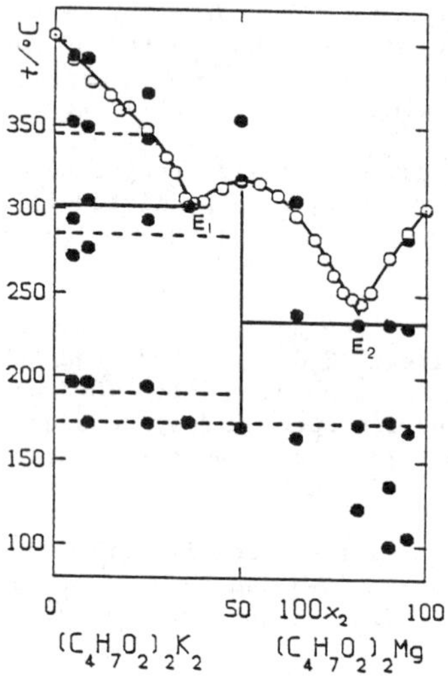

a T/K values calculated by the compiler.
b Differential thermal analysis.
c Initial crystallization.
d First eutectic stop.
e Second eutectic stop.
f First transition of the system.
g Second transition of the system.
h Third transition of the system.
i Fourth transition of the system.
j Fifth transition of the system.
k Sixth transition of the system.
l Seventh transition of the system (no explanation if offered by the author for the occurrence of this point above the liquidus, compiler).

COMPONENTS:

(1) Potassium butanoate (potassium butyrate); $(C_4H_7O_2)_2K_2$; [589-39-9]
(2) Magnesium butanoate (magnesium butyrate); $(C_4H_7O_2)_2Mg$; [556-45-6]

ORIGINAL MEASUREMENTS:

Pochtakova, E.I.
Zh. Obshch. Khim. 1974, **44**, 241-248.

VARIABLES:

Temperature.

PREPARED BY:

Baldini, P.

EXPERIMENTAL VALUES: (continued)

Characteristic point(s):

Eutectic, E_1, at 300 °C (302 °C by D.T.A.), and $100x_2$= 36.0 (author).
Eutectic, E_2, at 235 °C (232 °C by D.T.A.), and $100x_2$= 81.5 (author).

Intermediate compound(s):

$(C_7H_4O_2)_4K_2Mg$, congruently melting at 318 °C.

AUXILIARY INFORMATION

METHOD/APPARATUS/PROCEDURE:

Visual polythermal analysis, supplemented with differential thermal analysis.

SOURCE AND PURITY OF MATERIALS:

Materials prepared (Ref. 1) by reacting the proper ("chemically pure") carbonate with a slight excess of **n**-butanoic acid of analytical purity. Component 1 undergoes phase transitions at $t_{trs}(1)/°C$ = 190, 280-285, 345 (Ref. 2).

ESTIMATED ERROR:

Temperature: accuracy probably ±2 K (compiler).

REFERENCES:

(1) Sokolov, N.M.
Zh. Obshch. Khim. 1954, **24**, 1581-1593.
(2) Sokolov, N.M.
Tezisy Dokl. X Nauch. Konf. S.M.I. 1956.

COMPONENTS:

(1) Potassium butanoate (potassium butyrate); $(C_4H_7O_2)K$; [589-39-9]
(2) Sodium butanoate (sodium butyrate); $(C_4H_7O_2)Na$; [156-54-7]

EVALUATOR:

Franzosini, P.,
Dipartimento di Chimica Fisica
Universita´ di Pavia (ITALY).

CRITICAL EVALUATION:

The visual polythermal method was employed by Sokolov and Pochtakova (Ref. 1), and by Dmitrevskaya (Ref. 2) to study the lower boundary of the isotropic liquid field: according to these authors, continuous series of solid solutions ought to exist.

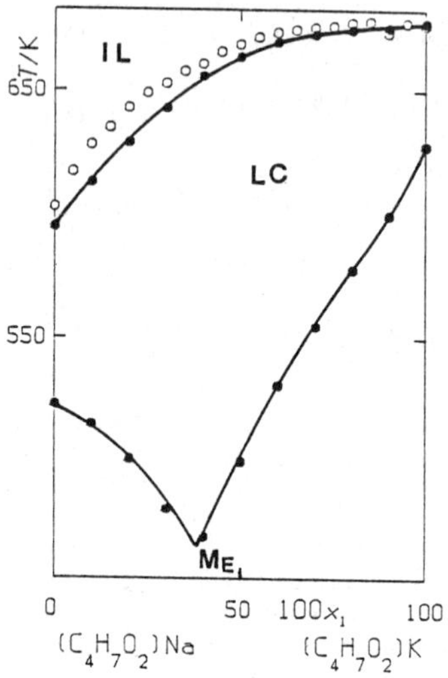

Both components, however, form liquid crystals. Consequently: (i) the fusion temperatures, $T_{fus}(1)$= 677 K (404 °C) and T_{fus} (2)= 603 K (330 °C), reported in Refs. 1, 2 should be identified with the clearing temperatures; and (ii) a continuous series of liquid crystal (and not of solid) solutions should be expected.

More recently, Prisyazhnyi et al. (Ref. 3) - to whom Refs. 1, 2 seem to be unknown - carried out a derivatographical re-investigation of the system, which allowed them to draw the lower boundaries of both the isotropic liquid, and the liquid crystal field. Their clearing [678 K (405 °C); 595 K (322 °C)] and fusion [628 K (355 °C); 523 K (250 °C)] temperatures substantially agree with the corresponding values from Preface, Table 1 (677.3$\pm$0.5, 600.4$\pm$0.2, and 626.1$\pm$0.7, 524.5$\pm$0.5 K, respectively).

Prisyazhnyi et al.´s, and Dmitrevskaya´s results (filled and empty circles, respectively) are compared in the figure (IL: isotropic liquid; LC: liquid crystals). The complete phase diagram ought to be similar to that reported in Scheme C.1, and the only invariant ought to be classified as an M_E point, at which equilibrium occurs among one liquid crystalline and two solid phases. The statements made in Refs. 1, 2 cannot be considered as correct, whereas Prisyazhnyi et al.´s measurements look as compatible with expectation.

The latter measurements can be further commented as follows: (i) the two-phase region pertinent to the liquid crystal - isotropic liquid equilibria might be so narrow as to prevent observation of two distinct sets of points in this region; (ii) the lack of information about eutectic fusion in the different samples submitted to derivatographical analysis remains, however, rather surprising.

(continued in the next page)

COMPONENTS:	EVALUATOR:
(1) Potassium butanoate (potassium butyrate); $(C_4H_7O_2)K$; [589-39-9] (2) Sodium butanoate (sodium butyrate); $(C_4H_7O_2)Na$; [156-54-7]	Franzosini, P., Dipartimento di Chimica Fisica Universita' di Pavia (ITALY).

CRITICAL EVALUATION: (continued)

Finally, the following two points deserve attention.

(i) Among the phase transformation temperatures of component 1 quoted in Refs. 1, 2 from Ref. 4 (i.e., 618, 553-558, and 463 K) the first one can be reasonably identified with the fusion temperature (626.1±0.7 K) listed in Preface, Table 1, whereas the second and third ones lie each halfway between the two pairs of solid state transition temperatures (i.e., 562.2±0.6 and 540.8±1.1, and 467.2±0.5 and 461.4±1.0, respectively) also reported in Table 1 of the Preface.

(ii) For component 2, Table 1 of the Preface [besides the clearing temperature] provides solid state transitions at 450.4±0.5, 489.8±0.2, 498.3±0.3, and 508.4±0.5, and fusion at 524.5±0.5. It is to be stressed that these phase relations, first stated on the basis of DSC records, were subsequently confirmed by Schiraldi and Chiodelli's conductometric results (Ref. 5). On the other hand, phase transformations are quoted in Refs. 1, 2 from Ref. 4 as occurring at 390, 505, 525, and 589 K, respectively. A comparison of the two sets of data allows one to identify conveniently the two intermediate transition temperatures from Ref. 4 with the first transition temperature and the fusion temperature from Table 1, whereas reasonable doubts can be cast about the actual existence of the highest and lowest transformations quoted in Refs. 1, 2.

REFERENCES:

(1) Sokolov, N.M.; Pochtakova, E.I.
Zh. Obshch. Khim. 1958, **28**, 1693-1700 (*); **Russ. J. Gen. Chem.(Engl. Transl.)** 1958, **28**, 1741-1747.

(2) Dmitrevskaya, O.I.
Zh. Obshch. Khim. 1958, **28**, 2007-2013 (*); **Russ. J. Gen. Chem. (Engl. Transl.)** 1958, **28**, 2046-2051.

(3) Prisyazhnyi, V.D.; Mirnyi, V.N.; Mirnaya, T.A.
Ukr. Khim. Zh. 1983, **49**, 659-660.

(4) Sokolov, N.M.
Tezisy Dokl. X Nauch. Konf. S.M.I. 1956.

(5) Schiraldi, A.; Chiodelli, G.
J. Phys. E: Sci. Instr. 1977, **10**, 596-599.

COMPONENTS:

(1) Potassium butanoate (potassium butyrate); $(C_4H_7O_2)K$; [589-39-9]
(2) Sodium butanoate (sodium butyrate); $(C_4H_7O_2)Na$; [156-54-7]

ORIGINAL MEASUREMENTS:

Sokolov, N.M.; Pochtakova, E.I.
Zh. Obshch. Khim. 1958, **28**, 1693-1700 (*);
Russ. J. Gen. Chem. (Engl. Transl.) 1958, **28**, 1741-1747.

VARIABLES:

Temperature.

PREPARED BY:

Baldini, P.

EXPERIMENTAL VALUES:

t/°C	T/K[a]	$100x_1$
330	603	0
339	612	5
348	621	10
356	629	15
364	637	20
370	643	25
375	648	30
380	653	35
385	658	40
389	662	45
393	666	50
396	669	55
399	672	60
402	675	65
405	678	70
406	679	75
406	679	80
405	678	85
405	678	90
404	677	95
404	677	100

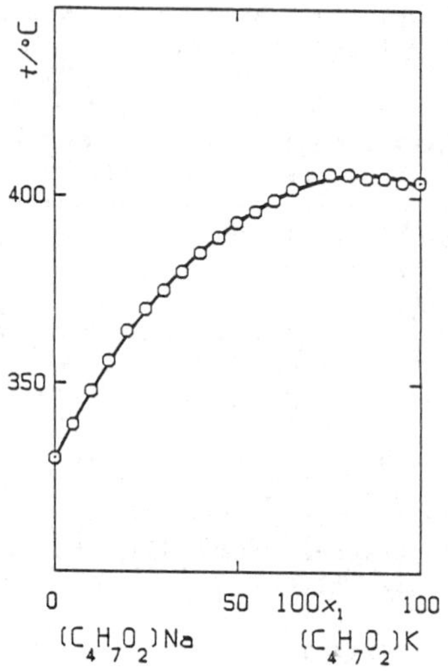

[a] T/K values calculated by the compiler.

Characteristic point(s): Continuous series of solid solutions (author).

AUXILIARY INFORMATION

METHOD/APPARATUS/PROCEDURE:

Visual polythermal analysis.
Temperatures measured with a Nichrome-Constantane thermocouple.

SOURCE AND PURITY OF MATERIALS:

Components synthetized from "chemically pure" potassium and sodium hydrogen carbonates, and **n**-butanoic acid (Ref. 2, where, however, carbonates instead of hydrogen carbonates are employed; compiler); the salts obtained were recrystallized from **n**-butanol. Component 1 undergoes phase transitions at $t_{trs}(1)$/°C= 190, 280-285, 345 (Ref. 2). Component 2 undergoes phase transitions at $t_{trs}(2)$/°C= 117, 232, 252, 316 (Ref. 2).

ESTIMATED ERROR:

Temperature: accuracy probably ±2 K (compiler).

REFERENCES:

(1) Sokolov, N.M.
Zh. Obshch. Khim. 1954, **24**, 1581-1593.
(2) Sokolov, N.M.
Tezisy Dokl. X Nauch. Konf. S.M.I. 1956.

COMPONENTS:

(1) Potassium butanoate (potassium butyrate); $(C_4H_7O_2)K$; [589-39-9]
(2) Sodium butanoate (sodium butyrate); $(C_4H_7O_2)Na$; [156-54-7]

ORIGINAL MEASUREMENTS:

Dmitrevskaya, O.I.
Zh. Obshch. Khim. 1958, 28, 2007-2013 (*);
Russ. J. Gen. Chem. (Engl. Transl.) 1958, **28**, 2046-2051.

VARIABLES:

Temperature.

PREPARED BY:

Baldini, P.

EXPERIMENTAL VALUES:

$t/^{\circ}C$	T/K^a	$100x_1$
330	603	0
344	617	5
355	628	10
362	635	15
370	643	20
376	649	25
380	653	30
385	658	35
388	661	40
393	666	45
396	669	50
399	672	55
401	674	60
402	675	65
403	676	70
403.5	676.5	75
405	678	80
405.5	678.5	85
401	674	90
405	678	95
404	677	100

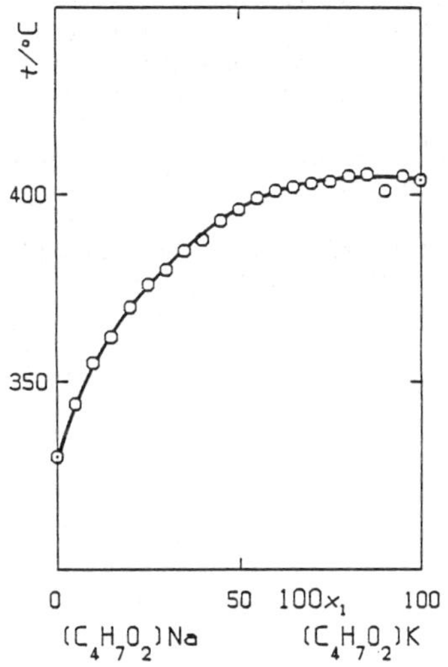

[a] T/K values calculated by the compiler.

Characteristic point(s): Continuous series of solid solutions (author).

AUXILIARY INFORMATION

METHOD/APPARATUS/PROCEDURE:

Visual polythermal analysis.
Temperatures measured with a Nichrome-Constantane thermocouple.

SOURCE AND PURITY OF MATERIALS:

Components synthetized from "chemically pure" potassium and sodium hydrogen carbonates, and **n**-butanoic acid twice distilled.
Component 1 undergoes phase transitions at $t_{trs}(1)/^{\circ}C$= 190, 280-285, 345 (Ref. 1).
Component 2 undergoes phase transitions at $t_{trs}(2)/^{\circ}C$= 117, 232, 252, 316 (Ref. 1).

ESTIMATED ERROR:

Temperature: accuracy probably ± 2 K (compiler).

REFERENCES:

(1) Sokolov, N.M.
Tezisy Dokl. X Nauch. Konf. S.M.I. 1956.

COMPONENTS:

(1) Potassium butanoate (potassium butyrate); $(C_4H_7O_2)K$; [589-39-9]
(2) Sodium butanoate (sodium butyrate); $(C_4H_7O_2)Na$; [156-54-7]

ORIGINAL MEASUREMENTS:

Prisyazhnyi, V.D.; Mirnyi, V.N.; Mirnaya, T.A.
Ukr. Khim. Zh. 1983, **49**, 659-660.

VARIABLES:

Temperature.

PREPARED BY:

Baldini, P.

EXPERIMENTAL VALUES:

The results are reported only in graphical form (see figure; data read with a digitizer by the compiler on Fig. 1 of the original paper; empty circles: liquid crystal - isotropic liquid equilibria; filled circles: solid - liquid crystal equilibria).

Characteristic point(s): Eutectic, E, at 194 °C and $100x_1$= 38 (authors).

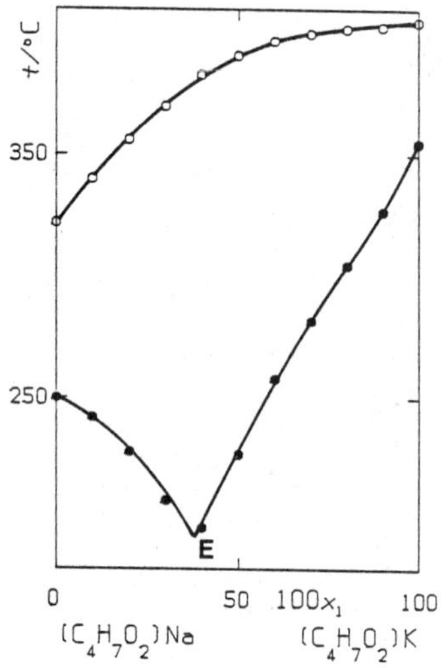

AUXILIARY INFORMATION

METHOD/APPARATUS/PROCEDURE:

The heating and cooling traces were recorded in an atmosphere of purified argon with an OD-102 derivatograph (MOM, Hungary) working at a rate of 6 K min^{-1}, and using Al_2O_3 as reference material. Temperatures were measured with a Pt/Pt-Rh thermocouple. A hot-stage Amplival polarizing microscope was employed to detect the transformation points from the liquid crystalline into the isotropic liquid phase. Supplementary information was obtained by conductometry.

SOURCE AND PURITY OF MATERIALS:

Not stated.
Component 1: $t_{fus}(1)/^oC$ about 355; $t_{clr}(1)/^oC$ about 405 (compiler).
Component 2: $t_{fus}(2)/^oC$ about 250; $t_{clr}(2)/^oC$ about 322 (compiler).

ESTIMATED ERROR:

Temperature: accuracy not evaluable (compiler).

COMPONENTS:	EVALUATOR:
(1) Lithium butanoate (lithium butyrate); $(C_4H_7O_2)Li$; [21303-03-7] (2) Sodium butanoate (sodium butyrate); $(C_4H_7O_2)Na$; [156-54-7]	Schiraldi, A., Dipartimento di Chimica Fisica Universita´ di Pavia (ITALY).

CRITICAL EVALUATION:

The visual polythermal analysis was employed by Tsindrik and Sokolov (Ref. 1) to study the lower boundary of the isotropic liquid field: according to these authors, a eutectic ought to exist at 495 K (222 °C), and $100\mathbf{x}_2$= 50.

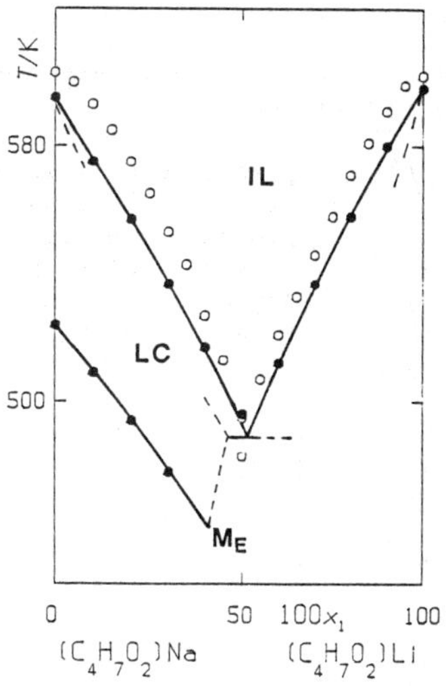

Component 2, however, forms liquid crystals. Consequently: (i) the fusion temperature, 603 K (330 °C) reported in Ref. 1 should be identified with the clearing temperature; (ii) the two branches of the curve refer to equilibria of different kind; and (iii) the intersection of the two branches cannot be classified as a eutectic.

More recently, Prisyazhnyi et al. (Ref. 2) - to whom Ref. 1 seems to be unknown - carried out a derivatographical re-investigation of the system, which allowed them to draw the lower boundaries of both the isotropic liquid, and the liquid crystal field. Their clearing [$\mathbf{T}_{clr}(2)$= 595 K (322 °C)] and fusion [$\mathbf{T}_{fus}(1)$= 598 K (325 °C); $T_{fus}(2)$= 524 K (251 °C)] temperatures substantiallly agree with the corresponding values from Table 1 of the Preface (600.4±0.2; 591.7±0.5, and 524.5±0.5 K, respectively).

Prisyazhnyi et al.´s, and Tsindrik and Sokolov´s results (filled and empty circles, respectively) are compared in the figure (IL: isotropic liquid; LC: liquid crystals). Assuming that limited solid solutions are present, the complete phase diagram ought to be similar to that reported in Preface, Scheme A.1. The upper invariant ought to be classified as an M'_E point, and the lower one as an M_E point.

Prisyazhnyi et al.´s measurements look as compatible with expectation, although the lack of information about eutectic fusion in the different samples studied by derivatographical analysis remains rather surprising. Instead, the narrowness of the two-phase region pertinent to the liquid crystal - isotropic liquid equilibria could have prevented the observation of two distinct sets of points in this region.

Finally, the following point requires attention. For component 2, Table 1 of the Preface [besides the $\mathbf{T}_{clr}(2)$ value] provides four solid state transitions at 450.4±0.5, 489.8±0.2, 498.3±0.3, and 508.4±0.5 K, and $\mathbf{T}_{fus}(2)$/K= 524.5±0.5. It is to be stressed that these phase relations, first stated on the basis of DSC records, were subsequently confirmed by Schiraldi and Chiodelli´s conductometric results (Ref. 3). On the other hand, phase transformations are quoted in Ref. 1 from Ref. 4 as occurring at 390, 505, 525, and 589 K, respectively. A comparison of the two sets of data allows one to identify conveniently the two intermediate transition temperatures from Ref. 4 with the first solid state transition and fusion temperatures from Table 1 of the Preface, whereas reasonable doubts can be cast about the actual existence of the highest and lowest transformations quoted in Ref. 1.

REFERENCES:

(1) Tsindrik, N.M.; Sokolov, N.M.
Zh. Obshch. Khim. 1958, **28**, 1728-1733 (*); **Russ. J. Gen. Chem. (Engl. Transl.)** 1958, **28**, 1775-1780.
(2) Prisyazhnyi, V.D.; Mirnyi, V.N.; Mirnaya, T.A.
Ukr. Khim. Zh. 1983, **49**, 659-660.
(3) Schiraldi, A.; Chiodelli, G.
J. Phys. E: Sci. Instr. 1977, **10**, 596-599.
(4) Sokolov, N.M.
Tezisy Dokl. X Nauch. Konf. S.M.I. 1956.

COMPONENTS:	ORIGINAL MEASUREMENTS:
(1) Lithium butanoate (lithium butyrate); $(C_4H_7O_2)Li$; [21303-03-7] (2) Sodium butanoate (sodium butyrate); $(C_4H_7O_2)Na$; [156-54-7]	Tsindrik, N.M.; Sokolov, N.M. **Zh. Obshch. Khim.** 1958, **28**, 1728-1733 (*); **Russ. J. Gen. Chem. (Engl. transl.)** 1958, **28**, 1775-1780.
VARIABLES:	PREPARED BY:
Temperature.	Baldini, P.

EXPERIMENTAL VALUES:

$t/^{o}C$	T/K^{a}	$100\mathbf{x}_2$
329	602	0
326	599	5
318	591	10
308	581	15
298	571	20
285	558	25
273	546	30
260	533	35
248	521	40
234	507	45
222	495	50
240	513	55
254	527	60
270	543	65
280	553	70
292	565	75
302	575	80
312	585	85
320	593	90
327	600	95
330	603	100

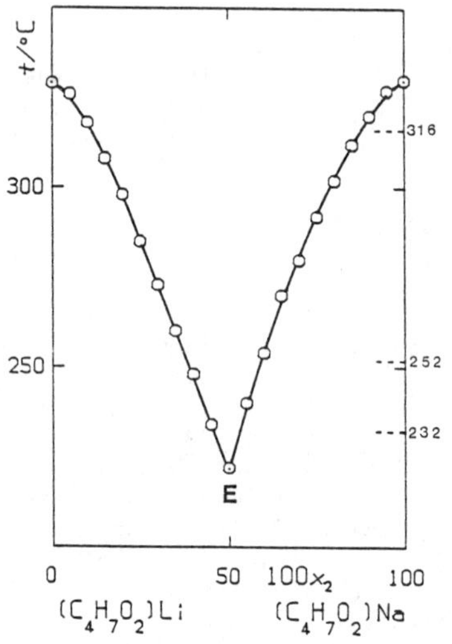

[a] T/K values calculated by the compiler.

Characteristic point(s): Eutectic, E, at 222 °C and $100\mathbf{x}_2$= 50 (authors).

AUXILIARY INFORMATION

METHOD/APPARATUS/PROCEDURE:

Visual polythermal analysis; temperatures of initial crystallization measured with a Nichrome-Constantane thermocouple and a millivoltmeter.

SOURCE AND PURITY OF MATERIALS:

Both components prepared from "chemically pure" carbonates and **n**-butanoic acid (Ref. 1); the solids recovered after evaporation were recrystallized from n-butanol. Component 2 undergoes phase transitions at $t_{trs}(2)/^{o}C$= 117, 232, 252, 316 (Ref. 2).

ESTIMATED ERROR:

Temperature: accuracy probably ±2 K (compiler).

REFERENCES:

(1) Sokolov, N.M. **Zh. Obshch. Khim.** 1954, 24, 1581-1593.
(2) Sokolov, N.M. **Tezisy Dokl.** X **Nauch. Konf. S.M.I.** 1956.

COMPONENTS:	ORIGINAL MEASUREMENTS:
(1) Lithium butanoate (lithium butyrate); $(C_4H_7O_2)Li$; [21303-03-7] (2) Sodium butanoate (sodium butyrate); $(C_4H_7O_2)Na$; [156-54-7]	Prisyazhnyi, V.D.; Mirnyi, V.N.; Mirnaya, T.A. **Ukr. Khim. Zh.** 1983, **49**, 659-660.
VARIABLES:	PREPARED BY:
Temperature.	Baldini, P.

EXPERIMENTAL VALUES:

The results are reported only in graphical form (see figure). Data read with a digitizer by the compiler on Fig. 1 of the original paper; empty circles: liquid crystal - isotropic liquid equilibria; filled circles: solid - liquid crystal or solid - isotropic liquid equilibria).

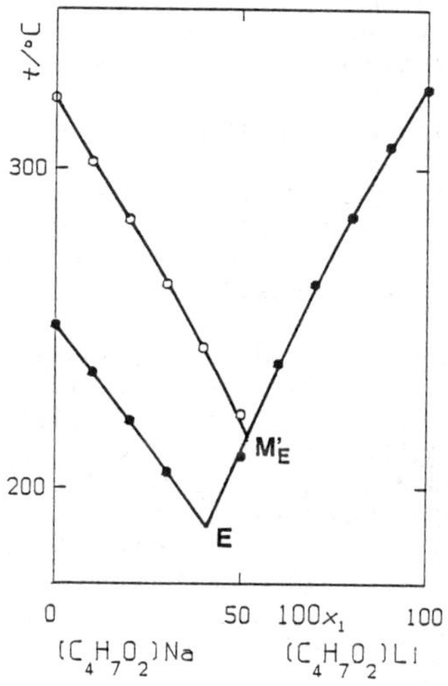

Characteristic point(s):

Eutectic, E, at 188 °C and $100\mathbf{x}_1$ = 41 (authors).
Invariant point, M'_E, at about 215 °C and $100\mathbf{x}_1$ about 52 (compiler).

AUXILIARY INFORMATION

METHOD/APPARATUS/PROCEDURE:	SOURCE AND PURITY OF MATERIALS:
The heating and cooling traces were recorded in an atmosphere of purified argon with an OD-102 derivatograph (MOM, Hungary) working at a rate of 6 K min^{-1}, and using Al_2O_3 as the reference material. Temperatures were measured with a Pt/Pt-Rh thermocouple. A hot-stage Amplival polarizing microscope was employed to detect the transformation points from the liquid crystalline into the isotropic liquid phase. Supplementary information was obtained by conductometry.	Not stated. Component 1: $t_{fus}(1)/^oC$ about 325 (compiler). Component 2: $t_{fus}(2)/^oC$ about 251; $t_{clr}(2)/^oC$ about 322 (compiler). ESTIMATED ERROR: Temperature: accuracy not evaluable (compiler). REFERENCES:

COMPONENTS:	EVALUATOR:
(1) Magnesium butanoate (magnesium butyrate); $(C_4H_7O_2)_2Mg$; [556-45-6] (2) Sodium butanoate (sodium butyrate); $(C_4H_7O_2)_2Na_2$; [156-54-7]	Franzosini, P. Dipartimento di Chimica fisica, Universita´ di Pavia (ITALY)

CRITICAL EVALUATION:

This binary was studied only by Pochtakova (Ref. 1) who (on the basis of visual polythermal and DTA results) claimed the occurrence of the congruently melting intermediate compound $(C_4H_7O_2)_7\ Mg_2Na_3$, able to give eutectics with either component.

Component 2, however, goes through the liquid crystalline state before transformation into a clear melt. Therefore the topology of the phase diagram at $0 \leq 100\mathbf{x}_1 \leq 57$ should be described more correctly with (probable) reference to Preface, Scheme A.1: in this case the invariant ought to be of the M'_E type.

The following points are still worth mentioning.

(i) Pochtakova's fusion temperature of component 1 (575 K) is noticeably higher than data by other authors reported in Ref. 2, whereas her $\mathbf{T}_{fus}(2)$ value (603 K) is in reasonable agreement with the clearing temperature (600.4$\pm$0.2 K) listed in Preface, Table 1 for component 2.

(ii) Again for component 2, Table 1 of the Preface provides four transition temperatures (450.4$\pm$0.5, 489.8$\pm$0.2, 498.3$\pm$0.3, and 508.4$\pm$0.5 K), and $\mathbf{T}_{fus}(2)/K = 524.5\pm0.5$. It is to be stressed that these phase relations, first stated on the basis of DSC records, were subsequently confirmed by Schiraldi and Chiodelli's conductometric results (Ref. 3). On the other hand, phase transformations are quoted in Ref. 1 from Ref. 4 as occurring at 390, 505, 525, and 589 K, respectively. A comparison of the two sets of data allows one to identity conveniently the two intermediate transition temperatures from Ref. 4 with the highest solid state transition and fusion, respectively, from Table 1 of the Preface, whereas reasonable doubts can be cast about the actual existence of the highest and lowest transformations quoted by Pochtakova.

(iii) In the DTA traces taken at $100\mathbf{x}_1 = 10$ and 35, Pochtakova observed discontinuities at 587 and 573 K, and at 528 and 507 K, respectively, which might correspond to the higher (587 and 528 K) and lower (573 and 507 K) boundary of a diphasic region, thus supporting an interpretation of the phase diagram based on Scheme A.1 of the Preface.

(iv) The author's explanation, that the discontinuities observed at temperatures corresponding to the lowest section of the subsolidus might be due to the transformation (at about 435 K) of the intermediate compound into a metastable phase turning to stable at about 410 K, should be more detailed and better supported.

In conclusion, it seems to the evaluator that the existence of an intermediate compound, the location of both eutectics, and the liquidus branch richest in component 1 are sufficiently well assessed, whereas other parts of the diagram need refinements.

REFERENCES:

(1) Pochtakova, E.I.
Zh. Obshch. Khim. 1974, **44**, 241-248.

(2) Sanesi, M.; Cingolani, A.; Tonelli, P.L.; Franzosini, P.
Thermal Properties, in **Thermodynamic and Transport Properties of Organic Salts**, IUPAC Chemical Data Series No. 28 (Franzosini, P.; Sanesi, M.; Editors), Pergamon Press, Oxford, 1980, 29-115.

(3) Schiraldi, A.; Chiodelli, G.
J. Phys. E: Sci. Instr. 1977, **10**, 596-599.

(4) Sokolov, N.M.
Tezisy Dokl. X Nauch. Konf. S.M.I. 1956.

COMPONENTS:	ORIGINAL MEASUREMENTS:
(1) Magnesium butanoate (magnesium butyrate); $(C_4H_7O_2)_2Mg$; [556-45-6] (2) Sodium butanoate (sodium butyrate); $(C_4H_7O_2)_2Na_2$; [156-54-7]	Pochtakova, E.I. **Zh. Obshch. Khim.** 1974, **44**, 241-248.
VARIABLES:	PREPARED BY:
Temperature.	Baldini, P.

EXPERIMENTAL VALUES:

$t/^{o}C$	T/K^{a}	$100x_1$	$t/^{o}C$	T/K^{a}	$100x_1$
330	603	0	217	490	65
318	591	5	220[bc]	493	65
305	578	10	206[be]	479	65
300[bc]	573	10	135[bf]	408	65
208[bd]	481	10	215	488	67.5
248[bi]	521	10	208[bc]	481	69
314[bj]	587	10	208[be]	481	69
288	561	15	140[bf]	413	69
278	551	20	162[bg]	435	69
268	541	25	218	491	70
258	531	30	226	499	72.5
248	521	35	234	507	75
255[bc]	528	35	230[bc]	503	75
214[bd]	487	35	205[be]	478	75
234[bh]	507	35	126[bf]	399	75
238	511	37.5	164[bg]	437	75
236	509	40	247	520	80
226	499	42.5	248[bc]	521	80
220	493	45	204[be]	477	80
220[bc]	493	45	133[bf]	406	80
220[bd]	493	45	158[bg]	431	80
223	496	47.5	266	539	85
224	497	50	275	548	90
225[bc]	498	50	270[bc]	543	90
216[bd]	489	50	202[be]	475	90
225	498	55	138[bf]	411	90
222[bc]	495	55	302	575	100
219	492	60			

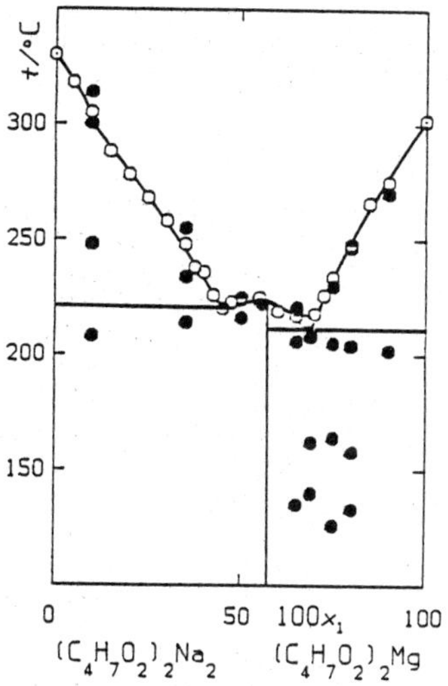

[a] T/K values calculated by the compiler.
[b] Differential thermal analysis (filled circles in the Figure).
All other data are from visual polythermal analysis and are represented as empty circles in the Figure.
[c] Initial crystallization.
[d] Eutectic stop (E_2).
[e] Eutectic stop (E_1).
[f] First transition of the system.
[g] Second transition of the system.
[h] Third transition of the system.
[i] Fourth transition of the system.
[j] Fifth transition of the system (no explanation if offered by the author for the occurrence of this point above the liquidus, compiler).

(continued in the next page)

COMPONENTS:

(1) Magnesium butanoate (magnesium butyrate); $(C_4H_7O_2)_2Mg$; [556-45-6]
(2) Sodium butanoate (sodium butyrate); $(C_4H_7O_2)_2Na_2$; [156-54-7]

ORIGINAL MEASUREMENTS:

Pochtakova, E.I.
Zh. Obshch. Khim. 1974, **44**, 241-248.

VARIABLES:

Temperature.

PREPARED BY:

Baldini, P.

EXPERIMENTAL VALUES: (continued)

Characteristic point(s):

Eutectic, E_1, at 210 ^{o}C (208 ^{o}C by DTA), and $100\mathbf{x}_1$= 69 (author).
Eutectic, E_2, at 220 ^{o}C and $100\mathbf{x}_1$= 45 (author).

Intermediate compound(s):

$(C_4H_7O_2)_7Mg_2Na_3$ (author), congruently melting at 225 ^{o}C (as reported in Ref. 1, Fig. 1, compiler).

AUXILIARY INFORMATION

METHOD/APPARATUS/PROCEDURE:

Visual polythermal analysis, (empty circles in the Figure) supplemented with DTA (filled circles).

SOURCE AND PURITY OF MATERIALS:

Materials prepared (Ref. 2) by reacting the proper ("chemically pure") carbonate with a slight excess of butanoic acid of analytical purity. Component 2 undergoes phase transitions at $t_{trs}(2)/\ ^{o}C$= 117, 232, 252, 316 (Ref. 3).

ESTIMATED ERROR:

Temperature: accuracy probably ± 2 K (compiler).

REFERENCES:

(1) Pochtakova, E.I.
Zh. Obshch. Khim. 1978, **48**, 1212-1214.
(2) Sokolov, N.M.
Zh. Obshch. Khim. 1954, **24**, 1581-1593.
(3) Sokolov, N.M.; **Tezisy Dokl. X Nauch. Konf. S.M.I.** 1956.

COMPONENTS:	EVALUATOR:
(1) Potassium **iso**.butanoate (potassium **iso**.butyrate); ($i.C_4H_7O_2$)K; [19455-20-0] (2) Sodium **iso**.butanoate (sodium **iso**.butyrate); ($i.C_4H_7O_2$)Na; [996-30-5]	Schiraldi, A., Dipartimento di Chimica Fisica Universita´ di Pavia (ITALY).

CRITICAL EVALUATION:

This system was studied only by Sokolov and Pochtakova (Ref. 1) who suggested the phase diagram to be of the eutectic type, the invariant point occurring at 521 K (248 ^oC) and $100\mathbf{x}_1 = 7.5$.

Component 1, however, forms liquid crystals. Therefore the temperature of 633 K (360 ^oC) given in Ref. 1 should be identified with the clearing (and not the fusion) temperature of this component, and compared with the $\mathbf{T}_{clr}(1)$ value (625.6 ± 0.8 K) reported in Table 2.

For the same component, three phase transition temperatures are quoted in Ref. 1 from Ref. 2, i.e., 621, 546, and 481 K, the second of which can be reasonably identified with the fusion temperature [$\mathbf{T}_{fus}(1) = 553.9 \pm 0.5$ K] listed in Preface, Table 2. Consequently: (i) the transition temperature at 621 K (if actually existing) might correspond to some kind of transformation (undetected by DSC, see Preface, Table 2) within the liquid crystal field; and (ii) only the transition at 481 K should correspond to a solid state transformation, although the latter figure is almost 60 K higher than the single $\mathbf{T}_{trs}(1)$ value (424 ± 3 K) listed in Table 2 of the Preface.

Concerning component 2, the fusion temperature of 535 K (262 ^oC; Ref. 1) is in reasonable agreement with that (526.9 ± 0.7 K) reported in Table 2 of the Preface. In this Table, however, no mention is made of other phase transformations, although three solid state transitions are quoted for this component in Ref. 1 (from Ref. 2), at 493, 364, and 340 K (220, 91, and 67 ^oC), respectively. Duruz et al. (Ref. 3) report in turn: fusion at 527 K (in agreement with the fusion temperature from Table 2), and solid state transitions at 493 K (in agreement with the highest transition temperature from Ref. 2), and at 468 K (a figure which has no correspondence in Ref. 2). Finally, Ferloni et al. (Ref. 4) are inclined to think that Sokolov´s transformation at 340 K (Ref. 2) actually represents a transition of a hydrated form of the salt.

In the evaluator´s opinion, a re-investigation of the phase relations in solid sodium **iso**.butanoate would be desirable. At any rate, the phase diagram suggested by Sokolov and Pochtakova (Ref. 1) has to be modified (due to the occurrence of liquid crystals in component 1) with reference to Schemes A.1, or A.3, of the Preface according to the kind of solid state miscibility between components, the eutectic point actually being an M'_E point.

REFERENCES:

(1) Sokolov, N.M.; Pochtakova, E.I.
Zh. Obshch. Khim. 1960, **30**, 1405-1410 (*); **Russ. J. Gen. Chem. (Engl. Transl.)** 1960, **30**, 1433-1437.

(2) Sokolov, N.M.
Tezisy Dokl. X Nauch. Konf. S.M.I. 1956.

(3) Duruz, J.J.; Michels, H.J.; Ubbelohde, A.R.
Proc. Roy. Soc. London 1971, **A 322**, 281-299.

(4) Ferloni, P.; Sanesi, M.; Tonelli, P.L.; Franzosini, P.
Z. Naturforsch. 1978, **A 33**, 240-242.

COMPONENTS:	ORIGINAL MEASUREMENTS:
(1) Potassium **iso**.butanoate (potassium **iso**.butyrate); $(i.C_4H_7O_2)K$; [19455-20-0] (2) Sodium **iso**.butanoate (sodium **iso**.butyrate); $(i.C_4H_7O_2)Na$; [996-30-5]	Sokolov, N.M.; Pochtakova, E.I. **Zh. Obshch. Khim.** 1960, **30**, 1405-1410 (*); **Russ. J. Gen. Chem. (Engl. Transl.)** 1960, **30**, 1433-1437.
VARIABLES:	PREPARED BY:
Temperature.	Baldini, P.

EXPERIMENTAL VALUES:

t/°C	T/K[a]	$100x_1$
262	535	0
254	527	5
248	521	7.5
255	528	10
266	539	15
277	550	20
285	558	25
293	566	30
302	575	35
308	581	40
315	588	45
320	593	50
325	598	55
331	604	60
335	608	65
338	611	70
342	615	75
345	618	80
348	621	85
353	626	90
357	630	95
360	633	100

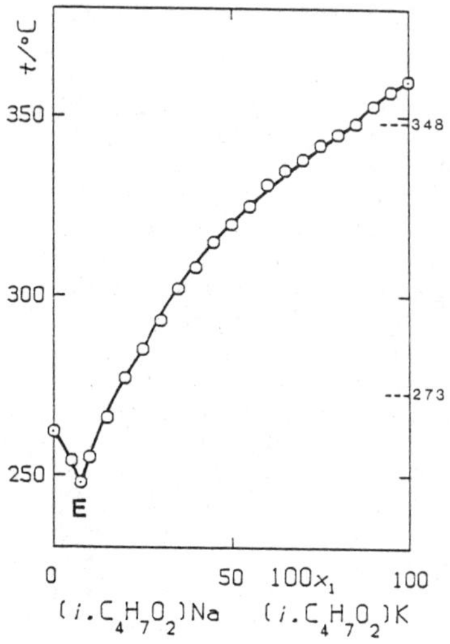

[a] **T**/K values calculated by the compiler.

Characteristic point(s): Eutectic, E, at 248 °C and $100x_1$= 7.5 (authors).

AUXILIARY INFORMATION

METHOD/APPARATUS/PROCEDURE:

Visual polythermal analysis.

SOURCE AND PURITY OF MATERIALS:

Both components were prepared from commercial "pure" grade **iso**.butanoic acid, distilled before use, and the proper "chemically pure" hydrogen carbonate (Ref. 1); then recrystallized from n-butanol. Component 1 undergoes phase transitions at $t_{trs}(1)$/°C= 208, 273, 348 (Ref. 2). Component 2 undergoes phase transitions at $t_{trs}(2)$/°C= 67, 91, 220 (Ref. 2).

REFERENCES:

(1) Sokolov, N.M. **Zh. Obshch. Khim.** 1954, **24**, 1150-1156.
(2) Sokolov, N.M. **Tezisy Dokl. X Nauch. Konf. S.M.I.** 1956 (this is Ref. 6 in the original paper, and not Ref. 5 as erroneously quoted in the text; compiler).

ESTIMATED ERROR:

Temperature: accuracy probably ±2 K (compiler).

COMPONENTS:

(1) Potassium pentanoate (potassium valerate);
$(C_5H_9O_2)K$; [19455-21-1]
(2) Sodium pentanoate (sodium valerate);
$(C_5H_9O_2)Na$; [6106-41-8]

EVALUATOR:

Schiraldi, A.,
Dipartimento di Chimica Fisica,
Universita' di Pavia (ITALY).

CRITICAL EVALUATION:

This system was studied only by Dmitrevskaya and Sokolov (Ref. 1), who claimed that continuous series of solid solutions exist.

Both components, however, form liquid crystals (see Preface, Table 1). Consequently: (i) the fusion temperatures, $T_{fus}(1)$= 717 K (444 °C) and $T_{fus}(2)$= 630 K (357 °C) given in Ref. 1, are actually to be identified with the clearing temperatures (the corresponding values from Preface, Table 1 being 716±2 K and 631±4 K, respectively); (ii) the transition temperatures $T_{trs}(1)$= 580 K (307 °C) and $T_{trs}(2)$= 489 K (216 °C) quoted in Ref. 1 from Ref. 2, are in turn to be identified with the actual fusion temperatures (the corresponding values from Table 1 of the Preface being 586.6±0.7 K and 498±2 K, respectively).

Continuous series of liquid crystal (instead of solid) solutions ought to form, and the phase diagram ought to be similar to that shown in Preface, Scheme C.1.

Moreover, the following point deserves attention. For component 2, Table 1 of the Preface reports no solid state transition, whereas Dmitrevskaya and Sokolov quote (again from Ref. 2) $T_{trs}(2)$/K= 482 and 453. It is, however, to be stressed that the single transition observed (at 479±1 K) with DTA in sodium n-pentanoate by Duruz et al. (Ref. 3) was not more mentioned in a subsequent DSC investigation by the same group (Ref. 4).

In conlusion, due to the lack of information about the boundaries of the mesomorphic liquid field, and to conflicting assertions about solid state transitions, a re-investigation of the system would be desirable.

REFERENCES:

(1) Dmitrevskaya, O.I.; Sokolov, N.M.
Zh. Obshch. Khim. 1965, **35**, 1905-1909.

(2) Sokolov, N.M.
Tezisy Dokl. X Nauch. Konf. S.M.I. 1956.

(3) Duruz, J.J.; Michels, H.J.; Ubbelohde, A.R.
Proc. Roy. Soc. London 1971, **A322**, 281-299.

(4) Michels, H.J.; Ubbelohde, A.R.
JCS Perkin II 1972, 1879-1881.

COMPONENTS:

(1) Potassium pentanoate (potassium valerate); $(C_5H_9O_2)K$; [19455-21-1]
(2) Sodium pentanoate (sodium valerate); $(C_5H_9O_2)Na$; [6106-41-8]

ORIGINAL MEASUREMENTS:

Dmitrevskaya, O.I.; Sokolov, N.M.
Zh. Obshch. Khim. 1965, **35**, 1905-1909.

VARIABLES:

Temperature.

PREPARED BY:

Baldini, P.

EXPERIMENTAL VALUES:

$t/^{o}C$	T/K^{a}	$100\mathbf{x}_1$
357	630	0
366	639	5
375	648	10
382	655	15
388	661	20
393	666	25
397	670	30
402	675	35
406	679	40
407	680	45
414	687	50
418	691	55
422	695	60
426	699	65
430	703	70
432	705	75
436	709	80
439	712	85
440	713	90
442	715	95
444	717	100

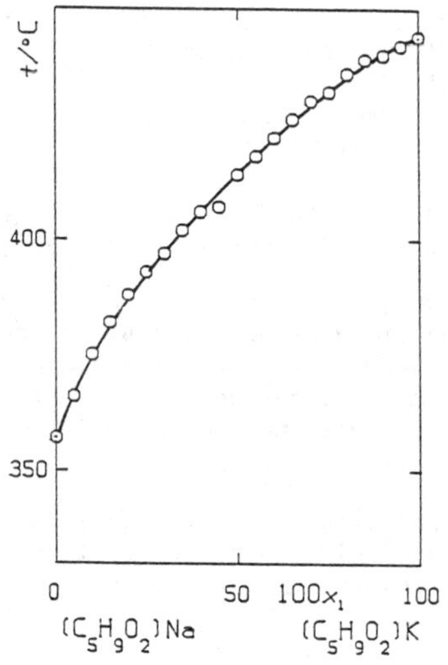

[a]T/K values calculated by the compiler.

Characteristic point(s): Continuous series of solid solutions.

AUXILIARY INFORMATION

METHOD/APPARATUS/PROCEDURE:

Visual polythermal analysis.

SOURCE AND PURITY OF MATERIALS:

Both components prepared from **n**-pentanoic acid and the proper carbonate (Ref. 1). Component 1 undergoes a phase transition at $t_{trs}(1)/^{o}C$= 307 (Ref. 2).
Component 2 undergoes phase transitions at $t_{trs}(2)/^{o}C$= 180, 209, 216 (Ref. 2).

ESTIMATED ERROR:

Temperature: accuracy probably ±2 K (compiler).

REFERENCES:

(1) Sokolov, N.M.
Zh. Obshch. Khim. 1954, **24**, 1581-1593.
(2) Sokolov, N.M.
Tezisy Dokl. X Nauch. Konf. S.M.I. 1956.

COMPONENTS:	EVALUATOR:
(1) Magnesium pentanoate (magnesium valerate); $(C_5H_9O_2)_2Mg$; [556-37-6] (2) Sodium pentanoate (sodium valerate); $(C_5H_9O_2)_2Na_2$; [6106-41-8]	Franzosini, P., Dipartimento di Chimica Fisica, Universita´ di Pavia (ITALY).

CRITICAL EVALUATION:

This binary was studied only by Pochtakova (Ref. 1) who employed the visual polythermal analysis to draw the lower boundary of the isotropic liquid region at $0 \leq 100\mathbf{x}_1 \leq 72.5$ (the investigation of mixtures richer in component 1 being prevented by their tendency to form glasses). She claimed the occurrence of an incongruently melting intermediate compound, i.e, $(C_5H_9O_2)_8Mg_3Na_2$, able to give a eutectic with component 2.

The latter component, however, forms liquid crystals. Consequently, the topology of the phase diagram at $0 \leq 100\mathbf{x}_1 \leq 55$ could be described more correctly with (possible) reference to Scheme D.1 of the Preface: accordingly, Pochtakova´s eutectic ought to be an M'_E point, and an invariant type M_E (undetected by visual polythermal analysis) ought to exist at $100\mathbf{x}_1 < 55$.

A few more points are worth mentioning.

(i) Pochtakova´s (extrapolated) fusion temperature of component 1 (537 K) seems reasonable, although somewhat higher than the only other value provided by the literature (531 K; Ref. 2), while her $\mathbf{T}_{fus}(2)$ value (630 K) agrees fairly with the clearing temperature (631$\pm$4 K) listed in Preface, Table 1 for component 2.

(ii) For the same component, Table 1 of the Preface provides also a $\mathbf{T}_{fus}(2)$= 498$\pm$2 K, a figure which can be identified (even if not fully satisfactorily) with that (489 K) corresponding to the highest phase transformation temperature quoted by Pochtakova from Ref. 3.

(iii) Once more for component 2, Table 1 of the Preface reports no solid state transition, whereas Pochtakova quotes (from Ref. 3) $\mathbf{T}_{trs}(2)/K$= 482 and 453. It is, however, to be stressed that the single transition observed (at 479$\pm$1 K) with DTA in sodium **n**-pentanoate by Duruz et al. (Ref. 4) was not more mentioned in a subsequent DSC investigation by the same group (Ref. 5).

In conlusion, due to the lack of information about the boundaries of the mesomorphic liquid field, and to conflicting assertions about solid state transitions, a re-investigation of the system would be desirable.

REFERENCES:

(1) Pochtakova, E.I.
Zh. Obshch. Khim. 1974, **44**, 241-248.

(2) Sanesi, M.; Cingolani, A.; Tonelli, P.L.; Franzosini, P.
Thermal Properties, in **Thermodynamic and Transport Properties of Organic Salts,** IUPAC Chemical Data Series No. 28 (Franzosini, P.; Sanesi, M.; Editors), Pergamon Press, Oxford, 1980, 29-115.

(3) Sokolov, N.M.
Tezisy Dokl. X Nauch. Konf. S.M.I. 1956.

(4) Duruz, J.J.; Michels, H.J.; Ubbelohde, A.R.
Proc. Roy. Soc. London 1971, **A322**, 281-299.

(5) Michels, H.J.; Ubbelohde, A.R.
JCS Perkin II 1972, 1879-1881.

COMPONENTS:	ORIGINAL MEASUREMENTS:
(1) Magnesium pentanoate (magnesium valerate); $(C_5H_9O_2)_2Mg$; [556-37-6] (2) Sodium pentanoate (sodium valerate); $(C_5H_9O_2)_2Na_2$; [6106-41-8]	Pochtakova, E.I. **Zh. Obshch. Khim.** 1974, **44**, 241-248.
VARIABLES:	PREPARED BY:
Temperature.	Baldini, P.

EXPERIMENTAL VALUES:

$t/^oC$	T/K^a	$100x_1$	$t/^oC$	T/K^a	$100x_1$
357	630	0	277	550	35
350	623	5	272	545	37.5
348	621	7.5	266	539	40
344	617	10	250	523	50
338	611	12.5	240	513	52.5
332	605	15	221	494	55
326	599	17.5	224	497	57.5
322	595	20	224	497	60
312	585	22.5	224	497	62.5
307	580	25	224	497	67.5
299	572	27.5	231	504	70
295	568	30	241	514	72.5
284	557	32.5			

[a] T/K values calculated by the compiler.

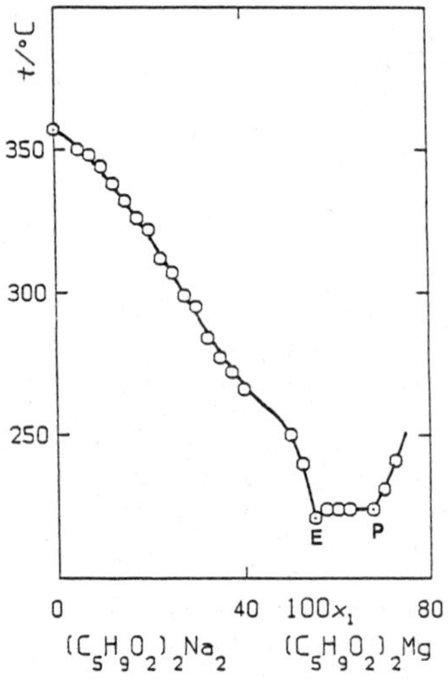

Characteristic point(s):

Eutectic, E, at 221 °C and $100x_1$= 55 (author).
Peritectic, P, at 224 °C and $100x_1$= 67.5 (author).

Intermediate compound(s):

$(C_5H_9O_2)_8Mg_3Na_2$ (probable composition), incongruently melting (author).

AUXILIARY INFORMATION

METHOD/APPARATUS/PROCEDURE:

Visual polythermal analysis.

NOTE:

The system was investigated at $0 \leq 100x_1 \leq 72.5$ due to the tendency of component 1 to form glasses.

SOURCE AND PURITY OF MATERIALS:

Materials prepared (Ref. 1) by reacting the proper ("chemically pure") carbonate with a slight excess of **n**-pentanoic acid of analytical purity.
Component 1: $t_{fus}/^oC$ (extrapolated) at 264.
Component 2 undergoes phase transitions at $t_{trs}(2)/^oC$= 180, 209, 216 (Ref. 2).

ESTIMATED ERROR:

Temperature: accuracy probably ±2 K (compiler).

REFERENCES:

(1) Sokolov, N.M.
Zh. Obshch. Khim. 1954, **24**, 1581-1593.
(2) Sokolov, N.M.
Tezisy Dokl. X Nauch. Konf. S.M.I. 1956.

COMPONENTS:	EVALUATOR:
(1) Potassium **iso**.pentanoate (potassium iso.valerate); ($\mathbf{i}.C_5H_9O_2$)K; [589-46-8] (2) Sodium **iso**.pentanoate (sodium **iso**.valerate); ($\mathbf{i}.C_5H_9O_2$)Na; [539-66-2]	Schiraldi, A., Dipartimento di Chimica Fisica, Universita´ di Pavia (ITALY).

CRITICAL EVALUATION:

This system was studied by Pochtakova (Ref. 1), and by Dmitrevskaya and Sokolov (Ref. 2): according to both papers, continuous series of solid solutions ought to be formed.

Both components, however, form liquid crystals (see Preface, Table 2). Consequently the fusion temperatures, $T_{fus}(1)$= 669 K (396 °C; Refs. 1, 2), and $T_{fus}(2)$= 533 K (260 °C; Ref. 1) or 535 K (262 °C; Ref. 2), are actually to be identified with the clearing temperatures, the corresponding values from Table 2 of the Preface being 679$\pm$2 K and 559$\pm$1 K, respectively. The latter figure is remarkably higher than those given by the Russian authors, although meeting rather satisfactorily those reported by Ubbelohde et al. (556 K; Ref. 3) and by Duruz et al. (553 K; Ref. 4).

No mention is made in Refs. 1, 2 of the actual fusion of component 1 which occurs at 531$\pm$3 K (Table 2): the latter figure is supported by the trend of the thermomagnetical curves plotted by Duruz and Ubbelohde (Ref. 5). As for the other phase transitions of the same component, Pochtakova quotes from Ref. 6 two T_{trs} values, i.e., 327 and 618 K (54 and 345 °C, respectively), for which no comparison is possible with the findings by other investigators, inasmuch as: (i) no transformation is reported in Table 2 as occurring below $T_{fus}(1)$= 531$\pm$3 K; and (ii) no transformation is reported in Table 2 or in Ref. 5 as occurring within the field of existence of the mesomorphic liquid. It is a bit puzzling the fact that for potassium **iso**.pentanoate Dmitrevskaya and Sokolov (Ref. 2) quote from the same source (Ref. 6) transitions at 618, 493, and 473 K (ignoring that quoted by Pochtakova at 327 K).

In the case of component 2, the transition at 451 K (178 °C; quoted in Refs. 1, 2 from Ref. 5) should be indentified with the actual fusion temperature (the corresponding value from Table 2 of the Preface being 461.5$\pm$0.6 K).

Taking into account the above remarks, the upper part of Dmitrevskaya and Sokolov´s diagram, Ref. 2, (to be compared with the upper part of Preface, Scheme C.1) supports the idea that continuous series of liquid crystal (instead of solid) solutions do form. Moreover, the left-hand side of the lower part of the same diagram might suggest that, at lower temperatures, solid solutions are also present.

REFERENCES:

(1) Pochtakova, E.I.
Zh. Obshch. Khim. 1963, **33**, 342-347.

(2) Dmitrevskaya, O.I.; Sokolov, N.M.
Zh. Obshch. Khim. 1967, **37**, 2160-2166 (*); **Russ. J. Gen. Chem. (Engl. Transl.)** 1967, **37**, 2050-2054.

(3) Ubbelohde, A.R.; Michels, H.J.; Duruz, J.J.
Nature 1970, **228**, 50-52.

(4) Duruz, J.J.; Michels, H.J.; Ubbelohde, A.R.
Proc. R. Soc. London 1971, **A 322**, 281-299.

(5) Duruz, J.J.; Ubbelohde, A.R.
Proc. R. Soc. London 1975, **A 342**, 39-49.

(6) Sokolov, N.M.
Tezisy Dokl. X Nauch. Konf. S.M.I. 1956.

COMPONENTS:	ORIGINAL MEASUREMENTS:
(1) Potassium **iso**.pentanoate (potassium **iso**.valerate); $(i.C_5H_9O_2)K$; [589-46-8] (2) Sodium **iso**.pentanoate (sodium **iso**.valerate); $(i.C_5H_9O_2)Na$; [539-66-2]	Pochtakova, E.I. **Zh. Obshch. Khim.** 1963, **33**, 342-347.
VARIABLES:	PREPARED BY:
Temperature.	Baldini, P.

EXPERIMENTAL VALUES:

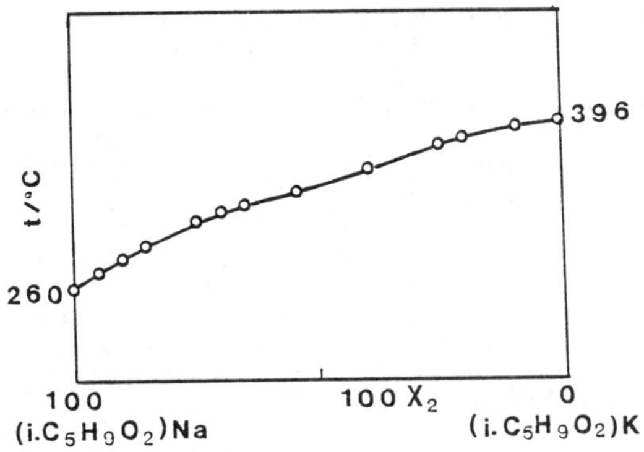

The results are reported only in graphical form (see figure).

Characteristic point(s): Continuous series of solid solutions.

AUXILIARY INFORMATION

METHOD/APPARATUS/PROCEDURE:	SOURCE AND PURITY OF MATERIALS:
Visual polythermal analysis.	Both components prepared from commercial **iso**.pentanoic acid (distilled twice before use) and the proper "chemically pure" hydrogen carbonate (Ref. 1, where, however, carbonates instead of hydrogen carbonates are employed; compiler). Component 1 undergoes phase transitions at $t_{trs}(1)/^oC$= 54, 345 (Ref. 2) and melts at $t_{fus}(1)/^oC$= 396. Component 2 undergoes phase transitions at $t_{trs}(2)/^oC$= 152, 178 (Ref. 2) and melts at $t_{fus}(2)/^oC$= 260.
ESTIMATED ERROR:	
Temperature: accuracy probably ± 2 K (compiler).	
REFERENCES:	
(1) Sokolov, N.M. **Zh. Obshch. Khim.** 1954, **24**, 1581-1593. (2) Sokolov, N.M. **Tezisy Dokl. X Nauch. Konf. S.M.I.** 1956.	

COMPONENTS:

(1) Potassium **iso**.pentanoate (potassium **iso**.valerate); $(i.C_5H_9O_2)K$; [589-46-8]
(2) Sodium **iso**.pentanoate (sodium **iso**.valerate); $(i.C_5H_9O_2)Na$; [539-66-2]

ORIGINAL MEASUREMENTS:

Dmitrevskaya, O.I.; Sokolov, N.M.
Zh. Obshch. Khim. 1967, **37**, 2160-2166 (*);
Russ. J. Gen. Chem. (Engl. Transl.) 1967, **37**, 2050-2054.

VARIABLES:

Temperature.

PREPARED BY:

Baldini, P.

EXPERIMENTAL VALUES:

$t/^{o}C$	T/K^{a}	$100x_1$
262	535	0
178[b]	451	0
152[b]	425	0
320	593	25
300[b]	573	25
200[b]	473	25
186[b]	459	25
350	623	50
340[b]	613	50
216[b]	489	50
384	657	75
370[b]	643	75
242[b]	515	75
396	669	100
345[b]	618	100
220[b]	493	100
200[b]	473	100

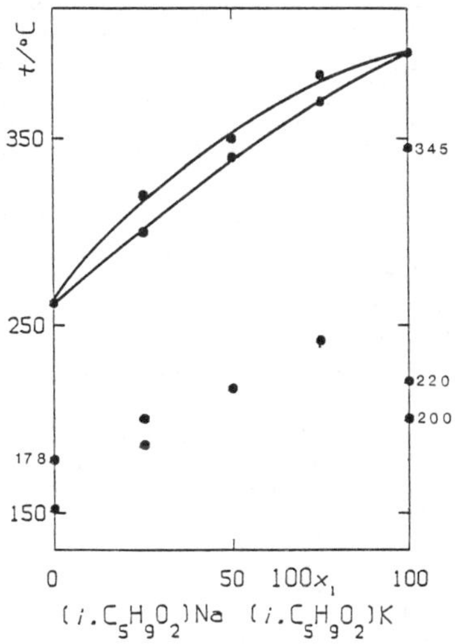

[a] T/K values calculated by the compiler.
[b] Transformation in phase.

Characteristic point(s): Continuous series of solid solutions.

AUXILIARY INFORMATION

METHOD/APPARATUS/PROCEDURE:

Thermographical investigation (heating curves recorded automatically).

NOTE:

The data tabulated (and plotted in the figure) refer to the thermographical investigation; other points of the liquidus, taken by visual polythermal analysis and consistent with the tabulated ones, are reported only in a graphical form (Fig. 2 of the original paper). For the latter, reference is made to a previous paper by Sokolov et al. (Ref. 1) where, however, the present binary is merely mentioned as a side of a reciprocal ternary.

SOURCE AND PURITY OF MATERIALS:

Both components synthetized from **iso**.butanoic acid and the proper carbonate (Ref. 2).
Component 1 undergoes phase transitions at $t_{trs}(1)/^{o}C$= 345, 220, 200 (Ref. 3).
Component 2 undergoes phase transitions at $t_{trs}(2)/^{o}C$= 152, 178 (Ref. 3).

ESTIMATED ERROR:

Temperature: accuracy probably ± 2 K (compiler).

REFERENCES:

(1) Sokolov, N.M.; Tsindrik, N.M.; Dmitrevskaya, O.I. **Zh. Obshch. Khim.** 1961, **31**, 1051-1056.
(2) Sokolov, N.M. **Zh. Obshch. Khim.** 1954, **24**, 1581-1593.
(3) Sokolov, N.M. **Tezisy Dokl. X Nauch. Konf. S.M.I.** 1956.

COMPONENTS:	EVALUATOR:
(1) Potassium hexanoate (potassium caproate); $(C_6H_{11}O_2)K$; [19455-00-6] (2) Sodium hexanoate (sodium caproate); $(C_6H_{11}O_2)Na$; [10051-44-2]	Schiraldi, A., Dipartimento di Chimica Fisica, Universita´ di Pavia (ITALY).

CRITICAL EVALUATION:

This system was studied only by Pochtakova (Ref. 1), who claimed the existence of a continuous series of solid solutions.

Both components, however, form liquid crystals (see Preface, Table 1). Consequently: (i) the fusion temperatures, $T_{fus}(1)$= 717.7 K (444.5 °C) and $T_{fus}(2)$= 638 K (365 °C) given in Ref. 1 are actually to be identified with the clearing temperatures (the corresponding values from Table 1 of the Preface being 725.8±0.8 K and 639.0±0.5 K, respectively); (ii) the transition temperatures $T_{trs}(1)$= 575 K (302 °C) and $T_{trs}(2)$= 499 K (226 °C), quoted in Ref. 1 from Ref. 2, are in turn to be identified with the fusion temperatures (the corresponding values from Table 1 of the Preface being 581.7±0.5 K and 499.6±0.6 K).

Finally, the following point deserves attention. Two more transitions are quoted in Ref. 1 from Ref. 2 as occurring in component 2 at 615 K (342 °C) and 476 K (203 °C), respectively. The latter one corresponds to that reported at 473±2 K in Table 1 of the Preface, whereas no evidence was obtained by subsequent investigators (Ref. 3) for a transition comparable with the former one: should it exist, it might mean that two different mesomorphic phases are present in sodium hexanoate.

As a conclusion, in the evaluator´s opinion Pochtakova´s data support reasonably the idea that continuous series of liquid crystal (instead of solid) solutions are formed, and the phase diagram ought to be not far from that shown in Preface, Scheme C.1.

REFERENCES:

(1) Pochtakova, E.I.
Zh. Obshch. Khim. 1959, **29**, 3183-3189 (*); **Russ. J. Gen. Chem. (Engl. Transl.)** 1959, **29**, 3149-3154.

(2) Sokolov, N.M.
Tezisy Dokl. X Nauch. Konf. S.M.I. 1956.

(3) Sanesi, M.; Cingolani, A.; Tonelli, P.L.; Franzosini, P.
Thermal Properties, in **Thermodynamic and Transport Properties of Organic Salts**, IUPAC Chemical Data Series No. 28 (Franzosini, P.; Sanesi, M.; Editors), Pergamon Press, Oxford, 1980, 29-115.

COMPONENTS:	ORIGINAL MEASUREMENTS:
(1) Potassium hexanoate (potassium caproate); $(C_6H_{11}O_2)K$; [19455-00-6] (2) Sodium hexanoate (sodium caproate); $(C_6H_{11}O_2)Na$; [10051-44-2]	Pochtakova, E.I. **Zh. Obshch. Khim.** 1959, **29**, 3183-3189 (*); **Russ. J. Gen. Chem. (Engl. Transl.)** 1959, **29**, 3149-3154.
VARIABLES:	PREPARED BY:
Temperature.	Baldini, P.

EXPERIMENTAL VALUES:

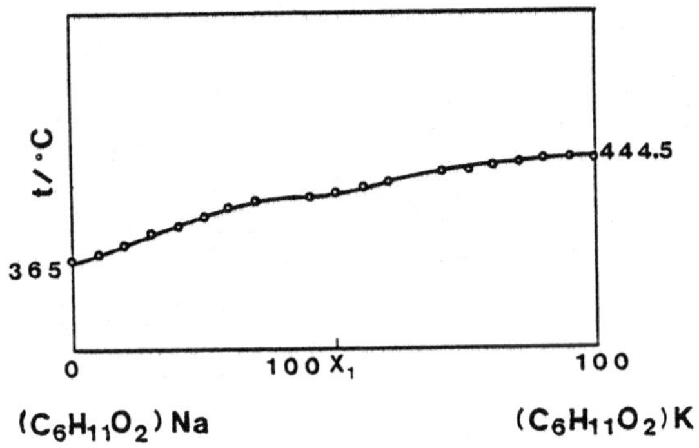

The results are reported only in graphical form (see figure).

Characteristic point(s): Continuous series of solid solutions.

AUXILIARY INFORMATION

METHOD/APPARATUS/PROCEDURE:	SOURCE AND PURITY OF MATERIALS:
Visual polythermal analysis.	Components prepared by reacting the proper carbonate with n-hexanoic acid (Ref. 1). Component 1 undergoes a phase transition at $t_{trs}(1)/^{o}C = 302$ (Ref. 2). Component 2 undergoes phase transitions at $t_{trs}(2)/^{o}C = 203, 226, 342$ (Ref. 2).
	ESTIMATED ERROR:
	Temperature: accuracy probably ± 2 K (compiler).
	REFERENCES:
	(1) Sokolov, N.M. **Zh. Obshch. Khim.** 1954, **24**, 1581-1593. (2) Sokolov, N.M. **Tezisy Dokl. X Nauch. Konf. S.M.I.** 1956.

SYSTEMS WITH COMMON CATION

COMPONENTS:	ORIGINAL MEASUREMENTS:
(1) Cesium ethanoate (cesium acetate); $CsC_2H_3O_2$; [3396-11-0] (2) Cesium nitrite; $CsNO_2$; [13454-83-6]	Diogenov, G.G.; Morgen, L.T. **Nekotorye Vopr. Khimii Rasplavlen. Solei i Produktov Destruktsii Sapropelitov, Irkutsk**, 1974, 32-34.
VARIABLES:	PREPARED BY:
Temperature.	Baldini, P.

EXPERIMENTAL VALUES:

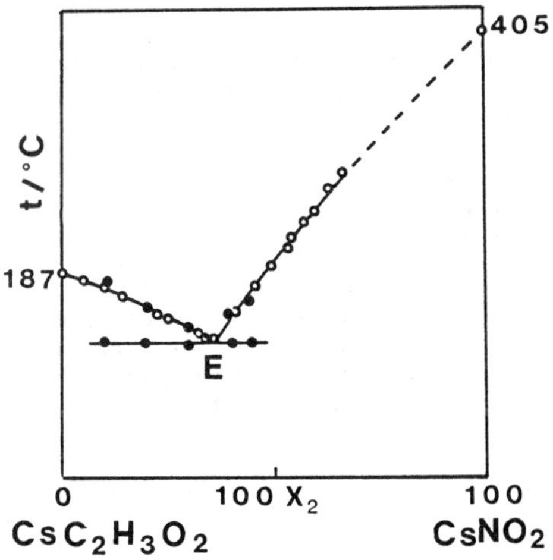

The results are reported only in graphical form (see figure).

Characteristic point(s):
Eutectic, E, at 125 °C and $100\mathbf{x}_2$= 36 (authors).

AUXILIARY INFORMATION

METHOD/APPARATUS/PROCEDURE:

Visual polythermal analysis; temperatures measured with a Chromel-Alumel thermocouple and a 15 mV millivoltmeter. Supplementary measurements (filled circles in the figure) were performed by thermographical analysis.

NOTE:

Concerning component 1, the value of the fusion temperature by Diogenov and Morgen (460 K) is not far from that (463±1 K) listed in Preface, Table 1. For the same component, Nurminskii and Diogenov reported previously (Ref. 1) a solid state transition at 447 K whose existence, however, was not confirmed by any subsequent investigator (Ref. 2).

ESTIMATED ERROR:

Temperature: accuracy probably ±2 K (compiler).

SOURCE AND PURITY OF MATERIALS:

Not stated.
Component 1: $t_{fus}(1)/^{\circ}C$= 187 (Fig. 1 of the original paper).
Component 2: $t_{fus}(2)/^{\circ}C$= 405 (Fig. 1).

REFERENCES:

(1) Nurminskii, N.N.; Diogenov, G.G. **Zh. Neorg. Khim.** 1960, 5, 2084-2087; **Russ. J. Inorg. Chem. (Engl. Transl.)** 1960, 5, 1011-1013.
(2) Sanesi, M.; Cingolani, A.; Tonelli, P.L.; Franzosini, P. **Thermal Properties**, in **Thermodynamic and Transport Properties of Organic Salts**, IUPAC Chemical Data Series No. 28 (Franzosini, P.; Sanesi, M.; Editors), Pergamon Press, Oxford, 1980, 29-115.

COMPONENTS:	EVALUATOR:
(1) Cesium ethanoate (cesium acetate); $CsC_2H_3O_2$; [3396-11-0] (2) Cesium nitrate; $CsNO_3$; [7789-18-6]	Schiraldi, A. Dipartimento di Chimica fisica, Universita´ di Pavia (ITALY)

CRITICAL EVALUATION:

This binary was studied with visual polythermal analysis by Nurminskii and Diogenov (as a side of the reciprocal ternary Cs, K/$C_2H_3O_2$, NO_3; Ref. 1), and by Gimel´shtein and Diogenov (as a side of the reciprocal ternary Cs, Na/$C_2H_3O_2$, NO_3; Ref. 2), with a substantially similar conclusion: the system is of the eutectic type, the invariant being at either 415 K (142 ^oC; Ref. 1), or 429 K (156 ^oC; Ref. 2), and $100\mathbf{x}_2$= 25 (Refs. 1,2).

In Ref. 1 the authors claim also the existence of a phase transition of component 1 at 447 K (174 ^oC) whose existence, however, was neither mentioned in Ref. 2, nor confirmed by other investigators (Ref. 3).

The fusion temperature of component 1 reported in both papers, i.e., 455 K (182 ^oC) represents the third lowest value among those listed in Ref. 3, which range between 453 and 467 K. It seems then likely that some impurity (possibly water) was present in the material used by Diogenov et al.

In the evaluator´s opinion, there is no reason to reject the assertion made in Refs. 1 and 2, that the diagram is of the eutectic type: however, due to the possibly inadequate purity of component 1, and to the large discrepancy in the eutectic temperature, a re-investigation of the system would be highly desirable.

REFERENCES:

(1) Nurminskii, N.N.; Diogenov, G.G.
Zh. Neorg. Khim. 1960, **5**, 2084-2087; **Russ. J. Inorg. Chem. (Engl. Transl.)** 1960, **5**, 1011-1013(*).

(2) Gimel´shtein, V.G.; Diogenov, G.G.
Tr. Irkutsk. Politekh. Inst., Ser. Khim., 1966, **27**, 69-75.

(3) Sanesi, M.; Cingolani, A.; Tonelli, P.L.; Franzosini, P.
Thermal Properties, in **Thermodynamic and Transport Properties of Organic Salts**, IUPAC Chemical Data Series No. 28 (Franzosini, P.; Sanesi, M.; Editors), Pergamon Press, Oxford, 1980, 29-115.

COMPONENTS:	ORIGINAL MEASUREMENTS:
(1) Cesium ethanoate (cesium acetate); $CsC_2H_3O_2$; [3396-11-0] **(2) Cesium nitrate;** $CsNO_3$; [7789-18-6]	Gimel´shtein, V.G.; Diogenov, G.G. **Tr. Irkutsk. Politekh. Inst., Ser. Khim., 1966, 27**, 69-75.
VARIABLES:	PREPARED BY:
Temperature.	Baldini, P.

EXPERIMENTAL VALUES:

Characteristic point(s): Eutectic, E, at 156 ^oC and $100\mathbf{x}_2$= 25 (authors).

AUXILIARY INFORMATION

METHOD/APPARATUS/PROCEDURE:	SOURCE AND PURITY OF MATERIALS:
Visual polythermal analysis. Temperatures measured with a Chromel-Alumel thermocouple and a 17 mV millivoltmeter.	Not stated. Component 1: $t_{fus}(1)/^oC$= 182 (Fig. 2 of the original paper). Component 2: $t_{fus}(2)/^oC$= 407 (Fig. 2).
ESTIMATED ERROR:	REFERENCES:
Temperature: accuracy probably ± 2 K (compiler).	

COMPONENTS:

(1) Cesium ethanoate (cesium acetate); $CsC_2H_3O_2$; [3396-11-0]
(2) Cesium nitrate; $CsNO_3$; [7789-18-6]

ORIGINAL MEASUREMENTS:

Nurminskii, N.N.; Diogenov, G.G.
Zh. Neorg. Khim. 1960, 5, 2084-2087; **Russ. J. Inorg. Chem. (Engl. Transl.)** 1960, 5, 1011-1013 (*).

VARIABLES:

Temperature.

PREPARED BY:

Baldini, P.

EXPERIMENTAL VALUES:

$t/^{o}C$	T/K^{a}	$100x_2$
180	453	0
176	449	2.5
172	445	8.0
164	437	14.0
156	429	19.0
148	421	23.0
147	420	26.0
176	449	31.0
195	468	35.0
211	484	39.0
236	509	45.0
263	536	52.5
284	557	60.0

[a] T/K values calculated by the compiler.

Characteristic point(s):

Eutectic, E, at 142 ^{o}C and $100x_2$= 25 (authors).

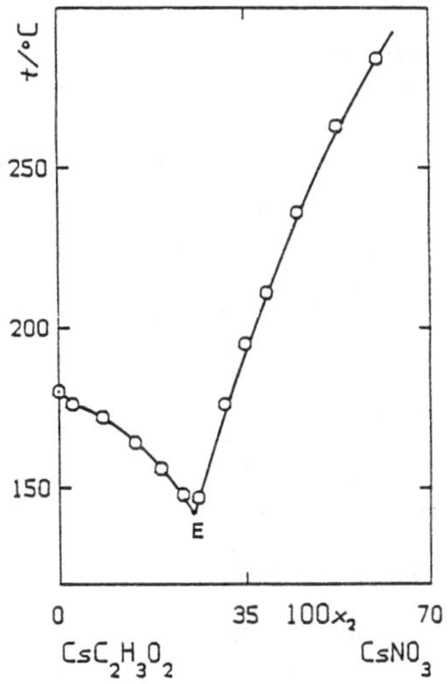

AUXILIARY INFORMATION

METHOD/APPARATUS/PROCEDURE:

Visual polythermal analysis. Temperatures measured with a Chromel-Alumel thermocouple and a 17 mV millivoltmeter.

SOURCE AND PURITY OF MATERIALS:

Not stated.
Component 1 undergoes a phase transition at $t_{trs}(1)/^{o}C$= 174 and melts at $t_{fus}(1)/^{o}C$ = 182 (Fig. 1 of the original paper), or 180 (table).
Component 2 undergoes a phase transition at $t_{trs}(2)/^{o}C$= 392 and melts at $t_{fus}(2)/^{o}C$ = 407 (Fig. 1).

ESTIMATED ERROR:

Temperature: accuracy probably ± 2 K (compiler).

REFERENCES:

COMPONENTS:

(1) Potassium bromide; KBr; [7758-02-3]
(2) Potassium methanoate (potassium formate); $KCHO_2$; [590-29-4]

ORIGINAL MEASUREMENTS:

Leonesi, D.; Braghetti, M.; Cingolani, A.; Franzosini, P.
Z. Naturforsch. 1970, **25a**, 52-55.

VARIABLES:

Temperature.

PREPARED BY:

Baldini, P.

EXPERIMENTAL VALUES:

t/°C	T/K[a]	$100x_1$
168.7	441.9	0
168.2	441.4	0.20
167.8	441.0	0.60
167.4	440.6	1.00
166.8	440.0	1.27
166.4	439.6	1.70
165.7	438.9	2.03
165.0	438.2	2.61
164.6	437.8	2.89
163.7	436.9	3.43
163.3	436.5	3.98
162.6	435.8	4.50
161.8	435.0	4.98
161.5	434.7	5.25
166.3	439.5	5.51
173.0	446.2	5.81
176.5	449.7	5.97
194.6	467.8	6.85
235.2	508.4	9.04
264.0	537.2	10.99
303.1	576.3	13.90

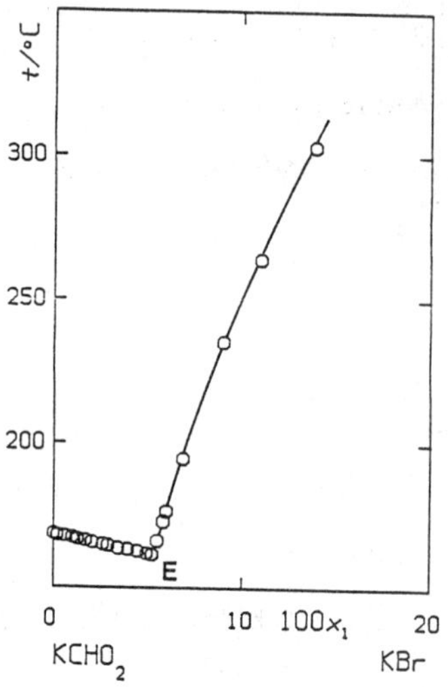

[a] T/K values calculated by the compiler.

Characteristic point(s): Eutectic, E, at 161.3 °C and $100x_1$= 5.3 (authors).

AUXILIARY INFORMATION

METHOD/APPARATUS/PROCEDURE:

A Pyrex device, suitable for work under an inert atmosphere, and allowing one to observe the system visually, was employed (for details, see Ref. 1). The initial crystallization temperatures were measured with a Chromel-Alumel thermocouple checked by comparison with a certified Pt resistance thermometer, and connected with a L&N Type K-3 potentiometer.

SOURCE AND PURITY OF MATERIALS:

C. Erba RP meterials, dried by heating under vacuum.

NOTES:

In the original paper the results were shown in graphical form. The above listed numerical values represent a private communication by one of the authors (F., P.) to the compiler.
The system could not be investigated above about 300 °C due to the thermal instability of the methanoate.
According to the authors, the trend of the liquidus branch richer in component 2 is close to ideal, and the formation of solid solutions in this region ought to be either insignificant, or at least contained within narrow limits.

ESTIMATED ERROR:

Temperature: accuracy probably ±0.1 K (compiler).

REFERENCES:

(1) Braghetti, M.; Leonesi, D.; Franzosini, P. **Ric. Sci.** 1968, **38**, 116-118.

COMPONENTS:

(1) Potassium bromide; KBr; [7758-02-3]
(2) Potassium ethanoate (potassium acetate); $KC_2H_3O_2$; [127-08-2]

ORIGINAL MEASUREMENTS:

Il'yasov. I.I.; Bergman, A.G.
Zh. Obshch. Khim. 1961, **31**, 368-370.

VARIABLES:

Temperature.

PREPARED BY:

Baldini, P.

EXPERIMENTAL VALUES:

The results are given only in graphical form (see figure). The system was investigated at $0 \leq 100x_1 \leq 25$.

Characteristic point(s):

Eutectic, E, at 290 °C and $100x_1 = 10$ (authors).

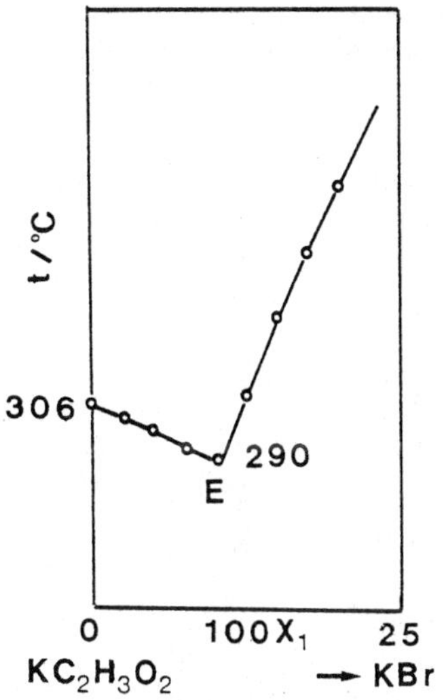

AUXILIARY INFORMATION

METHOD/APPARATUS/PROCEDURE:

Visual polythermal analysis; temperatures measured with a Nichrome-Constantane thermocouple and a millivoltmeter (Ref. 1).

NOTE:

See the NOTE relevant to the results obtained by Piantoni et al. (Ref. 2) on the same system (next Table).

SOURCE AND PURITY OF MATERIALS:

Not stated.
Component 1: $t_{fus}(1)/°C = 740$.
Component 2: $t_{fus}(2)/°C = 306$.

ESTIMATED ERROR:

Temperature: accuracy probably ±2 K.

REFERENCES:

(1) Il'yasov, I.I.; Bergman, A.G. **Zh. Obshch. Khim.** 1960, **30**, 355-358.
(2) Piantoni, G.; Leonesi, D.; Braghetti, M.; Franzosini, P. **Ric. Sci.** 1968, **38**, 127-132.

COMPONENTS:

(1) Potassium bromide;
KBr; [7758-02-3]
(2) Potassium ethanoate (potassium acetate);
$KC_2H_3O_2$; [127-08-2]

ORIGINAL MEASUREMENTS:

Piantoni, G.; Leonesi, D.; Braghetti, M.; Franzosini, P.
Ric. Sci. 1968, **38**, 127-132.

VARIABLES:

Temperature.

PREPARED BY:

Baldini, P.

EXPERIMENTAL VALUES:

The results are given only in graphical form (see figure). The system was investigated at $0 \leq 100\mathbf{x}_1 \leq 13$.

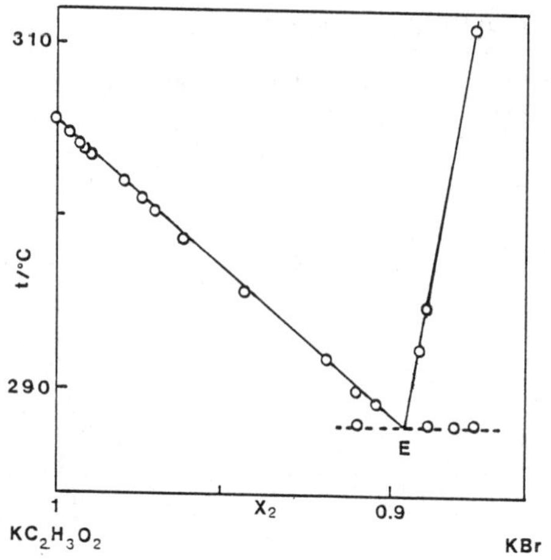

Characteristic point(s):

Eutectic, E, at 287.9 °C and $100\mathbf{x}_2$= 89.5 (authors).

AUXILIARY INFORMATION

METHOD/APPARATUS/PROCEDURE:

A Pyrex device, suitable for work under an inert atmosphere, and allowing one to observe the system visually, was employed (for details, see Ref. 1). The initial crystallization temperatures were measured with a Chromel-Alumel thermocouple checked by comparison with a certified Pt resistance thermometer, and connected with a L&N Type K-3 potentiometer.

NOTE:

Higher accuracy, and satisfactory mutual consistency of the results obtained by Piantoni et al. for the three binaries $K/C_2H_3O_2$, (Br,Cl,I) suggest to prefer here the data by these authors to those by Il´yasov and Bergman (Ref. 2). Increasingly positive deviation from ideality was observed by Piantoni et al. for the liquidus branch richer in the halide when KCl, KBr, and KI were successively taken into account. This is consistent with the (cryometric) limiting values:

[$\lim_{m_2 \to 0} (\Delta T/m_2)$ = 17.7, 17.4, and 16.0 K molality^{-1}, respectively] previously found by Braghetti et al. (Ref. 1) when the same halides were employed as solutes in molten potassium ethanoate (whose cryometric constant is: $\mathbf{K}_1$= 18.0$\pm$0.3 K molality^{-1}; Ref. 1).

SOURCE AND PURITY OF MATERIALS:

C. Erba RP materials, dried by heating under vacuum (private communication by the authors to the compiler).

ESTIMATED ERROR:

Temperature: accuracy probably $\pm$0.1 K.

REFERENCES:

(1) Braghetti, M.; Leonesi, D.; Franzosini, P.
Ric. Sci. 1968, **38**, 116-118.

COMPONENTS:

(1) Potassium methanoate (potassium formate); $KCHO_2$; [590-29-4]
(2) Potassium ethanoate (potassium acetate); $KC_2H_3O_2$; [127-08-2]

ORIGINAL MEASUREMENTS:

Sokolov, N.M.
Zh. Obshch. Khim. 1965, 35, 1897-1902.

VARIABLES:

Temperature.

PREPARED BY:

Baldini, P.

EXPERIMENTAL VALUES:

$t/^{\circ}C$	T/K^a	$100x_2$
167	440	0
157	430	5
154	427	10
157	430	15
168	441	20
179	452	25
188	461	30
199	472	35
209	482	40
217	490	45
224	497	50
232	505	55
240	513	60
249	522	65
257	530	70
266	539	75
273	546	80
282	555	85
290	563	90
297	570	95
302	575	100

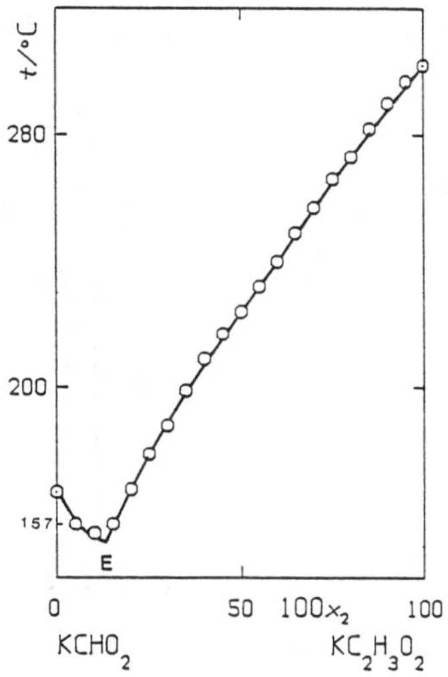

[a] T/K values calculated by the compiler.

Characteristic point(s): Eutectic, E, at 151 °C and $100x_2$= 13 (author).

AUXILIARY INFORMATION

METHOD/APPARATUS/PROCEDURE:

Visual polythermal analysis.

NOTE:

The fusion temperatures found by Sokolov are in reasonable agreement with those reported in Preface, Table 1 [$T_{fus}(1)$= 441.9±0.5 K, and $T_{fus}(2)$= 578.7±0.5 K]. Disagreement, on the contrary, exists about the number and location of the solid state transitions. As an example, for component 1 Table 1 of the Preface reports a single transition at a temperature (418±1 K) halfway between the highest (430 K) and second highest (408 K) values by Sokolov; the literature unfortunately provides no other data (Ref. 2).

ESTIMATED ERROR:

Temperature: accuracy probably ±2 K (compiler).

SOURCE AND PURITY OF MATERIALS:

Not stated.
Component 1 undergoes phase transitions at $t_{trs}(1)/^{\circ}C$= 60, 135, 157 (Ref. 1).
Component 2 undergoes phase transitions at $t_{trs}(2)/^{\circ}C$= 58, 150 [Ref. 1; the figure 150, however, is probably a misprint, because in several other papers the same author, quoting the same source (unavailable to the compiler), reports the figure 155; compiler].

REFERENCES:

(1) Sokolov, N.M. **Tezisy Dokl. X Nauch. Konf. S.M.I.** 1956.
(2) Sanesi, M.; Cingolani, A.; Tonelli, P.L.; Franzosini, P. **Thermal Properties,** in **Thermodynamic and Transport Properties of Organic Salts,** IUPAC Chemical Data Series No. 28 (Franzosini, P.; Sanesi, M.; Editors) Pergamon Press, Oxford, 1980, 29-115.

COMPONENTS:	ORIGINAL MEASUREMENTS:
(1) Potassium methanoate (potassium formate); $KCHO_2$; [590-29-4] (2) Potassium propanoate (potassium propionate); $KC_3H_5O_2$ [327-62-8]	Sokolov, N.M.; Minchenko, S.P. **Zh. Obshch. Khim.** 1971, **41**, 1656-1659.
VARIABLES:	PREPARED BY:
Temperature.	Baldini, P.

EXPERIMENTAL VALUES:

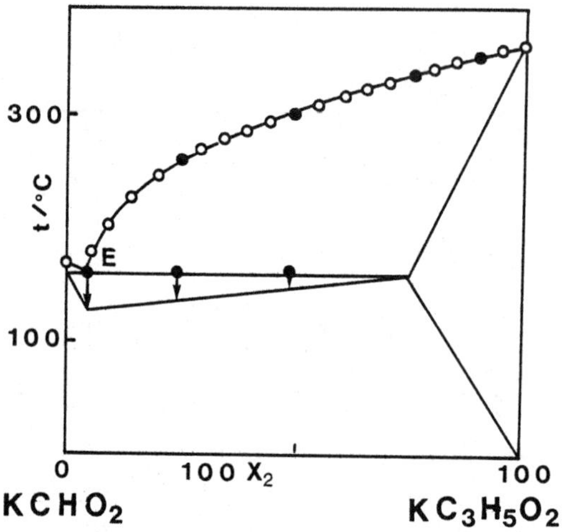

The results are reported only in graphical form (see figure; empty circles: visual polythermal analysis; filled circles: thermographical analysis).

Characteristic point(s): Eutectic, E, at 160 °C and $100\mathbf{x}_2$= 5 (authors).

AUXILIARY INFORMATION

METHOD/APPARATUS/PROCEDURE:

Visual polythermal analysis supplemented with thermographical analysis.

NOTE:

The fusion temperatures by Sokolov and Minchenko (440 and 638 K for components 1 and 2, respectively) almost coincide with those listed in Preface, Table 1 (respectively 441.9±0.5 K and 638.3±0.5 K). An approximate agreement exists also on the solid state transition of component 2. On the contrary, there in disagreement about solid state transitions of component 1, inasmuch as Table 1 of the Preface reports a single transformation at a temperature (418±1 K) halfway between the highest (430 K) and second highest (408 K) values by Sokolov and Minchenko; the literature, unfortunately, provides no other data (Ref. 3).

ESTIMATED ERROR:
Temperature: accuracy probably ±2 K (compiler).

SOURCE AND PURITY OF MATERIALS:

Component 1 (commercial material recrystallized from methanoic acid) melts at $t_{fus}(1)/^{o}C$= 167 and undergoes phase transitions at $t_{trs}(1)/^{o}C$= 60, 135, 157 (Ref. 1).
Component 2 (prepared from propanoic acid and carbonate, Ref.2) melts at 365 °C and undergoes a phase transition at 68 °C (Ref. 1).

REFERENCES:

(1) Sokolov, N.M. **Tezisy Dokl. X Nauch. Konf. S.M.I.** 1956,
(2) Sokolov, N.M. **Zh. Obshch. Khim.** 1954, **24**, 1581-1593.
(3) Sanesi, M.; Cingolani, A.; Tonelli, P.L.; Franzosini, P.; **Thermal Properties**, in **Thermodynamic and Transport Properties of Organic Salts**, IUPAC Chemical Data Series No. 28 (Franzosini, P.; Sanesi, M.; Editors), Pergamon Press, Oxford, 1980, 29-115.

COMPONENTS:	EVALUATOR:
(1) Potassium methanoate (potassium formate); $KCHO_2$; [590-29-4] (2) Potassium butanoate (potassium butyrate); $KC_4H_7O_2$ [589-39-9]	Spinolo, G., Dipartimento di Chimica Fisica, Universita´ di Pavia (ITALY).

CRITICAL EVALUATION:

This binary was studied only by Sokolov and Minchenko (Ref. 1), who employed the visual polythermal analysis to outline the lower boundary of the isotropic liquid field, and claimed the existence of a single invariant, i.e., a eutectic at 439 K and $100x_2$= 0.9.

However, taking into account that component 2 forms liquid crystals and that its actual fusion temperature is 626.1±0.7 K (see Preface, Table 1), the topology of the phase diagram ought to be described more correctly with reference to Schemes B.1 or B.2 of the Preface. An invariant type M'_p (undetected by Sokolov and Minchenko) should also exist: accordingly, the main branch of Sokolov and Minchenko's diagram should represent solid-liquid equilibria only at temperatures lower than that corresponding to M'_p.

It can be further noted that a reasonable agreement exists: (i) between the fusion temperature reported for component 1 in Ref. 1 (440 K) and in Table 1 of the Preface (441.9±0.5 K); and (ii) between Sokolov and Minchenko's fusion temperature of component 2 (677 K) and the clearing temperature (677.3±0.5 K) listed in Preface, Table 1 for the same component.

REFERENCES:

(1) Sokolov, N.M.; Minchenko, S.P.; **Zh. Obshch. Khim.**, 1974, **44**, 1429-1431.

COMPONENTS:	ORIGINAL MEASUREMENTS:
(1) Potassium methanoate (potassium formate); $KCHO_2$; [590-29-4] (2) Potassium butanoate (potassium butyrate); $KC_4H_7O_2$ [589-39-9]	Sokolov, N.M.; Minchenko, S.P. **Zh. Obshch. Khim.**, 1974, 44, 1429-1431.
VARIABLES:	PREPARED BY:
Temperature.	Baldini, P.

EXPERIMENTAL VALUES:

The results are reported only in graphical form (see figure).

Characteristic point(s): Eutectic, E, at 166 °C and $100x_2$= 0.9 (authors).

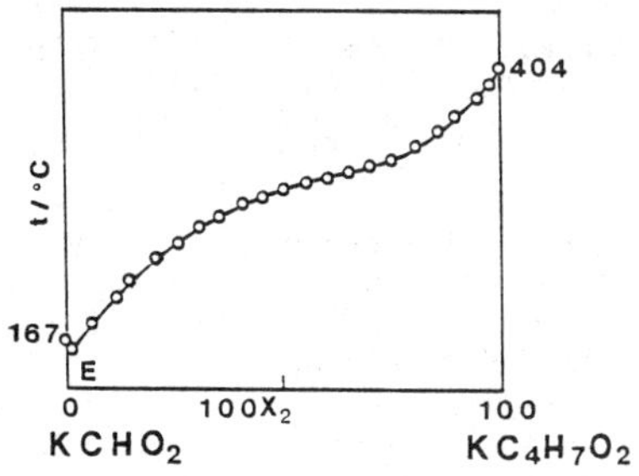

METHOD/APPARATUS/PROCEDURE:

Visual polythermal analysis.

SOURCE AND PURITY OF MATERIALS:

Component 1: commercial material recrystallized; it melts at 167 °C. Component 2: prepared from n-butanoic acid and the carbonate (Ref. 1); it melts at $t_{fus}(2)/^oC$= 404.

ESTIMATED ERROR:

Temperature: accuracy probably ±2 K (compiler).

REFERENCES:

(1) Sokolov, N.M.; **Zh. Obshch. Khim.** 1954, **24**, 1581-1593.

COMPONENTS:	EVALUATOR:
(1) Potassium methanoate (potassium formate); $KCHO_2$; [590-29-4] (2) Potassium **iso.**butanoate (potassium **iso.**butyrate); $Ki.C_4H_7O_2$ [19455-20-0]	Spinolo, G., Dipartimento di Chimica Fisica, Universita´ di Pavia (ITALY).

CRITICAL EVALUATION:

This binary was studied only by Sokolov and Minchenko (Ref. 1), who employed the visual polythermal analysis to outline the lower boundary of the isotropic liquid field, and claimed the existence of a single invariant, i.e., a eutectic at 437 K and $100\mathbf{x}_2 = 1$.

Component 2, however, forms liquid crystals. Therefore the topology of the phase diagram ought to be described more correctly with reference to Schemes B.1 or B.2 of the Preface, and an invariant type M'_p (undetected by Sokolov and Minchenko) should also exist. Accordingly, the main branch of Sokolov and Minchenko´s diagram should represent solid-liquid equilibria only at temperatures lower than that corresponding to M'_p.

It can be further noted that a reasonable agreement exists: (i) between the fusion temperature reported for component 1 in Ref. 1 (440 K) and in Preface, Table 1 (441.9 ± 0.5 K); and (ii) between Sokolov and Minchenko´s fusion temperature of component 2 (629 K) and the clearing temperature (625.6 ± 0.8 K) listed in Table 2 of the Preface for the same component.

Disagreement, on the contrary, exists about the remaining phase transformations. For component 1, Table 1 of the Preface reports a single solid state transition occurring at a temperature (418 ± 1 K) halfway between the highest (430 K) and second highest (408 K) values by Sokolov and Minchenko; the literature, unfortunately, provides no other data (Ref. 2).

For component 2, three phase transition temperatures are mentioned in Ref. 1, i.e., 621, 546, and 481 K, the second of which can be reasonably identified with the fusion temperature [$\mathbf{T}_{fus}(2) = 553.9 \pm 0.5$ K] listed in Preface, Table 2. Consequently: (i) the transition temperature at 621 K (if actually existing) might correspond to some kind of transformation (undetected by DSC, see Preface, Table 2) within the liquid crystal field; and (ii) only the transition at 481 K should correspond to a solid state transformation, although the latter figure is almost 60 K higher than the single $\mathbf{T}_{trs}(2)$ value (424 ± 3 K) listed in Table 2 of the Preface.

REFERENCES:

(1) Sokolov, N.M.; Minchenko, S.P.
Zh. Obshch. Khim., 1977, **47**, 740-742.

(2) Sanesi, M.; Cingolani, A.; Tonelli, P.L.; Franzosini, P.
Thermal Properties, in **Thermodynamic and Transport Properties of Organic Salts**, IUPAC Chemical Data Series No. 28 (Franzosini, P.; Sanesi, M.; Editors), Pergamon Press, Oxford, 1980, 29-115.

COMPONENTS:

(1) Potassium methanoate (potassium formate); $KCHO_2$; [590-29-4]
(2) Potassium **iso.**butanoate (potassium **iso.**butyrate); $Ki.C_4H_7O_2$ [19455-20-0]

ORIGINAL MEASUREMENTS:

Sokolov, N.M.; Minchenko, S.P.
Zh. Obshch. Khim., 1977, **47**, 740-742.

VARIABLES:

Temperature.

PREPARED BY:

Baldini, P.

EXPERIMENTAL VALUES:

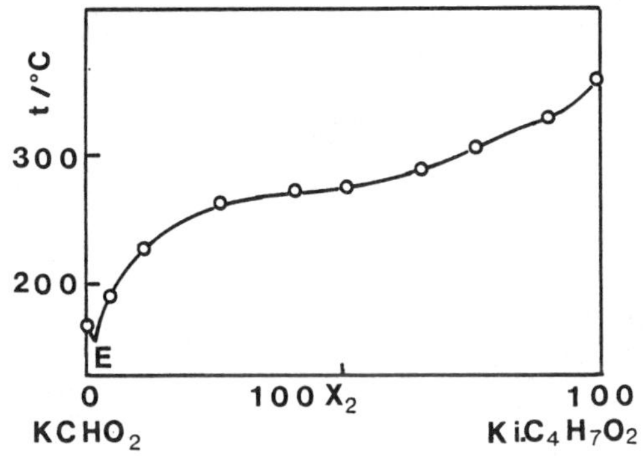

The results are reported only in graphical form (see figure).

Characteristic point(s):

Eutectic, E, at 164 °C and $100x_2$= 1 (authors).

AUXILIARY INFORMATION

METHOD/APPARATUS/PROCEDURE:

Visual polythermal analysis.

SOURCE AND PURITY OF MATERIALS:

Component 1: commercial material recrystallized from methanoic acid; it melts at $t_{fus}(1)/^{o}C$= 167 and undergoes phase transitions at $t_{trs}(1)/^{o}C$= 60, 135, 157.
Component 2: prepared from **i.**butanoic acid and the carbonate (Ref. 1); it melts at $t_{fus}(2)/^{o}C$= 356 and undergoes phase transitions at $t_{trs}(2)/^{o}C$= 208, 273, 348.

ESTIMATED ERROR:

Temperature: accuracy probably $\pm$2 K (compiler).

REFERENCES:

(1) Sokolov, N.M.
Zh. Obshch. Khim. 1954, **24**, 1581-1593.

COMPONENTS:

(1) Potassium methanoate (potassium formate);
$KCHO_2$; [590-29-4]
(2) Potassium chloride;
KCl; [7447-40-7]

ORIGINAL MEASUREMENTS:

Leonesi, D.; Braghetti, M.; Cingolani, A.; Franzosini, P.
Z. Naturforsch. 1970, **25a**, 52-55.

VARIABLES:

Temperature.

PREPARED BY:

Baldini, P.

EXPERIMENTAL VALUES:

$t/^{\circ}C$	T/K^a	$100x_2$
168.7	441.9	0
168.3	441.5	0.20
168.1	441.3	0.43
167.7	440.9	0.60
167.4	440.6	0.84
166.7	439.9	1.06
165.7	438.9	1.83
165.5	438.7	1.98
165.0	438.2	2.55
164.3	437.5	2.95
163.8	437.0	3.46
163.6	436.8	3.67
163.6	436.8	3.75
166.6	439.8	3.78
172.1	445.3	4.02
176.6	449.8	4.16
179.2	452.4	4.22
183.5	456.7	4.45
195.8	469.0	4.96
218.4	491.6	6.09
244.0	517.2	7.50
270.1	543.3	9.02
299.7	572.9	11.16

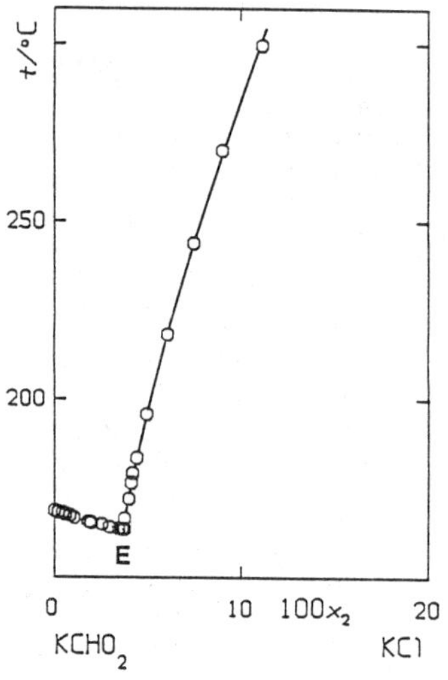

[a]T/K values calculated by the compiler.

Characteristic point(s): Eutectic, E, at 163.5 °C and $100x_2$= 3.7 (authors).

AUXILIARY INFORMATION

NOTES:

In the original paper the results were shown in graphical form. The above listed numerical values represent a private communication by one of the authors (F., P.) to the compiler.

According to the authors, the trend of the liquidus branch richer in component 1 is close to ideal, and the formation of solid solutions in this region ought to be either insignificant, or at least contained within very narrow limits. Indeed, previous investigations by the same group (Ref. 2) stated that the cryometric constant of potassium methanoate was $K = 11.5 \pm 0.1$ K molality^{-1}, and that $\lim_{m \to 0} (\Delta T/m) = 11.6$ K molality^{-1} (ΔT : experimental freezing point depression; m: molality of the solute) when KCl was the solute.

SOURCE AND PURITY OF MATERIALS:

C. Erba RP meterials, dried by heating under vacuum.

METHOD/APPARATUS/PROCEDURE:

A Pyrex device, suitable for work under an inert atmosphere, and allowing one to observe the system visually, was employed (for details, see Ref. 1). The initial crystallization temperatures were measured with a Chromel-Alumel thermocouple checked by comparison with a certified Pt resistance thermometer, and connected with a L&N Type K-3 potentiometer.

The system could not be investigated above 300 °C due to the thermal instability of the methanoate.

ESTIMATED ERROR:

Temperature: accuracy probably ± 0.1 K (compiler).

REFERENCES:

(1) Braghetti, M.; Leonesi, D.; Franzosini, P.; **Ric. Sci.** 1968, **38**, 116-118.
(2) Leonesi, D.; Piantoni, G.; Berchiesi, G.; Franzosini, P.
Ric. Sci. 1968, **38**, 702-705.

COMPONENTS:	EVALUATOR:
(1) Potassium methanoate (potassium formate); $KCHO_2$; [590-29-4] (2) Potassium thiocyanate; KCNS; [333-20-0]	Franzosini, P., Dipartimento di Chimica Fisica, Universita´ di Pavia (ITALY).

CRITICAL EVALUATION:

The liquidus of this binary was studied with visual methods by Sokolov and Pochtakova (Ref. 1), and by Berchiesi and Laffitte [Ref. 2, where reference is made to a previous investigation by Braghetti et al. (Ref. 3) for what concerns the branch richer in component 2, and the T_{trs} and $\Delta_{trs}H_m$ values of either components]. According to both papers, a single eutectic exists whose coordinates should be either 356 K and $100x_2$= 47.5 (Ref. 1), or 351.7 K and $100x_2$= 46 (Ref. 2).

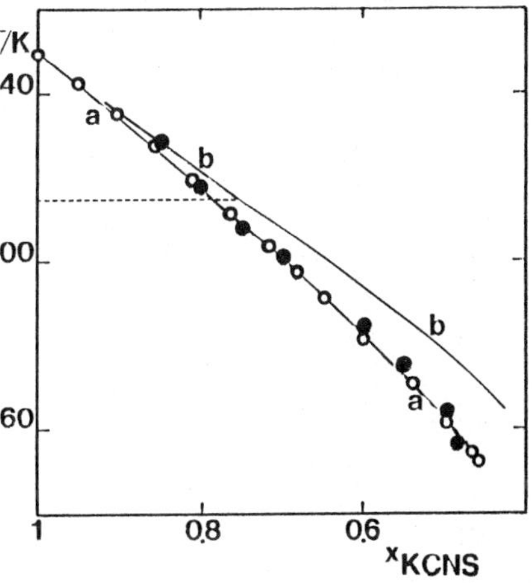

Substantially agreeing figures are reported for $T_{fus}(1)$ [440 K (Ref. 1); 441.85 K (Ref. 2)], $T_{fus}(2)$ [450 K (Ref. 1); 449.15 K (Ref. 2)], and $T_{trs}(2)$ [415.7 K (Ref. 4, quoted in Ref. 1); 415 K (Ref. 3, quoted in Ref. 2)]. Conversely, disagreement exists about the number and location of the solid state transitions of component 1, which ought to be three (at 430, 408, and 333 K, respectively) according to Ref. 5, quoted in Ref. 1, and only one (at 418 K) according to Ref. 3, quoted in Ref. 2. The latter information, however, ought to be looked at as more trustworthy being based on DSC records.

According once more to Ref. 3, solid solutions ought to be absent (or at least contained within very narrow limits) in the composition range between pure component 2 and the eutectic, as suggested by the DSC traces. The authors could thus employ the well known equation

$$T(2) = \frac{[\Delta_{fus}(2)H_m/R + \Delta_{trs}(2)H_m/R] + (A/R)(x_1)^2}{[\Delta_{fus}(2)S_m/R + \Delta_{trs}(2)S_m/R] - \ln x_2}$$

to calculate the solid-liquid equilibrium temperatures, $T(2)/K$, relevant to the liquidus branch richer in component 2 (see curve a of Fig. 1), assuming the following numerical values: $\Delta_{fus}(2)H_m/R$= 1545 K; $\Delta_{trs}(2)H_m/R$= 186 K (to be introduced only when $T(2) \leq T_{trs}$); and $A/K = -800/R + (360/R)x_1$ (A/K: empirical factor introduced to take into account the non-ideal behavior of the mixtures; ideality represented by curve b).

In the figure, the filled and empty circles correspond to data from Ref. 1 and Ref. 2, respectively. It is apparent that spreading is larger in the first set than in the second one, which, moreover, gives [at $T_{trs}(2)/K$= 415] a better evidence of the expected change of slope. Accordingly, and taking also into account the poor reliability of the $T_{trs}(1)$ values quoted in Ref. 1, the evaluator recommends Berchiesi and Laffitte´s presentation (Ref. 2), although regretting that information was not extended to the solidus.

REFERENCES:

(1) Sokolov, N.M.; Pochtakova, E.I.; **Zh. Obshch. Khim.** 1958, **28**, 1391-1397 (*); **Russ. J. Gen. Chem. (Engl. Transl.)** 1958, **28**, 1449-1454.

(2) Berchiesi, G.; Laffitte, M.; **J. Chim. Phys.** 1971, 877-881.

(3) Braghetti, M.; Berchiesi, G.; Franzosini, P.; **Ric. Sci.** 1969, **39**, 576-584.

(4) Ravich, M.I.; Ketkovich, V.I.; Rassonskaya, I.S.; **Izv. Sektora Fiz.-Khim. Anal.** 1949, **17**, 254.

(5) Sokolov, N.M.; **Tezisy Dokl. X Nauch. Konf. S.M.I.** 1956.

COMPONENTS:	ORIGINAL MEASUREMENTS:
(1) Potassium methanoate (potassium formate); $KCHO_2$; [590-29-4] (2) Potassium thiocyanate; KCNS; [333-20-0]	Sokolov, N.M.; Pochtakova, E.I. **Zh. Obshch. Khim.** 1958, **28**, 1391-1397 (*); **Russ. J. Gen. Chem. (Engl. Transl.)** 1958, **28**, 1449-1454.
VARIABLES:	PREPARED BY:
Temperature.	Baldini, P.

EXPERIMENTAL VALUES:

$t/^{o}C$	T/K^{a}	$100\mathbf{x}_1$
177	450	0
168	441	5
162	435	10
156	429	15
145	418	20
135	408	25
128	401	30
119	392	35
111	384	40
102	375	45
91	364	50
83	356	52.5
89	362	55
96	369	57.5
104	377	60
112	385	65
119	392	70
129	402	75
137	410	80
147	420	85
155	428	90
162	435	95
167	440	100

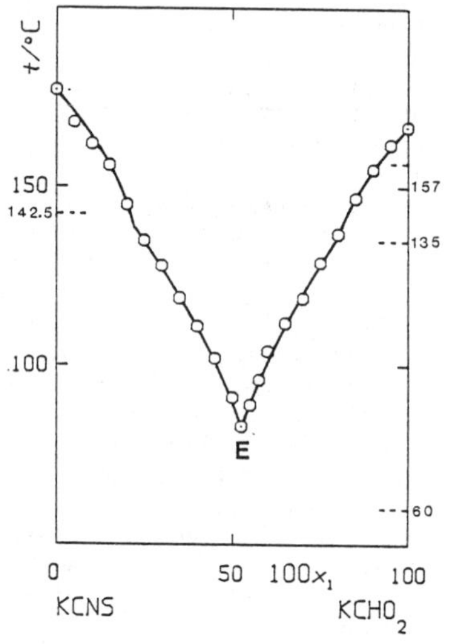

[a] **T**/K values calculated by the compiler.

Characteristic point(s): Eutectic, E, at 83 ^{o}C and $100\mathbf{x}_1$= 52.5 (authors).

AUXILIARY INFORMATION

METHOD/APPARATUS/PROCEDURE:

Visual polythermal analysis.

SOURCE AND PURITY OF MATERIALS:

Not stated.
Component 1: commercial material recrystallized from water; it undergoes phase transitions at $t_{trs}(1)/^{o}C$= 60, 135, 157 (Ref. 1).
Component 2: commercial material recrystallized from alcohol; it undergoes a phase transition at $t_{trs}(2)/^{o}C$= 142.5 (Ref. 2).

ESTIMATED ERROR:

Temperature: accuracy probably ±2 K (compiler).

REFERENCES:

(1) Sokolov, N.M. **Tezisy Dokl. X Nauch. Konf. S.M.I.** 1956.
(2) Ravich, M.I.; Ketkovich, V.I.; Rassonskaya, I.S.; **Izv. Sektora Fiz.-Khim. Anal.** 1949, **17**, 254.

COMPONENTS:

(1) Potassium methanoate (potassium formate); $KCHO_2$; [590-29-4]
(2) Potassium thiocyanate; KCNS; [333-20-0]

ORIGINAL MEASUREMENTS:

Berchiesi, G.; Laffitte, M.
J. Chim. Phys. 1971, 877-881.

VARIABLES:

Temperature.

PREPARED BY:

Baldini, P.

EXPERIMENTAL VALUES:

$t/^{o}C$	T/K^{a}	$100x_2$	$t/^{o}C$	T/K^{a}	$100x_2$
176.00	449.15	100	84.25	357.40	42.18
169.10	442.25	95.10	87.55	360.70	40.52
161.98	435.13	90.43	89.70	362.85	39.28
154.50	427.65	85.79	93.88	367.03	37.25
146.50	419.65	81.03	96.20	369.35	35.88
138.28	411.43	76.41	105.48	378.63	32.19
130.28	403.43	71.76	114.40	387.55	28.41
123.90	397.05	68.20	124.20	397.35	24.37
117.85	391.00	64.85	133.90	407.05	19.91
108.22	381.37	59.85	143.92	417.07	15.12
97.75	370.90	54.00	152.95	426.10	10.16
88.15	361.30	50.03	161.65	434.80	4.97
81.03	354.18	46.91	168.70	441.85	0
79.85	353.00	44.45			

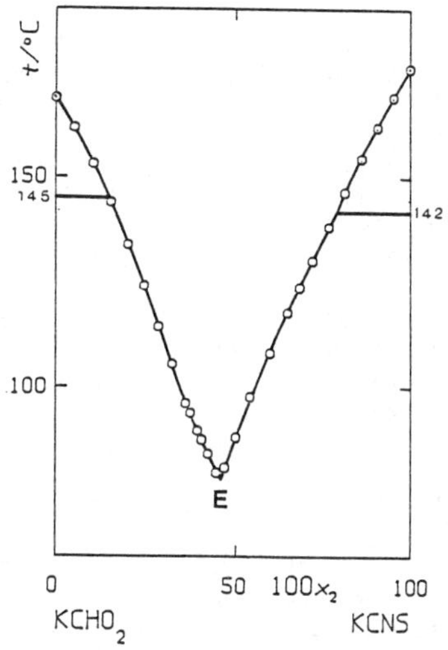

[a] T/K values calculated by the compiler.

Characteristic point(s):

Eutectic, E, at 78.5 °C and $100x_2$= 46 (compiler).

Note - The data relevant to the liquidus branch richer in component 2 were already published in Ref. 1.

AUXILIARY INFORMATION

METHOD/APPARATUS/PROCEDURE:

The liquidus was determined with a visual method, (details in Ref. 2), supplemented with DSC measurements. The enthalpy changes (not to be listed here) associated with various thermodynamic processes were measured with differential flux calorimetry (using a modified Tian-Calvet calorimeter).

SOURCE AND PURITY OF MATERIALS:

Not stated.
Component 1 undergoes a phase transition at $T_{trs}(1)/K$= 418 (Ref. 1).
Component 2 undergoes a phase transition at $T_{trs}(2)/K$= 415 (Ref. 1).

ESTIMATED ERROR:

Temperature: accuracy probably ±0.05 K (compiler).

REFERENCES:

(1) Braghetti, M.; Berchiesi, G.; Franzosini, P. **Ric. Sci.** 1969, **39**, 576-584.
(2) Braghetti, M.; Leonesi, D.; Franzosini, P. **Ric. Sci.** 1968, **38**, 116-118.

COMPONENTS:	ORIGINAL MEASUREMENTS:
(1) Potassium methanoate (potassium formate); $KCHO_2$; [590-29-4] (2) Potassium iodide; KI; [7681-11-0]	Leonesi, D.; Braghetti, M.; Cingolani, A.; Franzosini, P. **Z. Naturforsch.** 1970, **25a**, 52-55.
VARIABLES:	PREPARED BY:
Temperature.	Baldini, P.

EXPERIMENTAL VALUES:

$t/^{o}C$	T/K^{a}	$100x_2$	$t/^{o}C$	T/K^{a}	$100x_2$
168.7	441.9	0	160.0	433.2	6.06
168.2	441.4	0.32	160.0	433.2	6.24
167.5	440.7	0.85	158.5	431.7	7.02
167.0	440.2	1.06	157.2	430.4	8.01
166.5	439.7	1.61	157.6	430.8	8.82
165.9	439.1	1.94	159.7	432.9	8.96
165.9	439.1	2.05	165.5	438.7	9.19
165.5	438.7	2.33	171.3	444.5	9.49
165.3	438.5	2.46	172.6	445.8	9.64
164.7	437.9	2.86	190.8	464.0	10.72
163.7	436.9	3.60	212.0	485.2	11.99
163.6	436.8	3.61	240.6	513.8	14.03
161.6	434.8	5.00	276.7	549.9	16.78
161.3	434.5	5.10	298.3	571.5	18.98

[a] **T**/K values calculated by the compiler.

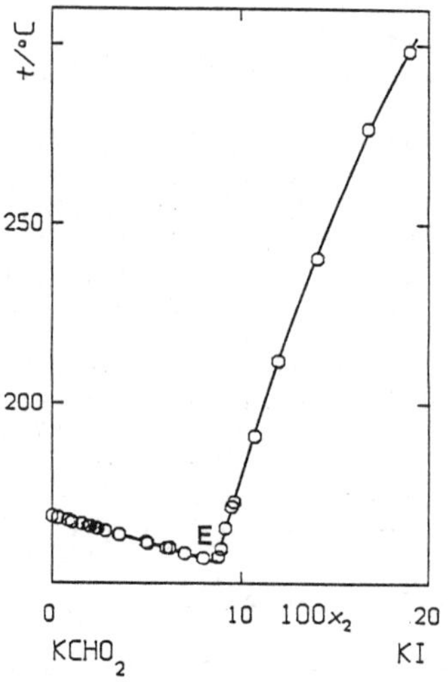

Note 1 - In the original paper the results were shown in graphical form. The above listed numerical values represent a private communication by one of the authors (F., P.) to the compiler.

Note 2 - The system could not be investigated above 300 °C due to the thermal instability of the methanoate.

Characteristic point(s): Eutectic, E, at 156.3 °C and $100x_2$= 8.7 (authors).

AUXILIARY INFORMATION

METHOD/APPARATUS/PROCEDURE:

A Pyrex device, suitable for work under an inert atmosphere, and allowing one to observe the system visually, was employed (for details, see Ref. 1). The initial crystallization temperatures were measured with a Chromel-Alumel thermocouple checked by comparison with a certified Pt resistance thermometer, and connected with a L&N Type K-3 potentiometer.

NOTE:

According to the authors, the trend of the liquidus branch richer in component 2 is close to ideal, and the formation of solid solutions in this region ought to be either insignificant, or at least contained within narrow limits.

SOURCE AND PURITY OF MATERIALS:

C. Erba RP materials, dried by heating under vacuum.

ESTIMATED ERROR:

Temperature: accuracy probably ±0.1 K (compiler).

REFERENCES:

(1) Braghetti, M.; Leonesi, D.; Franzosini, P.
Ric. Sci. 1968, **38**, 116-118.

COMPONENTS:

(1) Potassium methanoate (potassium formate); $KCHO_2$; [590-29-4]
(2) Potassium nitrite; KNO_2; [7758-09-0]

ORIGINAL MEASUREMENTS:

Sokolov, N.M.; Minich, M.A.
Zh. Neorg. Khim. 1961, 6, 2558-2562 (*);
Russ. J. Inorg. Chem. (Engl. Transl.) 1961, 6, 1293-1295.

VARIABLES:

Temperature.

PREPARED BY:

Baldini, P.

EXPERIMENTAL VALUES:

$t/^oC$	T/K^a	$100x_2$	$t/^oC$	T/K^a	$100x_2$
168	441	0	223	496	55
163	436	5	253	526	60
155	428	10	278	551	65
147	420	15	305	578	70
135	408	20	330	603	75
130	403	25	353	626	80
119	392	30	376	649	85
114	387	35	399	672	90
137	410	40	420	693	95
169	442	45	436	709	100
199	472	50			

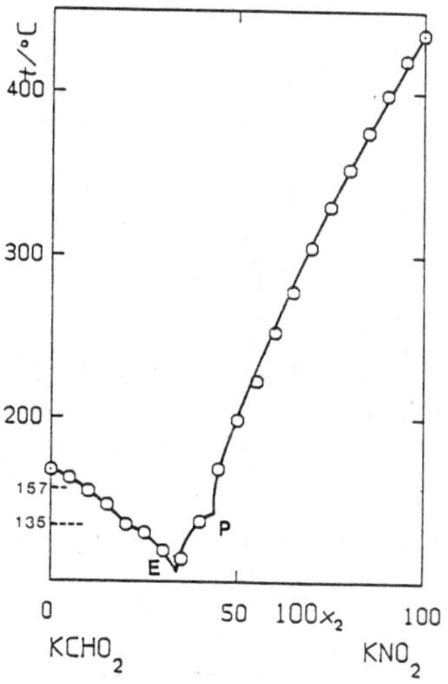

[a] T/K values calculated by the compiler.

Characteristic point(s):

Eutectic, E, at 107 °C and $100x_2$= 33.5 (authors).
Peritectic, P, at 143 °C and $100x_2$= 44 (authors).

Intermediate compound(s):

$K_2CHO_2NO_2$ (tentative composition; authors) incongruently melting.

AUXILIARY INFORMATION

METHOD/APPARATUS/PROCEDURE:

Visual polythermal analysis.

NOTE:

Solid state transitions of component 1 should be three (occurring at 430, 408, and 333 K, respectively) according to Ref. 1, and only one (at 418$\pm$1 K) according to Table 1 of the Preface. Unfortunately, no information from other sources is available. It can be noted, however, that, e.g., the trend of the liquidus branch richer in $KCHO_2$ of the binary K/CHO_2, NO_3 studied by Berchiesi et al. (Ref. 3) supports Table 1 statement. Moreover, the existence (and composition) of the intermediate compound ought to be more convincingly proved.

SOURCE AND PURITY OF MATERIALS:

Component 1: commercial "chemically pure" material recrystallized from methanoic acid; it undergoes phase transitions at $t_{trs}(1)/^oC$= 60, 135, 157 (Ref. 1).
Component 2: material prepared by reducing potassium nitrate with lead, melting at $t_{fus}(2)/^oC$= 436 after three recrystalliza tions; it undergoes a phase transition at $t_{trs}(2)/^oC$= 45 (Ref. 2).

ESTIMATED ERROR:

Temperature: accuracy probably $\pm$2 K (compiler).

REFERENCES:

(1) Sokolov, N.M. **Tezisy Dokl. X Nauch. Konf. S.M.I.** 1956.
(2) Berul', S.I.; Bergman, A.G. **Izv. Sektora Fiz.-Khim. Anal.** 1952, 21, 178-183.
(3) Berchiesi, G.; Cingolani, A.; Leonesi, D. **Z. Naturforsch.** 1970, 25a, 1766-1767.

COMPONENTS:	EVALUATOR:
(1) Potassium methanoate (potassium formate) $KCHO_2$; [590-29-4] (2) Potassium nitrate; KNO_3; [7757-79-1]	Franzosini, P., Dipartimento di Chimica Fisica, Universita' di Pavia (ITALY).

CRITICAL EVALUATION:

This system was studied first (Ref. 1) by Dmitrevskaya who, on the basis of her visual polythermal investigation, claimed the existence of the congruently melting intermediate compound $K_3(CHO_2)_2NO_3$, able to give eutectics with either component, at 423 K and $100\mathbf{x}_2$= 32.5, and 419 K and $100\mathbf{x}_2$= 44, respectively.

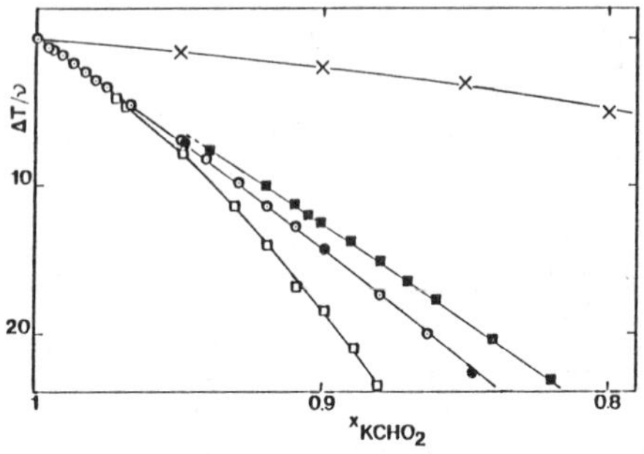

Berchiesi et al. (Ref. 2) re-investigated the binary (employing again a visual method) as a side of the ternary K/CHO_2, CNS, NO_3, and found an incongruently melting intermediate compound [whose composition, argued from auxiliary DSC measurements, should be $K_5(CHO_2)_4NO_3$], a peritectic at 399.7 K and $100\mathbf{x}_2$= 26.2, and a eutectic at 387 K and $100\mathbf{x}_2$= 37.9.

The following considerations can help to evaluate the trustworthiness of these far different results.

Leonesi et al. (Ref. 3) performed cryometric measurements in molten $KCHO_2$, stating that the cryometric constant was $\mathbf{K}_1$= 11.5±0.1 K molality^{-1} which corresponds to $\Delta_{fus}(1)\mathbf{H}_m$= 11.9 kJ mol^{-1} (2.84±0.03 kcal mol^{-1} in the original text). The latter value is, in turn, in satisfactory agreement with those subsequently determined with DSC by Braghetti et al. (11.8 kJ mol^{-1}; Ref. 4), and with Calvet microcalorimetry by Berchiesi and Laffitte (11.5±0.1 kJ mol^{-1}; Ref. 5).

In particular, Leonesi et al. (Ref. 3) found limiting values

$$[\lim_{m_2 \to 0} (\Delta T/m_2)]/\nu = 11.5, \quad 11.55, \quad \text{and} \quad 11.4 \text{ K molality}^{-1}$$

for KNO_3 (ν= 1; ν : number of cryometrically active foreign species), $LiNO_3$ (ν= 2), and $CsNO_3$ (ν= 2), respectively, which implies that a solubility of these solutes in $KCHO_2$ in the solid state should be either absent, or negligible. The three sets of $(\Delta T/\nu)$ vs. $\mathbf{x}_1$ data from Ref. 3 (KNO_3: empty circles; $LiNO_3$: empty squares; $CsNO_3$: filled squares), which exhibit a satisfactory mutual consistency, are compared in the figure with the data taken in K/CHO_2, NO_3 mixtures rich in component 1 by Berchiesi et al. (filled circles; Ref. 2), and by Dmitrevskaya (crosses; Ref. 1), respectively: it is apparent that the results from Ref. 1 are inconsistent with those from both Ref. 2 and Ref. 3.

Concerning solid state transformations of component 1, three transitions are quoted in Ref. 1 from Ref. 6 as occurring at 333, 408, and 430 K, respectively, whereas a single transition (at 418±1 K) is listed in Table 1 of the Preface. Berchiesi et al. (Ref. 2) make no explicit reference to any transition, but an inspection of their liquidus branch richest in component 1 allows one to observe a single change of slope around 418 K, i.e., in correspondence with the value from Table 1.

In conclusion, in the evaluator's opinion the data by Berchiesi et al. (Ref. 2) are to be recommended, although a better knowledge of the solidus would be desirable.

REFERENCES:

(1) Dmitrevskaya, O.I.; **Zh. Obshch. Khim.** 1958, **28**, 299-304 (*); **Russ. J. Gen. Chem. (Engl. Transl.)** 1958, **28**, 295-300.
(2) Berchiesi, G.; Cingolani, A.; Leonesi, D. **Z. Naturforsch.** 1970, **25a**, 1766-1767.
(3) Leonesi, D.; Piantoni, G.; Berchiesi, G.; Franzosini, P. **Ric. Sci.** 1968, **38**, 702-705.
(4) Braghetti, M.; Berchiesi, G.; Franzosini, P. **Ric. Sci.** 1969, **39**, 576-584.
(5) Berchiesi, G.; Laffitte, M.; **J. Chim. Fis.** 1971, 877-881.
(6) Sokolov, N.M.; **Tezisy Dokl. X Nauch. Konf. S.M.I.** 1956.

COMPONENTS:

(1) Potassium methanoate (potassium formate); $KCHO_2$; [590-29-4]
2) Potassium nitrate; KNO_3; [7757-79-1]

ORIGINAL MEASUREMENTS:

Dmitrevskaya, O.I.
Zh. Obshch. Khim. 1958, **28**, 299-304 (*);
Russ. J. Gen. Chem. (Engl. Transl.) 1958, **28**, 295-300.

VARIABLES:

Temperature.

PREPARED BY:

Baldini, P.

EXPERIMENTAL VALUES:

$t/^{o}C$	T/K^a	$100x_2$	$t/^{o}C$	T/K^a	$100x_2$
167	440	0	161	434	47.5
166	439	5	172	445	50
165	438	10	193	466	55
164	437	15	212	485	60
162	435	20	230	503	65
159	432	25	248	521	70
155	428	30	264	537	75
150	423	32.5	279	552	80
153	426	35	294	567	85
154	427	37.5	309	582	90
150.5	423.5	40	323	596	95
146	419	42	337	610	100
153	426	45			

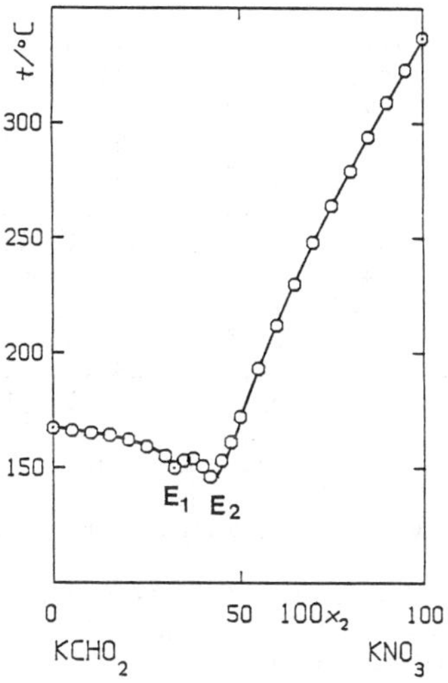

[a] T/K values calculated by the compiler.

Characteristic point(s):

Eutectic, E_1, at 150 °C and $100x_1$= 67.5 (author).
Eutectic, E_2, at 146 °C and $100x_2$= 44 (author).

Intermediate compound(s):

$K_3(CHO_2)_2NO_3$, congruently melting (author).

AUXILIARY INFORMATION

METHOD/APPARATUS/PROCEDURE:

Visual polythermal analysis; temperatures measured with a Nichrome-Constantane thermocouple.

SOURCE AND PURITY OF MATERIALS:

"Chemically pure" materials, recrystallized and dried to constant mass.
Component 1 undergoes phase transitions at $t_{trs}(1)/^{o}C$= 60, 135, 157 (Ref. 1).
Component 2 undergoes phase transitions at $t_{trs}(2)/^{o}C$= 124, 316 (current literature).

ESTIMATED ERROR:

Temperature: accuracy probably ±2 K (compiler).

REFERENCES:

(1) Sokolov, N.M.
Tezisy Dokl. X Nauch. Konf. S.M.I. 1956.

COMPONENTS:	ORIGINAL MEASUREMENTS:
(1) Potassium methanoate (potassium formate); $KCHO_2$; [590-29-4] (2) Potassium nitrate; KNO_3; [7757-79-1]	Berchiesi, G.; Cingolani, A.; Leonesi, D. **Z. Naturforsch.** 1970, **25a**, 1766-1767.
VARIABLES:	PREPARED BY:
Temperature.	Baldini, P.

EXPERIMENTAL VALUES:

$t/^{\circ}C$	T/K^a	$100x_2$	$t/^{\circ}C$	T/K^a	$100x_2$
168.7	441.9	0	115.9	389.1	36.5
161.6	434.8	5.3	114.3	387.5	37.6
154.3	427.5	10.2	118.6	391.8	38.8
146.0	419.2	15.5	124.5	397.7	40.0
142.3	415.5	17.9	138.0	411.2	42.6
138.0	411.2	20.4	163.4	436.6	47.9
133.9	407.1	22.5	173.4	446.6	50.0
131.2	404.4	23.8	196.2	469.4	55.4
128.5	401.7	25.2	212.5	485.7	60.0
125.9	399.1	27.5	221.6	494.8	63.0
123.3	396.5	30.3	232.9	506.1	66.0
122.2	395.4	31.5	245.5	518.7	69.7
120.9	394.1	32.5	261.0	534.2	74.6
118.1	391.3	35.0	278.1	551.3	80.0

[a] T/K values calculated by the compiler.

Note - Measurements at $t/^{\circ}C \geq 280$ could not be taken due to the thermal instability of the melts (authors).

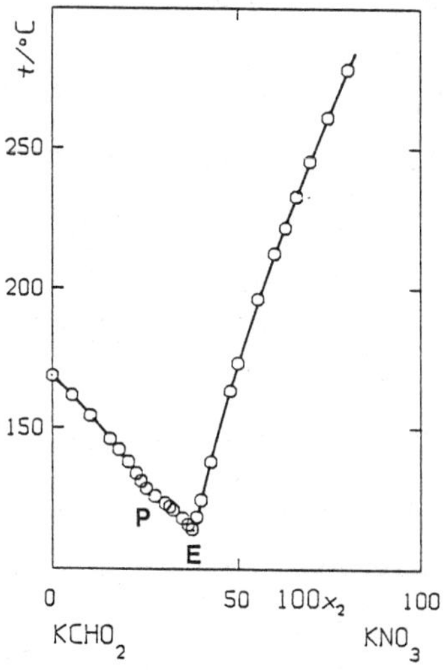

Characteristic point(s):
Eutectic, E, at 114 °C and $100x_2$ = 37.9 (authors).
Peritectic, P, at 126.5 °C and $100x_2$ = 26.2 (authors).

Intermediate compound(s):
$K_5(CHO_2)_4NO_3$, incongruently melting (authors).

AUXILIARY INFORMATION

METHOD/APPARATUS/PROCEDURE:	SOURCE AND PURITY OF MATERIALS:
Visual method (for details, see Ref. 1) supplemented with DSC measurements.	C. Erba (Milan, Italy) materials dried before use.

ESTIMATED ERROR:

Temperature: accuracy probably ± 0.1 K (compiler).

REFERENCES:

(1) Braghetti, M.; Leonesi, D.; Franzosini, P. **Ric. Sci.** 1968, **38**, 116-118.

COMPONENTS:

(1) Potassium ethanoate (potassium acetate); $KC_2H_3O_2$; [127-08-2]
(2) Potassium propanoate (potassium propionate); $KC_3H_5O_2$; [327-62-8]

ORIGINAL MEASUREMENTS:

Sokolov, N.M.; Pochtakova, E.I.
Zh. Obshch. Khim. 1958, **28**, 1397-1404.

VARIABLES:

Temperature.

PREPARED BY:

Baldini, P.

EXPERIMENTAL VALUES:

t/°C	T/K[a]	$100x_2$	t/°C	T/K[a]	$100x_2$
301	574	0	342	615	55
310	583	5	346	619	60
311	584	10	348	621	65
312	585	15	352	625	70
317	590	20	356	629	75
320	593	25	358	631	80
322	595	30	362	635	85
328	601	35	364	637	90
331	604	40	365	638	95
334	607	45	365	638	100
339	612	50			

[a] T/K values calculated by the compiler.

Characteristic point(s):

Continuous series of solid solutions.

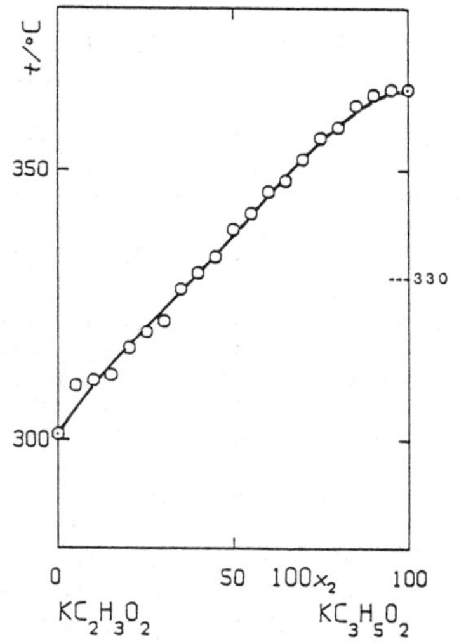

AUXILIARY INFORMATION

METHOD/APPARATUS/PROCEDURE:

Visual polythermal analysis.

NOTE:

The occurrence of a continuous series of solid solutions in this binary seems likely.
The fusion temperatures of both components [$T_{fus}(1)$= 574 K, and $T_{fus}(2)$= 638 K] are in reasonable agreement with the corresponding data listed in Preface, Table 1 (578.7±0.5 K, and 638.3±0.5 K, respectively). Conversely, there is no correspondence between the solid state transition temperature of component 2 quoted from Ref. 1 (603 K) and that of Table 1 (352.5±0.5 K). It is, however, to be noted that in other papers by the same group (see, e.g., Ref. 2) a transition of component 2 - ignored here - is quoted from the same Ref. 1 as occurring at 341 K.

SOURCE AND PURITY OF MATERIALS:

Component 1: "chemically pure" material.
Component 2: prepared from commercial propanoic acid (distilled before use) and "chemically pure" potassium carbonate; the recovered solid was recrystallized from n-butanol; it undergoes a phase transition at $t_{trs}(2)$/°C= 330 (Ref. 1).

ESTIMATED ERROR:

Temperature: accuracy probably ±2 K (compiler).

REFERENCES:

(1) Sokolov, N.M.
Tezisy Dokl. X Nauch. Konf. S.M.I. 1956.
(2) Sokolov, N.M.; Tsindrik, N.M.
Zh. Neorg. Khim. 1969, 14, 584-590 (*);
Russ. J. Inorg. Chem. (Engl. Transl.) 1969, 14, 302-306.

COMPONENTS:	EVALUATOR:
(1) Potassium ethanoate (potassium acetate); $KC_2H_3O_2$; [127-08-2] (2) Potassium butanoate (potassium butyrate); $KC_4H_7O_2$; [589-39-9]	Franzosini, P., Dipartimento di Chimica Fisica, Universita' di Pavia (ITALY).

CRITICAL EVALUATION:

The visual polythermal analysis was employed by Sokolov and Pochtakova (Ref. 1) to study the lower boundary of the isotropic liquid field. According to these authors, an intermediate compound of presumable composition $K_7C_2H_3O_2(C_4H_7O_2)_6$ ought to form, and two invariants, i.e., a eutectic, E [at 546 K (273 °C), and $100\mathbf{x}_1$= 85.5], and a "perekhodnaya tochka", P [at 623 K (350 °C), and $100\mathbf{x}_1$= 20.5], ought to exist.

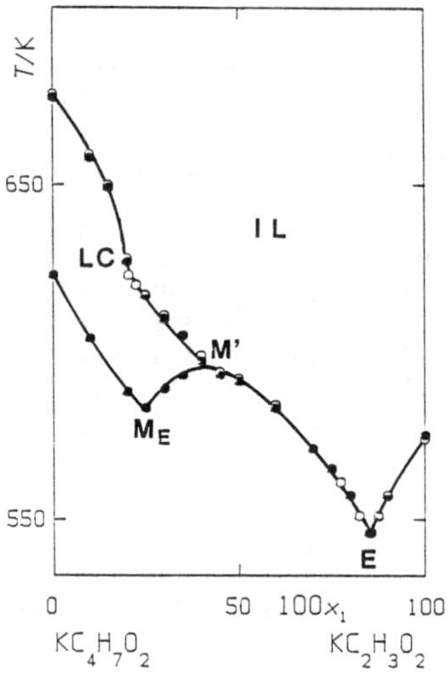

Component 2, however, forms liquid crystals, which causes the statements about the composition of the intermediate compound and the occurrence of the invariant P to become inconsistent, as explained below. Sokolov and Pochtakova's fusion temperature [677 K (404 °C)], and solid state transition at 618 K (345 °C; quoted from Ref. 2) should be identified with the clearing and fusion temperatures of component 2, respectively.

More recently, Prisyazhnyi et al. (Ref. 3) - to whom Ref. 1 seems to be unknown - carried out a derivatographical re-investigation of the system, which allowed them to draw the lower boundaries of both the isotropic liquid, and the liquid crystal field. Their clearing [$\mathbf{T}_{clr}(2)$= 676 K (403 °C)] and fusion [$\mathbf{T}_{fus}(1)$= 575 K (302 °C); $\mathbf{T}_{fus}(2)$= 623 K (350 °C)] temperatures substantially agree with the corresponding values from Table 1 of the Preface (677.3±0.5; 578.7±0.5, and 626.1±0.7 K, respectively). Prisyazhnyi et al.'s, and Sokolov and Pochtakova's results (filled and empty circles, respectively) are compared in the figure (IL: isotropic liquid; LC: liquid crystals), an inspection of which allows one to remark that: (i) the correct composition of the intermediate compound ought to be $K_5(C_2H_3O_2)_2(C_4H_7O_2)_3$ (Ref. 3) and not $K_7C_2H_3O_2(C_4H_7O_2)_6$ (Ref. 1); (ii) point P mentioned in Ref. 1 cannot be an invariant, but corresponds merely to an inflection (on the origin of which, however, no sure explanation can be offered by the evaluator) of the pertinent curve; and (iii) besides the eutectic, E, two more invariants exist, i.e., an M_E point, and an M' point. The abscissa of the latter being known only approximately, it can be hardly decided if this M' point is actually of the M'_E or of the M'_P type: in the former case, the complete phase diagram should be similar to Scheme D.1 of the Preface; in the latter case, to Scheme D.3.
The two-phase region pertinent to the liquid crystal - isotropic liquid equilibria might be so narrow as to have prevented Prisyazhnyi et al. to observe two distinct sets of points in this region, whereas the lack of information about eutectic fusion in the different samples submitted to derivatographical analysis remains rather surprising.

Finally, the following two points require attention.
(i) In Ref. 1 solid state transitions of component 1 are quoted from Ref. 2 as occurring at 428 and 331 K (155 and 58 °C, respectively), whereas mention is made in Preface (Table 1) of a single transition at 422.2±0.5 K.
(ii) Again in Ref. 1 (and from the same source), two more transformation temperatures, i.e., 558 and 463 K, respectively, are quoted for component 2 which lie each halfway between the two pairs of solid state transition temperatures (i.e., 562.2±0.6 and 540.8±1.1 K, and 467.2±0.5 and 461.4±1.0 K, respectively) also reported in Table 1 of the Preface.

REFERENCES:

(1) Sokolov, N.M.; Pochtakova, E.I.; **Zh. Obshch. Khim.** 1960, **30**, 1401-1405 (*); **Russ. J. Gen. Chem. (Engl. Transl.)** 1960, **30**, 1429-1433.
(2) Sokolov, N.M.; **Tezisy Dokl. X Nauch. Konf. S.M.I.** 1956.
(3) Prisyazhnyi, V.D.; Mirnyi, V.N.; Mirnaya, T.A.; **Zh. Neorg. Khim.** 1983, **28**, 253-255; **Russ. J. Inorg. Chem. (Engl. Transl.)** 1983, **28**, 140-141 (*).

COMPONENTS:

(1) Potassium ethanoate (potassium acetate); $KC_2H_3O_2$; [127-08-2]
(2) Potassium butanoate (potassium butyrate); $KC_4H_7O_2$; [589-39-9]

ORIGINAL MEASUREMENTS:

Sokolov, N.M.; Pochtakova, E.I.
Zh. Obshch. Khim. 1960, **30**, 1401-1405 (*);
Russ. J. Gen. Chem. (Engl. Transl.) 1960, **30**, 1429-1433.

VARIABLES:

Temperature.

PREPARED BY:

Baldini, P.

EXPERIMENTAL VALUES:

$t/^oC$	T/K^a	$100\mathbf{x}_1$	$t/^oC$	T/K^a	$100\mathbf{x}_1$
404	677	0	321	594	45
386	659	10	319	592	50
377	650	15	311	584	60
355	628	20	288	561	77.5
350	623	20.5	278	551	82.5
347	620	22.5	273	546	85.5
344	617	25	278	551	87.5
338	611	30	284	557	90
326	599	40	301	574	100

[a] **T**/K values calculated by the compiler.

Characteristic point(s):

Eutectic, E, at 273 ^oC and $100\mathbf{x}_1$= 85.5 (authors).

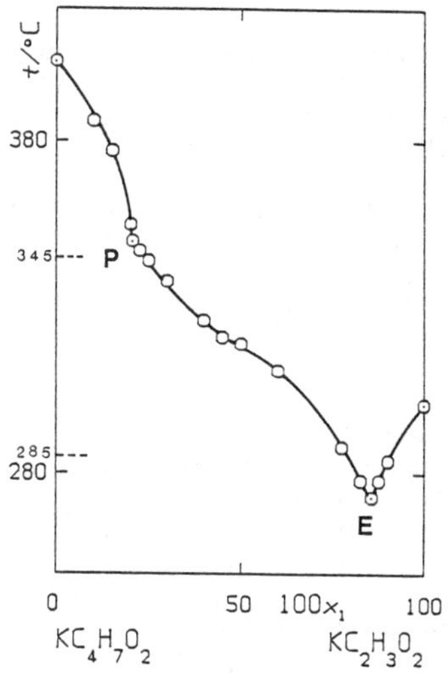

Characteristic point, P (**perekhodnaya tochka** in the original text; see the Introduction) at 350 ^oC and $100\mathbf{x}_1$= 20.5 (authors).

Intermediate compound(s):

$K_7C_2H_3O_2(C_4H_7O_2)_6$ (presumable composition; authors) incongruently melting.

AUXILIARY INFORMATION

METHOD/APPARATUS/PROCEDURE:

Visual polythermal analysis.

SOURCE AND PURITY OF MATERIALS:

Component 1: "chemically pure" material recrystallized; it undergoes phase transitions at $t_{trs}(1)/^oC$= 58, 155 (Ref. 1).
Component 2: prepared by reacting $KHCO_3$ with n-butanoic acid, and recrystallized from n-butanol (Ref. 2, where, however, carbonate instead of hydrogen carbonate was employed; compiler); it undergoes phase transitions at $t_{trs}(2)/^oC$= 190, 285, 345 (Ref. 1).

ESTIMATED ERROR:

Temperature: accuracy probably ±2 K (compiler).

REFERENCES:

(1) Sokolov, N.M.
Tezisy Dokl. X Nauch. Konf. S.M.I. 1956.
(2) Sokolov, N.M.
Zh. Obshch. Khim. 1954, **24**, 1581-1593.

COMPONENTS:

(1) Potassium ethanoate (potassium acetate); $KC_2H_3O_2$; [127-08-2]
(2) Potassium butanoate (potassium butyrate); $KC_4H_7O_2$; [589-39-9]

ORIGINAL MEASUREMENTS:

Prisyazhnyi, V.D.; Mirnyi, V.N.; Mirnaya, T.A.
Zh. Neorg. Khim. 1983, **28**, 253-255; **Russ. J. Inorg. Chem. (Engl. Transl.)** 1983, **28**, 140-141 (*).

VARIABLES:

Temperature.

PREPARED BY:

Baldini, P.

EXPERIMENTAL VALUES:

The results are reported only in graphical form (see figure; data read with a digitizer by the compiler on Fig. 1 of the original paper; empty circles: liquid crystal - isotropic liquid equilibria; filled circles: solid - liquid crystal or solid - isotropic liquid equilibria).

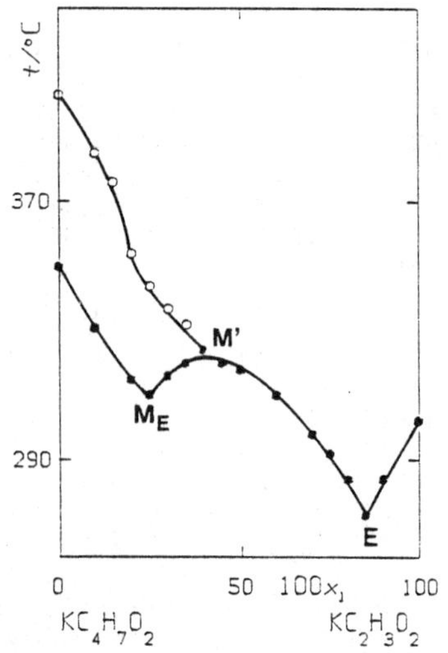

Characteristic point(s):

Invariant point, M'_E, at about 310 °C and $100x_1$ about 25 (compiler).
Eutectic, E, at about 273 °C and $100x_1$ about 85 (compiler).
Invariant point, M′, at about 323 °C and $100x_1$ about 40 (compiler).

Intermediate compound(s):

$K_5(C_2H_3O_2)_2(C_4H_7O_2)_3$, melting at about 323 °C (compiler).

AUXILIARY INFORMATION

METHOD/APPARATUS/PROCEDURE:

The heating and cooling traces were recorded in an atmosphere of purified argon with an OD-102 derivatograph (MOM, Hungary) working at a rate of 6-8 K min^{-1}, and using Al_2O_3 as the reference material. Temperatures were measured with a Pt/Pt-Rh thermocouple. A hot-stage Amplival polarizing microscope was employed to detect the transformation points from the liquid crystalline into the isotropic liquid phase.

SOURCE AND PURITY OF MATERIALS:

Not stated.
Component 1: $t_{fus}(1)/°C$ about 302 (compiler).
Component 2: $t_{fus}(2)/°C$ about 350; $t_{clr}(2)/°C$ about 403 (compiler).

ESTIMATED ERROR:

Temperature: accuracy not evaluable (compiler).

REFERENCES:

COMPONENTS:	EVALUATOR:
(1) Potassium ethanoate (potassium acetate); $KC_2H_3O_2$; [127-08-2] (2) Potassium **iso**.butanoate (potassium **iso**.butyrate); $K\mathbf{i}.C_4H_7O_2$; [19455-20-0]	Schiraldi, A., Dipartimento di Chimica Fisica, Universita´ di Pavia (ITALY).

CRITICAL EVALUATION:

This system was studied only by Sokolov and Pochtakova (Ref. 1) who claimed the existence of: (i) a eutectic, E_1, at 564 K (291 ^oC) and $100\mathbf{x}_1$= 86.5; (ii) a eutectic, E_2, at 567 K (294 ^oC) and $100\mathbf{x}_1$= 32; and (iii) an intermediate compound, $K_5(C_2H_3O_2)_3(\mathbf{i}.C_4H_7O_2)_2$, congruently melting at 578 K (305 ^oC).

Component 2, however, forms liquid crystals. Therefore the temperature of 633 K (360 ^oC) given in Ref. 1 should be identified with the clearing (and not the fusion) temperature of this component, and compared with the $\mathbf{T}_{clr}(2)$ value (625.6$\pm$0.8 K) reported in Preface, Table 2.

For the same component, three more phase transition temperatures are quoted in Ref. 1 from Ref. 2, i.e., 621, 546, and 481 K, the second of which can be reasonably identified with the fusion temperature [$\mathbf{T}_{fus}(2)$= 553.9$\pm$0.5 K] listed in Table 2 of the Preface. Consequently: (i) the transition temperature at 621 K (if actually existing) might correspond to some kind of transformation (undetected by DSC, see Table 2) within the liquid crystal field; and (ii) only the transition at 481 K should correspond to a solid state transformation, although the latter figure is almost 60 K higher than the single $\mathbf{T}_{trs}(2)$ value (424$\pm$3 K) listed in Preface, Table 2.

The fusion temperature of component 1, $\mathbf{T}_{fus}(1)$= 574 K (301 ^oC; Ref. 1), and the transition temperature [428 K (155 ^oC; quoted in Ref. 1 from Ref. 2)] satisfactorily correspond with the values listed in Table 1 of the Preface (578.7$\pm$0.5 K, and 422.2$\pm$0.5 K, respectively), whereas the other solid-solid transition quoted by the authors from Ref. 2 as occurring at 331 K (58 ^oC) has no correspondence in Table 1.

In conclusion, the phase diagram ought to be similar to that shown in Scheme D.1 of the Preface. Accordingly, the eutectic E_2 should actually be an M'_E point. The existence of the intermediate compound $K_5(C_2H_3O_2)_3(\mathbf{i}.C_4H_7O_2)_2$ seems reasonably supported by the available data.

REFERENCES:

(1) Sokolov, N.M.; Pochtakova, E.I.
Zh. Obshch. Khim. 1960, 30, 1405-1410 (*); **Russ. J. Gen. Chem. (Engl. Transl.)** 1960, **30**, 1433-1437.

(2) Sokolov, N.M.
Tezisy Dokl. X Nauch. Konf. S.M.I. 1956 .

COMPONENTS:

(1) Potassium ethanoate (potassium acetate); $KC_2H_3O_2$; [127-08-2]
(2) Potassium **iso**.butanoate (potassium **iso**.butyrate); $K\mathbf{i}.C_4H_7O_2$; [19455-20-0]

ORIGINAL MEASUREMENTS:

Sokolov, N.M.; Pochtakova, E.I.
Zh. Obshch. Khim. 1960, **30**, 1405-1410 (*);
Russ. J. Gen. Chem. (Engl. Transl.) 1960, **30**, 1433-1437.

VARIABLES:

Temperature.

PREPARED BY:

Baldini, P.

EXPERIMENTAL VALUES:

$t/^oC$	T/K^a	$100\mathbf{x}_1$	$t/^oC$	T/K^a	$100\mathbf{x}_1$
360	633	0	304	577	50
352	625	5	304	577	55
343	616	10	305	578	60
320	593	20	302	575	70
310	583	25	298	571	75
304	577	27.5	294	567	85
298	571	30	291	564	86.5
294	567	32	293	566	87.5
296	569	32.5	294	567	90
297	570	35	298	571	95
301	574	40	301	574	100

[a] **T**/K values calculated by the compiler.

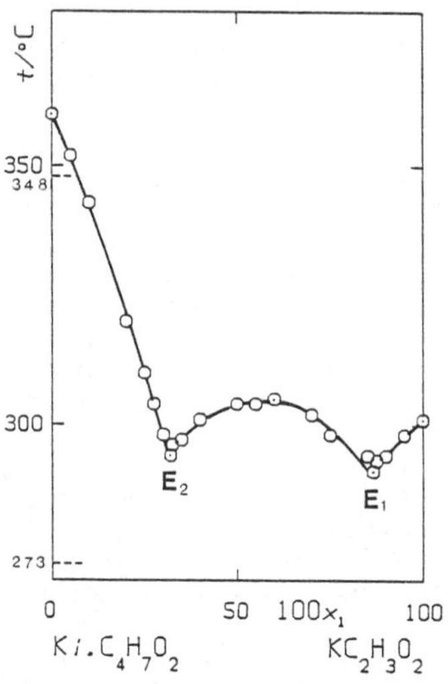

Characteristic point(s):

Eutectic, E_1, at 291 ^oC and $100\mathbf{x}_1$= 86.5 (authors).
Eutectic, E_2, at 294 ^oC and $100\mathbf{x}_1$= 32 (authors).

Intermediate compound(s):

$K_5(C_2H_3O_2)_3(\mathbf{i}.C_4H_7O_2)_2$ (probable composition) congruently melting at 305 ^oC.

AUXILIARY INFORMATION

METHOD/APPARATUS/PROCEDURE:

Visual polythermal analysis.

SOURCE AND PURITY OF MATERIALS:

Component 1: "chemically pure" material recrystallized.
Component 2: prepared from commercial "pure" grade **iso**.butanoic acid, distilled before use, and potassium "chemically pure" hydrogen carbonate (Ref. 1); then recrystallized from **n**-butanol.
Component 1 undergoes phase transitions at $\mathbf{t}_{trs}(1)/^oC$= 58, 155 (Ref. 2). Component 2 undergoes phase transitions at $\mathbf{t}_{trs}(2)/^oC$= 208, 273, 348 (Ref. 2).

ESTIMATED ERROR:

Temperature: accuracy probably ± 2 K (compiler).

REFERENCES:

(1) Sokolov, N.M.
Zh. Obshch. Khim. 1954, **24**, 1150-1156.
(2) Sokolov, N.M.
Tezisy Dokl. X Nauch. Konf. S.M.I. 1956.
(this is Ref. 6 in the original paper, and not Ref. 5 as erroneously quoted in the text; compiler).

COMPONENTS:	EVALUATOR:
(1) Potassium ethanoate (potassium acetate); $KC_2H_3O_2$; [127-08-2] (2) Potassium pentanoate (potassium valerate); $KC_5H_9O_2$; [19455-21-1]	Schiraldi, A., Dipartimento di Chimica Fisica, Universita´ di Pavia (ITALY).

CRITICAL EVALUATION:

This system was studied only by Pochtakova (Ref. 1) who claimed the existence of: (i) a eutectic point at 553 K (280 °C) and $100\mathbf{x}_2$= 12.5; (ii) a "perekhodnaya tochka" (likely a peritectic) at 607 K (334 °C) and $100\mathbf{x}_2$= 52.5; and (iii) an incongruently melting intermediate compound, $K_5(C_2H_3O_2)_2(C_5H_9O_2)_3$.

Component 2, however, forms liquid crystals. Therefore, Pochtakova´s fusion temperature, $\mathbf{T}_{fus}(2)$= 717 K (444 °C), should be identified with the clearing temperature, the corresponding value from Table 1 of the Preface being 716±2 K. The phase transition quoted by the author from Ref. 2 as occurring in the same component at 580 K (307 °C; Ref. 2) can be reasonably identified with the actual fusion temperature, the value from Preface, Table 1 being $\mathbf{T}_{fus}(2)$= 586.6±0.7 K. No mention is made of further transformations, although Table 1 reports a solid state transition at 399.5±0.9 K.

Among the phase transition temperatures mentioned by Pochtakova for component 1, the fusion at 575 K (302 °C; Ref. 1), and the solid state transition at 428 K (155 °C; quoted from Ref. 2), can be satisfactorily identified with the corresponding values of Table 1 of the Preface, viz., 578.7±0.5 K and 422.2±0.5 K, respectively. On the contrary, the lower solid-solid transition quoted from Ref. 2 as occurring at 331 K (58 °C) has no correspondence in Table 1.

In conclusion, it can be asserted that in Pochtakova´s phase diagram the branch whose ends are $\mathbf{T}_{clr}(2)$ and point P is relevant to isotropic liquid - liquid crystal equilibria, whereas it is hard to decide, on the basis of the available data, whether or not an intermediate compound is formed.

The existence of the intermediate compound might be argued from analogy with the topology of the binary potassium ethanoate -potassium **iso.**butanoate (Ref. 3) where evidence was obtained for the formation of a 3:2 compound. Accordingly, the phase diagram might be similar to Scheme D.3 of the Preface with an M'_p point at about 588 K (315 °C) and $100\mathbf{x}_2$ about 40. In this case, Pochtakova´s P point should be a mere inflection in the relevant branch.

Conversely, if the existence of the compound is not accepted, the phase diagram might be interpreted with reference to Scheme B.2.

REFERENCES:

(1) Pochtakova, E.I.
Zh. Obshch. Khim. 1966, **36**, 3-8.

(2) Sokolov, N.M.
Tezisy Dokl. X Nauch. Konf. S.M.I. 1956.

(3) Sokolov, N.M.; Pochtakova, E.I.
Zh. Obshch. Khim. 1960, **30**, 1405-1410 (*); **Russ. J. Gen. Chem. (Engl. Transl.)** 1960, **30**, 1433-1437.

COMPONENTS:

(1) Potassium ethanoate (potassium acetate); $KC_2H_3O_2$; [127-08-2]
(2) Potassium pentanoate (potassium valerate); $KC_5H_9O_2$; [19455-21-1]

ORIGINAL MEASUREMENTS:

Pochtakova, E.I.
Zh. Obshch. Khim. 1966, **36**, 3-8.

VARIABLES:

Temperature.

PREPARED BY:

Baldini, P.

EXPERIMENTAL VALUES:

t/°C	T/K[a]	$100x_2$	t/°C	T/K[a]	$100x_2$
302	575	0	320	593	45
288	561	5	325	598	47.5
283	556	10	330	603	50
280	553	12.5	334	607	52.5
286	559	15	345	618	55
290	563	17.5	355	628	57.5
295	568	20	364	637	60
302	575	25	381	654	65
308	581	30	409	682	75
313	586	35	418	691	80
315[b]	588	40	426	699	85
318	591	42.5	444	717	100

[a] T/K values calculated by the compiler.
[b] 415 in the original table (compiler).

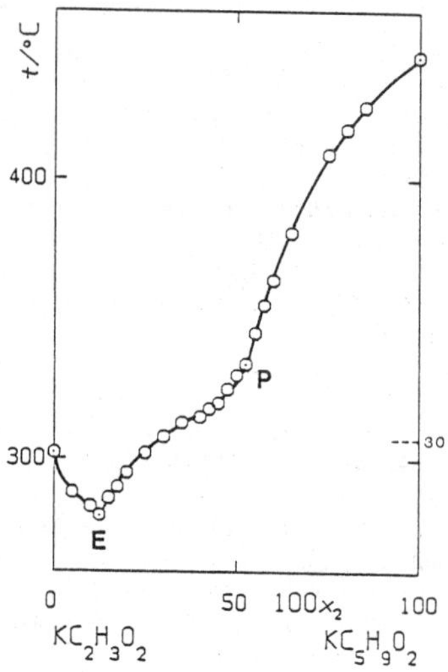

Characteristic point(s):

Eutectic, E, at 280 °C and $100x_2$= 12.5 (author).
Characteristic point, P (**perekhodnaya tochka** in the original text; see the Introduction), at 334 °C and $100x_2$= 52.5 (author).

Intermediate compound(s):

$K_5(C_2H_3O_2)_2(C_5H_9O_2)_3$ (probable composition), incongruently melting.

AUXILIARY INFORMATION

METHOD/APPARATUS/PROCEDURE:

Visual polythermal analysis.

SOURCE AND PURITY OF MATERIALS:

Component 1: "chemically pure" material.
Component 2: prepared from n-pentanoic acid and hydrogen carbonate (Ref. 1, where, however, carbonate instead of hydrogen carbonate was employed).
Component 1 undergoes phase transitions at $t_{trs}(1)$/°C= 58, 155 (Ref. 2).
Component 2 undergoes a phase transition at $t_{trs}(2)$/°C= 307 (Ref. 2).

ESTIMATED ERROR:

Temperature: accuracy probably ±2 K (compiler).

REFERENCES:

(1) Sokolov, N.M.
Zh. Obshch. Khim. 1954, **24**, 1581-1593.
(2) Sokolov, N.M.
Tezisy Dokl. X Nauch. Konf. S.M.I. 1956.

COMPONENTS:	EVALUATOR:
(1) Potassium ethanoate (potassium acetate); $KC_2H_3O_2$; [127-08- 2] (2) Potassium **iso**.pentanoate (potassium iso.valerate); $K\mathbf{i}.C_5H_9O_2$; [589-46-8]	Schiraldi, A., Dipartimento di Chimica Fisica, Universita´ di Pavia (ITALY).

CRITICAL EVALUATION:

This system was studied only by Pochtakova (Ref. 1), who claimed the existence of: (i) a eutectic, E, at 542 K (269 °C) and $100\mathbf{x}_2$= 50; (ii) a peritectic, P, at 543 K (270 °C) and $100\mathbf{x}_2$= 18.5; and (iii) an incongruently melting compound, of probable composition $K_8(C_2H_3O_2)_7\mathbf{i}.C_5H_9O_2$.

Component 2, however, forms liquid crystals. Therefore the fusion temperature, $\mathbf{T}_{fus}(2)$= 669 K (396 °C) reported by the author should be identified with the clearing temperature, the corresponding value from Preface, Table 2 being 679±2 K. No mention is made by the author of the actual fusion which occurs at 531±3 K according to Table 2: the latter figure is supported by the trend of the thermomagnetical curves plotted by Duruz and Ubbelohde (Ref. 2).

As for the other phase transitions quoted by Pochtakova from Ref. 3 at 327 and 618 K (54 and 345 °C, respectively), no identification is possible with the findings by other investigators, inasmuch as: (i) no transformation is reported in Table 2 as occurring below $\mathbf{T}_{fus}(2)$= 531±3 K; and (ii) no transformation is reported in Table 2 or in Ref. 2 as occurring within the field of existence of the mesomorphic liquid. It is a bit puzzling the fact that for potassium **iso**.pentanoate Dmitrevskaya and Sokolov (Ref. 4) quote from Ref. 3 (unavailable to the evaluator) transitions at 618, 493, and 473 K (ignoring that quoted by Pochtakova at 327 K), and Pochtakova quotes from the same source transitions at 618 and 327 K (ignoring those quoted by Dmitrevskaya and Sokolov at 493 and 473 K).

Component 1, as quoted in Ref. 1 from Ref. 3, undergoes phase transitions at 331 and 428 K (58 and 155 °C, respectively), the latter figure being in reasonable agreement with the $\mathbf{T}'_{trs}$ value (422.2±0.5 K) from Table 1 of the Preface.

The available data do not seem sufficient to prove unambiguously the existence of any intermediate compound. Should it exist, the phase relations at $50 \le 100\mathbf{x}_2 \le 100$ could be reasonably interpreted with reference to Scheme D.1: Pochtakova´s eutectic could be actually an M'_E point, and a further invariant of the M_E type should exist.

REFERENCES:

(1) Pochtakova, E.I.
Zh. Obshch. Khim. 1963, **33**, 342-347.

(2) Duruz, J.J.; Ubbelohde, A.R.
Proc. Roy. Soc. London 1975, **A 342**, 39-49.

(3) Sokolov, N.M.
Tezisy Dokl. X Nauch. Konf. S.M.I. 1956.

(4) Dmitrevskaya, O.I.; Sokolov, N.M.
Zh. Obshch. Khim. 1967, **37**, 2160-2166; **Russ. J. Gen. Chem. (Engl. Transl.)** 1967, **37**, 2050-2054.

COMPONENTS:	ORIGINAL MEASUREMENTS:
(1) Potassium ethanoate (potassium acetate); $KC_2H_3O_2$; [127-08- 2] (2) Potassium **iso**.pentanoate (potassium **iso**.valerate); $K\mathbf{i}.C_5H_9O_2$; [589-46-8]	Pochtakova, E.I. **Zh. Obshch. Khim.** 1963, 33, 342-347.
VARIABLES:	PREPARED BY:
Temperature.	Baldini, P.

EXPERIMENTAL VALUES:

The results are reported only in graphical form (see figure).

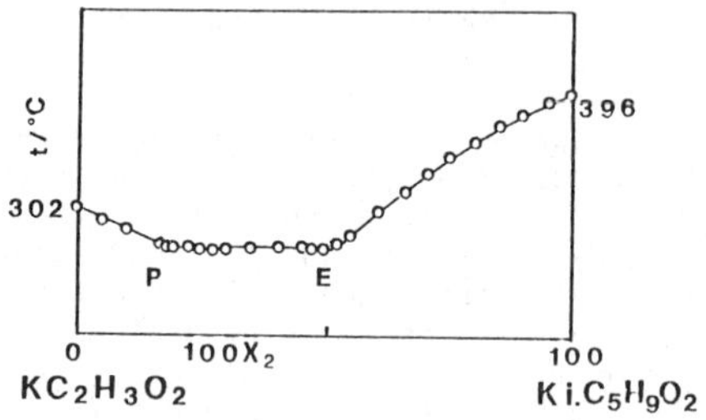

Characteristic point(s):

Eutectic, E, at 269 °C and $100\mathbf{x}_2$= 50.0.
Peritectic, P (**perekhodnaya tochka** in the original text; see the Introduction), at 270 °C and $100\mathbf{x}_2$= 18.5.

Intermediate compound(s):

$K_8(C_2H_3O_2)_7\mathbf{i}.C_5H_9O_2$ (probable composition).

AUXILIARY INFORMATION

METHOD/APPARATUS/PROCEDURE:

Visual polythermal analysis.

SOURCE AND PURITY OF MATERIALS:

Component 1: "chemically pure" material.
Component 2: prepared from commercial **iso**.pentanoic acid (distilled twice before use) and "chemically pure" hydrogen carbonate (Ref. 1).
Component 1 undergoes phase transitions at $t_{trs}(1)/^{\circ}C$= 58, 155 (Ref. 2) and melts at $t_{fus}(1)/^{\circ}C$= 302.
Component 2 undergoes phase transitions at $t_{trs}(2)/^{\circ}C$= 54, 345 (Ref. 2) and melts at $t_{fus}(2)/^{\circ}C$= 396.

ESTIMATED ERROR:

Temperature: accuracy probably ± 2 K (compiler).

REFERENCES:

(1) Sokolov, N.M.
Zh. Obshch. Khim. 1954, 24, 1581-1593.
(2) Sokolov, N.M.
Tezisy Dokl. X Nauch. Konf. S.M.I. 1956.

COMPONENTS:

(1) Potassium ethanoate (potassium acetate); $KC_2H_3O_2$; [127-08-2]
(2) Potassium hexanoate (potassium caproate); $KC_6H_{11}O_2$; [19455-00-6]

EVALUATOR:

Schiraldi, A.,
Dipartimento di Chimica Fisica,
Universita´ di Pavia (ITALY).

CRITICAL EVALUATION:

This system was studied only by Pochtakova (Ref. 1) who suggested the existence of: (i) a eutectic, E, at 560 K (287 °C), and $100\mathbf{x}_2$= 11.0; (ii) a "perekhodnaya tochka" (likely a peritectic) at 592 K (319 °C) and $100\mathbf{x}_2$= 39.0; and (iii) an incongruently melting intermediate compound, $K_5(C_2H_3O_2)_3(C_6H_{11}O_2)_2$.

Component 2, however, forms liquid crystals. Therefore the fusion temperature, $T_{fus}(2)$= 717.7 K (444.5 °C; Ref. 1), should be identified with the clearing temperature, the corresponding value from Preface, Table 1 being 725.8±0.8 K. For the same component, the phase transition quoted in Ref. 1 from Ref. 2 as occurring at 575 K (302 °C) can be identified with the actual fusion temperature, $T_{fus}(2)$= 581.7±0.5 K (Preface, Table 1).

Concerning component 1, fusion occurs at 574 K (301 °C; Ref. 1), and solid state transitions occur at 428 K (155 °C; Ref. 2), and 331 K (58 °C; Ref. 2). Only the former two values, however, find a direct identification with data listed in Table 1 of the Preface, i.e., 578.7±0.5 K and 422.2±0.5 K, respectively.

In conclusion, it can be asserted that in Pochtakova´s phase diagram the branch whose ends are $T_{clr}(2)$ and point P is relevant to isotropic liquid - liquid crystal equilibria, whereas it is hard to decide, on the basis of the available data, whether or not an intermediate compound is formed.

The existence of the intermediate compound might be argued from analogy with the topology of the binary potassium ethanoate -potassium **iso**.butanoate (Ref. 3) where evidence was obtained for the formation of a 3:2 compound. Accordingly, the phase diagram might be similar to Scheme D.3 of the Preface. In this case, Pochtakova´s P point should be a mere inflection in the relevant branch.

Conversely, if the existence of the compound is not accepted, the phase diagram might be interpreted with reference to Scheme B.2.

REFERENCES:

(1) Pochtakova, E.I.
Zh. Obshch. Khim. 1959, **29**, 3183-3189 (*); **Russ. J. Gen. Chem. (Engl. Transl.)** 1959, **29**, 3149-3154.

(2) Sokolov, N.M.
Tezisy Dokl. X Nauch. Konf. S.M.I. 1956.

(3) Sokolov, N.M.; Pochtakova, E.I.
Zh. Obshch. Khim. 1960, 30, 1405-1410 (*); **Russ. J. Gen. Chem. (Engl. Transl.)** 1960, **30**, 1433-1437.

COMPONENTS:

(1) Potassium ethanoate (potassium acetate); $KC_2H_3O_2$; [127-08-2]
(2) Potassium hexanoate (potassium caproate); $KC_6H_{11}O_2$; [19455-00-6]

ORIGINAL MEASUREMENTS:

Pochtakova, E.I.
Zh. Obshch. Khim. 1959, **29**, 3183-3189 (*);
Russ. J. Gen. Chem. (Engl. Transl.) 1959, **29**, 3149-3154.

VARIABLES:

Temperature.

PREPARED BY:

Baldini, P.

EXPERIMENTAL VALUES:

The results are reported only in graphical form (see figure).

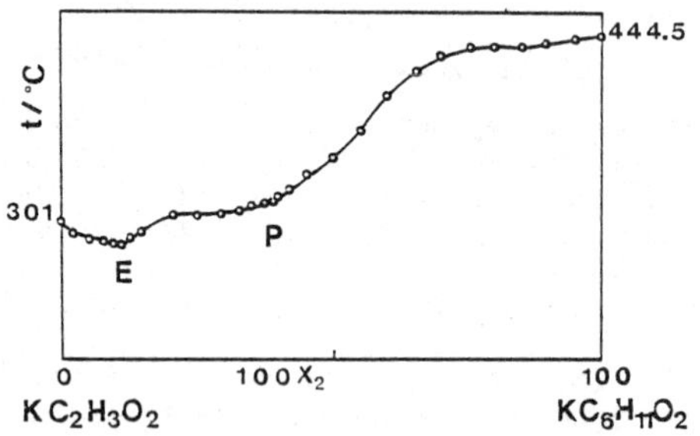

Characteristic point(s):

Eutectic, E, at 287 °C and $100\mathbf{x}_2$= 11.0 (author).
Characteristic point, P (**perekhodnaya tochka** in the original text; see the Introduction), at 319 °C and $100\mathbf{x}_2$= 39.0.

Intermediate compound(s):

$K_5(C_2H_3O_2)_3(C_6H_{11}O_2)_2$ (approximate composition), incongruently melting.

AUXILIARY INFORMATION

METHOD/APPARATUS/PROCEDURE

Visual polythermal analysis.

SOURCE AND PURITY OF MATERIALS:

"Chemically pure" $KC_2H_3O_2$, and $KC_6H_{11}O_2$ prepared by reacting K_2CO_3 with n-hexanoic acid (Ref. 1).
Component 1 undergoes phase transitions at $\mathbf{t}_{trs}(1)$/°C= 58, 155 (Ref. 2).
Component 2 undergoes a phase transition at $\mathbf{t}_{trs}(2)$/°C= 302 (Ref. 2).

ESTIMATED ERROR:

Temperature: accuracy probably ±2 K (compiler).

REFERENCES:

(1) Sokolov, N.M.
Zh. Obshch. Khim. 1954, **24**, 1581-1593.
(2) Sokolov, N.M.
Tezisy Dokl. X Nauch. Konf. S.M.I. 1956.

COMPONENTS:	ORIGINAL MEASUREMENTS:
(1) Potassium ethanoate (potassium acetate); $KC_2H_3O_2$; [127-08-2] (2) Potassium chloride; KCl; [7447-40-7]	Il'yasov, I.I.; Bergman, A.G. **Zh. Obshch. Khim.** 1960, **30**, 355-358.
VARIABLES:	PREPARED BY:
Temperature.	Baldini, P.

EXPERIMENTAL VALUES:

t/°C	T/K[a]	$100x_2$
306	579	0
305	578	2.5
304	577	5.0
295	568	10.0
300	573	11.0
328	601	13.0
366[b]	639	16.0
383[c]	656	17.5

[a] T/K values calculated by the compiler.
[b] Erroneously reported as 266 in Table 1 of the original paper (compiler).
[c] Erroneously reported as 283 in Table 1 of the original paper (compiler).

Characteristic point(s):

Eutectic, E, at 293 °C and $100x_2$= 10.5 (authors).

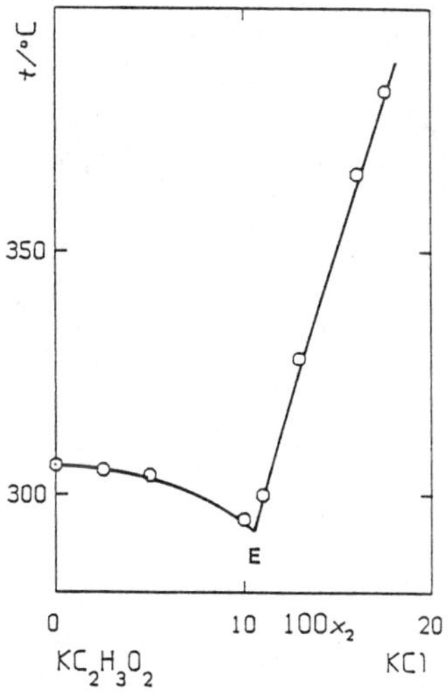

AUXILIARY INFORMATION

METHOD/APPARATUS/PROCEDURE:

Visual polythermal analysis; temperatures measured with a Nichrome-Constantane thermocouple and a millivoltmeter.

NOTES:

The system was investigated at $0 \leq 100x_2 \leq 17.5$ due to thermal instability of component 1.

See also the note relevant to the results obtained by Piantoni et al. (Ref. 1) on the same system (next Table).

SOURCE AND PURITY OF MATERIALS:

Not stated.

ESTIMATED ERROR:
Temperature: accuracy probably ±2 K (compiler).

REFERENCES:

(1) Piantoni, G.; Leonesi, D.; Braghetti, M.; Franzosini, P.
Ric. Sci. 1968, **38**, 127-132.

COMPONENTS:	ORIGINAL MEASUREMENTS:
(1) Potassium ethanoate (potassium acetate); $KC_2H_3O_2$; [127-08-2] (2) Potassium chloride; KCl; [7447-40-7]	Piantoni, G.; Leonesi, D.; Braghetti, M.; Franzosini, P. **Ric. Sci.** 1968, **38**, 127-132.
VARIABLES:	PREPARED BY:
Temperature.	Baldini, P.

EXPERIMENTAL VALUES:

The results are given only in graphical form (see figure). The system was investigated at $0 \leq 100x_2 \leq 8$.

Characteristic point(s):
Eutectic, E, at 293.6 °C and $100x_1 = 93.3$ (authors).

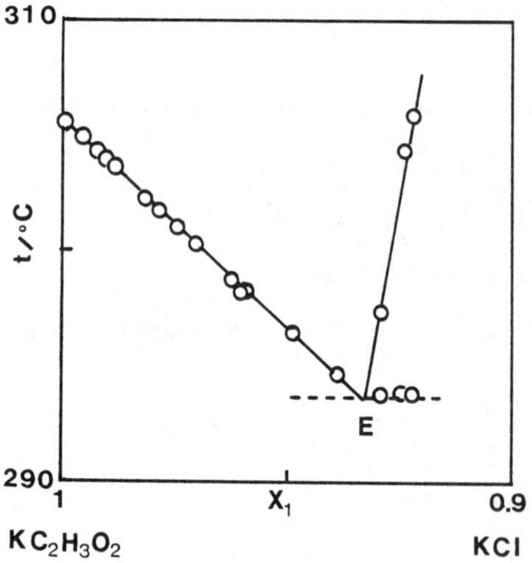

AUXILIARY INFORMATION

METHOD/APPARATUS/PROCEDURE:

A Pyrex device, suitable for work under an inert atmosphere, and allowing one to observe the system visually, was employed (for details, see Ref. 1). The initial crystallization temperatures were measured with a Chromel-Alumel thermocouple checked by comparison with a certified Pt resistance thermometer, and connected with a L&N Type K-3 potentiometer.

NOTE:
Higher precision, and satisfactory mutual consistency of the results obtained by Piantoni et al. for the three binaries $K/C_2H_3O_2$, (Br,Cl,I) suggest to prefer here the data by these authors to those by Il'yasov and Bergman (Ref. 2). Increasingly positive deviation from ideality was observed by Piantoni et al. for the liquidus branch richer in the halide when KCl, KBr, and KI were successively taken into account. This is consistent with the (cryometric) limiting values [$\lim_{m_2 \to 0} (\Delta T/m_2) = 17.7$, 17.4, and 16.0 K molality^{-1}, respectively] previously found by Braghetti et al. (Ref. 1) when the same halides were employed as solutes in molten potassium ethanoate (the cryometric constant of which is: $K_1 = 18.0 \pm 0.3$ K molality^{-1}; Ref. 1).

SOURCE AND PURITY OF MATERIALS:

C. Erba RP materials, dried by heating under vacuum (private communication by the authors to the compiler).

ESTIMATED ERROR:

Temperature: accuracy probably ±0.1 K.

REFERENCES:

(1) Braghetti, M.; Leonesi, D.; Franzosini, P. **Ric. Sci.** 1968, **38**, 116-118.
(2) Il'yasov. I.I.; Bergman, A.G. **Zh. Obshch. Khim.** 1960, **30**, 355-358.

COMPONENTS:	EVALUATOR
(1) Potassium ethanoate (potassium acetate); $KC_2H_3O_2$; [127-08-2] (2) Potassium thiocyanate; KCNS; [333-20-0]	Spinolo, G., Dipartimento di Chimica Fisica, Universita´ di Pavia (ITALY).

CRITICAL EVALUATION:

The binary $K/C_2H_3O_2$, CNS was studied by Golubeva et al. (Ref. 1), and by Sokolov (Ref. 2). In both papers, the visual polythermal analysis was employed to draw the lower boundary of the isotropic liquid field.

Concerning the thermal behavior of component 1, it can be noted that a reasonable agreement exists: (i) between the fusion temperatures from Refs. 1, 2, and that listed in Table 1 of the Preface (578.7±0.5 K); and (ii) between Sokolov´s (Ref. 2) higher transition temperature (428 K), and the single $T_{trs}(1)$ value (422.2±0.5 K) from Table 1. No correspondence with Table 1 can be found for Sokolov´s lower transition (331 K). No solid state transformation of this component is mentioned in Ref. 1.

The main features of the phase diagram given in either source exhibit rather close similarities, as shown here:

	Ref. 1	Ref. 2
$T_{fus}(1)/K$:	579	575
$T_{fus}(2)/K$:	449	450
Intermediate compound	$K_3C_2H_3O_2(CNS)_2$	$K_3C_2H_3O_2(CNS)_2$
Eutectic E_1; T/K:	405	410-412
Eutectic E_1; $100x_1$:	42.5	39
Eutectic E_2; T/K:	403	408
Eutectic E_2; $100x_1$:	27	22.5

It is, however, to be stressed that Sokolov´s graphical presentation of the diagram is somewhat conflicting with the few numerical data reported in the text. Accordingly, the evaluator is inclined to prefer the values listed under the heading "Ref. 1", although regretting that no tabulation of the experimental points is supplied by the authors.

REFERENCES:

(1) Golubeva, M.S.; Aleshkina, N.N.; Bergman, A.G.
Zh. Neorg. Khim. 1959, 4, 2606-2610; **Russ. J. Inorg. Chem. (Engl. Transl.)** 1959, **4**, 1201-1203 (*).

(2) Sokolov, N.M.
Zh. Obshch. Khim. 1966, **36**, 577-582.

COMPONENTS:

(1) Potassium ethanoate (potassium acetate); $KC_2H_3O_2$; [127-08-2]
(2) Potassium thiocyanate; KCNS; [333-20-0]

ORIGINAL MEASUREMENTS:

Golubeva, M.S.; Aleshkina, N.N.; Bergman, A.G.
Zh. Neorg. Khim. 1959, 4, 2606-2610; **Russ. J. Inorg. Chem. (Engl. Transl.)** 1959, **4**, 1201-1203 (*).

VARIABLES:

Temperature.

PREPARED BY:

Baldini, P.

EXPERIMENTAL VALUES:

The results are reported only in graphical form (see figure).

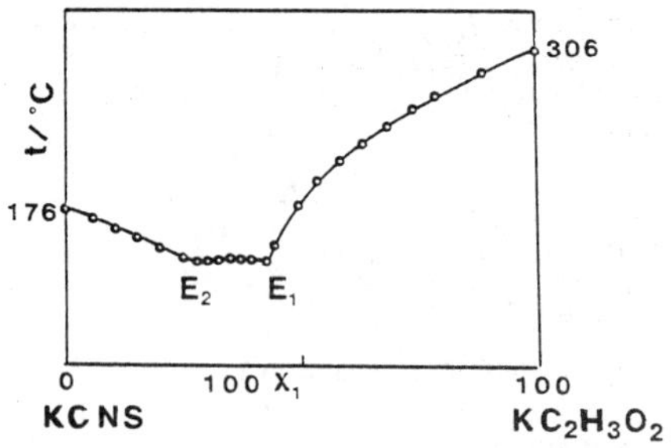

Characteristic point(s):

Eutectic, E_1, at 132 °C and $100x_1 = 42.5$ (authors).
Eutectic, E_2, at 130 °C and $100x_1 = 27$ (authors).

Intermediate compound(s):

$K_3C_2H_3O_2(CNS)_2$ congruently melting at 134 °C (authors).

AUXILIARY INFORMATION

METHOD/APPARATUS/PROCEDURE:

Visual observation of fusion of the salt mixtures contained in a glass tube surrounded by a wider tube to secure a more uniform heating. Temperatures measured with a Chromel-Alumel thermocouple.

SOURCE AND PURITY OF MATERIALS:

Materials of analytical purity recrystallized twice.

ESTIMATED ERROR:

Temperature: accuracy probably ±2 K (compiler).

REFERENCES:

COMPONENTS:

(1) Potassium ethanoate (potassium acetate); $KC_2H_3O_2$; [127-08-2]
(2) Potassium thiocyanate; KCNS; [333-20-0]

ORIGINAL MEASUREMENTS:

Sokolov, N.M.
Zh. Obshch. Khim. 1966, 36, 577-582.

VARIABLES:

Temperature.

PREPARED BY:

Baldini, P.

EXPERIMENTAL VALUES:

The results are reported only in graphical form (see figure).

Characteristic point(s):

Eutectic, E_1, at 137 °C (text and Fig. 1 of the original paper) or 139 °C (Fig. 2 and Fig. 3) and $100\mathbf{x}_1$= 39 (author).
Eutectic, E_2, at 135 °C and $100\mathbf{x}_1$= 22.5 (author).

Intermediate compound(s):
$K_3C_2H_3O_2(CNS)_2$, congruently melting.

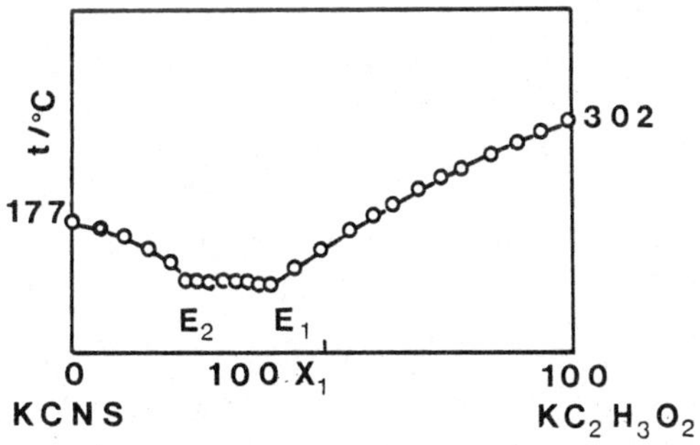

AUXILIARY INFORMATION

METHOD/APPARATUS/PROCEDURE:

Visual polythermal analysis.

NOTE:

Curve 1 of Fig. 1 of the original paper, which is reproduced in the figure, is somewhat unsatisfactory inasmuch E_2 seems higher than E_1 (compiler).

SOURCE AND PURITY OF MATERIALS:

Not stated.
Component 1 melts at $t_{fus}(1)/°C$= 302 and undergoes phase transitions at $t_{trs}(1)/°C$= 58, 155 (Ref. 1).
Component 2 melts at $t_{fus}(2)/°C$= 177 and undergoes a phase transition at $t_{trs}(2)/°C$= 143 (Ref. 2).

ESTIMATED ERROR:

Temperature: accuracy probably ±2 K (compiler).

REFERENCES:

(1) Sokolov, N.M.
Tezisy Dokl. X Nauch. Konf. S.M.I. 1956.
(2) Vrzhesnevskij, I.B.
Zh. Russk. Fiz.-Khim. Obshch. 1911, 43, 1368.

COMPONENTS:	ORIGINAL MEASUREMENTS:
(1) Potassium ethanoate (potassium acetate); $KC_2H_3O_2$; [127-08-2] (2) Potassium iodide KI; [7681-11-0]	Diogenov, G.G.; Erlykov, A.M. **Nauch. Dokl. Vysshei Shkoly, Khim. i Khim. Tekhnol.** 1958, **No. 3**, 413-416.
VARIABLES:	PREPARED BY:
Temperature.	Baldini, P.

EXPERIMENTAL VALUES:

$t/^{\circ}C$	T/K^a	$100x_1$
310.5	583.5	100
305	578	97.0
296	569	94.9
295	568	93.3
293	566	90.0
290	563	87.2
287	560	86.0
286	559	85.0
282	555	83.5
277	550	82.0
282	555	81.3
314	587	76.4
350	623	70.8

[a] T/K values calculated by the compiler.

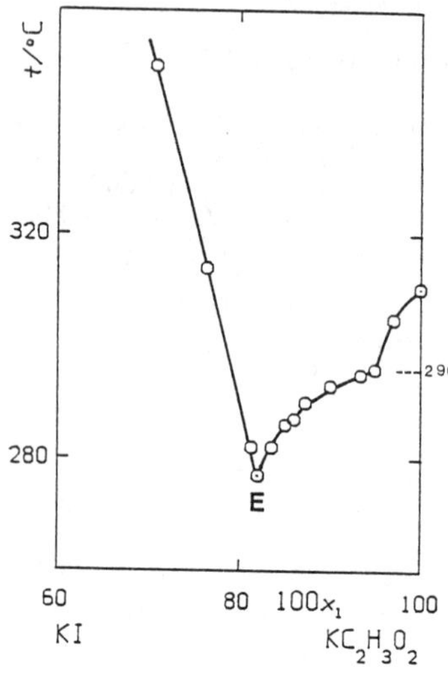

Characteristic point(s):

Eutectic, E, at 277 °C and $100x_1$ = 82.0.

AUXILIARY INFORMATION

METHOD/APPARATUS/PROCEDURE:	SOURCE AND PURITY OF MATERIALS:
Visual polythermal analysis. NOTE: The system was investigated at $100 \geq 100x_1 \geq 70.8$. See also the Note relevant to the results obtained by Piantoni et al. (Ref. 1) on the same system (next Table).	Not stated. Component 1 undergoes a phase transition at $t_{trs}(1)/^{\circ}C$ = 296. Component 2 melts at $t_{fus}(2)/^{\circ}C$ = 683.
	ESTIMATED ERROR:
	Temperature: accuracy probably ±2 K (compiler).
	REFERENCES:
	(1) Piantoni, G.; Leonesi, D.; Braghetti, M.; Franzosini, P. **Ric. Sci.** 1968, **38**, 127-132.

COMPONENTS:	ORIGINAL MEASUREMENTS:
(1) Potassium ethanoate (potassium acetate); $KC_2H_3O_2$; [127-08-2] (2) Potassium iodide; KI; [7681-11-0]	Piantoni, G.; Leonesi, D.; Braghetti, M.; Franzosini, P. **Ric. Sci.** 1968, **38**, 127-132.
VARIABLES:	PREPARED BY:
Temperature.	Baldini, P.

EXPERIMENTAL VALUES:

The results are given only in graphical form (see figure).
The system was investigated only at $0 \leq 100x_2 \leq 20$.

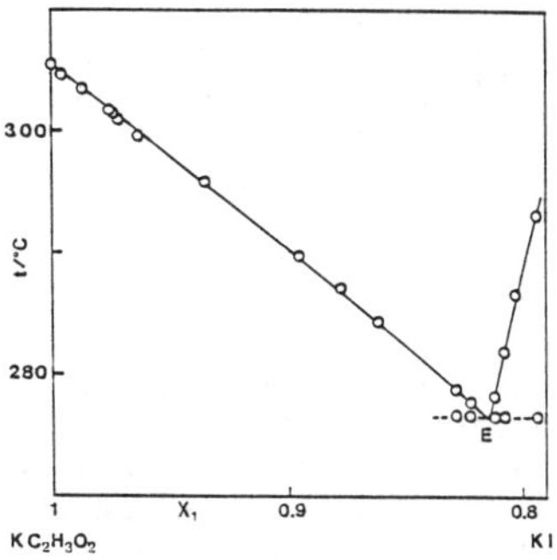

Characteristic point(s):
Eutectic, E, at 276.5 °C and $100x_1$= 81.4 (authors).

NOTE:
Higher precision, and satisfactory mutual consistency of the results obtained by Pianto ni et al. for the three binaries $K/C_2H_3O_2$, (Br,Cl,I) suggest to prefer here the data by these authors to those by Diogenov and Erlykov (Ref. 2), whose solid state transition of component 1 at 569 K, moreover, was not confirmed in more recent literature (Ref. 3). Increasingly positive deviation from ideality was observed by Piantoni et al. for the liquidus branch richer in the halide when KCl, KBr, and KI were successively taken into account. This is coherent with the (cryometric) limiting values $[\lim_{m_2 \to 0} (\Delta T/m_2)$= 17.7, 17.4, and 16.0 K molality^{-1}, respectively] previously found by Braghetti et al. (Ref. 1) when the same halides were employed as solutes in molten potassium ethanoate (whose cryometric constant is: K_1= 18.0±0.3 K molality^{-1}; Ref. 1).

AUXILIARY INFORMATION

METHOD/APPARATUS/PROCEDURE:	SOURCE AND PURITY OF MATERIALS:
A Pyrex device, suitable for work under an inert atmosphere, and allowing one to ob serve the system visually, was employed (for details, see Ref. 1). The initial crystallization temperatures were measured with a Chromel-Alumel thermocouple checked by comparison with a certified Pt resistan ce thermometer, and connected with a L&N Type K-3 potentiometer.	C. Erba RP materials, dried by heating under vacuum (private communication by the authors to the compiler). ESTIMATED ERROR: Temperature: accuracy probably ±0.1 K. REFERENCES: (1) Braghetti, M.; Leonesi, D.; Franzosini, P. **Ric. Sci.** 1968, **38**, 116-118. (2) Diogenov, G.G.; Erlykov, A.M. **Nauch. Dokl. Vysshei Shkoly, Khim. i Khim. Tekhnol.** 1958, **No. 3**, 413-416. (3) Sanesi, M.; Cingolani, A.; Tonelli, P.L.; Franzosini, P. **Thermal Properties**, in **Thermodynamic and Transport Properties of Organic Salts**, IUPAC Chemical Data Series No. 28 (Franzosini, P.; Sanesi, M.; Editors), Pergamon Press, Oxford, 1980, 29-115.

COMPONENTS:	EVALUATOR:
(1) Potassium ethanoate (potassium acetate); $KC_2H_3O_2$; [127-08-2] (2) Potassum nitrite; KNO_2; [7758-09-0]	Spinolo, G., Dipartimento di Chimica Fisica, Universita´ di Pavia (ITALY).

CRITICAL EVALUATION:

This binary was studied by Bergman and Evdokimova (Ref. 1), and by Sokolov and Minich (Ref. 2): in both papers, the visual polythermal analysis was employed to draw the lower boundary of the isotropic liquid field.

Concerning the thermal behavior of component 1, it can be noted that a reasonable agreement exists: (i) between the fusion temperature (575 K) from Refs. 1, 2, and that listed in Table 1 of the Preface (578.7±0.5 K); and (ii) between Sokolov and Minich´s (Ref. 2) higher transition temperature (428 K), and the single $T_{trs}(1)$ value (422.2±0.5 K) from Table 1. No correspondence with Table 1 can be found for Sokolov and Minich´s lower transition (331 K). No solid state transformation of this component is mentioned in Ref. 1.

The experimental points from both papers exhibit rather similar trends; a discrepancy, however, exists about interpretation of the results. Indeed, in Sokolov and Minich´s opinion (Ref. 2), the system ought to be characterized by a eutectic and a peritectic, and accordingly by the presence of an incongruently melting intermediate compound. Conversely, in Bergman and Evdokimova´s opinion (Ref. 1), the system shows a single invariant, i.e., a eutectic at 573 K and $100\mathbf{x}_2 = 50$. It is worth mentioning that, in the evaluator´s opinion, the existence of a third (intermediate) branch of the liquidus - if any - might be supported rather by the experimental data from Ref. 1 than by those from Ref. 2. Moreover, the composition of the intermediate compound suggested in Ref. 2 is not compatible with Sokolov and Minich´s experimental values.

In conlusion, the evaluator is inclined to think that the actual existence of an intermediate compound is poorly supported by the available data, and therefore to prefer the picture of the system drawn in Ref. 1.

REFERENCES:

(1) Bergman, A.G.; Evdokimova, K.A.
Izv. Sektora Fiz.-Khim. Anal., Inst. Obshchei i Neorg. Khim. Akad. Nauk SSSR 1956, **27**, 296-314.

(2) Sokolov, N.M.; Minich, M.A.
Zh. Neorg. Khim. 1961, **6**, 2558-2562 (*); **Russ. J. Inorg. Chem. (Engl. Transl.)** 1961, **6**, 1293-1295.

COMPONENTS:

(1) Potassium ethanoate (potassium acetate); $KC_2H_3O_2$; [127-08-2]
(2) Potassum nitrite; KNO_2; [7758-09-0]

ORIGINAL MEASUREMENTS:

Bergman, A.G.; Evdokimova, K.A.
Izv. Sektora Fiz.-Khim. Anal., Inst. Obshchei i Neorg. Khim. Akad. Nauk SSSR 1956, 27, 296-314.

VARIABLES:

Temperature.

PREPARED BY:

Baldini, P.

EXPERIMENTAL VALUES:

$t/^{\circ}C$	T/K^a	$100x_2$
302	575	0
294	567	4.0
288	561	8.4
278	551	13.0
273	546	17.8
264	537	22.6
258	531	27.3
240	513	36.1
227	500	40.3
218	491	44.3
208	481	47.9
206	479	51.2
206^b	479	54.4
208^b	481	57.3
252	525	60.1
264	537	62.5
282	555	65.8
300	573	70.0

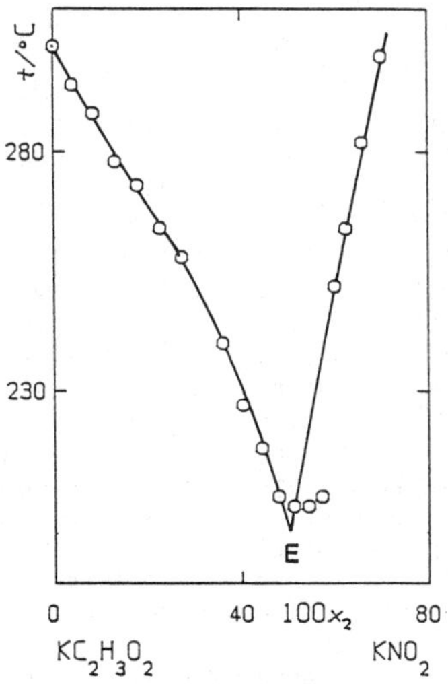

[a] T/K values calculated by the compiler.
[b] Point not considered in Fig. 2 of the original paper in order to draw the fusibility curve (compiler)

Characteristic point(s):
Eutectic, E, at 200 °C and $100x_2$= 50 (authors).

Note - The system was investigated at $0 \leq 100x_2 \leq 70$.

AUXILIARY INFORMATION

METHOD/APPARATUS/PROCEDURE:

Visual polythermal analysis: the temperatures of initial crystallization were measured with a Nichrome-Constantane thermocouple and a 17 mV full-scale millivoltmeter.

SOURCE AND PURITY OF MATERIALS:

Source not stated.
Component 2: $t_{fus}(2)/^{\circ}C$= 440.

ESTIMATED ERROR:

Temperature: accuracy probably ±2 K (compiler).

REFERENCES:

COMPONENTS:	ORIGINAL MEASUREMENTS:
(1) Potassium ethanoate (potassium acetate); $KC_2H_3O_2$; [127-08-2] (2) Potassium nitrite; KNO_2; [7758-09-0]	Sokolov, N.M.; Minich, M.A. **Zh. Neorg. Khim.** 1961, 6, 2558-2562 (*); **Russ. J. Inorg. Chem. (Engl. Transl.)** 1961, 6, 1293-1295.
VARIABLES:	PREPARED BY:
Temperature.	Baldini, P.

EXPERIMENTAL VALUES:

t/°C	T/K[a]	$100x_2$	t/°C	T/K[a]	$100x_2$
302	575	0	221	494	55
295	568	5	245	518	60
287	560	10	271	544	65
276	549	15	292	565	70
265	538	20	315	588	75
252	525	25	339	612	80
245	518	30	360	633	85
233	506	35	381	654	90
220	493	40	408	681	95
208	481	45	436	709	100
210	483	50			

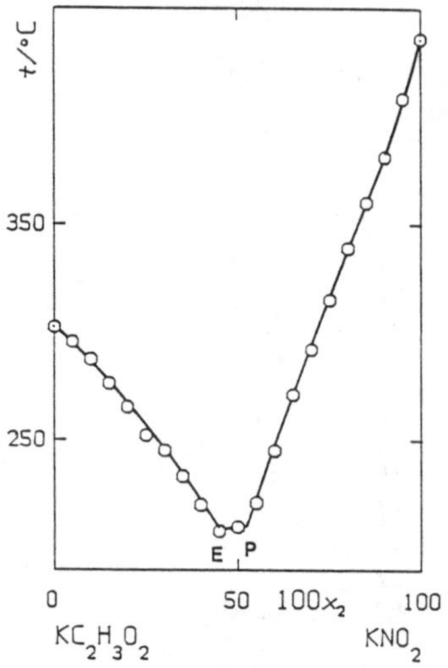

[a] T/K values calculated by the compiler.

Characteristic point(s):
Eutectic, E, at 208 °C and $100x_2$= 45 (compiler).
Peritectic, P, at 210 °C and $100x_2$= 52 (compiler).

Note - The assignement of the invariants as reported in Table 2 of the original paper, i.e., E at 210 °C and $100x_2$= 45, and P at 208 °C and $100x_2$= 52, is nonsensical and not compatible with the experimental values (compiler).

Intermediate compound(s):
$K_2C_2H_3O_2NO_2$ (tentative composition; authors) incongruently melting (the tentative compo sition by the authors is not compatible with the above assignement of the invariants; compiler).

AUXILIARY INFORMATION

METHOD/APPARATUS/PROCEDURE:

Visual polythermal analysis.

SOURCE AND PURITY OF MATERIALS:

Component 1: commercial "chemically pure" material recrystallized from water; it undergoes phase transitions at $t_{trs}(1)$/°C= 58, 155 (Ref. 1).
Component 2: material prepared by reducing potassium nitrate with lead, melting at $t_{fus}(2)$/°C= 436 after three recrystalliza tions; it undergoes a phase transition at $t_{trs}(2)$/°C= 45 (Ref. 2).

ESTIMATED ERROR:

Temperature: accuracy probably ±2 K (compiler).

REFERENCES:

(1) Sokolov, N.M.
Tezisy Dokl. X Nauch. Konf. S.M.I. 1956.
(2) Berul', S.I.; Bergman, A.G.
Izv. Sektora Fiz.-Khim. Anal. 1952, 21, 178-183.

COMPONENTS:	EVALUATOR:
(1) Potassium ethanoate (potassium acetate); $KC_2H_3O_2$; [127-08-2] (2) Potassium nitrate; KNO_3; [7757-79-1]	Spinolo, G., Dipartimento di Chimica Fisica, Universita´ di Pavia (ITALY).

CRITICAL EVALUATION:

This binary was studied by Bergman and Evdokimova (Ref. 1), Diogenov et al. (Ref. 2), Gimel´shtein (Ref. 3), and Diogenov and Chumakova (Ref. 4). In Ref. 3, the automatic record of the heating curves with a DTA device allowed the author to gain a complete picture of the phase diagram in the superambient region, whereas in Refs. 1, 2, and 4 the visual polythermal analysis was employed to draw merely the lower boundary of the isotropic liquid field.

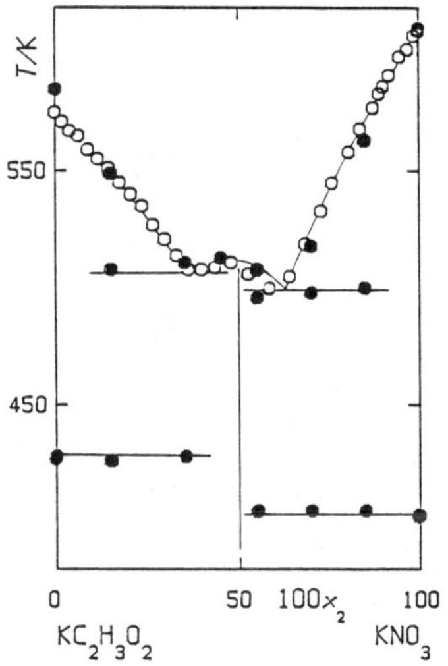

Concerning component 1, the fusion temperatures from Refs. 1-4 (575, 583, 586, and 575 K, respectively) fluctuate (rather widely) around the $T_{fus}(1)$ value (578.7±0.5 K) listed in Preface, Table 1. Moreover, a reasonable agreement exists between the (single) solid state transition temperature reported in Ref. 3 and Table 1 (428 K and 422.2±0.5 K, respectively), whereas, in the evaluator´s opinion, poor reliability is to be attached to Diogenov et al.´s (Ref. 2) assertion that a transition occurs at 565 K, because no support to it is provided by the findings of any author foreign to Diogenov´s group (Ref. 5).

The main features of the phase diagram reported in Refs. 1-4 appear to be rather similar, so that the following points can be taken as unambiguously stated: (i) a 1:1 intermediate compound is formed; (ii) it melts congruently, and, accordingly, two eutectics separate its crystallization branches from those relevant to the pure components; and (iii) a fair agreement exists among the coordinates of the invariants provided by Refs. 1, 3, and 4 (see below), whereas the temperature values from Ref. 2 appear to be systematically too low.

	Ref. 1	Ref. 2	Ref. 3	Ref. 4
Eutectic E_1; T/K:	507	493	507	507
Eutectic E_1; $100\mathbf{x}_2$:	36	39	35.5	35
Eutectic E_2; T/K:	495	485	497	497
Eutectic E_2; $100\mathbf{x}_2$:	61.5	61	62.5	62
Int. comp.; T_{fus}/K:	511	502	511	511

A direct comparison of the visual polythermal (empty circles) and derivatographical (filled circles) data from Refs. 1 and 3, respectively, is made in the figure.

REFERENCES:

(1) Bergman, A.G.; Evdokimova, K.A.; **Iz. Sektora Fiz.-Khim. Anal., Inst. Obshchei i Neorg. Khim. Akad. Nauk SSSR** 1956, 27, 296-314.
(2) Diogenov, G.G.; Nurminskii, N.N.; Gimel´shtein, V.G.; **Zh. Neorg. Khim.** 1957, 2, 1596-1600 (*); **Russ. J. Inorg. Chem. (Engl. Transl.)** 1957, 2(7), 237-245.
(3) Gimel´shtein, V.G.; **Tr. Irkutsk. Politekh. Inst.** 1971, No. 66, 80-100.
(4) Diogenov, G.G.; Chumakova, V.P. **Fiz.-Khim. Issled. Rasplavov Solei, Irkutsk,** 1975, 7-12.
(5) Sanesi, M.; Cingolani, A.; Tonelli, P.L.; Franzosini, P. **Thermal Properties,** in **Thermodynamic and Transport Properties of Organic Salts,** IUPAC Chemical Data Series No. 28 (Franzosini, P.; Sanesi, M.; Editors), Pergamon Press, Oxford, 1980, 29-115.

COMPONENTS:	ORIGINAL MEASUREMENTS:
(1) Potassium ethanoate (potassium acetate); $KC_2H_3O_2$; [127-08-2] (2) Potassium nitrate; KNO_3; [7757-79-1]	Bergman, A.G.; Evdokimova, K.A. **Iz. Sektora Fiz.-Khim. Anal., Inst. Obshchei i Neorg. Khim. Akad. Nauk SSSR** 1956, **27**, 296-314.
VARIABLES:	PREPARED BY:
Temperature.	Baldini, P.

EXPERIMENTAL VALUES:

t/°C	T/K[a]	$100x_2$	t/°C	T/K[a]	$100x_2$
302	575	0	233	506	52.7
298	571	2.0	227	500	58.5
294	567	4.0	232	505	64.2
292	565	6.4	246	519	68.5
286	559	9.0	260	533	72.9
282	555	11.7	272	545	76.0
278	551	14.6	285	558	80.7
272	545	17.5	295	568	83.9
267	540	20.5	304	577	87.5
262	535	23.5	310	583	89.1
254	527	26.5	313	586	90.2
248	521	29.6	318[b]	591	91.9
241	514	32.8	326[c]	599	94.8
235	508	36.1	329[d]	602	97.0
235	508	39.8	335	608	98.8
236	509	43.6	337	610	100
238	511	48.0			

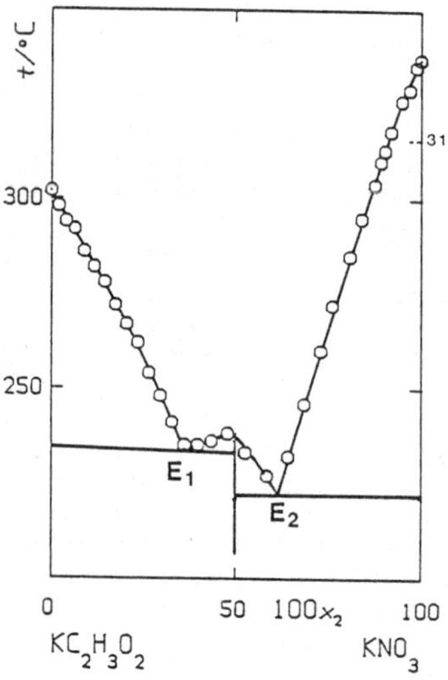

[a] T/K values calculated by the compiler.
[b] Erroneously reported as 218 in table 5 of the original paper (compiler).
[c] Erroneously reported as 226 in table 5 of the original paper (compiler).
[d] Erroneously reported as 229 in table 5 of the original paper (compiler).

Characteristic point(s):

Eutectic, E_1, at 234 °C and $100x_1$= 64 (authors).
Eutectic, E_2, at 222 °C and $100x_2$= 61.5 (authors).

Intermediate compound(s):

$K_2C_2H_3O_2NO_3$, congruently melting at 238 °C (authors).

AUXILIARY INFORMATION

METHOD/APPARATUS/PROCEDURE:	SOURCE AND PURITY OF MATERIALS:
Visual polythermal analysis: the temperatures of initial crystallization were measured with a Nichrome-Constantane thermocouple and a 17 mV full-scale millivoltmeter.	Source not stated. Component 2: in the temperature field of interest it undergoes a phase transition at $t_{trs}(2)$/°C= 316-318 (Ref. 1).
ESTIMATED ERROR:	REFERENCES:
Temperature: accuracy probably ±2 K (compiler).	(1) Bergman, A.G.; Berul', S.I. **Izv. Sektora Fiz.-Khim. Anal.** 1952, **21**, 178-183.

COMPONENTS:	ORIGINAL MEASUREMENTS:
(1) Potassium ethanoate (potassium acetate); $KC_2H_3O_2$; [127-08-2] (2) Potassium nitrate; KNO_3; [7757-79-1]	Diogenov, G.G.; Nurminskii, N.N.; Gimel'shtein, V.G. **Zh. Neorg. Khim.** 1957, 2, 1596-1600 (*); **Russ. J. Inorg. Chem. (Engl. Transl.)** 1957, 2(7), 237-245.
VARIABLES:	PREPARED BY:
Temperature.	Baldini, P.

EXPERIMENTAL VALUES:

t/°C	T/K[a]	$100x_2$	t/°C	T/K[a]	$100x_2$
310	583	0	225	498	54.5
303	576	3	223	496	55.5
292	565	6	220	493	57.5
288	561	8.5	217	490	59
276	549	15	216	489	62.5
264	537	22	221	494	65
250	523	28	235	508	70
245	518	31	250	523	74
242	515	32	265	538	77.5
238	511	33.5	291	564	84.5
230	503	36	301	574	87.5
225	498	37.5	307	580	89
220	493	39	311	584	91
225	498	41.5	318	591	93
228	501	44	332	605	98
228	501	49.5	337	610	100
228	501	51			

[a] T/K values calculated by the compiler.

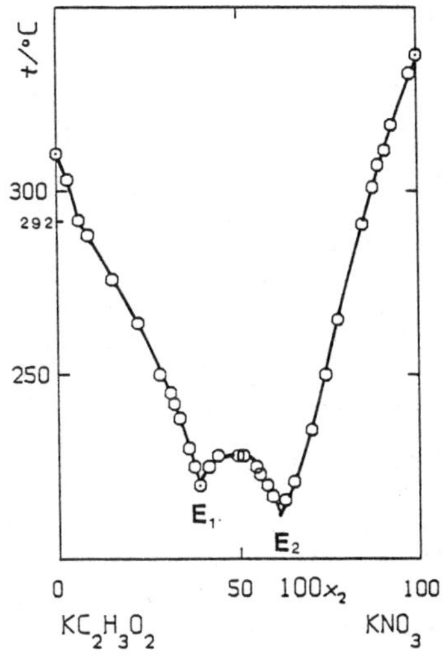

Characteristic point(s):
Eutectic, E_1, at 220 °C and $100x_1$= 61 (authors).
Eutectic, E_2, at 212 °C and $100x_2$= 61 (authors).

Intermediate compound(s):
$K_2C_2H_3O_2NO_3$, congruently melting at 229 °C (authors).

AUXILIARY INFORMATION

METHOD/APPARATUS/PROCEDURE:	SOURCE AND PURITY OF MATERIALS:
Visual polythermal analysis.	Source not stated. Component 1 undergoes a phase transition at $t_{trs}(1)$/°C= 292.
	ESTIMATED ERROR:
	Temperature: accuracy probably ±2 K (compiler).
	REFERENCES:

COMPONENTS:

(1) Potassium ethanoate (potassium acetate); $KC_2H_3O_2$; [127-08-2]
(2) Potassium nitrate; KNO_3; [7757-79-1]

ORIGINAL MEASUREMENTS:

Gimel´shtein, V.G.
Tr. Irkutsk. Politekh. Inst. 1971, No. 66, 80-100.

VARIABLES:

Temperature

PREPARED BY:

Baldini, P.

EXPERIMENTAL VALUES:

t/°C	T/K[a]	$100x_2$	t/°C	T/K[a]	$100x_2$
312	585	0	132	405	55.0
155	428	0	245	518	70.0
276	549	15.0	225	498	70.0
235	508	15.0	132	405	70.0
153	426	15.0	290	563	85.0
238	511	35.0	227	500	85.0
155	428	35.0	132	405	85.0
240	513	45.0	338	611	100
235	508	55.0	130	403	100
223	496	55.0			

[a] T/K values calculated by the compiler.

Characteristic point(s):

Eutectic, E_1, at 234 °C and $100x_2$= 35.5 (author).
Eutectic, E_2, at 224 °C, and $10x_2$= 62.5 (author).

Intermediate compound(s):

$K_2C_2H_3O_2NO_3$, congruently melting at 238 °C (author).

AUXILIARY INFORMATION

METHOD/APPARATUS/PROCEDURE:

Differential thermal analysis (using a derivatograph with automatic recording of the heating curves) was employed.

NOTE:

The meaning of the data listed in the table becomes apparent by observing the figure reported in the critical evaluation.
The coordinates of the characteristic points were stated by the author on the basis of his own DTA measurements, and of previous literature data (Ref. 1).

SOURCE AND PURITY OF MATERIALS:

Not stated.
Component 1 melts at $t_{fus}(1)$/°C= 312 (310 °C according to Fig. 13 of the original paper; compiler), and undergoes a phase transition at $t_{trs}(1)$/°C= 155.
Component 2 melts at $t_{fus}(2)$/°C= 338 (337 °C according to Fig. 13 of the original paper; compiler), and undergoes a phase transition at $t_{trs}(2)$/°C= 130.

ESTIMATED ERROR:

Temperature: accuracy probably ±2 K (compiler).

REFERENCES:

(1) Bergman, A.G.; Evdokimova, K.A. **Iz. Sektora Fiz.-Khim. Anal., Inst. Obshchei i Neorg. Khim. Akad. Nauk SSSR** 1956, **27**, 296-314.

COMPONENTS:

(1) Potassium ethanoate (potassium acetate); $KC_2H_3O_2$; [127-08-2]
(2) Potassium nitrate; KNO_3; [7757-79-1]

ORIGINAL MEASUREMENTS:

Diogenov, G.G.; Chumakova, V.P.
Fiz.-Khim. Issled. Rasplavov Solei, Irkutsk, 1975, 7-12.

VARIABLES:

Temperature.

PREPARED BY:

Baldini, P.

EXPERIMENTAL VALUES:

Eutectic, E_1, at 234 ^oC (Fig. 1 of the original paper); composition not stated ($100\mathbf{x}_1$ about 65 in compiler's graphical estimation).
Eutectic, E_2, at 224 ^oC (Fig. 1 of the original paper); composition not stated ($100\mathbf{x}_1$ about 38 in compiler's graphical estimation).

Intermediate compound(s):

$K_2C_2H_3O_2NO_3$, congruently melting at 238 ^oC (Fig. 1 of the original paper).

AUXILIARY INFORMATION

METHOD/APPARATUS/PROCEDURE:

Visual polythermal analysis.

SOURCE AND PURITY OF MATERIALS:

Not stated.
Component 1: $t_{fus}(1)/^oC$= 302;
component 2: $t_{fus}(2)/^oC$= 337 (Fig. 1 of the original paper).

ESTIMATED ERROR:

Temperature: accuracy probably ± 2 K (compiler).

REFERENCES:

COMPONENTS:

(1) Potassium propanoate (potassium propionate); $KC_3H_5O_2$; [327-62-8]
(2) Potassium thiocyanate; KCNS; [333-20-0]

ORIGINAL MEASUREMENTS:

Sokolov, N.M.
Zh. Obshch. Khim. 1966, **36**, 577-582.

VARIABLES:

Temperature.

PREPARED BY:

Baldini, P.

EXPERIMENTAL VALUES:

The results are reported only in graphical form (see figure).

Characteristic point(s):

Eutectic, E, at 157 °C and $100\mathbf{x}_1$= 14 (author).

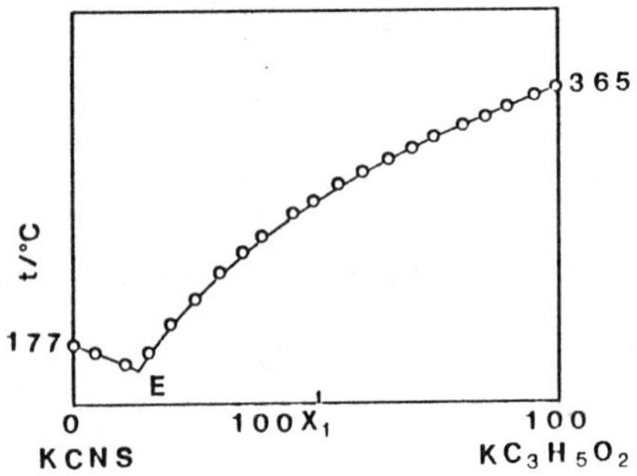

AUXILIARY INFORMATION

METHOD/APPARATUS/PROCEDURE:

Visual polythermal analysis.

NOTE:

A substantial agreement exists between the solid state transition and fusion temperatures reported by Sokolov [341 (instead of 350) and 638 K, respectively] for component 1, and those listed in Table 1 of the Preface (352.5±0.5 and 638.3±0.5 K, respectively).

SOURCE AND PURITY OF MATERIALS:

Not stated. Component 1 melts at $t_{fus}(1)$/°C= 365, and undergoes a phase transition at $t_{trs}(1)$/°C= 77 (Ref. 1; in the compiler´s opinion the correct figure ought to be 68, as quoted in several papers by the same author from the same source; according to Ref. 1, 77 °C is the temperature at which a transition occurs in $NaC_3H_5O_2$). Component 2 melts at $t_{fus}(2)$/°C= 177 and undergoes a phase transition at $t_{trs}(2)$/°C= 143 (Ref. 2).

ESTIMATED ERROR:

Temperature: accuracy probably ±2 K (compiler).

REFERENCES:

(1) Sokolov, N.M. **Tezisy Dokl. X Nauch. Konf. S.M.I.** 1956.
(2) Vrzhesnevskij, I.B. **Zh. Russk. Fiz.-Khim. Obshch.** 1911, **43**, 1368.

COMPONENTS:	ORIGINAL MEASUREMENTS:
(1) Potassium propanoate (potassium propionate); $KC_3H_5O_2$; [327-62-8] (2) Potassium nitrite; KNO_2; [7758-09-0]	Sokolov, N.M.; Minich, M.A. **Zh. Neorg. Khim.** 1961, **6**, 2558-2562 (*); **Russ. J. Inorg. Chem. (Engl. Transl.)** 1961, **6**, 1293-1295.
VARIABLES:	PREPARED BY:
Temperature.	Baldini, P.

EXPERIMENTAL VALUES:

t/°C	T/K[a]	$100x_2$	t/°C	T/K[a]	$100x_2$
366	639	0	285	558	55
357	630	5	309	582	60
349	622	10	322	595	65
341	614	15	336	609	70
336	609	20	347	620	75
327	600	25	356	629	80
319	592	30	369	642	85
309	582	35	382	655	90
299	572	40	401	674	95
293	566	45	436	709	100
289	562	50			

[a] T/K values calculated by the compiler.

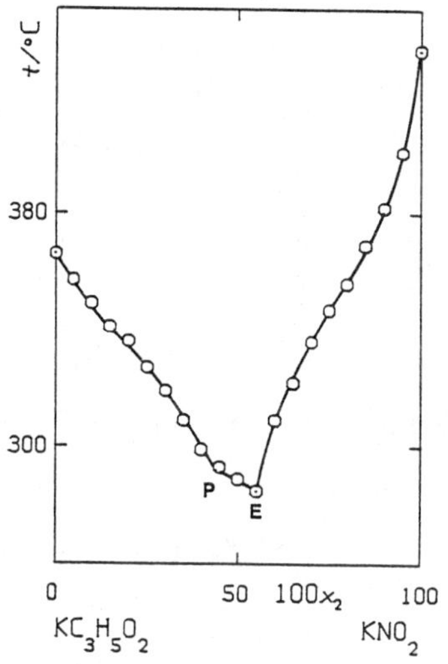

Characteristic point(s):

Eutectic, E, at either 285 °C (according to the tabulated data; compiler), or 283 °C (according to table 2 of the original paper; authors), and $100x_2$= 55 (authors).
Peritectic, P, at 292 °C (figure in poor agreement with the tabulated data; compiler), and $100x_2$= 44 (authors).

Intermediate compound(s):

$K_5(C_3H_5O_2)_3(NO_2)_2$ (tentative composition; authors) incongruently melting.

AUXILIARY INFORMATION

METHOD/APPARATUS/PROCEDURE:

Visual polythermal analysis.

NOTE:

A substantial agreement exists between the solid state transition and fusion temperatures reported by Sokolov and Minich (341 and 639 K, respectively) for component 1, and those listed in Table 1 (352.5±0.5 and 638.3±0.5 K, respectively). The actual existence (and composition) of the intermediate compound ought to be more convincingly proved.

SOURCE AND PURITY OF MATERIALS:

Component 1: prepared from "chemically pure" $KHCO_3$ and the fatty acid, and recrystallized from n-butanol after having been deposited from the aqueous solution and dried; it undergoes a phase transition at $t_{trs}(1)$/°C= 68 (Ref. 1).
Component 2: material prepared by reducing potassium nitrate with lead, melting at $t_{fus}(2)$/°C= 436 after three recrystallizations; it undergoes a phase transition at $t_{trs}(2)$/°C= 45 (Ref. 2).

ESTIMATED ERROR:

Temperature: accuracy probably ±2 K (compiler).

REFERENCES:

(1) Sokolov, N.M.
Tezisy Dokl. X Nauch. Konf. S.M.I. 1956.
(2) Berul', S.I.; Bergman, A.G.
Izv. Sektora Fiz.-Khim. Anal. 1952, **21**, 178-183.

COMPONENTS:	ORIGINAL MEASUREMENTS:
(1) Potassium propanoate (potassium propionate); $KC_3H_5O_2$; [327-62-8] (2) Potassium nitrate; KNO_3; [7757-79-1]	Dmitrevskaya, O.I.; Sokolov, N.M. **Zh. Obshch. Khim.** 1958, **28**, 2920-2926 (*); **Russ. J. Gen. Chem. (Engl. Transl.)** 1958, **28**, 2949-2954.
VARIABLES:	PREPARED BY:
Temperature.	Baldini, P.

EXPERIMENTAL VALUES:

$t/^{o}C$	T/K^{a}	$100\mathbf{x}_2$	$t/^{o}C$	T/K^{a}	$100\mathbf{x}_2$
365	638	0	281	554	55
360	633	5	272	545	60
354	627	10	267	540	62.5
348	621	15	264	537	65
341	614	20	268	541	67.5
335	608	25	274	547	70
328	601	30	282	555	75
321	594	35	292	565	80
310	583	40	301	574	85
300	573	45	311	584	90
292	565	50	337	610	100

[a] **T**/K values calculated by the compiler.

Characteristic point(s):

Eutectic, E, at 264 ^{o}C and $100\mathbf{x}_2$= 65 (authors).

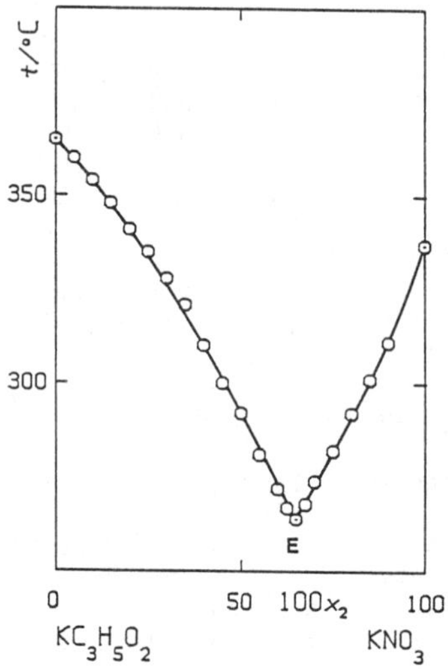

AUXILIARY INFORMATION

METHOD/APPARATUS/PROCEDURE:	SOURCE AND PURITY OF MATERIALS:
Visual polythermal analysis. Temperature of initial crystallization measured with a Nichrome-Constantane thermocouple checked at the boiling point of water, and at the fusion points of benzoic acid, mannitol, succinic acid, silver nitrate, tin, potassium nitrate, and potassium dichromate. Mixtures melted in a glass tube inserted into a wider tube to ensure uniform heating. Glass fiber stirrer used.	Component 1 prepared by adding a small excess of distilled commercial propanoic acid to a solution of the "chemically pure" hydrogen carbonate; the solid recovered after evaporation of the solvent was recrystallized from **n**-butanol. Component 2: "chemically pure" material recrystallized. Component 1 undergoes a phase transition at $\mathbf{t}_{trs}(1)/^{o}C$= 68 (Ref. 1). Component 2 undergoes phase transitions at $\mathbf{t}_{trs}(2)/^{o}C$= 124, 316 (current literature).
NOTE:	ESTIMATED ERROR:
A substantial agreement exists between the solid state transition and fusion temperatures reported by Dmitrevskaya and Sokolov (341 and 638 K, respectively) for component 1, and those listed in Preface, Table 1 (352.5$\pm$0.5 and 638.3$\pm$0.5 K, respectively).	Temperature: accuracy probably $\pm$2 K (compiler). REFERENCES: (1) Sokolov, N.M. **Tezisy Dokl. X Nauch. Konf. S.M.I.** 1956.

COMPONENTS:	EVALUATOR:
(1) Potassium butanoate (potassium butyrate); $KC_4H_7O_2$; [589-39-9] (2) Potassium thiocyanate; KCNS; [333-20-0]	Franzosini, P., Dipartimento di Chimica Fisica, Universita' di Pavia (ITALY).

CRITICAL EVALUATION:

This system was studied only by Sokolov and Pochtakova (Ref. 1), who suggested the existence of: (i) an intermediate compound of probable composition $K_7(C_4H_7O_2)_6CNS$; (ii) a "perekhodnaya tochka" (likely a peritectic), P, at 608 K (335 °C) and $100\mathbf{x}_1$= 82; and (iii) a eutectic, E, at 443 K (170 °C) and $100\mathbf{x}_1$= 6.5.

Component 1, however, forms liquid crystals. Therefore, the fusion temperature, $\mathbf{T}_{fus}(1)$= 677 K (404 °C), reported in Ref. 1 should be identified with the clearing temperature, the corresponding value from Table 1 of the Preface being $\mathbf{T}_{clr}(1)$= 677.3±0.5 K.

For the same component, the phase transition temperatures quoted (from Ref. 2) in Ref. 1, viz., 618 K (345 °C), 553-558 K (280-285 °C), and 463 K (190 °C), might correspond respectively to the fusion temperature (626.1±0.7 K) and to the first and third solid state transition temperatures (562.2±0.6 K, and 467.2±0.5 K) of Table 1 of the Preface. No mention is made by the authors of other phase transitions, although in Table 1 two more $\mathbf{T}_{trs}$ values are reported (540.8±1.1 K and 461.4±1.0 K).

The phase diagram as suggested by the authors can be considered as adequate only for the region (rich in component 2) including the eutectic, whereas it does not seem reliable in the remaining part.

In particular:

(i) the "perekhodnaya tochka", P, should rather be an M'_p point, at which the equilibria involving the isotropic liquid and the liquid crystals might be those described in Preface, Scheme B.1;

(ii) the available data cannot be considered as sufficient to support the existence of any intermediate compound.

REFERENCES:

(1) Sokolov, N.M.; Pochtakova, E.I.
Zh. Obshch. Khim. 1958, **28**, 1693-1700 (*); **Russ. J. Gen. Chem. (Engl. Transl.)** 1958, **28**, 1741-1747.

(2) Sokolov, N.M.
Tezisy Dokl. X Nauch. Konf. S.M.I. 1956.

COMPONENTS:

(1) Potassium butanoate (potassium butyrate); $KC_4H_7O_2$; [589-39-9]
(2) Potassium thiocyanate; KCNS; [333-20-0]

ORIGINAL MEASUREMENTS:

Sokolov, N.M.; Pochtakova, E.I.
Zh. Obshch. Khim. 1958, **28**, 1693-1700 (*);
Russ. J. Gen. Chem. (Engl. Transl.) 1958, **28**, 1741-1747.

VARIABLES:

Temperature.

PREPARED BY:

Baldini, P.

EXPERIMENTAL VALUES:

$t/^oC$	T/K^a	$100x_1$	$t/^oC$	T/K^a	$100x_1$
177	450	0	301	574	65
176	449	2.5	310	583	70
174	447	5	322	595	75
170	443	6.5	327	600	77.5
173	446	7.5	335	608	80
183	456	10	335	608	82
204	477	15	337	610	82.5
215	488	20	342	615	85
224	497	25	364	637	90
232	505	30	379	652	95
260	533	45	404	677	100
278	551	55			

[a] T/K values calculated by the compiler.

Characteristic point(s):

Eutectic, E, at 170 °C and $100x_1$= 6.5 (authors).
Characteristic point, P (perekhodnaya tochka in the original text; see the Introduction), at 335 °C and $100x_1$= 82.

Intermediate compound(s):

$K_7(C_4H_7O_2)_6CNS$ (proposed by the authors).

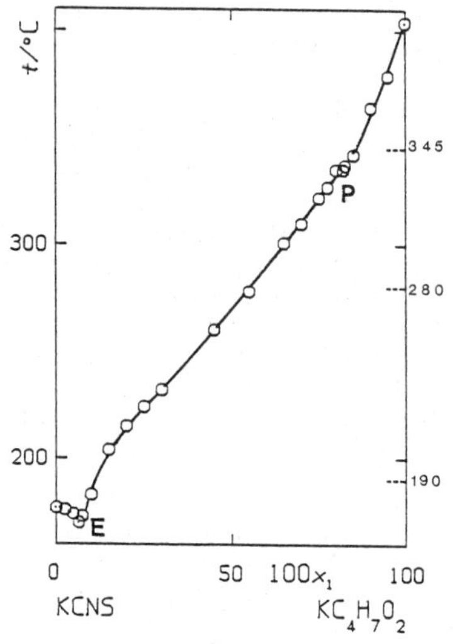

AUXILIARY INFORMATION

METHOD/APPARATUS/PROCEDURE:

Visual polythermal analysis. Temperatures measured with a Nichrome-Constantane thermocouple.

SOURCE AND PURITY OF MATERIALS:

Component 1: synthetized from "chemically pure" potassium hydrogen carbonate and n-butanoic acid (Ref. 1, where, however, carbonates instead of hydrogen carbonates are employed; compiler), and recrystallized from n-butanol; it undergoes phase transitions at $t_{trs}(1)/^oC$= 190, 280-285, 345 (Ref. 2). Component 2: commercial material recrystallized once from water and once from alcohol.

ESTIMATED ERROR:

Temperature: accuracy probably ±2 K (compiler).

REFERENCES:

(1) Sokolov, N.M.
Zh. Obshch. Khim. 1954, **24**, 1581-1593.
(2) Sokolov, N.M.
Tezisy Dokl. X Nauch. Konf. S.M.I. 1956.

COMPONENTS:	EVALUATOR:
(1) Potassium butanoate (potassium butyrate); $KC_4H_7O_2$; [589-39-9] (2) Potassium nitrite; KNO_2; [7758-09-0]	Franzosini, P., Dipartimento di Chimica Fisica, Universita' di Pavia (ITALY).

CRITICAL EVALUATION:

The visual polythermal analysis was employed by Sokolov and Minich (Ref. 1) to study the lower boundary of the isotropic liquid field: they claimed the occurrence of a congruently melting intermediate compound [of tentative composition $K_8(C_4H_7O_2)_5(NO_2)_3$], able to give eutectics with either component.

Component 1, however, forms liquid crystals. Therefore, the phase diagram has to be re-interpreted, possibly with reference to Preface, Scheme D.1. In this case, Sokolov and Minich's eutectic E_1 should be an M'_E point, and a further (still undetected) invariant type M_E ought to exist.

The fusion temperature, $T_{fus}(1)$= 677 K, reported in Ref. 1, should be identified with the clearing temperature of component 1, and agrees fairly with the $T_{clr}(1)$ value (677.3 ± 0.5 K) listed in Preface, Table 1.

Neither of the phase transformation temperatures, i.e., 553-558 and 463 K, quoted in Ref. 1 from Ref. 2 for the same component correspond to the $T_{fus}(1)$ value (626.1 ± 0.7 K) given in Table 1, inasmuch as they lie each halfway between the two pairs of solid state transition temperatures (i.e., 562.2 ± 0.6 and 540.8 ± 1.1 K, and 467.2 ± 0.5 and 461.4 ± 1.0 K, respectively) also reported in Table 1. It is, however, to be noted that in other papers by the same group (see, e.g., Ref. 3) a phase transformation occurring at 618 K, i.e., close to the $T_{fus}(1)$ value of Table 1, is also mentioned.

REFERENCES:

(1) Sokolov, N.M.; Minich, M.A.
Zh. Neorg. Khim. 1961, **6**, 2558-2562 (*); **Russ. J. Inorg. Chem. (Engl. Transl.)** 1961, **6**, 1293-1295.

(2) Sokolov, N.M.
Tezisy Dokl. X Nauch. Konf. S.M.I. 1956.

(3) Sokolov, N.M.; Pochtakova, E.I.
Zh. Obshch. Khim. 1958, **28**, 1693-1700 (*); **Russ. J. Gen. Chem. (Engl. Transl.)** 1958, **28**, 1741-1747.

COMPONENTS:

(1) Potassium butanoate (potassium butyrate); $KC_4H_7O_2$; [589-39-9]
(2) Potassium nitrite; KNO_2; [7758-09-0]

ORIGINAL MEASUREMENTS:

Sokolov, N.M.; Minich, M.A.
Zh. Neorg. Khim. 1961, 6, 2558-2562 (*);
Russ. J. Inorg. Chem. (Engl. Transl.) 1961, 6, 1293-1295.

VARIABLES:

Temperature.

PREPARED BY:

Baldini, P.

EXPERIMENTAL VALUES:

$t/^oC$	T/K^a	$100x_2$	$t/^oC$	T/K^a	$100x_2$
404	677	0	334	607	55
396	669	5	347	620	60
382	655	10	357	630	65
365	638	15	368	641	70
352	625	20	374	647	75
339	612	25	383	656	80
325	598	30	392	665	85
317	590	35	403	676	90
316	589	40	413	686	95
306	579	45	436	709	100
319	592	50			

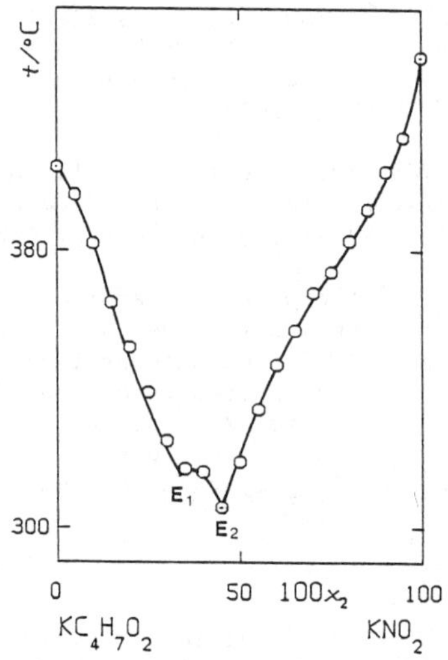

[a] T/K values calculated by the compiler.

Characteristic point(s):

Eutectic, E_1, 315 ^oC and $100x_2$= 33.5 (authors).
Eutectic, E_2, 306 ^oC and $100x_2$= 45 (authors).

Intermediate compound(s):

$K_8(C_4H_7O_2)_5(NO_2)_3$ (tentative composition; authors) congruently melting (at 317 ^oC; compiler).

AUXILIARY INFORMATION

METHOD/APPARATUS/PROCEDURE:

Visual polythermal analysis.

SOURCE AND PURITY OF MATERIALS:

Component 1: prepared from "chemically pure" $KHCO_3$ and the fatty acid, and recrystallized from n-butanol after having been deposited from the aqueous solution and dried; it undergoes phase transitions at $t_{trs}(1)/^oC$= 190, 280-285 (Ref. 1).
Component 2: material prepared by reducing potassium nitrate with lead, melting at $t_{fus}(2)/^oC$= 436 after three recrystallizations; it undergoes a phase transition at $t_{trs}(2)/^oC$= 45 (Ref. 2).

ESTIMATED ERROR:

Temperature: accuracy probably ± 2 K (compiler).

REFERENCES:

(1) Sokolov, N.M. **Tezisy Dokl. X Nauch. Konf. S.M.I.** 1956.
(2) Berul´, S.I.; Bergman, A.G. **Izv. Sektora Fiz.-Khim. Anal.** 1952, 21, 178-183.

COMPONENTS:	EVALUATOR:
(1) Potassium butanoate (potassium butyrate); $KC_4H_7O_2$; [589-39-9] (2) Potassium nitrate; KNO_3; [7757-79-1]	Schiraldi, A., Dipartimento di Chimica Fisica, Universita' di Pavia (ITALY).

CRITICAL EVALUATION:

This system was studied only by Dmitrevskaya (Ref. 1), who, on the basis of her visual polythermal investigation, suggested the phase diagram to be of the eutectic type, the invariant point being at 556 K (283 °C) and $100\mathbf{x}_2$= 58.

Component 1, however, forms liquid crystals. Therefore, Dmitrevskaya's fusion temperature, $T_{fus}(1)$= 677 K (404 °C), should be identified with the clearing temperature of potassium butanoate, the corresponding value from Preface, Table 1 being $\mathbf{T}_{clr}(1)$= 677.3±0.5 K.

Accordingly, it seems likely that the actual phase diagram of this system should correspond to Preface, Scheme B.1 or B.2.

Among the phase transformation temperatures of component 1 quoted in Ref. 1 from Ref. 2 (i.e., 618, 553-558, and 463 K) the first one can be reasonably identified with the fusion temperature (626.1±0.7 K) listed in Table 1, whereas the second and third ones lie each halfway between the two pairs of solid state transition temperatures (i.e., 562.2±0.6 and 540.8±1.1 K, and 467.2±0.5 and 461.4±1.0 K, respectively) also reported in Table 1.

REFERENCES:

(1) Dmitrevskaya, O.I.
Zh. Obshch. Khim. 1958, **28**, 2007-2013 (*); **Russ. J. Gen. Chem. (Engl. Transl.)** 1958, **28**, 2046-2051.

(2) Sokolov, N.M.
Tezisy Dokl. X Nauch. Konf. S.M.I. 1956.

COMPONENTS:

(1) Potassium butanoate (potassium butyrate); $KC_4H_7O_2$; [589-39-9]
(2) Potassium nitrate; KNO_3; [7757-79-1]

ORIGINAL MEASUREMENTS:

Dmitrevskaya, O.I.
Zh. Obshch. Khim. 1958, **28**, 2007-2013 (*);
Russ. J. Gen. Chem. (Engl. Transl.) 1958, **28**, 2046-2051.

VARIABLES:

Temperature.

PREPARED BY:

Baldini, P.

EXPERIMENTAL VALUES:

$t/^{o}C$	T/K^a	$100x_2$	$t/^{o}C$	T/K^a	$100x_2$
404	677	0	289	562	55
387	660	5	283	556	58
370	643	10	286	559	60
356	629	15	291	564	65
342	615	20	294	567	70
330	603	25	300	573	75
320	593	30	306	579	80
313	586	35	314	587	85
306	579	40	320	593	90
300	573	45	328	601	95
294	567	50	337	610	100

[a] T/K values calculated by the compiler.

Characteristic point(s):

Eutectic, E, at 283 ^{o}C and $100x_2$= 58 (author).

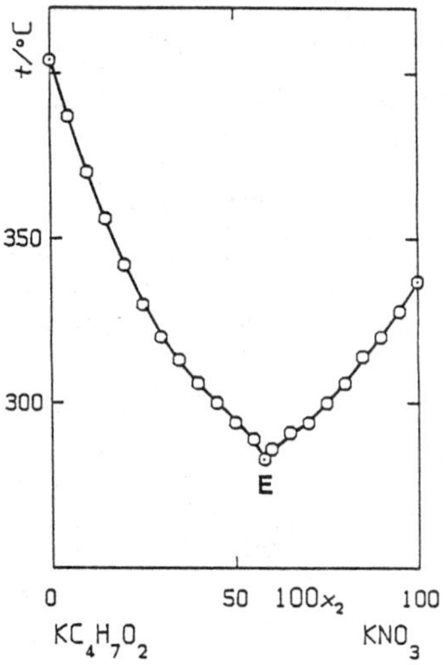

AUXILIARY INFORMATION

METHOD/APPARATUS/PROCEDURE:

Visual polythermal analysis.
Temperatures measured with a Nichrome-Constantane thermocouple.

SOURCE AND PURITY OF MATERIALS:

Component 1 synthetized from "chemically pure" potassium hydrogen carbonate and **n**-butanoic acid twice distilled. "Chemically pure" component 2 recrystallized and dried to constant mass.
Component 1 undergoes phase transitions at $t_{trs}(1)/^{o}C$= 190, 280-285, 345 (Ref. 1).
Component 2 undergoes phase transitions at $t_{trs}(2)/^{o}C$= 124, 316 (current literature).

ESTIMATED ERROR:

Temperature: accuracy probably ±2 K (compiler).

REFERENCES:

(1) Sokolov, N.M.
Tezisy Dokl. X Nauch. Konf. S.M.I. 1956.

COMPONENTS:	EVALUATOR:
(1) Potassium **iso**.butanoate (potassium **iso**.butyrate); $Ki.C_4H_7O_2$; [19455-20-0] (2) Potassium nitrite; KNO_2; [7758-09-0]	Schiraldi, A., Dipartimento di Chimica Fisica, Universita´ di Pavia (ITALY).

CRITICAL EVALUATION:

The visual polythermal analysis was employed by Sokolov and Minich (Ref. 1) to study the lower boundary of the isotropic liquid field: they claimed the formation of a continuous series of solid solutions with a minimum at 535 K (262 °C) and $100\mathbf{x}_2$= 32.5.

Component 1, however, goes through the liquid crystalline state before to turn into a clear melt. Accordingly, the topology of the system has to be re-interpreted, a possibility (not very convincing, however) being that shown in Preface, Scheme B.3 which is based on the assumption that continuous solutions do form between solid KNO_2 and solid $Ki.C_4H_7O_2$.

Sokolov and Minich´s fusion temperature of component 1, i.e., 638 K (365 °C), should be identified with the $T_{clr}(1)$ value (625.6±0.8 K) listed in Preface, Table 2. The discrepancy between the two figures is noticeable: in previous papers by Sokolov´s group, however, lower values, i.e., 629 K (Ref. 2) and 633 K (Ref. 3), were reported.

It is further to be noted that three phase transition temperatures are quoted in Ref. 1 from Ref. 4 for component 1, i.e., 621, 546, and 481 K, the second of which can be reasonably identified with the fusion temperature [$T_{fus}(1)$= 553.9±0.5 K] listed in Table 2 of the Preface. Consequently: (i) the transition temperature at 621 K (if actually existing) might correspond to some kind of transformation (undetected by DSC, see Table 2) within the liquid crystal field; and (ii) only the transition at 481 K should correspond to a solid state transformation, although the latter figure is almost 60 K higher than the single $T_{trs}(1)$ value (424±3 K) listed in Table 2.

REFERENCES:

(1) Sokolov, N.M.; Minich, M.A.
Zh. Neorg. Khim. 1961, 6, 2558-2562 (*); **Russ. J. Inorg. Chem. (Engl. Transl.)** 1961, 6, 1293-1295.

(2) Dmitrevskaya, O.I.; Sokolov, N.M.
Zh. Obshch. Khim. 1960, 30, 20-25 (*); **Russ. J. Gen. Chem. (Engl. Transl.)** 1960, 30, 19-24.

(3) Sokolov, N.M.; Pochtakova, E.I.
Zh. Obshch. Khim. 1960, 30, 1405-1410 (*); **Russ. J. Gen. Chem. (Engl. Transl.)** 1960, 30, 1433-1437.

(4) Sokolov, N.M.
Tezisy Dokl. X Nauch. Konf. S.M.I. 1956.

COMPONENTS:	ORIGINAL MEASUREMENTS:
(1) Potassium **iso**.butanoate (potassium **iso**.butyrate); $Ki.C_4H_7O_2$; [19455-20-0] (2) Potassium nitrite; KNO_2; [7758-09-0]	Sokolov, N.M.; Minich, M.A. **Zh. Neorg. Khim.** 1961, 6, 2558-2562 (*); **Russ. J. Inorg. Chem. (Engl. Transl.)** 1961, 6, 1293-1295.
VARIABLES:	**PREPARED BY:**
Temperature.	Baldini, P.

EXPERIMENTAL VALUES:

$t/^oC$	T/K^a	$100x_2$	$t/^oC$	T/K^a	$100x_2$
365	638	0	333	606	55
339	612	5	345	618	60
326	599	10	356	629	65
310	583	15	366	639	70
292	565	20	374	647	75
276	549	25	381	654	80
264	537	30	392	665	85
266	539	35	401	674	90
283	556	40	415	688	95
301	574	45	436	709	100
319	592	50			

[a] T/K values calculated by the compiler.

Characteristic point(s):

Continuous series of solid solutions with a minimum at 262 °C and $100x_2$= 32.5 (authors).

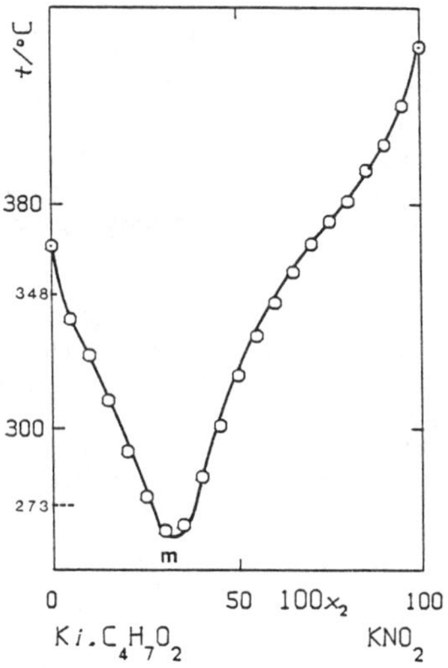

AUXILIARY INFORMATION

METHOD/APPARATUS/PROCEDURE:	SOURCE AND PURITY OF MATERIALS:
Visual polythermal analysis.	Component 1: prepared from "chemically pure" $KHCO_3$ and the fatty acid, and recrystallized from n-butanol after having been deposited from the aqueous solution and dried; it undergoes phase transitions at $t_{trs}(1)/^oC$= 208, 273, 348 (Ref. 1). Component 2: material prepared by reducing potassium nitrate with lead, melting at $t_{fus}(2)/^oC$= 436 after three recrystallizations; it undergoes a phase transition at $t_{trs}(2)/^oC$= 45 (Ref. 2).
	ESTIMATED ERROR:
	Temperature: accuracy probably ±2 K (compiler).
	REFERENCES:
	(1) Sokolov, N.M. **Tezisy Dokl. X Nauch. Konf. S.M.I.** 1956. (2) Berul´, S.I.; Bergman, A.G. **Izv. Sektora Fiz.-Khim. Anal.** 1952, 21, 178-183.

COMPONENTS:	EVALUATOR:
(1) Potassium **iso**.butanoate (potassium **iso**.butyrate); K**i**.$C_4H_7O_2$; [19455-20-0] (2) Potassium nitrate; KNO_3; [7757-79-1]	Spinolo, G., Dipartimento di Chimica Fisica, Universita´ di Pavia (ITALY).

CRITICAL EVALUATION:

This binary was studied only by Dmitrevskaya and Sokolov (Ref. 1). On the basis of their visual polythermal results, they claimed the existence of: (i) an incongruently melting intermediate compound of supposed composition K_2**i**.$C_4H_7O_2NO_3$; (ii) a "perekhodnaya tochka" (likely a peritectic), P, at 529 K (256 °C) and $100\mathbf{x}_2$= 47.5; and (iii) a eutectic at 526 K (253 °C) and $100\mathbf{x}_2$= 32.5.

Component 1, however, goes through the liquid crystalline state before to turn into a clear melt. Therefore, the authors´ fusion temperature [$\mathbf{T}_{fus}$(1)= 629 K (356 °C)] should be identified with the clearing temperature, the corresponding value from Table 2 of the Preface being $\mathbf{T}_{clr}$(1)= 625.6±0.8 K.

Moreover, three phase transition temperatures are quoted in Ref. 1 from Ref. 2 for the same component, i.e., 621, 546, and 481 K, the second of which can be reasonably identified with the fusion temperature [$\mathbf{T}_{fus}$(1)= 553.9±0.5 K] listed in Preface, Table 2. Consequently: (i) the transition temperature at 621 K (if actually existing) might correspond to some kind of transformation (undetected by DSC, see Table 2) within the liquid crystal field; and (ii) only the transition at 481 K should correspond to a solid state transformation, although the latter figure is almost 60 K higher than the single $\mathbf{T}_{trs}$(1) value (424±3 K) listed in Table 2.

In conclusion, the authors´ interpretation of the topology of this system is to be modified. In the evaluator´s opinion, it seems reasonable to assume that the phase diagram could be similar to that shown in Preface, Scheme D.1, allowance being made for the fact that in the present case the intermediate compound is incongruently (instead of congruently) melting. Dmitrevskaya and Sokolov´s eutectic should actually be an M'_E point, and a further invariant, type M_E, ought to exist. At any rate, a re-investigation of the system would be desirable, in order to obtain information on the solidus, and to assess unambiguously the composition of the intermediate compound.

REFERENCES:

(1) Dmitrevskaya, O.I.; Sokolov, N.M.
Zh. Obshch. Khim. 1960, 30, 20-25 (*); **Russ. J. Gen. Chem. (Engl. Transl.)** 1960, 30, 19-24.

(2) Sokolov, N.M.
Tezisy Dokl. X Nauch. Konf. S.M.I. 1956.

COMPONENTS:

(1) Potassium **iso**.butanoate (potassium **iso**.butyrate); $K\mathbf{i}.C_4H_7O_2$; [19455-20-0]
(2) Potassium nitrate; KNO_3; [7757-79-1]

ORIGINAL MEASUREMENTS:

Dmitrevskaya, O.I.; Sokolov, N.M.
Zh. Obshch. Khim. 1960, **30**, 20-25 (*);
Russ. J. Gen. Chem. (Engl. Transl.) 1960, **30**, 19-24.

VARIABLES:

Temperature.

PREPARED BY:

Baldini, P.

EXPERIMENTAL VALUES:

$t/^{\circ}C$	T/K^a	$100\mathbf{x}_2$	$t/^{\circ}C$	T/K^a	$100\mathbf{x}_2$
356	629	0	258	531	50
346	619	5	267	540	55
335	608	10	276	549	60
321	594	15	286	559	65
304	577	20	293	566	70
285	558	25	300	573	75
263	536	30	308	581	80
253	526	32.5	316	589	85
254	527	35	322	595	90
255	528	40	330	603	95
255.5	528.7	45	337	610	100
256	529	47.5			

[a] T/K values calculated by the compiler.

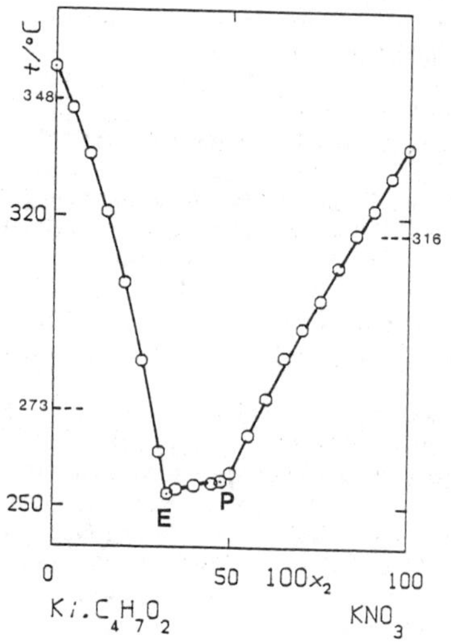

Characteristic point(s):

Eutectic, E, at 253 °C and $100\mathbf{x}_2$= 32.5 (authors).
Characteristic point, P (perekhodnaya tochka in the original text; see the Introduction), at 256 °C and $100\mathbf{x}_2$= 47.5.

Intermediate compound(s):

$K_2\mathbf{i}.C_4H_7O_2NO_3$ (supposed composition; authors) incongruently melting.

AUXILIARY INFORMATION

METHOD/APPARATUS/PROCEDURE:

Visual polythermal analysis.

SOURCE AND PURITY OF MATERIALS:

Component 1 synthetized from **iso**.butanoic acid and K_2CO_3 (Ref. 1). "Chemically pure" component 2 recrystallized.
Component 1 undergoes phase transitions at $t_{trs}(1)/^{\circ}C$= 208, 273, 348 (Ref 2).
Component 2 undergoes phase transitions at $t_{trs}(2)/^{\circ}C$= 124, 316 (current literature).

ESTIMATED ERROR:

Temperature: accuracy probably ±2 K (compiler).

REFERENCES:

(1) Sokolov, N.M.
Zh. Obshch. Khim. 1954, **24**, 1581-1593.
(2) Sokolov, N.M.
Tezisy Dokl. X Nauch. Konf. S.M.I. 1956.

COMPONENTS:	EVALUATOR:
(1) Potassium pentanoate (potassium valerate); $KC_5H_9O_2$; [19455-21-1] (2) Potassium nitrite; KNO_2; [7758-09-0]	Schiraldi, A., Dipartimento di Chimica Fisica, Universita´ di Pavia (ITALY).

CRITICAL EVALUATION:

The visual polythermal analysis was employed by Sokolov and Minich (Ref. 1) to study the lower boundary of the isotropic liquid field: they claimed the occurrence of a congruently melting intermediate compound [of tentative composition $K_7(C_5H_9O_2)_4(NO_2)_3$], able to give eutectics with either component.

Component 1, however, forms liquid crystals. Therefore, the phase diagram has to be re-interpreted, possibly with reference to Preface, Scheme D.1. In this case, Sokolov and Minich´s eutectic E_1 should be an M'_E point, and a further (still undetected) invariant type M_E ought to exist.

The fusion temperature, $T_{fus}(1)$= 717 K, reported in Ref. 1, should be identified with the clearing temperature of component 1, and agrees fairly with the $T_{clr}(1)$ value (716±2 K) listed in Preface, Table 1. Moreover, the transition temperature $T_{trs}(1)$= 580 K (307 °C) quoted in Ref. 1 from Ref. 2 should in turn be identified with the actual fusion temperature, the corresponding value from Table 1 of the Preface being 586.6±0.7 K.

REFERENCES:

(1) Sokolov, N.M.; Minich, M.A.
Zh. Neorg. Khim. 1961, **6**, 2558-2562 (*); **Russ. J. Inorg. Chem. (Engl. Transl.)** 1961, **6**, 1293-1295.

(2) Sokolov, N.M.
Tezisy Dokl. X Nauch. Konf. S.M.I. 1956.

COMPONENTS:	ORIGINAL MEASUREMENTS:
(1) Potassium pentanoate (potassium valerate); $KC_5H_9O_2$; [19455-21-1] (2) Potassium nitrite; KNO_2; [7758-09-0]	Sokolov, N.M.; Minich, M.A. **Zh. Neorg. Khim.** 1961, 6, 2558-2562 (*); **Russ. J. Inorg. Chem. (Engl. Transl.)** 1961, 6, 1293-1295.
VARIABLES:	PREPARED BY:
Temperature.	Baldini, P.

EXPERIMENTAL VALUES:

$t/°C$	T/K^a	$100x_2$	$t/°C$	T/K^a	$100x_2$
444	717	0	350	623	55
432	705	5	363	636	60
421	694	10	373	646	65
413	686	15	381	654	70
401	674	20	387	660	75
385	658	25	395	668	80
363	636	30	401	674	85
339	612	35	407	680	90
325	598	40	416	689	95
325	598	45	436	709	100
335	608	50			

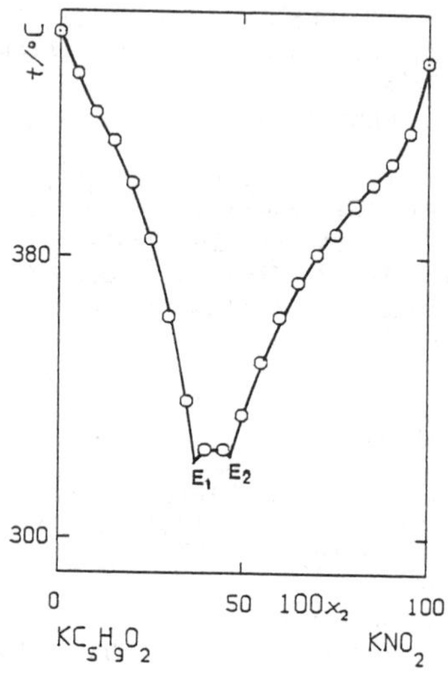

[a] T/K values calculated by the compiler.

Characteristic point(s):

Eutectic, E_1, at 321 °C and $100x_2$= 37 (authors).
Eutectic, E_2, at 323 °C and $100x_2$= 47 (authors).

Intermediate compound(s):

$K_7(C_5H_9O_2)_4(NO_2)_3$ (tentative composition; authors), congruently melting (at 325 °C; compiler).

AUXILIARY INFORMATION

METHOD/APPARATUS/PROCEDURE:

Visual polythermal analysis.

SOURCE AND PURITY OF MATERIALS:

Component 1: prepared from "chemically pure" $KHCO_3$ and the fatty acid, and recrystallized from n-butanol after having been deposited from the aqueous solution and dried; it undergoes a phase transition at $t_{trs}(1)/°C$= 307 (Ref. 1).
Component 2: material prepared by reducing potassium nitrate with lead, melting at $t_{fus}(2)/°C$= 436 after three recrystallizations; it undergoes a phase transition at $t_{trs}(2)/°C$= 45 (Ref. 2).

ESTIMATED ERROR:

Temperature: accuracy probably ±2 K (compiler).

REFERENCES:

(1) Sokolov, N.M.
Tezisy Dokl. X Nauch. Konf. S.M.I. 1956.
(2) Berul´, S.I.; Bergman, A.G.
Izv. Sektora Fiz.-Khim. Anal. 1952, 21, 178-183.

COMPONENTS:	EVALUATOR:
(1) Potassium pentanoate (potassium valerate); $KC_5H_9O_2$; [19455-21-1] (2) Potassium nitrate; KNO_3; [7757-79-1]	Ferloni, P., Dipartimento di Chimica Fisica, Universita' di Pavia (ITALY).

CRITICAL EVALUATION:

This system was studied only by Dmitrevskaya and Sokolov (Ref. 1), who suggested (on the basis of their visual polythermal observations) the phase diagram to be of the eutectic type, the invariant being at 583 K (310 ^oC) and $100\mathbf{x}_2$= 49.

Component 1, however, forms liquid crystals. Therefore, the fusion temperature, $T_{fus}(1)$= 717 K (444 ^oC), reported by the authors, should be identified with the clearing temperature, the corresponding value from Table 1 of the Preface being 716$\pm$2 K.

For the same component, the phase transition at 580 K (307 ^oC), quoted in Ref. 1 from Ref. 2, can be identified with the actual fusion temperature, $T_{fus}(1)$= 586.6$\pm$0.7 K, reported in Preface, Table 1.

Accordingly, the available experimental data justify a phase diagram possibly similar to Scheme A.1 in the Preface, the invariant point given in Ref. 1 being consequently an M'_E point and not a usual eutectic.

The slope change apparent in the liquidus branch richer in component 2 is consistent with the occurrence in KNO_3 of the solid state transition at 589 K (316 ^oC) mentioned by the authors.

REFERENCES:

(1) Dmitrevskaya, O.I.; Sokolov, N.M.
Zh. Obshch. Khim. 1965, 35, 1905-1909.

(2) Sokolov, N.M.
Tezisy Dokl. X Nauch. Konf. S.M.I. 1956.

COMPONENTS:

(1) Potassium pentanoate (potassium valerate); $KC_5H_9O_2$; [19455-21-1]
(2) Potassium nitrate; KNO_3; [7757-79-1]

ORIGINAL MEASUREMENTS:

Dmitrevskaya, O.I.; Sokolov, N.M.
Zh. Obshch. Khim. 1965, **35**, 1905-1909.

VARIABLES:

Temperature.

PREPARED BY:

Baldini, P.

EXPERIMENTAL VALUES:

$t/^oC$	T/K^a	$100x_2$	$t/^oC$	T/K^a	$100x_2$
444	717	0	312	585	55
430	703	5	313	586	60
415	688	10	314	587	65
400	673	15	316	589	70
384	657	20	319	592	75
365	638	25	322	595	80
350	623	30	326	599	85
336	609	35	329	602	90
325	598	40	332	605	95
314	587	45	337	610	100
311	584	50			

[a] T/K values calculated by the compiler.

Characteristic point(s):

Eutectic, E, at 310 ^oC and $100x_2$= 49 (authors).

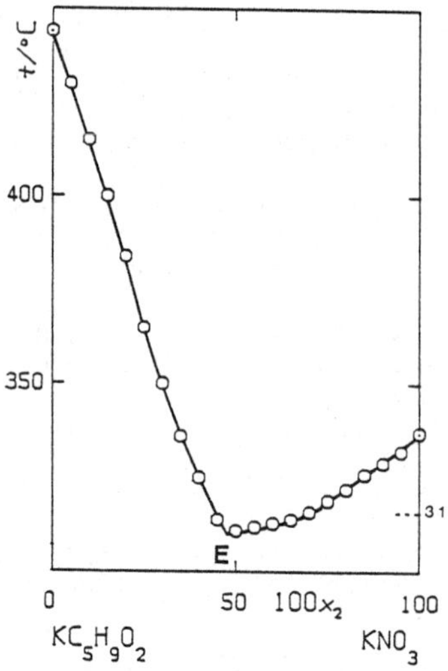

AUXILIARY INFORMATION

METHOD/APPARATUS/PROCEDURE:

Visual polythermal analysis.

SOURCE AND PURITY OF MATERIALS:

Component 1: prepared from n-pentanoic acid and the carbonate (Ref. 1); it undergoes a phase transition at $t_{trs}(1)/^oC$= 307 (Ref. 2).
Component 2: "chemically pure" material recrystallized; it undergoes phase transitions at $t_{trs}(2)/^oC$= 124, 316 (current literature).

ESTIMATED ERROR:

Temperature: accuracy probably ± 2 K (compiler).

REFERENCES:

(1) Sokolov, N.M. **Zh. Obshch. Khim.** 1954, **24**, 1581-1593.
(2) Sokolov, N.M. **Tezisy Dokl. X Nauch. Konf. S.M.I.** 1956.

COMPONENTS:	EVALUATOR:
(1) Potassium **iso**.pentanoate (potassium **iso**.valerate); $Ki.C_5H_9O_2$; [589-46-8] (2) Potassium nitrite; KNO_2; [7758-09-0]	Franzosini, P., Dipartimento di Chimica Fisica, Universita´ di Pavia (ITALY).

CRITICAL EVALUATION:

The visual polythermal analysis was employed by Sokolov and Minich (Ref. 1) to study the lower boundary of the isotropic liquid field: they claimed the formation of a continuous series of solid solutions with a minimum at 562 K (289 $^{\circ}$C) and $100\mathbf{x}_2$= 37.5.

Component 1, however, goes through the liquid crystalline state before to turn into a clear melt. Therefore, the fusion temperature, $T_{fus}(1)$= 669 K (396 $^{\circ}$C), reported by the authors should be identified with the clearing temperature, the corresponding value from Table 2 in the Preface being $T_{clr}(1)$= 679$\pm$2 K. No mention is made by the authors of the actual fusion which occurs at 531$\pm$3 K (Table 2): the latter figure is supported by the trend of the thermomagnetical curves plotted by Duruz and Ubbelohde (Ref. 2). Accordingly, the topology of the system has to be re-interpreted, a possibility (not very convincing, however) being that shown in Preface, Scheme B.3, which is based on the assumption that continuous solutions do form between solid KNO_2 and solid $Ki.C_5H_9O_2$.

As for the other phase transitions quoted by the authors for the same component from Ref. 3, at 327, and 618 K (54, and 345 $^{\circ}$C, respectively), no identification is possible with the findings by other investigators, inasmuch as: (i) no transformation is reported in Table 2 of the Preface as occurring below $T_{fus}(1)$= 531$\pm$3 K; and (ii) no transformation is reported either in Table 2 or in Ref. 2 as occurring within the field of existence of the mesomorphic liquid. It is, however, to be stressed that the transition temperatures mentioned by Sokolov and Minich do not seem to be trustworthy: indeed, it is a bit puzzling the fact that for potassium **iso**.pentanoate Dmitrevskaya and Sokolov (Ref. 4) quote from Ref. 3 transitions at 618, 493, and 473 K, ignoring that quoted by Sokolov and Minich at 327 K.

REFERENCES:

(1) Sokolov, N.M.; Minich, M.A.
Zh. Neorg. Khim. 1961, **6**, 2558-2562 (*); **Russ. J. Inorg. Chem. (Engl. Transl.)** 1961, **6**, 1293-1295.

(2) Duruz, J.J.; Ubbelohde, A.R.
Proc. Roy. Soc. London 1975, **A342**, 39-49.

(3) Sokolov, N.M.
Tezisy Dokl. X Nauch. Konf. S.M.I. 1956.

(4) Dmitrevskaya, O.I.; Sokolov, N.M.
Zh. Obshch. Khim. 1967, **37**, 2160-2166 (*); **Russ. J. Gen. Chem. (Engl. Transl.)** 1967, **37**, 2050-2054.

COMPONENTS:	ORIGINAL MEASUREMENTS:
(1) Potassium **iso**.pentanoate (potassium **iso**.valerate); $Ki.C_5H_9O_2$; [589-46-8] (2) Potassium nitrite; KNO_2; [7758-09-0]	Sokolov, N.M.; Minich, M.A. **Zh. Neorg. Khim.** 1961, 6, 2558-2562 (*); **Russ. J. Inorg. Chem. (Engl. Transl.)** 1961, 6, 1293-1295.
VARIABLES:	PREPARED BY:
Temperature.	Baldini, P.

EXPERIMENTAL VALUES:

$t/^{\circ}C$	T/K^a	$100x_2$	$t/^{\circ}C$	T/K^a	$100x_2$
396	669	0	343	616	55
370	643	5	357	630	60
352	625	10	368	641	65
339	612	15	377	650	70
326	599	20	385	658	75
313	586	25	395	668	80
300	573	30	400	673	85
291	564	35	409	682	90
292	565	40	420	693	95
310	583	45	436	709	100
327	600	50			

[a] T/K values calculated by the compiler.

Characteristic point(s):

Continuous series of solid solutions with a minimum at 289 $^{\circ}C$ (erroneously reported as 389 both in Fig. 2 of the original paper and in the text; compiler) and $100x_2$= 37.5.

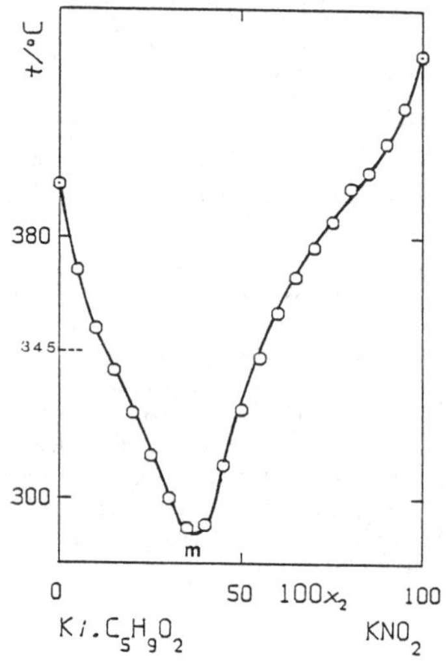

AUXILIARY INFORMATION

METHOD/APPARATUS/PROCEDURE:

Visual polythermal analysis.

SOURCE AND PURITY OF MATERIALS:

Component 1: prepared from "chemically pure" $KHCO_3$ and the fatty acid, and recrystallized from **n**-butanol after having been deposited from the aqueous solution and dried; it undergoes phase transitions at $t_{trs}(1)/^{\circ}C$= 54, 345 (Ref. 1).
Component 2: material prepared by reducing potassium nitrate with lead, melting at $t_{fus}(2)/^{\circ}C$= 436 after three recrystallizations; it undergoes a phase transition at $t_{trs}(2)/^{\circ}C$= 45 (Ref. 2).

ESTIMATED ERROR:

Temperature: accuracy probably ± 2 K (compiler).

REFERENCES:

(1) Sokolov, N.M.
Tezisy Dokl. X Nauch. Konf. S.M.I. 1956.
(2) Berul', S.I.; Bergman, A.G.
Izv. Sektora Fiz.-Khim. Anal. 1952, 21, 178-183.

COMPONENTS:	EVALUATOR:
(1) Potassium **iso**.pentanoate (potassium **iso**.valerate); $K\textbf{i}.C_5H_9O_2$; [589-46-8] (2) Potassium nitrate; KNO_3; [7757-79-1]	Ferloni, P., Dipartimento di Chimica Fisica, Universita' di Pavia (ITALY).

CRITICAL EVALUATION:

This system was studied only by Dmitrevskaya and Sokolov (Ref. 1), who claimed the existence of: (i) a eutectic, E_1, at 557 K (284 °C) and $100\mathbf{x}_2$= 27.5; (ii) a eutectic, E_2, at either 553 K (280 °C; according to visual polythermal determinations), or 549 K (276 °C; according to thermographical analysis), and $100\mathbf{x}_2$= 46.0; and (iii) an intermediate compound $K_3(\mathbf{i}.C_5H_9O_2)_2NO_3$, congruently melting at 557$\pm$2 K (284$\pm$2 °C).

Component 1, however, forms liquid crystals. Therefore, the fusion temperature, $\mathbf{T}_{fus}(1)$= 669 K (396 °C), reported by the authors should be identified with the clearing temperature, the corresponding value from Preface, Table 2 being $\mathbf{T}_{clr}(1)$= 679$\pm$2 K. No mention is made by the authors of the actual fusion which occurs at 531$\pm$3 K (Table 2): the latter figure is supported by the trend of the thermomagnetical curves plotted by Duruz and Ubbelohde (Ref. 2).

As for the other phase transitions quoted by the authors for component 1 from Ref. 3, at 473, 493, and 618 K (200, 220, and 345 °C, respectively), no identification is possible with the findings by other investigators, inasmuch as: (i) no transformation is reported in Table 2 of the Preface as occurring below $\mathbf{T}_{fus}(1)$= 531$\pm$3 K; and (ii) no transformation is reported either in Table 2 or in Ref. 2 as occurring within the field of existence of the mesomorphic liquid. It is, however, to be stressed that the transition temperatures mentioned by Dmitrevskaya and Sokolov do not seem to be trustworthy: indeed, it is a bit puzzling the fact that for potassium **iso**.pentanoate Dmitrevskaya and Sokolov (Ref. 1) quote from Ref. 3 transitions at 618, 493, and 473 K, whereas, e.g., Pochtakova (Ref. 4) quotes from the same source transitions at 618 and 327 K (ignoring those quoted by Dmitrevskaya and Sokolov at 493 and 473 K).

The interpretation of the phase diagram should be modified in the region rich in component 1. The evaluator is inclined to think that: (i) the transition reported (for component 1) in Ref. 3 at 618 K is erratic; (ii) despite the absence of thermographical evidence for the occurrence of fusion at about 530 K, this part of the diagram ought to be similar to that shown in Preface, Scheme D.1, the eutectic E_1 actually being an M'_E point. Accordingly, a further invariant of the M_E type should exist at lower temperature.

The composition of the intermediate compound could coincide with that suggested by the authors, viz., $100\mathbf{x}_2$= 33.3, and the remaining part of the diagram seems reliable.

REFERENCES:

(1) Dmitrevskaya, O.I.; Sokolov, N.M.
Zh. Obshch. Khim. 1967, **37**, 2160-2166 (*); **Russ. J. Gen. Chem. (Engl. Transl.)** 1967, **37**, 2050-2054.

(2) Duruz, J.J.; Ubbelohde, A.R.
Proc. Roy. Soc. London 1975, **A342**, 39-49.

(3) Sokolov, N.M.
Tezisy Dokl. X Nauch. Konf. S.M.I. 1956.

(4) Pochtakova, E.I.
Zh. Obshch. Khim. 1963, **33**, 342-347.

COMPONENTS:	ORIGINAL MEASUREMENTS:
(1) Potassium **iso**.pentanoate (potassium **iso**.valerate); $Ki.C_5H_9O_2$; [589-46-8] (2) Potassium nitrate; KNO_3; [7757-79-1]	Dmitrevskaya, O.I.; Sokolov, N.M. **Zh. Obshch. Khim.** 1967, **37**, 2160-2166 (*); **Russ. J. Gen. Chem. (Engl. Transl.)** 1967, **37**, 2050-2054.
VARIABLES:	PREPARED BY:
Temperature.	Baldini, P.

EXPERIMENTAL VALUES:

t/°C	T/K[a]	$100x_2$	t/°C	T/K[a]	$100x_2$	t/°C	T/K[a]	$100x_2$
396	669	0	284	557	27.5	130[c]	403	60
396[b]	669	0	284[b]	557	27.5	305	578	65
345[c]	618	0	284[d]	557	27.5	308	581	70
220[c]	493	0	208[c]	481	27.5	314	587	75
200[c]	473	0	284.5	557.7	30	316[b]	589	75
386	659	5	282[b]	555	30	276[d]	549	75
382[b]	655	5	283	556	40	127[c]	400	75
275[d]	548	5	286[b]	559	40	317	590	80
336[c]	609	5	280[d]	553	40	323	596	85
205[c]	478	5	130[c]	403	40	328	601	90
365	638	10	282	555	45	328[b]	601	90
344	617	15	280	553	46	275[d]	548	90
320	593	20	276[b]	549	46	130[c]	403	90
324[b]	597	20	276[d]	549	46	335	608	95
274[d]	547	20	126[c]	399	46	337	610	100
208[c]	481	20	286	559	50	337[b]	610	100
296	569	25	294	567	55	316[c]	589	100
302[b]	575	25	300	573	60	127[c]	400	100
278[d]	551	25	306[b]	579	60			
200[c]	473	25	275[d]	548	60			

[a] T/K values calculated by the compiler.
[b] Liquidus from thermographical analysis.
[c] Transformation in the solid state.
[d] Eutectic temperature.

(continued on next page)

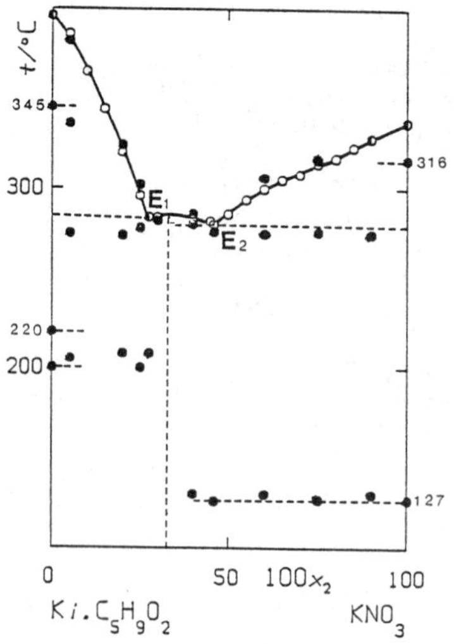

COMPONENTS:

(1) Potassium **iso**.pentanoate (potassium **iso**.valerate);
$Ki.C_5H_9O_2$; [589-46-8]
(2) Potassium nitrate;
KNO_3; [7757-79-1]

ORIGINAL MEASUREMENTS:

Dmitrevskaya, O.I.; Sokolov, N.M.
Zh. Obshch. Khim. 1967, **37**, 2160-2166 (*);
Russ. J. Gen. Chem. (Engl. Transl.) 1967, **37**, 2050-2054.

VARIABLES:

Temperature.

PREPARED BY:

Baldini, P.

EXPERIMENTAL VALUES: (continued)

Characteristic point(s):

Eutectic, E_1, at 284 ^oC and $100x_2$= 27.5.
Eutectic, E_2, at 280 ^oC (visual polythermal analysis) or 276 ^oC (thermographical analysis) and $100x_2$= 46.0.

Intermediate compound(s):

$K_3(i.C_5H_9O_2)_2NO_3$ (authors), congruently melting at 284$\pm$2 ^oC (compiler).

Note - In the figure the filled circles refer to thermographical analysis.

AUXILIARY INFORMATION

METHOD/APPARATUS/PROCEDURE:

Visual polythermal analysis supplemented with thermographical analysis (heating curves recorded automatically).

SOURCE AND PURITY OF MATERIALS:

Component 1: synthetized from **iso**.butanoic acid and the carbonate (Ref. 1).
Component 2: "chemically pure" material recrystallized.
Component 1 undergoes phase transitions at $t_{trs}(1)/^oC$= 345, 220, 200 (Ref. 2).
Component 2 undergoes phase transitions at $t_{trs}(2)/^oC$= 316, 127 (current literature).

ESTIMATED ERROR:

Temperature: accuracy probably $\pm$2 K (compiler).

REFERENCES:

(1) Sokolov, N.M.
Zh. Obshch. Khim. 1954, **24**, 1581-1593.

(2) Sokolov, N.M.
Tezisy Dokl. X Nauch. Konf. S.M.I. 1956.

COMPONENTS:

(1) Potassium hexanoate (potassium caproate); $KC_6H_{11}O_2$; [19455-00-6]
(2) Potassium nitrite; KNO_2; [7758-09-0]

ORIGINAL MEASUREMENTS:

Sokolov, N.M.; Minich, M.A.
Zh. Neorg. Khim. 1961, 6, 2558-2562 (*);
Russ. J. Inorg. Chem. (Engl. Transl.) 1961, 6, 1293-1295.

VARIABLES:

Temperature.

PREPARED BY:

Baldini, P.

EXPERIMENTAL VALUES:

$t/^{\circ}C$	T/K^a	$100x_2$	$t/^{\circ}C$	T/K^a	$100x_2$
444.4	717.6	0	365	638	55
425	698	5	365	638	60
414	687	10	383	656	65
405	678	15	395	668	70
396	669	20	399	672	75
392	665	25	397	670	80
389	662	30	406	679	85
387	660	35	414	687	90
386	659	40	424	697	95
385	658	45	436	709	100
377	650	50			

[a] T/K values calculated by the compiler.

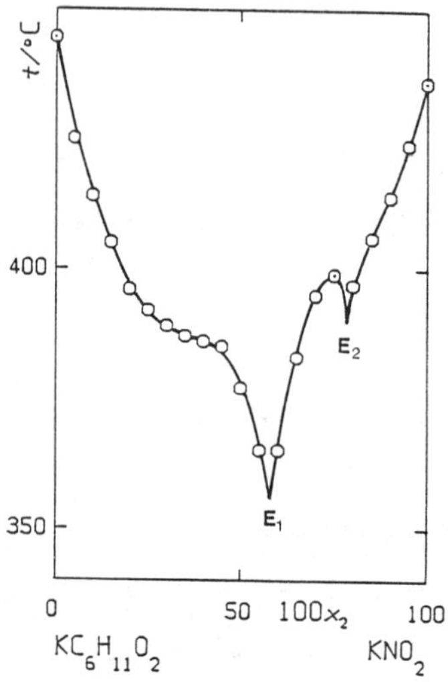

Characteristic point(s):

Eutectic, E_1, at 356 °C and $100x_2$= 58 (authors).
Eutectic, E_2, at 390 °C and $100x_2$= 78.5 (authors).

Intermediate compound(s):

$K_4C_6H_{11}O_2(NO_2)_3$ (tentative composition; authors) congruently melting (at 399 °C; compiler).

AUXILIARY INFORMATION

METHOD/APPARATUS/PROCEDURE:

Visual polythermal analysis.

NOTE:

Component 1 forms liquid crystals. Accordingly, the fusion temperature reported here, viz., 717.6 K (444.4 °C), should be identified with the clearing temperature (725.8±0.8 K) listed in Preface, Table 1, the actual fusion occurring at $T_{fus}(1)$= 581.7±0.5 K (Table 1). The latter figure, in turn, might be identified with the phase transition temperature quoted here from Ref. 1, viz., 575 K (302 °C). The diagram could be re-interpreted with reference to Scheme D.1 of the Preface, the authors' eutectic E_1 being possibly an M'_E point.

SOURCE AND PURITY OF MATERIALS:

Component 1: prepared from "chemically pure" $KHCO_3$ and the fatty acid, and recrystallized from butanol after having been deposited from the aqueous solution and dried; it undergoes a phase transition at $t_{trs}(1)/^{\circ}C$= 302 (Ref. 1).
Component 2: material prepared by reducing potassium nitrate with lead, melting at $t_{fus}(2)/^{\circ}C$= 436 after three recrystallizations; it undergoes a phase transition at $t_{trs}(2)/^{\circ}C$= 45 (Ref. 2).

ESTIMATED ERROR:

Temperature: accuracy probably ±2 K (compiler).

REFERENCES:

(1) Sokolov, N.M.
Tezisy Dokl. X Nauch. Konf. S.M.I. 1956.
(2) Berul', S.I.; Bergman, A.G.
Izv. Sektora Fiz.-Khim. Anal. 1952, 21, 178-183.

COMPONENTS:

(1) Potassium heptanoate (potassium enanthate); $KC_7H_{13}O_2$; [16761-12-9]
(2) Potassium nitrite; KNO_2; [7758-09-0]

ORIGINAL MEASUREMENTS:

Sokolov, N.M.; Minich, M.A.
Zh. Neorg. Khim. 1961, 6, 2558-2562 (*);
Russ. J. Inorg. Chem.(Engl. Transl.) 1961, 6, 1293-1295.

VARIABLES:

Temperature.

PREPARED BY:

Baldini, P.

EXPERIMENTAL VALUES:

$t/^{o}C$	T/K^{a}	$100x_2$	$t/^{o}C$	T/K^{a}	$100x_2$
452	725	0	403	676	55
444	717	5	404	677	60
439	712	10	400	673	65
432	705	15	395	668	70
425	698	20	392	665	75
419	692	25	401	674	80
413	686	30	405	678	85
407	680	35	410	683	90
401	674	40	417	690	95
395	668	45	436	709	100
395	668	50			

[a] T/K values calculated by the compiler.

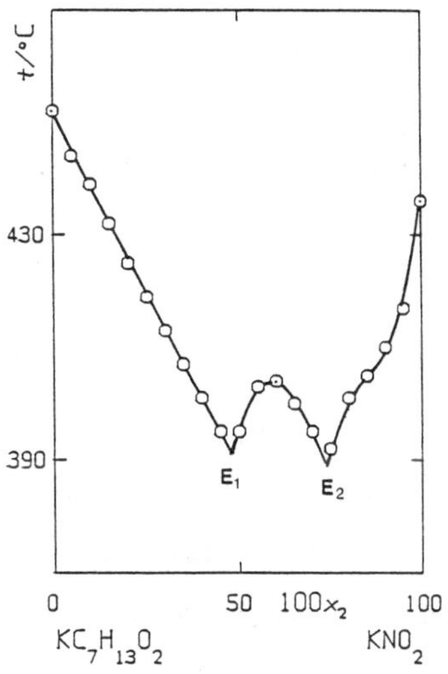

Characteristic point(s):

Eutectic, E_1, at 391 ^{o}C and $100x_2$= 47.5 (authors).
Eutectic, E_2, at 389 ^{o}C and $100x_2$= 74 (authors).

Intermediate compound(s):

$K_5(C_7H_{13}O_2)_2(NO_2)_3$ (tentative composition; authors) congruently melting (at 404 ^{o}C; compiler).

AUXILIARY INFORMATION

METHOD/APPARATUS/PROCEDURE:

Visual polythermal analysis.

NOTE:

Component 1 forms liquid crystals. Accordingly, the fusion temperature reported here, viz., 725 K (452 ^{o}C), should be identified with the clearing temperature (722$\pm$3 K) listed in Preface, Table 1, the actual fusion occurring at $T_{fus}(1)$= 571.3$\pm$0.9 K (Table 1). The diagram could be re-interpreted with reference to Scheme D.1, of the Preface, the authors′ eutectic E_1 possibly being an M'_E point.

SOURCE AND PURITY OF MATERIALS:

Component 1: prepared from "chemically pure" $KHCO_3$ and the fatty acid, evaporated on a steam-bath, dissolved in ethanol, and precipitated with ether.
Component 2: material prepared by reducing potassium nitrate with lead, melting at $t_{fus}(2)/^{o}C$= 436 after three recrystallizations; it undergoes a phase transition at $t_{trs}(2)/^{o}C$= 45 (Ref. 1).

ESTIMATED ERROR:

Temperature: accuracy probably $\pm$2 K (compiler).

REFERENCES:

(1) Berul′, S.I.; Bergman, A.G.
Izv. Sektora Fiz.-Khim. Anal. 1952, 21, 178-183.

COMPONENTS:	ORIGINAL MEASUREMENTS:
(1) Potassium octanoate (potassium caprylate); $KC_8H_{15}O_2$; [764-71-6] (2) Potassium nitrite; KNO_2; [7758-09-0]	Sokolov, N.M.; Minich, M.A. **Zh. Neorg. Khim.** 1961, 6, 2558-2562 (*); **Russ. J. Inorg. Chem. (Engl. Transl.)** 1961, 6, 1293-1295.
VARIABLES:	PREPARED BY:
Temperature.	Baldini, P.

EXPERIMENTAL VALUES:

t/°C	T/K[a]	$100x_2$	t/°C	T/K[a]	$100x_2$
444	717	0	366	639	55
419	692	5	356	629	60
396	669	10	345	618	65
368	641	15	360	633	70
347	620	20	373	646	75
324	597	25	387	660	80
335	608	30	399	672	85
346	619	35	411	684	90
358	631	40	426	699	95
364	637	45	436	709	100
369	642	50			

[a] T/K values calculated by the compiler.

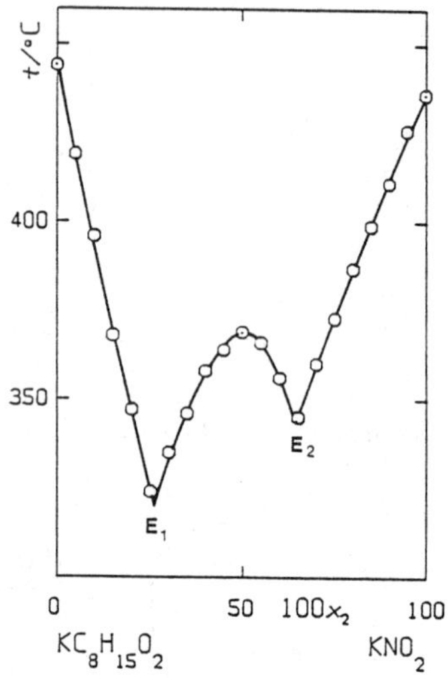

Characteristic point(s):

Eutectic, E_1, at 320 °C and $100x_2$= 26 (authors).
Eutectic, E_2, at 344 °C (authors) and $100x_2$= 64.5 (compiler: the figure 60.5 reported in Table 2 and in Fig. 2 of the original paper is not consistent with the tabulated data).

Intermediate compound(s):

$K_2C_8H_{15}O_2NO_2$ (tentative composition; authors) congruently melting (at 369 °C; compiler).

AUXILIARY INFORMATION

METHOD/APPARATUS/PROCEDURE:

Visual polythermal analysis.

NOTE:

Component 1 forms liquid crystals. Accordingly, the fusion temperature reported here, viz., 717 K (444 °C), should be identified with the clearing temperature (712±2 K) listed in Preface, Table 1, the actual fusion occurring at $T_{fus}(1)$= 560.6±0.8 K (Preface, Table 1). The diagram could be re-interpreted with reference to Scheme D.1, the authors′ eutectic E_1 possibly being an M'_E point.

SOURCE AND PURITY OF MATERIALS:

Component 1: prepared from "chemically pure" $KHCO_3$ and the fatty acid, evaporated on a steam-bath, dissolved in ethanol, and precipitated with ether.
Component 2: material prepared by reducing potassium nitrate with lead, melting at $t_{fus}(2)$/°C= 436 after three recrystallizations; it undergoes a phase transition at $t_{trs}(2)$/°C= 45 (Ref. 1).

ESTIMATED ERROR:

Temperature: accuracy probably ±2 K (compiler).

REFERENCES:

(1) Berul′, S.I.; Bergman, A.G. **Izv. Sektora Fiz.-Khim. Anal.** 1952, 21, 178-183.

COMPONENTS:

(1) Potassium nonanoate (potassium pelargonate); $KC_9H_{17}O_2$; [23282-34-0]
(2) Potassium nitrite; KNO_2; [7758-09-0]

ORIGINAL MEASUREMENTS:

Sokolov, N.M.; Minich, M.A.
Zh. Neorg. Khim. 1961, 6, 2558-2562 (*);
Russ. J. Inorg. Chem. (Engl. Transl.) 1961, 6, 1293-1295.

VARIABLES:

Temperature.

PREPARED BY:

Baldini, P.

EXPERIMENTAL VALUES:

$t/^{o}C$	T/K^{a}	$100x_2$
421	694	0
370^{b}	643	5
370	643	10
..	...	...
370	643	95
436	709	100

[a] T/K values calculated by the compiler.
[b] Figure not compatible with curve VI in Fig. 1 of the original paper (compiler).

Characteristic point(s):

Eutectic, E, at 332 ^{o}C and $100x_2 = 6.5$ (compiler: the figure 7.5 reported in table 2 of the original paper is not compatible with curve VI in Fig. 1 of the original paper).

Note - Liquid layering occurs at $7.5 \leq 100x_2 \leq 99$ at $t/^{o}C = 370$ (see the figure which is a reproduction of curve VI in Fig. 1 of the original paper, and not a plot of the data tabulated; compiler).

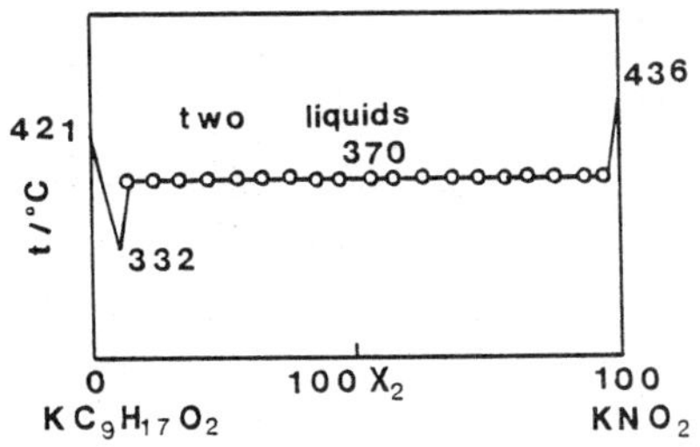

AUXILIARY INFORMATION

METHOD/APPARATUS/PROCEDURE:

Visual polythermal analysis.

NOTE:

Component 1 forms liquid crystals. Accordingly, the fusion temperature reported here, viz., 694 K (421 ^{o}C), should be identified with the clearing temperature (707.4 ± 0.8 K) listed in Table 1 of the Preface, the actual fusion occurring at $T_{fus}(1) = 549.1 \pm 0.8$ K (Table 1). A possible re-interpretation of the phase diagram might be done with reference to Scheme A.1 of the Preface, modified as shown in Fig. 2, the authors´ eutectic being in this case an M'_E point.

SOURCE AND PURITY OF MATERIALS:

Component 1: prepared from "chemically pure" $KHCO_3$ and the fatty acid, evaporated on a steam-bath, dissolved in ethanol, and precipitated with ether.
Component 2: material prepared by reducing potassium nitrate with lead, melting at $t_{fus}(2)/^{o}C = 436$ after three recrystallizations; it undergoes a phase transition at $t_{trs}(2)/^{o}C = 45$ (Ref. 1).

ESTIMATED ERROR:

Temperature: accuracy probably ± 2 K (compiler).

REFERENCES:

(1) Berul´, S.I.; Bergman, A.G. **Izv. Sektora Fiz.-Khim. Anal.** 1952, 21, 178-183.

COMPONENTS:	ORIGINAL MEASUREMENTS:
(1) Lithium methanoate (lithium formate); $LiCHO_2$; [556-63-8] (2) Lithium ethanoate (lithium acetate); $LiC_2H_3O_2$; [546-89-4]	Pochtakova, E.I. **Zh. Obshch. Khim.** 1975, **45**, 503-505.
VARIABLES:	PREPARED BY:
Temperature.	Baldini, P.

EXPERIMENTAL VALUES:

The results are reported only in graphical form (see figure).

Characteristic point(s):

Eutectic, E, at 240 °C and $100x_1$= 37.5 (author).

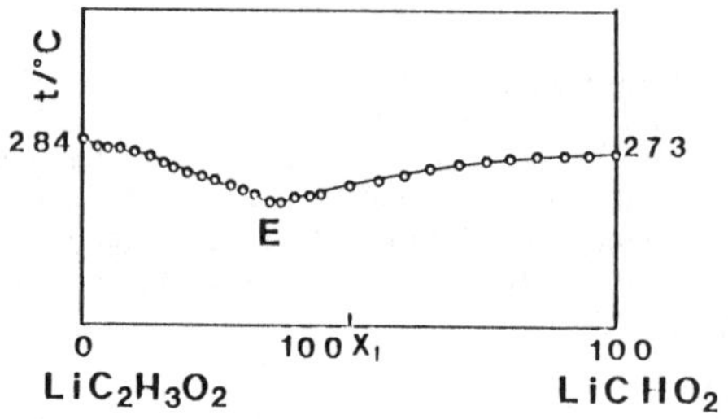

AUXILIARY INFORMATION

METHOD/APPARATUS/PROCEDURE:	SOURCE AND PURITY OF MATERIALS:
Visual polythermal analysis.	Not stated. Component 1 melts at $t_{fus}(1)/^{o}C$= 273. Component 2 melts at $t_{fus}(2)/^{o}C$= 284.
	ESTIMATED ERROR:
	Temperature: accuracy probably ± 2 K (compiler).
	REFERENCES:

COMPONENTS:	ORIGINAL MEASUREMENTS:
(1) Lithium methanoate (lithium formate); $LiCHO_2$; [556-63-8] (2) Lithium thiocyanate; LiCNS; [556-65-0]	Sokolov, N.M.; Dmitrevskaya, O.I. **Zh. Neorg. Khim.** 1969, **14**, 286-296 (*); **Russ. J. Inorg. Chem. (Engl. Transl.)** 1969, 14, 148-155.
VARIABLES:	PREPARED BY:
Temperature.	Baldini, P.

EXPERIMENTAL VALUES:

$t/^{o}C$	T/K^{a}	$100x_2$	$t/^{o}C$	T/K^{a}	$100x_2$
273	546	0	156	429	48.5
259	532	5	157	430	50
247	520	10	167	440	55
235	508	15	180	453	60
222	495	20	192	465	65
210	483	25	204	477	70
198	471	30	216	489	75
187	460	35	227	500	80
176	449	40	238	511	85
163	436	45	266	539	100

[a] T/K values calculated by the compiler.

Characteristic point(s):

Eutectic, E, at 156 ^{o}C and $100x_2$= 48.5 (authors).

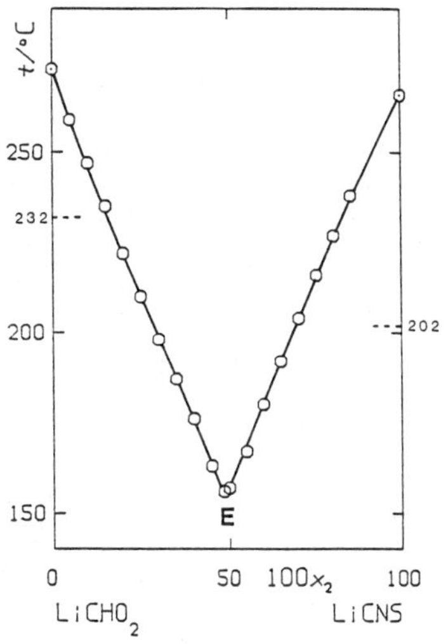

AUXILIARY INFORMATION

METHOD/APPARATUS/PROCEDURE:	SOURCE AND PURITY OF MATERIALS:
Visual polythermal analysis; solid state transition temperatures drawn from the heating curves obtained with automatic recording. NOTE: The fusion temperature of component 1 (546 K) coincides with that listed in Preface, Table 1 where, however, a single solid state transformation of the same component is mentioned as occurring at 496±2 K (i.e., some 10 K lower than the highest Sokolov and Dmitrevskaya's transition).	Not stated. Component 1 undergoes phase transitions at $t_{trs}(1)/^{o}C$= 87, 115, 232. Component 2 undergoes a phase transition at $t_{trs}(2)/^{o}C$= 202. ESTIMATED ERROR: Temperature: accuracy probably ±2 K (compiler). REFERENCES:

COMPONENTS:

(1) Lithium methanoate (lithium formate); $LiCHO_2$; [556-63-8]
(2) Lithium nitrate; $LiNO_3$; [7790-69-4]

ORIGINAL MEASUREMENTS:

Tsindrik, N.M.
Zh. Obshch. Khim. 1958, **28**, 830-834.

VARIABLES:

Temperature.

PREPARED BY:

Baldini, P.

EXPERIMENTAL VALUES:

t/°C	T/K[a]	$100x_1$	t/°C	T/K[a]	$100x_1$
256	529	0	180	453	45
247	520	5	170	443	50
238	511	10	162	435	55
228	501	15	178	451	60
220	493	20	194	467	65
212	485	25	208	481	70
206	479	30	220	493	75
196	469	35	232	505	80
188	461	40			

[a] T/K values calculated by the compiler.

Characteristic point(s):

Eutectic, E, at 162 °C and $100x_1$= 55 (author).

Note - The system was investigated at $0 \leq 100x_1 \leq 80$.

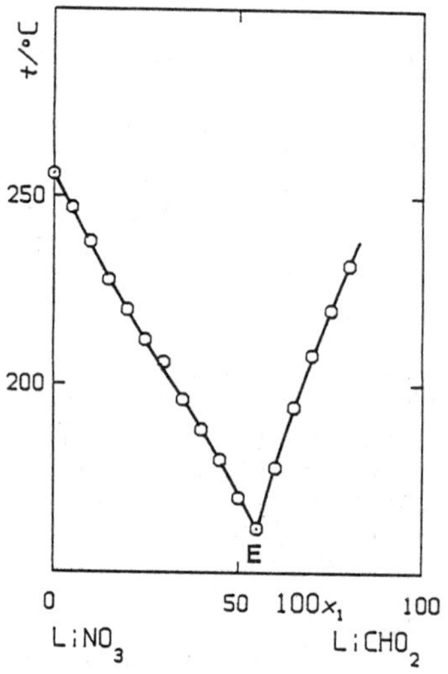

AUXILIARY INFORMATION

METHOD/APPARATUS/PROCEDURE:

Visual polythermal analysis; temperatures measured with a Nichrome-Constantane thermocouple.

NOTE:

The extrapolated $T_{fus}(1)$ reported by the author (546 K) coincides with that listed in Table 1.

SOURCE AND PURITY OF MATERIALS:

Materials of analytical purity recrystallized twice (extrapolated $t_{fus}(1)/°C$= 273; author).

ESTIMATED ERROR:

Temperature: accuracy probably ±2 K (compiler).

REFERENCES:

COMPONENTS:

(1) Lithium ethanoate (lithium acetate); $LiC_2H_3O_2$; [546-89-4]
(2) Lithium thiocyanate; LiCNS; [556-65-0]

ORIGINAL MEASUREMENTS:

Sokolov, N.M.; Dmitrevskaya, O.I.
Zh. Neorg. Khim. 1969, 14, 286-296 (*);
Russ. J. Inorg. Chem. (Engl. Transl.) 1969, **14**, 148-155.

VARIABLES:

Temperature.

PREPARED BY:

Baldini, P.

EXPERIMENTAL VALUES:

t/°C	T/K[a]	$100x_1$	t/°C	T/K[a]	$100x_1$
266	539	0	168	441	50
248	521	10	176	449	55
244	517	15	187	460	60
234	507	20	198	471	65
220	493	25	208	481	70
210	483	30	220	493	75
200	473	35	233	506	80
188	461	40	243	516	85
174	447	45	284[b]	557	100
166	439	49			

[a] T/K values calculated by the compiler.
[b] The figure 266 given in the original table is apparently a misprint, being the fusion temperature of component 2; the figure 284 is taken from Fig. 1 of the original paper (compiler).

Characteristic point(s):

Eutectic, E, at 166 °C (authors) and $100x_2$= 51 (compiler).

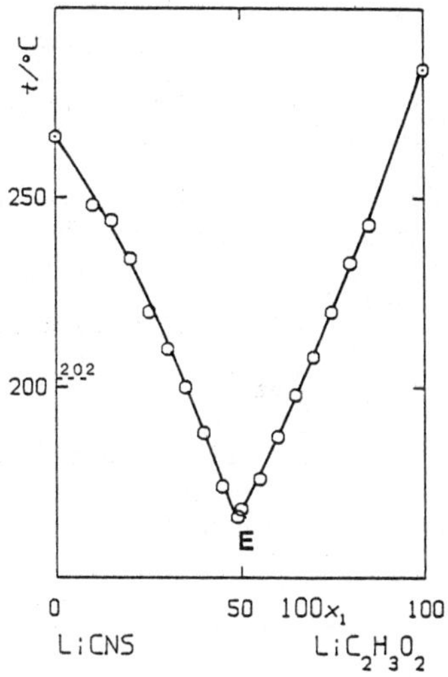

AUXILIARY INFORMATION

METHOD/APPARATUS/PROCEDURE:

Visual polythermal analysis; solid state transition temperatures drawn from the heating curves obtained with automatic recording.

NOTE:

The eutectic composition is given as $100x_2$= 49 in the original text, but this figure is conflicting with both the data tabulated and Fig. 1b of the original paper.

SOURCE AND PURITY OF MATERIALS:

Not stated.
Component 2 undergoes a phase transition at $t_{trs}(2)$/°C= 202.

ESTIMATED ERROR:

Temperature: accuracy probably ±2 K (compiler).

REFERENCES:

COMPONENTS:	EVALUATOR:
(1) Lithium ethanoate (lithium acetate); $LiC_2H_3O_2$; [546-89-4] (2) Lithium nitrate; $LiNO_3$; [7790-69-4]	Ferloni, P., Dipartimento di Chimica Fisica, Universita´ di Pavia (ITALY).

CRITICAL EVALUATION:

This binary was submitted to visual polythermal analysis by Diogenov (Ref. 1) as a side of the reciprocal ternary Li, $Na/C_2H_3O_2$, NO_3, and by Diogenov et al. (Ref. 2), and Sokolov and Tsindrik (Ref. 3) as a side of the reciprocal ternary K, $Li/C_2H_3O_2$, NO_3. All investigations were restricted to the liquidus.

The fusion temperature of component 1 given in Refs. 1, 2 (564 K) is 7 K higher than that (557 K) reported both in Ref. 3 and Table 1 of the Preface. Again for component 1, a solid state transformation is mentioned in Refs. 1, 2 as occurring at 536-538 K, whereas, in a subsequent paper by the same group (Ref. 4), a far different temperature (405 K) is reported. No information about the existence of any solid-solid transition in lithium ethanoate is known to the evaluator from any source (included Ref. 3 and Table 1), but Diogenov´s group.

The diagrams shown in Refs. 1-3 are qualitatively similar, and characterized by the presence of a single eutectic at $100\mathbf{x}_2$ about 51. It is, however, a bit surprising that neither Sokolov and Tsindrik (Ref. 3, where Ref. 1 is quoted), nor Diogenov et al. (Ref. 2, where Ref. 1 is not quoted) have commented on the unusually large discrepancies existing between the eutectic temperatures they found (463 K and 449 K, respectively) and the previous value (418 K) by Diogenov (Ref. 1). These discrepancies might be related to the fact that component 1 tends to form glasses.

At any rate, the evaluator - due to the apparent lack of internal consistency of the measurements by Diogenov´s group - is inclined to attach more reliability to the data from Ref. 3, although regretting that they are reported only in graphical form.

REFERENCES:

(1) Diogenov, G.G.
Zh. Neorg. Khim. 1956, **1**, 799-805 (*); **Russ. J. Inorg. Chem. (Engl. Transl.)** 1956, **1** (4), 199-205.

(2) Diogenov, G.G.; Nurminskii, N.N.; Gimel´shtein, V.G.
Zh. Neorg. Khim. 1957, **2**, 1596-1600 (*); **Russ. J. Inorg. Chem. (Engl. Transl.)** 1957, **2**(7), 237-245.

(3) Sokolov, N.M.; Tsindrik, N.M.
Zh. Neorg. Khim. 1969, **14**, 584-590 (*); **Russ. J. Inorg. Chem. (Engl. Transl.)** 1969, **14**, 302-306.

(4) Diogenov, G.G.; Erlykov, A.M.; Gimel´shtein, V.G.
Zh. Neorg. Khim. 1974, **19**, 1955-1960; **Russ. J. Inorg. Chem. (Engl. Transl.)** 1974, **19**, 1069-1073 (*).

COMPONENTS:

(1) Lithium ethanoate (lithium acetate); $LiC_2H_3O_2$; [546-89-4]
(2) Lithium nitrate; $LiNO_3$; [7790-69-4]

ORIGINAL MEASUREMENTS:

Diogenov, G.G.
Zh. Neorg. Khim. 1956, 1, 799-805 (*);
Russ. J. Inorg. Chem. (Engl. Transl.) 1956, **1** (4), 199-205.

VARIABLES:

Temperature.

PREPARED BY:

Baldini, P.

EXPERIMENTAL VALUES:

t/°C	T/K[a]	$100x_1$	t/°C	T/K[a]	$100x_1$
259	532	0	170	443	56
254	527	5	183	456	60
248	521	9	190	463	62.5
241	514	14	206	479	68
234	507	18.5	219	492	72.5
227	500	22.5	233	506	77.5
216	489	27.5	242	515	81
204	477	32	250	523	84.5
188	461	37.5	259	532	90
176	449	41	264	537	92.5
160	433	45	276	549	94
150	423	50	283	556	96
160	433	53	291	564	100

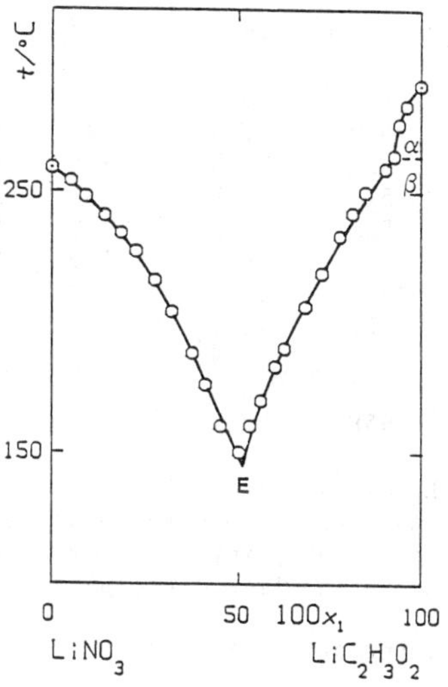

[a] T/K values calculated by the compiler.

Characteristic point(s):

Eutectic, E, at 145 °C and $100x_1$ = 51 (author).

Note - On the branch richer in component 1 the inflexion at 263 °C indicates a phase transition of $LiC_2H_3O_2$ (author).

AUXILIARY INFORMATION

METHOD/APPARATUS/PROCEDURE:

Visual polythermal analysis.

SOURCE AND PURITY OF MATERIALS:

Not stated.

ESTIMATED ERROR:

Temperature: accuracy probably ±2 K (compiler).

REFERENCES:

COMPONENTS:	ORIGINAL MEASUREMENTS:
(1) Lithium ethanoate (lithium acetate); $LiC_2H_3O_2$; [546-89-4] (2) Lithium nitrate; $LiNO_3$; [7790-69-4]	Diogenov, G.G.; Nurminskii, N.N.; Gimel'shtein, V.G. **Zh. Neorg. Khim.** 1957, 2, 1596-1600 (*); **Russ. J. Inorg. Chem. (Engl. Transl.)** 1957, **2**(7), 237-245.
VARIABLES:	PREPARED BY:
Temperature.	Baldini, P.

EXPERIMENTAL VALUES:

t/°C	T/K[a]	$100x_1$	t/°C	T/K[a]	$100x_1$
259	532	0	178[b]	451	55
257	530	1.5	196	469	60
249	522	8.5	214	487	68
240	513	15.5	230	503	75
232	505	21	239	512	80
221	494	28.5	250	523	85
209	482	36	259	532	90
198	471	41.5	265	538	92.5
188	461	46.5	277	550	94
180	453	48	291	564	100
185	458	52.5			

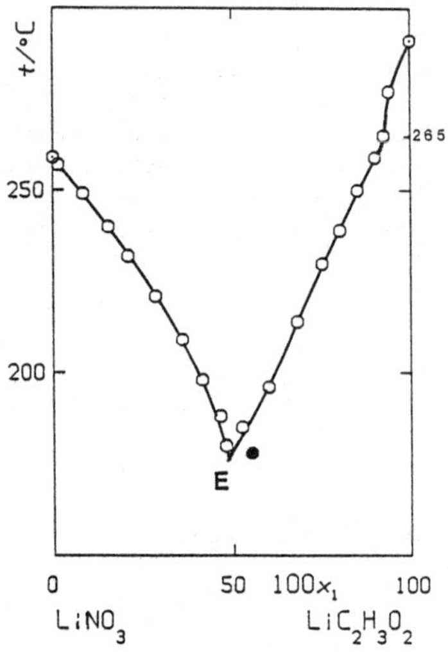

[a] T/K values calculated by the compiler.
[b] This figure seems to be a misprint: the corresponding point is reported as a filled circle in the figure (compiler).

Characteristic point(s):

Eutectic, E, at 176 °C (authors) and $100x_2$= 51 (compiler).

Note - The eutectic composition reported in the original paper ($100x_1$= 51) is not coherent with the tabulated data: in compiler's opinion, this might be due to a misprint.

AUXILIARY INFORMATION

METHOD/APPARATUS/PROCEDURE:	SOURCE AND PURITY OF MATERIALS:
Visual polythermal analysis.	Source not stated. Component 1 undergoes a phase transition at $t_{trs}(1)$/°C= 265.
	ESTIMATED ERROR:
	Temperature: accuracy probably ±2 K (compiler).
	REFERENCES:

COMPONENTS:	ORIGINAL MEASUREMENTS:
(1) Lithium ethanoate (lithium acetate); $LiC_2H_3O_2$; [546-89-4] (2) Lithium nitrate; $LiNO_3$; [7790-69-4]	Sokolov, N.M.; Tsindrik, N.M. **Zh. Neorg. Khim.** 1969, 14, 584-590 (*); **Russ. J. Inorg. Chem. (Engl. Transl.)** 1969, 14, 302-306.
VARIABLES:	PREPARED BY:
Temperature	Baldini, P.

EXPERIMENTAL VALUES:

The results are reported only in graphical form (see figure).

Characteristic point(s):

Eutectic, E, at 190 ^{o}C (authors) and $100x_2$ about 51 (compiler).

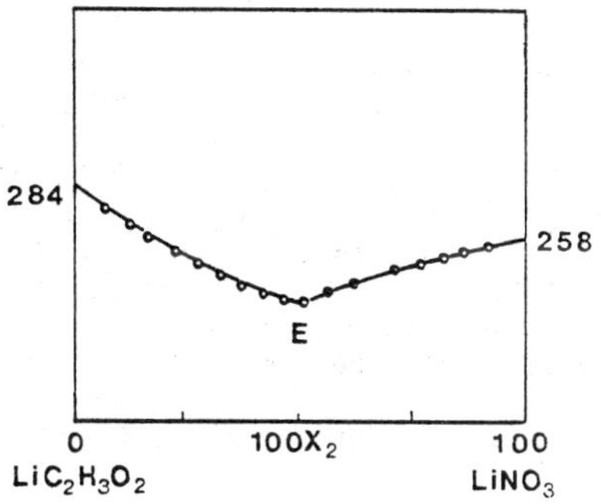

AUXILIARY INFORMATION

METHOD/APPARATUS/PROCEDURE:	SOURCE AND PURITY OF MATERIALS:
Visual polythermal analysis (compiler).	Commercial materials recrystallized (compiler). Component 1: $t_{fus}(1)/^{o}C$= 284. Component 2: $t_{fus}(2)/^{o}C$= 258.
	ESTIMATED ERROR:
	Temperature: accuracy probably ± 2 K (compiler).
	REFERENCES:

COMPONENTS:

(1) Lithium propanoate (lithium propionate); $LiC_3H_5O_2$; [6531-45-9]
(2) Lithium thiocyanate; LiCNS; [556-65-0]

ORIGINAL MEASUREMENTS:

Sokolov, N.M. and Dmitrevskaya, O.I.
Zh. Neorg. Khim. 1969, 14, 286-296 (*);
Russ. J. Inorg. Chem. (Engl. Transl.) 1969, **14**, 148-155.

VARIABLES:

Temperature.

PREPARED BY:

EXPERIMENTAL VALUES:

$t/^{o}C$	T/K^{a}	$100x_2$	$t/^{o}C$	T/K^{a}	$100x_2$
329	602	0	196	469	45
326	599	5	197	470	50
313	586	10	196	469	55
294	567	15	193	466	60
268	541	20	210	483	65
245	518	25	224	497	70
220	493	30	237	510	75
204	477	35	243	516	80
194	467	37.5	260	533	90
195	468	40	266	539	100

[a] T/K values calculated by the compiler.

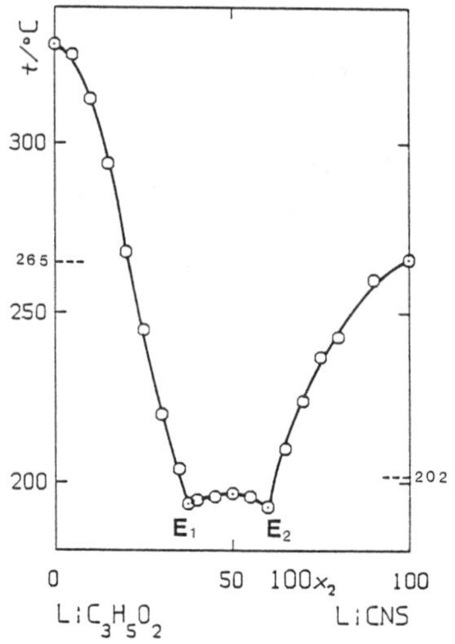

Characteristic point(s):

Eutectic, E_1, at 194 ^{o}C and $100x_2$= 37.5 (authors).
Eutectic, E_2, at 193 ^{o}C and $100x_2$= 60 (authors).

Intermediate compound(s):

$Li_2C_3H_5O_2CNS$ congruently melting at 197 ^{o}C (authors).

AUXILIARY INFORMATION

METHOD/APPARATUS/PROCEDURE:

Visual polythermal analysis; solid state transition temperatures drawn from the heating curves obtained with automatic recording.

SOURCE AND PURITY OF MATERIALS:

Not stated.
Component 1 undergoes a solid state transition at $t_{trs}(1)/^{o}C$= 265.
Component 2 undergoes a solid state transition at $t_{trs}(2)/^{o}C$= 202.

ESTIMATED ERROR:

Temperature: accuracy probably ± 2 K (compiler).

REFERENCES:

COMPONENTS:

(1) Lithium propanoate (lithium propionate); $LiC_3H_5O_2$; [6531-45-9]
(2) Lithium nitrate; $LiNO_3$; [7790-69-4]

ORIGINAL MEASUREMENTS:

Tsindrik, N.M.; Sokolov, N.M.
Zh. Obshch. Khim., 1958, **28**, 1404-1410 (*);
Russ. J. Gen. Chem. (Engl. Transl.) 1958, **28**, 1462-1467.

VARIABLES:

Temperature.

PREPARED BY:

Baldini, P.

EXPERIMENTAL VALUES:

$t/^{o}C$	T/K^a	$100x_1$	$t/^{o}C$	T/K^a	$100x_1$
256	529	0	253	526	45
252	525	5	252	525	50
246	519	10	252	525	55
240	513	15	252	525	60
234	507	20	250	523	65
232	505	21.5	244	517	70
234	507	22.5	266	539	75
238	511	25	280	553	80
244	517	30	307	580	90
248	521	35	329	602	100
252	525	40			

[a] T/K values calculated by the compiler.

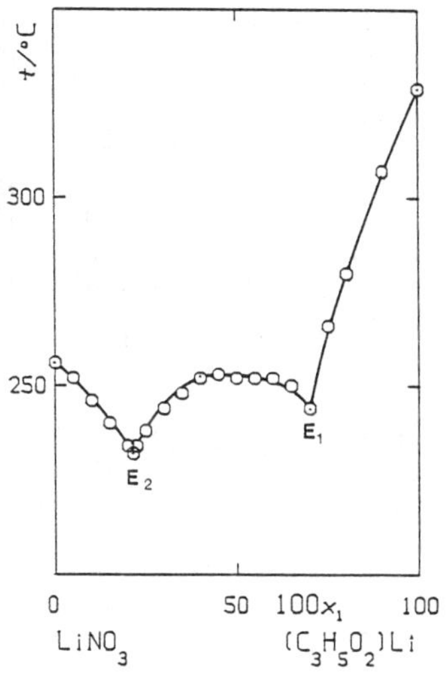

Characteristic point(s):

Eutectic, E_1, at 244 ^{o}C and $100x_1$= 70 (authors).
Eutectic, E_2, at 232 ^{o}C and $100x_1$= 21.5 (according to the table, although reported as 22.5 in the text; compiler).

Intermediate compound(s):

$Li_5(C_3H_5O_2)_2(NO_3)_3$, congruently melting at 252-253 ^{o}C.

AUXILIARY INFORMATION

METHOD/APPARATUS/PROCEDURE:

Visual polythermal analysis.

SOURCE AND PURITY OF MATERIALS:

Component 1: prepared from propanoic acid and lithium hydrogen carbonate (Ref. 1), and recrystallized from n-butanol.
Component 2: material of analytical grade recrystallized twice.

ESTIMATED ERROR:

Temperature: accuracy probably ± 2 K (compiler).

REFERENCES:

(1) Sokolov, N.M.
Zh. Obshch. Khim. 1954, **24**, 1150-1156.

COMPONENTS:

(1) Lithium butanoate (lithium butyrate); $LiC_4H_7O_2$; [21303-03-7]
(2) Lithium thiocyanate; LiCNS; [556-65-0]

ORIGINAL MEASUREMENTS:

Sokolov, N.M.; Dmitrevskaya, O.I.
Zh. Neorg. Khim. 1969, 14, 286-296 (*);
Russ. J. Inorg. Chem. (Engl. Transl.) 1969, 14, 148-155.

VARIABLES:

Temperature.

PREPARED BY:

Baldini, P.

EXPERIMENTAL VALUES:

t/°C	T/K[a]	$100x_2$	t/°C	T/K[a]	$100x_2$
329	602	0	215	488	45
316	589	5	208	481	50
308	581	10	215	488	55
290	563	15	224	497	60
278	551	20	233	506	65
265	538	25	241	514	70
251	524	30	250	523	75
238	511	35	262	535	85
225	498	40	266	539	100

[a] T/K values calculated by the compiler.

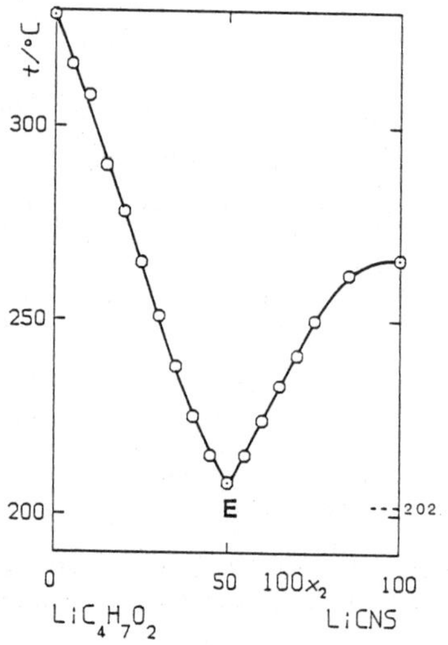

Characteristic point(s):

Eutectic, E, at 208 °C and $100x_2$= 50 (authors).

AUXILIARY INFORMATION

METHOD/APPARATUS/PROCEDURE:

Visual polythermal analysis; solid state transition temperatures drawn from the heating curves obtained with automatic recording.

NOTE:

The fusion temperature of component 1 given by the authors (602 K) is noticeably higher than that (591.7±0.5 K) listed in Preface, Table 1 where, moreover, no solid state transformation is reported for lithium n-butanoate.

SOURCE AND PURITY OF MATERIALS:

Not stated.
Component 1 undergoes a phase transition at $t_{trs}(1)$/°C= 98.
Component 2 undergoes a phase transition at $t_{trs}(2)$/°C= 202.

ESTIMATED ERROR:

Temperature: accuracy probably ±2 K (compiler).

REFERENCES:

COMPONENTS:

(1) Lithium butanoate (lithium butyrate); $LiC_4H_7O_2$; [21303-03-7]
(2) Lithium nitrate; $LiNO_3$; [7790-69-4]

ORIGINAL MEASUREMENTS:

Tsindrik, N.M.; Sokolov, N.M.
Zh. Obshch. Khim. 1958, **28**, 1728-1733 (*);
Russ. J. Gen. Chem. (Engl. Transl.) 1958, **28**, 1775-1780.

VARIABLES:

Temperature.

PREPARED BY:

Baldini, P.

EXPERIMENTAL VALUES:

$t/^{o}C$	T/K^{a}	$100x_1$	$t/^{o}C$	T/K^{a}	$100x_1$
256	529	0	216	489	45
248	521	5	228	501	50
242	515	10	238	511	55
238	511	12.5	248	521	60
232	505	15	258	531	65
232	505	17.5	268	541	70
232	505	20	278	551	75
230	503	25	288	561	80
230	503	30	298	571	85
226	499	35	308	581	90
224	497	40	329	602	100
220	493	42.5			

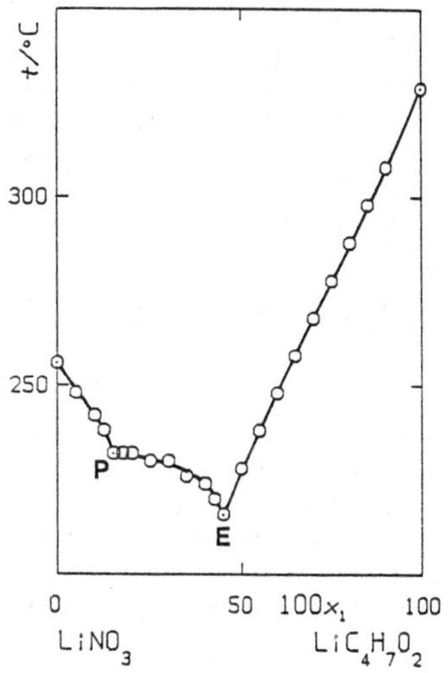

[a] T/K values calculated by the compiler.

Characteristic point(s):

Eutectic, E, at 216 °C and $100x_1$= 45 (authors).
Peritectic, P, at 232 °C and $100x_1$= 15 (authors).

Intermediate compound(s):

$Li_8C_4H_7O_2(NO_3)_7$ (probable composition; authors), incongruently melting.

AUXILIARY INFORMATION

METHOD/APPARATUS/PROCEDURE:

Visual polythermal analysis; temperatures of initial crystallization measured with a Nichrome-Constantane thermocouple and a millivoltmeter.

SOURCE AND PURITY OF MATERIALS:

Component 1: prepared from "chemically pure" carbonate and n-butanoic acid (Ref. 1); the solid recovered after evaporation was recrystallized from n-butanol.
Component 2: source not stated.

ESTIMATED ERROR:

Temperature: accuracy probably ±2 K (compiler).

REFERENCES:

(1) Sokolov, N.M.
Zh. Obshch. Khim. 1954, **24**, 1581-1593.

COMPONENTS:	ORIGINAL MEASUREMENTS:
(1) Sodium bromide; NaBr; [7647-15-6] (2) Sodium methanoate (sodium formate); $NaCHO_2$; [141-53-7]	Leonesi, D.; Braghetti, M.; Cingolani, A.; Franzosini, P. **Z. Naturforsch.** 1970, **25a**, 52-55.
VARIABLES:	PREPARED BY:
Temperature.	Baldini, P.

EXPERIMENTAL VALUES:

$t/^{o}C$	T/K^a	$100x_1$
257.5	530.7	0
254.5	527.7	2.02
251.8	525.0	3.91
250.4	523.6	4.99
248.8	522.0	6.01
245.7	518.9	8.00
244.2	517.4	8.98
250.9	524.1	10.00
265.1	538.3	11.04
274.0	547.2	11.76
296.1	569.3	13.55

[a] **T**/K values calculated by the compiler.

Note 1 - In the original paper the results were shown in a graphical form. The above listed numerical values represent a personal communication by one of the authors (F., P.) to the compiler.

Note 2 - The system could not be investigated above 300 ^{o}C due to the thermal instability of the methanoate.

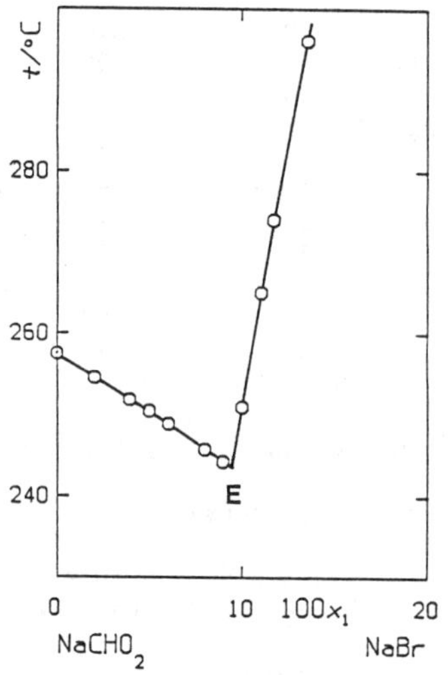

Characteristic point(s):

Eutectic, E, at 243.5 ^{o}C and $100x_1$= 9.5 (authors).

AUXILIARY INFORMATION

METHOD/APPARATUS/PROCEDURE:

A Pyrex device, suitable for work under an inert atmosphere, and allowing one to observe the system visually, was employed (for details, see Ref. 1). The initial crystallization temperatures were measured with a Chromel-Alumel thermocouple checked by comparison with a certified Pt resistance thermometer, and connected with a L&N Type K-3 potentiometer.

SOURCE AND PURITY OF MATERIALS:

C. Erba RP meterials, dried by heating under vacuum.

ESTIMATED ERROR:

Temperature: accuracy probably ± 0.1 K (compiler).

REFERENCES:

(1) Braghetti, M.; Leonesi, D.; Franzosini, P. **Ric. Sci.** 1968, **38**, 116-118.

COMPONENTS:

(1) Sodium bromide;
$NaBr$; [7647-15-6]
(2) Sodium ethanoate (sodium acetate);
$NaC_2H_3O_2$; [127-09-3]

ORIGINAL MEASUREMENTS:

Il'yasov. I.I.; Bergman, A.G.
Zh. Obshch. Khim. 1961, 31, 368-370.

VARIABLES:

Temperature.

PREPARED BY:

D'Andrea, G.

EXPERIMENTAL VALUES:

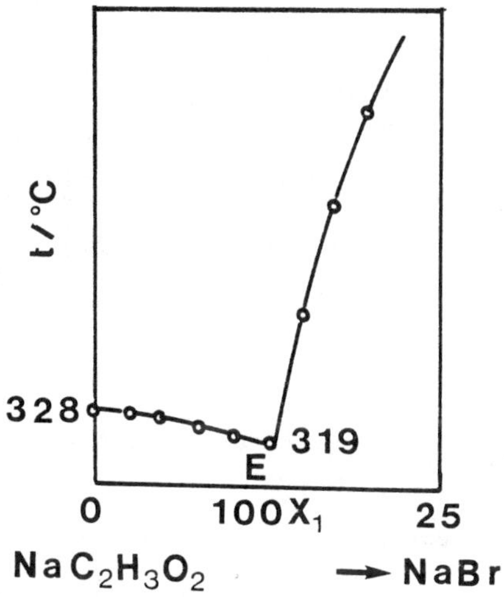

The results are given only in graphical form (see figure). The system was investigated at $0 \leq 100\mathbf{x}_1 \leq 25$.

Characteristic point(s): Eutectic, E, at 319 °C and $100\mathbf{x}_1 = 12.5$ (authors).

AUXILIARY INFORMATION

METHOD/APPARATUS/PROCEDURE:

Visual polythermal analysis; temperatures measured with a Nichrome-Constantane thermocouple and a millivoltmeter (Ref. 1).

SOURCE AND PURITY OF MATERIALS:

Not stated.
Component 1: $t_{fus}(1)/^{o}C = 755$.
Component 2: $t_{fus}(1)/^{o}C = 328$.

ESTIMATED ERROR:

Temperature: accuracy probably ± 2 K (compiler).

REFERENCES:

(1) Il'yasov, I.I.; Bergman, A.G.
Zh. Obshch. Khim. 1960, 30, 355-358.

COMPONENTS:

(1) Sodium bromide;
$NaBr$; [7647-15-6]
(2) Sodium ethanoate (sodium acetate);
$NaC_2H_3O_2$; [127-09-3]

ORIGINAL MEASUREMENTS:

Piantoni, G.; Leonesi, D.; Braghetti, M.; Franzosini, P.
Ric. Sci., 1968, **38**, 127-132.

VARIABLES:

Temperature.

PREPARED BY:

D'Andrea, G.

EXPERIMENTAL VALUES:

$t/^oC$	T/K^a	$100x_2$
328.1	601.3	100
327.7	600.9	99.6
327.1	600.3	98.9
326.5	599.7	98.2
326.1	599.3	97.8
325.4	598.6	97.1
325.4	598.6	97.1
325.4	598.6	97.0
323.0	596.2	94.4
323.1	596.3	94.4
321.0	594.2	92.2
318.8	592.0	89.8
322.3	595.5	88.5
325.9	599.1	88.2
331.8	605.0	87.6

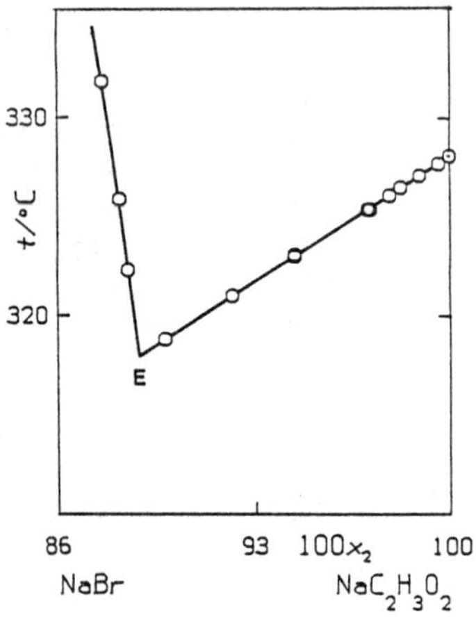

[a] T/K values calculated by the compiler.

Note 1 - In the original paper the results were shown in graphical form. The above listed numerical values represent a private communication by one of the authors (F., P.) to the compiler.

Note 2 - The system was investigated at $0 \leq 100x_1 \leq 12.5$.

Characteristic point(s): Eutectic, E, at 317.9 °C and $100x_2$= 88.9 (authors).

AUXILIARY INFORMATION

METHOD/APPARATUS/PROCEDURE:

A Pyrex device, suitable for work under an inert atmosphere, and allowing one to observe the system visually, was employed (for details, see Ref. 1). The initial crystallization temperatures were measured with a Chromel-Alumel thermocouple checked by comparison with a certified Pt resistance thermometer, and connected with a L&N Type K-3 potentiometer.

SOURCE AND PURITY OF MATERIALS:

C. Erba RP materials, dried by heating under vacuum.

NOTE:

The authors discuss their own results in comparison with both the expected ideal behaviour of the molten mixtures and the previous data from Ref. 2. Extension of this comparison to the cryometric constant at null molality for different solutes in molten sodium ethanoate allowed them to argue that sodium bromide and sodium ethanoate show a remarkable tendency to give mixed crystals.

ESTIMATED ERROR:

Temperature: accuracy probably ±0.1 K.

REFERENCES:

(1) Braghetti, M.; Leonesi, D.; Franzosini, P.
Ric. Sci. 1968, **38**, 116-118.
(2) Il'yasov. I.I.; Bergman, A.G.
Zh. Obshch. Khim. 1961, 31, 368-370.

COMPONENTS:	ORIGINAL MEASUREMENTS:
(1) Sodium methanoate (sodium formate); $NaCHO_2$; [141-53-7] (2) Sodium ethanoate (sodium acetate); $NaC_2H_3O_2$; [127-09-3]	Sokolov, N.M. **Zh. Obshch. Khim.** 1954, **24**, 1581-1593.
VARIABLES:	PREPARED BY:
Temperature.	Baldini, P.

EXPERIMENTAL VALUES:

t/°C	T/K[a]	$100x_2$	t/°C	T/K[a]	$100x_2$
258	531	0	296	569	50
252	525	5	300	573	55
244	517	10	303	576	60
242	515	10.5	308	581	65
252	525	15	313	586	70
260	533	20	316	589	75
267	540	25	320	593	80
270	543	30	323	596	85
278	551	35	326	599	90
284	557	40	330	603	95
291	564	45	331	604	100

[a] T/K values calculated by the compiler.

Characteristic point(s):

Eutectic, E, at 242 °C and $100x_2$= 10.5 (author).

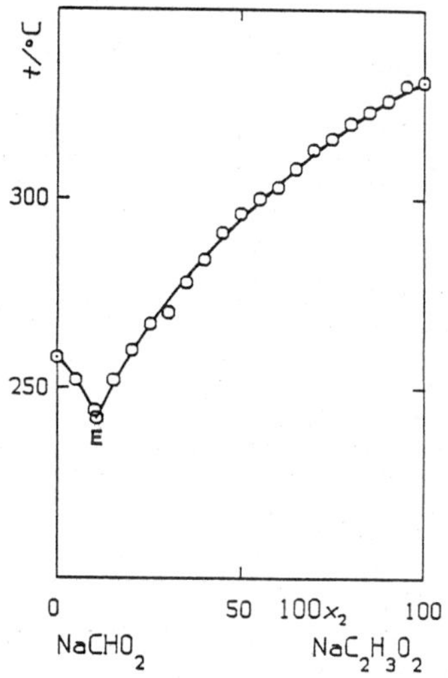

AUXILIARY INFORMATION

METHOD/APPARATUS/PROCEDURE:

Visual polythermal analysis.
Melts contained in a glass tube and stirred. Temperatures measured with a Nichrome-Constantane thermocouple and a 17 mV full scale millivoltmeter. The temperature readings refer to the disappearance of isotropicity in the melt on cooling.

SOURCE AND PURITY OF MATERIALS:

Component 1: prepared by reacting aqueous ("chemically pure") Na_2CO_3 with a slight excess of methanoic acid of analytical purity. The solvent and excess acid were removed by heating to 160 °C.
Component 2: "chemically pure" material.

ESTIMATED ERROR:

Temperature: accuracy probably ±2 K (compiler).

REFERENCES:

COMPONENTS:

(1) Sodium methanoate (sodium formate); $NaCHO_2$; [141-53-7]
(2) Sodium propanoate (sodium propionate); $NaC_3H_5O_2$ [137-40-6]

ORIGINAL MEASUREMENTS:

Sokolov, N.M.; Minchenko, S.P.
Zh. Obshch. Khim. 1971, 41, 1656-1659.

VARIABLES:

Temperature.

PREPARED BY:

Baldini, P.

EXPERIMENTAL VALUES:

The results are reported only in graphical form (see figure; empty circles: visual polythermal analysis; filled circles: thermographical analysis).

Characteristic point(s):

Eutectic, E_1, at 255 ^oC and $100x_2$= 6 (authors).
Eutectic, E_2, at 293 ^oC and $100x_2$= 98 (authors).

Intermediate compound(s):

$Na_3CHO_2(C_3H_5O_2)_2$, congruently melting.

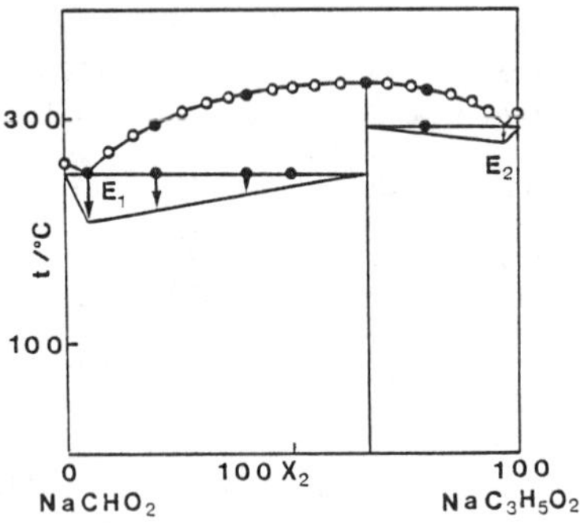

AUXILIARY INFORMATION

METHOD/APPARATUS/PROCEDURE:

Visual polythermal analysis supplemented with thermographical analysis.

NOTE:

The fusion temperature reported for component 1 (531 K) coincides with that listed in Preface, Table 1 (530.7 ± 0.5 K), whereas the T_{trs} values from Ref. 2 and Table 1 are significantly discrepant. Concerning component 2, the fusion temperature (571 K) looks as somewhat too high; moreover, doubts are to be cast about the reliability of the highest transition temperature (560 K) quoted by the authors from Ref. 2, inasmuch as both DSC (Table 1) and adiabatic calorimetry (Table 3 of the Preface) proved the occurrence of solid state transformations only at 467-470 and 491-494 K, respectively.

SOURCE AND PURITY OF MATERIALS:

Both components prepared from the proper acid and the carbonate (Ref. 1).
Component 1 melts at $t_{fus}(1)/^oC$= 258 and undergoes a phase transition at $t_{trs}(1)/^oC$= 242 (Ref. 2).
Component 2 melts at $t_{fus}(2)/^oC$= 298 (according to Fig.s 3, 4, of the original paper; compiler) or 300 (Fig. 1), and undergoes phase transitions at $t_{trs}(2)/^oC$= 195, 217, 287 (Ref. 2).

ESTIMATED ERROR:

Temperature: accuracy probably ± 2 K (compiler).

REFERENCES:

(1) Sokolov, N.M.
Zh. Obshch. Khim. 1954, **24**, 1581-1593.
(2) Sokolov, N.M.
Tezisy Dokl. X Nauch. Konf. S.M.I. 1956,

COMPONENTS:	EVALUATOR:
(1) Sodium methanoate (sodium formate); $NaCHO_2$; [141-53-7] (2) Sodium butanoate (sodium butyrate); $NaC_4H_7O_2$; [156-54-7]	Franzosini, P., Dipartimento di Chimica Fisica, Universita´ di Pavia (ITALY).

CRITICAL EVALUATION:

This system was studied only by Sokolov (Ref. 1), who suggested the existence of: (i) a eutectic, E_1, at 525 K (252 °C) and $100\mathbf{x}_2$= 2.5; (ii) a eutectic, E_2, at 581 K (308 °C) and $100\mathbf{x}_2$= 89; and (iii) an intermediate compound, $Na_2CHO_2C_4H_7O_2$, congruently melting at 614 K (341 °C).

Component 2, however, forms liquid crystals. Therefore, the fusion temperature, $\mathbf{T}_{fus}(2)$= 603 K (330 °C; Ref. 1), should be identified with the clearing temperature, the corresponding value from Table 1 of the Preface being $\mathbf{T}_{clr}(2)$= 600.4±0.2 K. No mention is made by the author of the other phase transitions occurring in component 2, including that corresponding to the actual fusion, viz., $\mathbf{T}_{fus}(2)$= 524.5±0.5 K (Table 1).

Conversely, the fusion temperature of component 1, $\mathbf{T}_{fus}(1)$= 531 K (258 °C; Ref. 1), satisfactorily corresponds to the value of Table 1, viz., $\mathbf{T}_{fus}(1)$= 530.7±0.5 K.

In conclusion, Sokolov´s assertion of the existence of the congruently melting intermediate compound is a reasonable interpretation of the trend of the available data. In this case, however, the phase diagram could be interpreted with reference to Scheme D.1 of the Preface: in particular, the eutectic E_2 could be actually identified with an M'_E point, Sokolov´s diagram likely being similar to that shown in Preface, Scheme D.1.

The unusual size of the dome and the absence of any information about the solidus does not allow one to exclude that Sokolov´s points might be at least in part relevant to liquid-liquid instead of solid-liquid equilibria. One might therefore take into account the occurrence of liquid layering as shown in the figure: in particular, the eutectic E_2 could be actually identified with an invariant at which equilibrium occurs among two isotropic liquid and one crystalline liquid phases.

REFERENCES:

(1) Sokolov, N.M.
Zh. Obshch. Khim. 1954, **24**, 1581-1593.

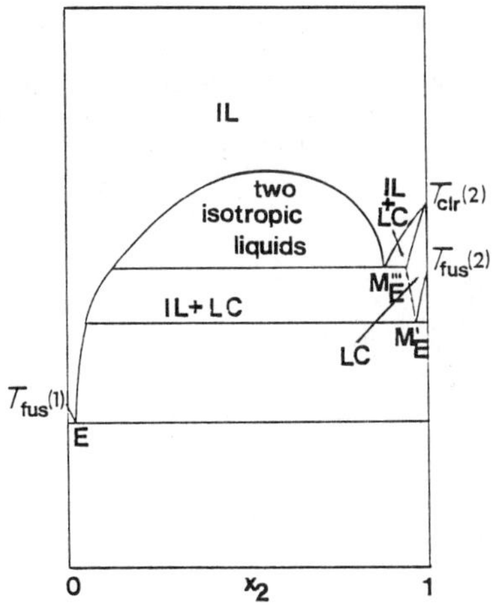

COMPONENTS:

(1) Sodium methanoate (sodium formate); $NaCHO_2$; [141-53-7]
(2) Sodium butanoate (sodium butyrate); $NaC_4H_7O_2$; [156-54-7]

ORIGINAL MEASUREMENTS:

Sokolov, N.M.
Zh. Obshch. Khim. 1954, **24**, 1581-1593.

VARIABLES:

Temperature.

PREPARED BY:

Baldini, P.

EXPERIMENTAL VALUES:

t/°C	T/K[a]	$100x_2$	t/°C	T/K[a]	$100x_2$
258	531	0	340	613	55
252	525	2.5	340	613	60
287	560	5	339	612	65
301	574	10	338	611	70
312	585	15	336	609	75
318	591	20	331	604	80
324	597	25	324	597	85
327	600	30	308	581	89
333	606	35	311	584	90
337	610	40	322	595	95
339	612	45	330	603	100
341	614	50			

[a] T/K values calculated by the compiler.

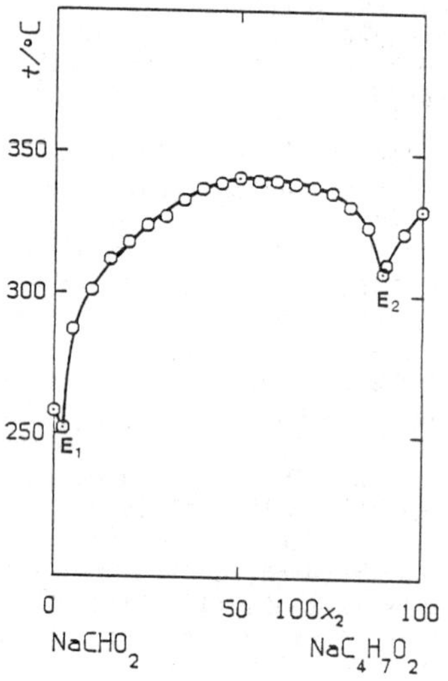

Characteristic point(s):

Eutectic, E_1, at 252 °C and $100x_2$ = 2.5 (author).
Eutectic, E_2, at 308 °C (erroneously reported as 318 °C in the text, compiler) and $100x_2$ = 89 (author).

Intermediate compound(s):

$Na_2CHO_2C_4H_7O_2$, congruently melting at 341 °C.

AUXILIARY INFORMATION

METHOD/APPARATUS/PROCEDURE:

Visual polythermal analysis.
Melts contained in a glass tube and stirred.
Temperatures measured with a Nichrome-Constantane thermocouple and a 17 mV full scale millivoltmeter. The temperature readings refer to the disappearance of isotropicity in the melt on cooling.

SOURCE AND PURITY OF MATERIALS:

Materials prepared by reacting aqueous ("chemically pure") Na_2CO_3 with a slight excess of the proper acid of analytical purity. The solvent and excess acid were removed by heating to 160 °C.

ESTIMATED ERROR:

Temperature: accuracy probably ±2 K (compiler).

REFERENCES:

COMPONENTS:	ORIGINAL MEASUREMENTS:
(1) Sodium methanoate (sodium formate); $NaCHO_2$; [141-53-7] (2) Sodium **iso**.butanoate (sodium **iso**.butyrate); $Nai.C_4H_7O_2$; [996-30-5]	Sokolov, N.M. **Zh. Obshch. Khim.** 1954, **24**, 1581-1593.
VARIABLES:	PREPARED BY:
Temperature.	Baldini, P.

EXPERIMENTAL VALUES:

t/°C	T/K[a]	$100x_2$	t/°C	T/K[a]	$100x_2$
258	531	0	329	602	55
252	525	1.3	327	600	60
290	563	5	325	598	65
305	578	10	320	593	70
314	587	15	314	587	75
319	592	20	306	579	80
321	594	25	296	569	85
324	597	30	282	555	90
326	599	35	258	531	95
327	600	40	250	523	96.5
329	602	45	260	533	100
330	603	50			

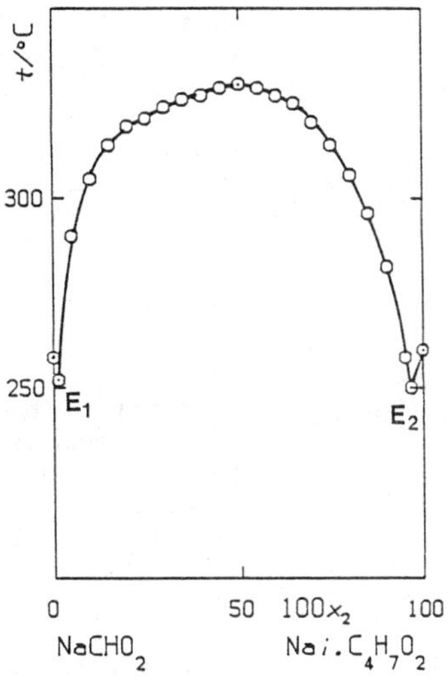

[a] T/K values calculated by the compiler.

Characteristic point(s):

Eutectic, E_1, at 252 °C and $100x_2$= 1.3 (author).
Eutectic, E_2, at 250 °C and $100x_2$= 96.5 (author).

Intermediate compound(s):

$Na_2CHO_2i.C_4H_7O_2$, congruently melting at 330 °C.

AUXILIARY INFORMATION

METHOD/APPARATUS/PROCEDURE:

Visual polythermal analysis.
Melts contained in a glass tube and stirred.
Temperatures measured with a Nichrome-Constantane thermocouple and a 17 mV full scale millivoltmeter. The temperature readings refer to the disappearance of isotropicity in the melt on cooling.

NOTE:

As an interpretation alternative to that by Sokolov, the large liquidus dome might be due to the occurrence of a miscibility gap in the liquid state.

SOURCE AND PURITY OF MATERIALS:

Materials prepared by reacting aqueous ("chemically pure") Na_2CO_3 with a slight excess of the proper acid of analytical purity. The solvent and excess acid were removed by heating to 160 °C.

ESTIMATED ERROR:

Temperature: accuracy probably ±2 K (compiler).

REFERENCES:

COMPONENTS:	EVALUATOR:
(1) Sodium methanoate (sodium formate); $NaCHO_2$; [141-53-7] (2) Sodium **iso**.pentanoate (sodium **iso**.valerate); $Na\mathbf{i}.C_5H_9O_2$; [539-66-2]	Franzosini, P., Dipartimento di Chimica Fisica, Universita´ di Pavia (ITALY).

CRITICAL EVALUATION:

This system was studied only by Sokolov (Ref. 1), who claimed the existence of:

(i) a eutectic, E_1, at 525 K (252 ^oC) and $100\mathbf{x}_2$= 0.75;
(ii) a eutectic, E_2, at 518 K (245 ^oC) and $100\mathbf{x}_2$= 94.5; and
(iii) an intermediate compound, $Na_5(CHO_2)_3(\mathbf{i}.C_5H_9O_2)_2$ congruently melting at 593 K (320 ^oC).

Component 2, however, forms liquid crystals. Therefore, the fusion temperature reported in Ref. 1, $\mathbf{T}_{fus}(2)$= 535 K (262 ^oC) is actually to be identified with the clearing temperature, the corresponding value from Table 2 of the Preface being $\mathbf{T}_{clr}(2)$= 559$\pm$1 K. The remarkable discrepancy between these values might be attributed to the presence of some impurity in Sokolov´s sample, inasmuch as the value from Table 2 meets rather satisfactorily those reported by Ubbelohde et al. (556 K; Ref. 2), and by Duruz et al. (553 K; Ref. 3). According to Table 2, component 2 melts at 461.5$\pm$0.6 K.

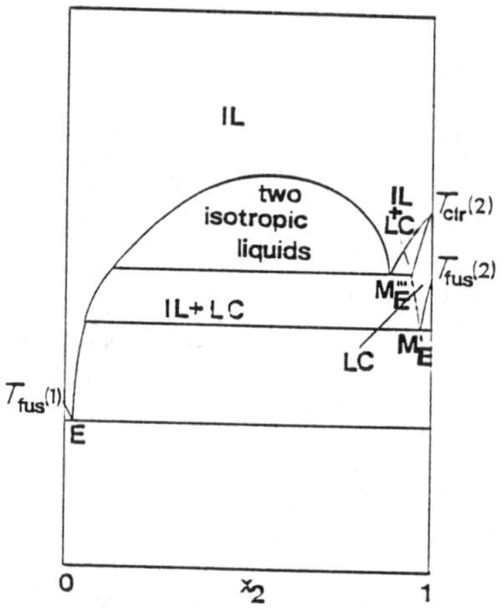

Conversely, the fusion temperature reported in Ref. 1 for component 1, $\mathbf{T}_{fus}(1)$= 531 K (238 ^oC) is in satisfactory agreement with the value from Table 1, viz., $\mathbf{T}_{fus}(1)$= 530.7$\pm$0.5 K.

In conclusion, Sokolov´s assertion of the existence of the congruently melting intermediate compound is a reasonable interpretation of the trend of the available data. In this case, however, the phase diagram should be modified as follows: the eutectic E_2 should be identified with an M'_E point, Sokolov´s diagram being likely similar to that shown in Scheme D.1.

The unusual size of the dome and the absence of any information about the solidus does not allow one to exclude that Sokolov´s points might be at least in part relevant to liquid-liquid instead of solid-liquid equilibria. One might therefore take into account the occurrence of liquid layering as shown in the figure: in particular, the eutectic E_2 could be actually identified with an invariant at which equilibrium occurs among two isotropic liquid and one crystalline liquid phases.

REFERENCES:

(1) Sokolov, N.M.
Zh. Obshch. Khim. 1954, **24**, 1581-1593.

(2) Ubbelohde, A.R.; Michels, H.J.; Duruz, J.J.
Nature 1970, **228**, 50-52.

(3) Duruz, J.J.; Michels, H.J.; Ubbelohde, A.R.
Proc. R. Soc. London 1971, **A322**, 281-299.

COMPONENTS:	ORIGINAL MEASUREMENTS:
(1) Sodium methanoate (sodium formate); $NaCHO_2$; [141-53-7] (2) Sodium **iso**.pentanoate (sodium **iso**.valerate); $Na\,i.C_5H_9O_2$; [539-66-2]	Sokolov, N.M. **Zh. Obshch. Khim.** 1954, **24**, 1581-1593.
VARIABLES:	PREPARED BY:
Temperature.	Baldini, P.

EXPERIMENTAL VALUES:

t/°C	T/K[a]	$100x_2$	t/°C	T/K[a]	$100x_2$
258	531	0	317	590	50
252	525	0.75	315	588	55
287	560	5	312	585	60
300	573	10	309	582	65
308	581	15	306	579	70
311	584	20	301	574	75
314	587	25	297	570	80
316	589	30	284	557	85
318	591	35	265	538	90
320	593	40	245	518	94.5
319	592	45	262	535	100

[a] T/K values calculated by the compiler.

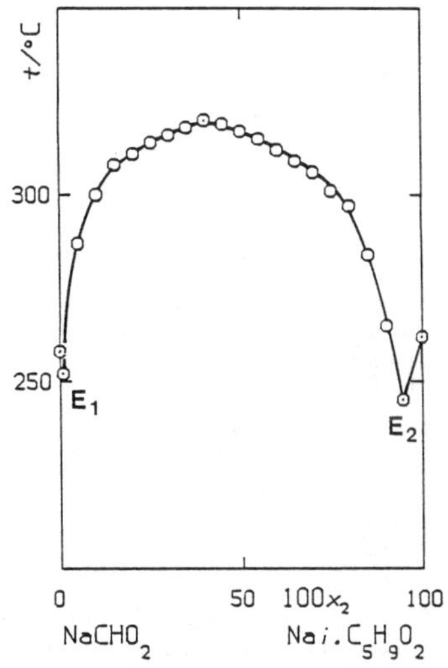

Characteristic point(s):

Eutectic, E_1, at 252 °C and $100x_2$= 0.75 (author).
Eutectic, E_2, at 245 °C and $100x_2$= 94.5 (author).

Intermediate compound(s):

$Na_5(CHO_2)_3(i.C_5H_9O_2)_2$, congruently melting at 320 °C.

AUXILIARY INFORMATION

METHOD/APPARATUS/PROCEDURE:

Visual polythermal analysis.
Melts contained in a glass tube and stirred.
Temperatures measured with a Nichrome-Constantane thermocouple and a 17 mV full scale millivoltmeter. The temperature readings refer to the disappearance of isotropicity in the melt on cooling.

SOURCE AND PURITY OF MATERIALS:

Materials prepared by reacting aqueous ("chemically pure") Na_2CO_3 with a slight excess of the proper acid of analytical purity. The solvent and excess acid were removed by heating to 160 °C.

ESTIMATED ERROR:

Temperature: accuracy probably ±2 K (compiler).

REFERENCES:

COMPONENTS:

(1) Sodium methanoate (sodium formate); $NaCHO_2$; [141-53-7]
(2) Sodium chloride; NaCl; [7647-14-5]

ORIGINAL MEASUREMENTS:

Leonesi, D.; Braghetti, M.; Cingolani, A.; Franzosini, P.
Z. Naturforsch. 1970, **25a**, 52-55.

VARIABLES:

Temperature.

PREPARED BY:

Baldini, P.

EXPERIMENTAL VALUES:

$t/^{o}C$	T/K^{a}	$100x_2$
257.5	530.7	0
256.0	529.2	0.95
254.5	527.7	2.10
253.2	526.4	3.00
251.6	524.8	3.94
250.7	523.9	4.60
249.8	523.0	5.14
252.5	525.7	5.22
259.5	532.7	5.63
260.5	533.7	5.70
266.8	540.0	5.98
275.4	548.6	6.48
286.3	559.5	7.13
293.6	566.8	7.55
298.9	572.1	7.83

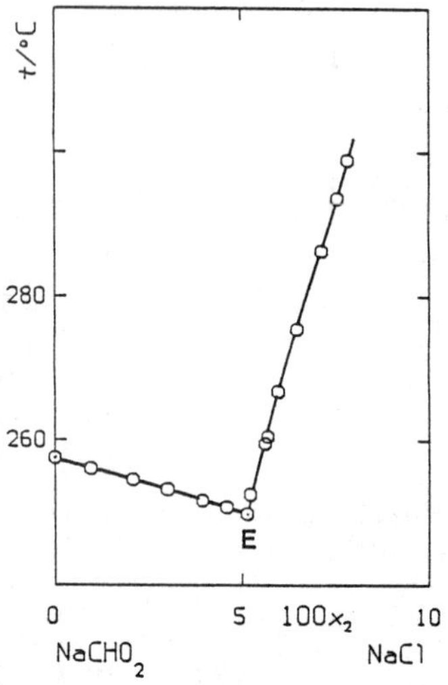

[a] T/K values calculated by the compiler.

Note 1 - In the original paper the results were shown in graphical form. The above listed numerical values represent a personal communication by one of the authors (F., P.) to the compiler.

Note 2 - The system could not be investigated above 300 °C due to the thermal instability of the methanoate.

Characteristic point(s):
Eutectic, E, at 249.8 °C and $100x_2$= 5.15 (authors).

AUXILIARY INFORMATION

METHOD/APPARATUS/PROCEDURE:

A Pyrex device, suitable for work under an inert atmosphere, and allowing one to observe the system visually, was employed (for details, see Ref. 1). The initial crystallization temperatures were measured with a Chromel-Alumel thermocouple checked by comparison with a certified Pt resistance thermometer, and connected with a L&N Type K-3 potentiometer.

NOTE:

Previous investigations by the same group (Ref. 2) stated that the cryometric constant of sodium methanoate was **K**= 9.4 ± 0.2 K $molality^{-1}$, and that

$$\lim_{m\to 0} (\Delta T/m) = 9.6 \text{ K molality}^{-1}$$

(where ΔT: experimental freezing point depression; m: molality of the solute) when NaCl was the solute. Consequently, the solubility of component 2 in component 1 in the solid state ought to be insignificant.

SOURCE AND PURITY OF MATERIALS:

C. Erba RP materials, dried by heating under vacuum.

ESTIMATED ERROR:

Temperature: accuracy probably ±0.1 K (compiler).

REFERENCES:

(1) Braghetti, M.; Leonesi, D.; Franzosini, P. **Ric. Sci.** 1968, **38**, 116-118.
(2) Leonesi, D.; Piantoni, G.; Berchiesi, G.; Franzosini, P. **Ric. Sci.** 1968, **38**, 702-705.

COMPONENTS:	EVALUATOR:
(1) Sodium methanoate (sodium formate); $NaCHO_2$; [141-53-7] (2) Sodium thiocyanate; NaCNS; [540-72-7]	Ferloni, P., Dipartimento di Chimica Fisica, Universita´ di Pavia (ITALY).

CRITICAL EVALUATION:

The system sodium methanoate - sodium thiocyanate was investigated by Sokolov, 1954 (Ref. 1), Sokolov and Pochtakova, 1958 (Ref. 2), Cingolani et al., 1971 (Ref. 3), and Storonkin et al., 1974 (Ref. 4).

The liquidus curve drawn on the basis of visual polythermal observations led Sokolov (Ref. 1) to express the opinion that the system was a eutectic one.

Sokolov and Pochtakova (Ref. 2) re-examined the system (as a side of the composition square of the reciprocal ternary K, Na/CHO_2, CNS) using the same method and came to parallel conclusions. It is however to be noted that: (i) differences up to 8 K exist between the fusion temperatures listed in either paper for mixtures of equal composition; and (ii) the coordinates of the eutectic are somewhat different, i.e., 460 K and $100\mathbf{x}_2$= 36 according to Ref. 1, and 462 K and $100\mathbf{x}_2$= 38 according to Ref. 2.

Cingolani et al. (Ref. 3), not aware of Ref.s 1, 2, found two invariants, viz. a eutectic at 462.7 K and $100\mathbf{x}_2$= 38.0 (in excellent agreement with Ref. 2) and the other one corresponding to the incongruent melting of the intermediate compound $Na_5(CHO_2)_4CNS$. They supplemented their visual observations (carried out at a cooling rate of about 0.25 K min^{-1}) with DSC analysis, and, in particular, asserted that the composition of the intermediate compound "was confirmed by DSC measurements". They could also observe in the composition triangle of each of the ternaries Na/Br, CHO_2, CNS, Na/CHO_2, Cl, CNS, and Na/CHO_2, CNS, I a crystallization region belonging to the binary intermediate compound and covering respectively 0.45, 0.80, and 1.80 % of the liquidus area.

Storonkin et al. (Ref. 4) employed DTA to investigate the ternary Na/CHO_2, CNS, NO_3, and once more found, for the binary system of interest here, just one eutectic at 443 K and $100\mathbf{x}_2$= 36; they also claimed the distribution coefficient of $NaCHO_2$ in NaCNS to be zero in the thiocyanate crystallization field. They were apparently not aware of Ref. 3.

Because of the better accuracy of the experimental approach, the evaluator is inclined to recommend (among those available so far) the data by Berchiesi et al. (Ref. 3). The fact that Storonkin et al. (Ref. 4), by employing a DTA technique, where not able to detect the intermediate compound still remains surprising. This fact, however, might be explained if the large supercooling effect found by the latter authors in the region of the ternary eutectic could not be prevented in the region of the binary eutectic. Efficient stirring and slow cooling rate have likely allowed Cingolani et al. (Ref. 3) to avoid this drawback. The presence of some impurity in Storonkin et al. (Ref. 4) methanoate is even possible, inasmuch as their $\mathbf{T}_{fus}(1)/K$ value (528) is some 3 K lower than those reported in Ref.s 1 (531), 2 (531), and 3 (530.65), and in Table 3 [530.46$\pm$0.04 (adiabatic calorimetry); 530.7$\pm$0.5 (DSC)].

REFERENCES:

(1) Sokolov, N.M.
Zh. Obshch. Khim. 1954, **24**, 1150-1156.

(2) Sokolov, N.M.; Pochtakova, E.I.
Zh. Obshch. Khim. 1958, **28**, 1391-1397 (*); **Russ. J. Gen. Chem. (Engl. Transl.)** 1958, **28**, 1449-1454.

(3) Cingolani, A.; Berchiesi, G; Piantoni, G.
J. Chem. Eng. Data 1971, **16**, 464-467.

(4) Storonkin, A.V.; Vasil´kova, I.V.; Potemin, S.S.
Vestn. Leningr. Univ., Fiz., Khim. 1974, (10), 84-88.

COMPONENTS:

(1) Sodium methanoate (sodium formate); $NaCHO_2$; [141-53-7]
(2) Sodium thiocyanate; NaCNS; [540-72-7]

ORIGINAL MEASUREMENTS:

Sokolov, N.M.
Zh. Obshch. Khim. 1954, **24**, 1150-1156.

VARIABLES:

Temperature.

PREPARED BY:

Baldini, P.

EXPERIMENTAL VALUES:

$t/^{o}C$	T/K^{a}	$100x_2$	$t/^{o}C$	T/K^{a}	$100x_2$
258	531	0	212	485	45
250	523	5	232	505	50
241	514	10	244	517	55
233	506	15	256	529	60
213	486	25	267	540	65
202	475	30	284	557	75
190	463	35	302	575	90
187	460	36	311	584	100
197	470	40			

[a] T/K values calculated by the compiler.

Characteristic point(s):

Eutectic, E, at 187 ^{o}C and $100x_2$= 36 (author).

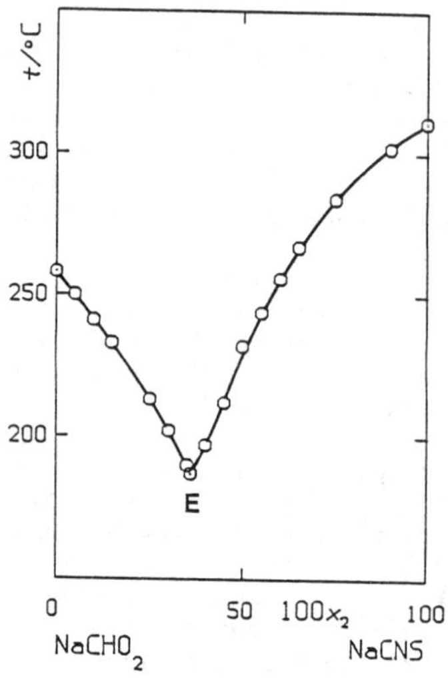

AUXILIARY INFORMATION

METHOD/APPARATUS/PROCEDURE:

Visual polythermal analysis.
Salt(s) melted in a test tube. Temperature measured with a Nichrome-Constantane thermocouple and a millivoltmeter with mirror reading to 17 mV.

SOURCE AND PURITY OF MATERIALS:

Component 1 synthetized from methanoic acid and $NaHCO_3$. Component 2 of analytical purity recrystallized once from water and once from ethanol.

ESTIMATED ERROR:

Temperature: accuracy probably ± 2 K (compiler).

REFERENCES:

COMPONENTS:

(1) Sodium methanoate (sodium formate); $NaCHO_2$; [141-53-7]
(2) Sodium thiocyanate; NaCNS; [540-72-7]

ORIGINAL MEASUREMENTS:

Sokolov, N.M.; Pochtakova, E.I.
Zh. Obshch. Khim. 1958, **28**, 1391-1397 (*);
Russ. J. Gen. Chem. (Engl. Transl.) 1958, **28**, 1449-1454.

VARIABLES:

Temperature.

PREPARED BY:

Baldini, P.

EXPERIMENTAL VALUES:

$t/^oC$	T/K^a	$100x_1$	$t/^oC$	T/K^a	$100x_1$
311	584	0	210	483	55
306	579	5	195	468	60
298	571	10	189	462	62
290	563	15	195	468	65
283	556	20	202	475	70
278	551	25	216	489	75
269	542	30	225	498	80
259	532	35	239	512	85
251	524	40	246	519	90
237	510	45	256	529	95
228	501	50	258	531	100

[a] T/K values calculated by the compiler.

Characteristic point(s):

Eutectic, E, at 189 ^oC and $100x_1$= 62 (authors).

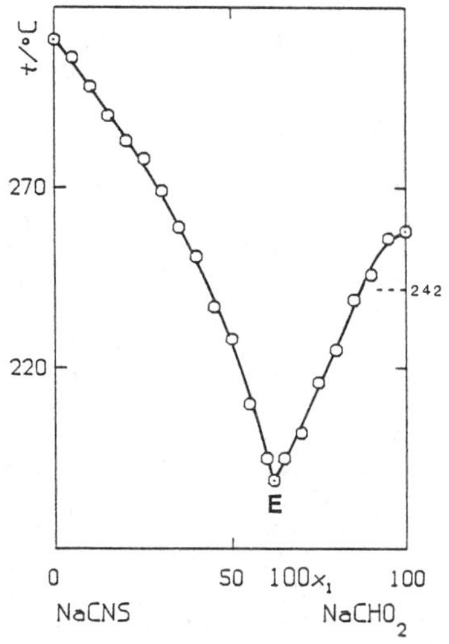

AUXILIARY INFORMATION

METHOD/APPARATUS/PROCEDURE:

Visual polythermal analysis.

SOURCE AND PURITY OF MATERIALS:

Component 1: commercial material recrystallized from water; it undergoes a phase transition at $t_{trs}(1)/^oC$= 242 (Ref. 1).
Component 2: commercial material recrystallized from alcohol.

ESTIMATED ERROR:

Temperature: accuracy probably ±2 K (compiler).

REFERENCES:

(1) Sokolov, N.M.
Tezisy Dokl. X Nauch. Konf. S.M.I. 1956.

COMPONENTS:	ORIGINAL MEASUREMENTS:
(1) Sodium methanoate (sodium formate); $NaCHO_2$; [141-53-7] (2) Sodium thiocyanate; NaCNS; [540-72-7]	Cingolani, A.; Berchiesi, G; Piantoni, G. **J. Chem. Eng. Data** 1971, **16**, 464-467.
VARIABLES:	PREPARED BY:
Temperature.	Baldini, P.

EXPERIMENTAL VALUES:

t/°C	T/K[a]	$100x_2$	t/°C	T/K[a]	$100x_2$
257.5	530.7	0	199.5	472.7	30.2
251.0	524.2	4.5	197.0	470.2	32.1
241.0	514.2	10.1	195.0	468.2	34.2
232.5	505.7	14.6	194.0	467.2	34.8
221.0	494.2	20.1	193.0	466.2	35.1
212.0	485.2	23.9	190.5	463.7	37.3
202.5	475.7	27.2	189.5	462.7	38.1
193.0	466.2	38.9	223.5	496.7	50.0
193.0	466.2	39.1	231.0	504.2	53.0
199.0	472.2	41.0	234.5	507.7	54.4
203.5	476.7	42.4	241.0	514.2	57.7
203.5	476.7	42.7	257.0	530.2	65.5
213.0	486.2	45.7	268.5	541.7	72.9
217.0	490.2	47.8			

[a] T/K values calculated by the compiler.

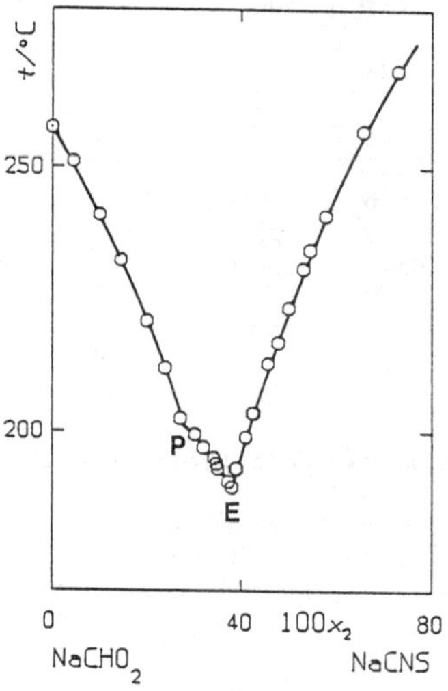

Note 1 - Measurements at t/°C $\geq$ 280 not taken due to instability of the melts (authors).
Note 2 - Despite the high accuracy of their temperature measurements, the authors chose to tabulate temperatures rounded at 0.5 K (compiler).

Characteristic point(s):
Eutectic, E, at 189.5 °C and $100x_2$= 38.0 (authors).
Peritectic, P, at 200.8 °C and $100x_2$= 28.6 (authors).
Intermediate compound(s):
$Na_5(CHO_2)_4CNS$, incongruently melting (authors).

AUXILIARY INFORMATION

METHOD/APPARATUS/PROCEDURE:

Visual method (for details, see Ref. 1). The melts contained in a Pyrex cryostat were cooled at a rate of about 0.25 K/min; the temperatures of initial crystallization were measured with a Chromel-Alumel thermocouple checked by comparison with a certified Pt resistance thermometer, and connected with a L&N potentiometer type K-3. Supplementary DSC measurements were also performed.

SOURCE AND PURITY OF MATERIALS:

Materials of stated purity $\geq$ 99 % were employed after careful drying.

ESTIMATED ERROR:

Temperature: accuracy $\pm$ 0.05 K (authors).

REFERENCES:

(1) Braghetti, M.; Leonesi, D.; Franzosini, P. **Ric. Sci.** 1968, **38**, 116-118.

COMPONENTS:

(1) Sodium methanoate (sodium formate); $NaCHO_2$; [141-53-7]
(2) Sodium thiocyanate; NaCNS; [540-72-7]

ORIGINAL MEASUREMENTS:

Storonkin, A.V.; Vasil'kova, I.V.; Potemin, S.S.
Vestn. Leningr. Univ., Fiz., Khim. 1974, (10), 84-88.

VARIABLES:

Temperature.

PREPARED BY:

Baldini, P.

EXPERIMENTAL VALUES:

t/°C	T/K[a]	$100x_2$
255	528	0
235	508	10
210	483	20
182	455	30
186	459	40
215	488	50
240	513	60
259	532	70
280	553	80
295	568	90
308	581	100

[a] T/K values calculated by the compiler.

Note - The data tabulated were drawn by the compiler from Fig. 1 of the original paper.

Characteristic point(s):

Eutectic, E, at 170 °C and $100x_2$= 36 (authors).

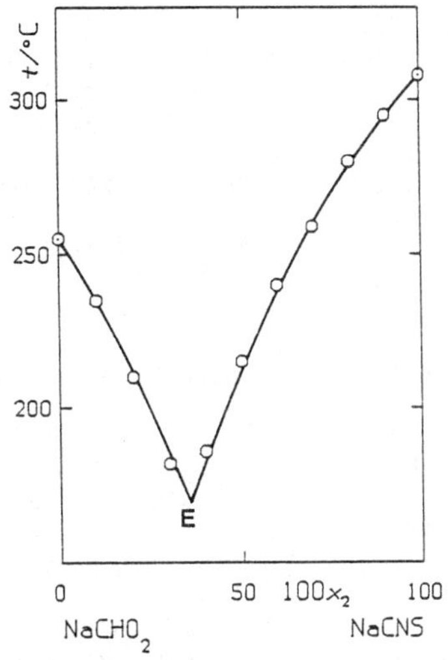

AUXILIARY INFORMATION

METHOD/APPARATUS/PROCEDURE:

DTA.
Thermograph with photorecorder.
Salt(s) sealed under vacuum in Pyrex ampoules.
No other information given.

SOURCE AND PURITY OF MATERIALS:

$NaCHO_2$ of analytical purity and "chemically pure" NaCNS, heated 10-15 h at temperatures 50-60 °C below their fusion temperatures, were employed.

ESTIMATED ERROR:

Temperature: accuracy probably ±2 K (compiler).

REFERENCES:

COMPONENTS:

(1) Sodium methanoate (sodium formate); $NaCHO_2$; [141-53-7]
(2) Sodium iodide; NaI; [7681-82-5]

ORIGINAL MEASUREMENTS:

Leonesi, D.; Braghetti, M.; Cingolani, A.; Franzosini, P.
Z. Naturforsch. 1970, **25a**, 52-55.

VARIABLES:

Temperature.

PREPARED BY:

Baldini, P.

EXPERIMENTAL VALUES:

$t/^{o}C$	T/K^{a}	$100x_2$	$t/^{o}C$	T/K^{a}	$100x_2$
257.5	530.7	0	243.5	516.7	9.00
256.8	530.0	0.42	241.6	514.8	9.98
256.3	529.5	0.73	237.7	510.9	12.01
255.9	529.1	0.98	234.0	507.2	13.99
255.4	528.6	1.34	232.2	505.4	15.00
254.7	527.9	1.79	230.0	503.2	15.99
254.3	527.5	2.03	236.5	509.7	17.99
251.2	524.4	4.02	248.4	521.6	18.98
248.3	521.5	6.00	270.9	544.1	20.99
244.9	518.1	7.99	306.3	579.5	24.61

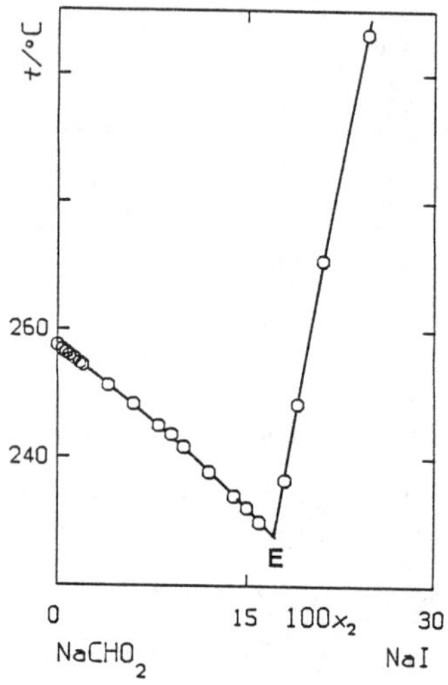

[a] T/K values calculated by the compiler.

Note 1 - In the original paper the results were shown in a graphical form. The above listed numerical values represent a personal communication by one of the authors (F., P.) to the compiler.

Note 2 - The system could not be investigated above about 300 °C due to the thermal instability of the methanoate.

Characteristic point(s):

Eutectic, E, at 227.7 °C and $100x_2$= 17.25 (authors).

AUXILIARY INFORMATION

METHOD/APPARATUS/PROCEDURE:

A Pyrex device, suitable for work under an inert atmosphere, and allowing one to observe the system visually, was employed (for details, see Ref. 1). The initial crystallization temperatures were measured with a Chromel-Alumel thermocouple checked by comparison with a certified Pt resistance thermometer, and connected with a L&N Type K-3 potentiometer.

SOURCE AND PURITY OF MATERIALS:

C. Erba RP meterials, dried by heating under vacuum.

ESTIMATED ERROR:

Temperature: accuracy probably $\pm$0.1 K (compiler).

REFERENCES:

(1) Braghetti, M.; Leonesi, D.; Franzosini, P. **Ric. Sci.** 1968, **38**, 116-118.

COMPONENTS:

(1) Sodium methanoate (sodium formate); $NaCHO_2$; [141-53-7]
(2) Sodium nitrite; $NaNO_2$; [7632-00-0]

ORIGINAL MEASUREMENTS:

Sokolov, N.M.
Zh. Obshch. Khim. 1957, 27, 840-844 (*);
Russ. J. Gen. Chem. (Engl. Transl.) 1957, 27, 917-920.

VARIABLES:

Temperature.

PREPARED BY:

Baldini, P.

EXPERIMENTAL VALUES:

t/°C	T/K[a]	$100x_2$
258	531	0
251	524	5
243	516	10
238	511	15
232	505	20
225	498	25
215	488	30
207	480	35
193	466	40
179	452	45
177	450	50
198	471	55
211	484	60
222	495	65
231	504	70
241	514	75
251	524	80
260	533	85
267	540	90
275	548	95
284	557	100

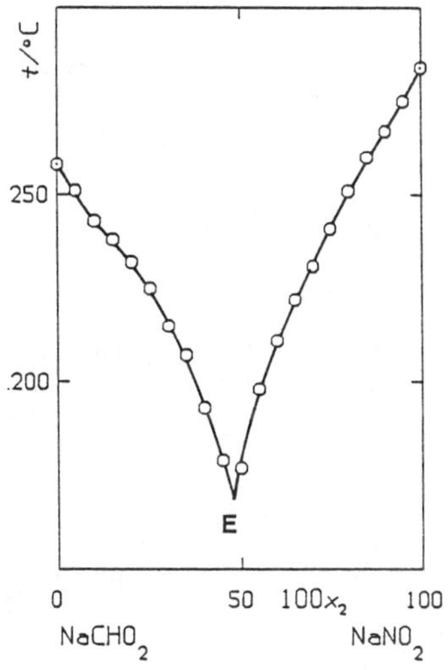

[a] T/K values calculated by the compiler.

Characteristic point(s): Eutectic, E, at 169 °C and $100x_2$= 48 (author).

AUXILIARY INFORMATION

METHOD/APPARATUS/PROCEDURE:

Visual polythermal analysis; salt mixtures melted in a glass tube (surrounded by a wider tube) and stirred with a glass thread. The temperatures of initial crystallization were measured with a Nichrome-Constantane thermocouple checked at the fusion points of water, benzoic acid, mannitol, $AgNO_3$, Cd, KNO_3, and $K_2Cr_2O_7$.

SOURCE AND PURITY OF MATERIALS:

"Chemically pure" materials recrystallized from water.

ESTIMATED ERROR:

Temperature: accuracy probably ±2 K (compiler).

NOTE:

The fusion temperature of component 1 (531 K) is in excellent agreement with the value (530.7±0.5 K) listed in Table 1 of the Preface, where a solid state transition (at 502±5 K), not mentioned by the author, is also reported.

REFERENCES:

COMPONENTS:	EVALUATOR:
(1) Sodium methanoate (sodium formate); $NaCHO_2$; [141-53-7] (2) Sodium nitrate; $NaNO_3$; [7631-99-4]	Ferloni, P., Dipartimento di Chimica Fisica, Universita´ di Pavia (ITALY).

CRITICAL EVALUATION:

The system sodium methanoate - sodium nitrate was investigated by Sokolov, 1954 (Ref. 1), Tsindrik, 1958 (Ref. 2), Berchiesi et al., 1972 (Ref. 3), and Storonkin et al., 1974 (Ref. 4).

The liquidus curve drawn on the basis of visual polythermal observations led Sokolov (Ref. 1) to express the opinion that the formation of any intermediate compound was to be excluded, and consequently the system was a eutectic one.

Tsindrik (Ref. 2), who belonged to the same Smolensk Medical Institute (S.M.I.) as Sokolov, re-examined the system (as a side of the composition square of the reciprocal ternary Li, Na/CHO_2, NO_3) using the same method and came to parallel conclusions. Significant discrepancies, however, exist in the trend of the liquidus curves given by either author; and for the coordinates of the eutectic, Tsindrik (Ref. 2) quoted figures (from a paper discussed in 1956 by Sokolov - Ref. 5) which coincide neither with those reported by Sokolov himself in his 1954 paper (Ref. 1) nor with those the evaluator could obtain by plotting Tsindrik´s experimental points (Ref. 2).

Berchiesi et al. (Ref. 3), being aware of Sokolov´s paper (Ref. 1), found two invariant points: a eutectic and one corresponding to the incongruent melting of the intermediate compound $Na_4(CHO_2)_3NO_3$. They supplemented their visual observations with DSC analysis of four mixtures. In the recorded traces they recognized: for x_1= 0.7926, "peaks corresponding to the peritectic transition (477 K) and to complete fusion"; for x_1= 0.7312, "peaks corresponding to the eutectic fusion (464 K), to the peritectic transition (477 K) and to complete fusion"; for x_1= 0.6560, "peaks corresponding to the eutectic fusion and to the peritectic transition"; for x_1= 0.5190, one "peak corresponding to the eutectic fusion". They could also observe in the composition triangle of the ternary Na/CHO_2, CNS, NO_3 a crystallization region belonging to the binary intermediate compound and covering 5.30 % of the liquidus area.

Storonkin et al. (Ref. 4) employed DTA to investigate the same ternary, and once more found, for the binary system of interest here, just one eutectic although at coordinates different from those reported by Sokolov (Ref. 1) and by Tsindrik (Ref. 2); they also claimed the distribution coefficient of $NaCHO_2$ in $NaNO_3$ to be zero in the nitrate crystallization field. Storonkin et al. (Ref. 4) were apparently aware only of a 1971 paper by Sokolov and Khaitina (Ref. 6), where in turn only Sokolov´s 1954 findings (Ref. 1) were quoted.

Finally, it is to be mentioned that the cryometric data of Leonesi et al., 1968 (Ref. 7), proved that the nitrate has no tendency (or at least a negligibly small tendency) to dissolve in the methanoate in the solid state.

In order to evaluate the consistency of the above sets of measurements, the following considerations may be useful.

In any binary system where solid solutions are absent, the branch of the liquidus curve rich in component 1 may often be represented satisfactorily by means of the approximate equation (Ref. 8)

$$\mathbf{T}(1) = \{\mathbf{H}(1)/\mathbf{R} + (\mathbf{A}/\mathbf{R})(\mathbf{x}_2)^2\} \;/\; \{\mathbf{S}(1)/\mathbf{R} - \ln\,(\mathbf{x}_2)\}$$

where A is an empirical constant which of course is zero for ideal systems, and

$$\mathbf{H}(1) = \Delta_{fus}(1)\mathbf{H}_m; \qquad \mathbf{S}(1) = \Delta_{fus}(1)\mathbf{S}_m\ .$$

When $\mathbf{T}(1)$ is between $[\mathbf{T}_{fus}(1)$ and $\mathbf{T}_{trs}(1)]$,

$$\mathbf{H}(1) = \Delta_{fus}(1)\mathbf{H}_m + \Delta_{trs}(1)\mathbf{H}_m; \qquad \mathbf{S}(1) = \Delta_{fus}(1)\mathbf{S}_m + \Delta_{trs}(1)\mathbf{S}_m\ .$$

Taking now the DSC numerical values listed in Table 3 of the Preface, which concern component 1, i.e. sodium methanoate, one obtains for the ideal behaviour the curve denoted as "ideal" in the Figure of the next page.

COMPONENTS:

(1) Sodium methanoate (sodium formate); $NaCHO_2$; [141-53-7]
(2) Sodium nitrate; $NaNO_3$; [7631-99-4]

EVALUATOR:

Ferloni, P.,
Dipartimento di Chimica Fisica,
Universita' di Pavia (ITALY).

CRITICAL EVALUATION (continued):

For the system K/CHO_2, NO_3 Leonesi et al. (Ref. 7) were able to fit their experimental points fairly well for the branch rich in methanoate, when A/R was assigned the value -175 K. In the present binary, formed with the common cation Na and the same pair of different anions, it seemed not unreasonable to expect analogous behavior. Introducing into Eq. (1) the above $\Delta(1)\mathbf{H}_m$ and $\Delta(1)\mathbf{S}_m$ values, and again A/R= -175 K, the "real" curve of the Figure is obtained. It can be seen that Berchiesi et al.'s (Ref. 3) points are the closest to this curves, whereas progressively increasing discrepancies are observed for the data of Tsindrik (Ref. 2), Sokolov (Ref. 1), and Storonkin et al. (Ref. 4) (each temperature being corrected in order to make allowance for the differences in the fusion temperatures of the methanoate given by the different authors).

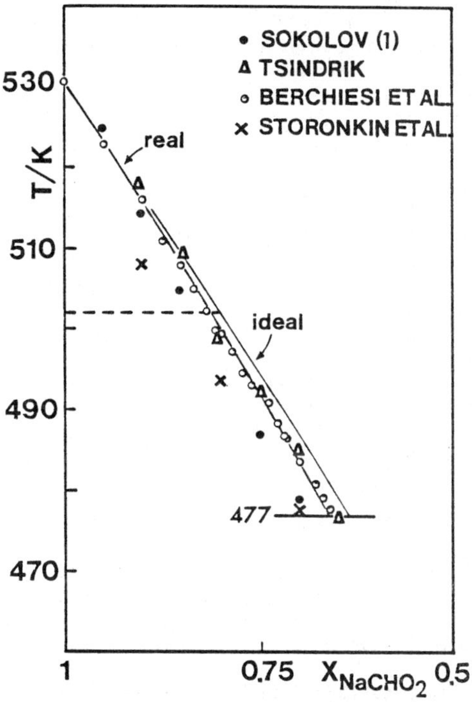

Thus, the evaluator is inclined to recommend (among those available so far) the data by Berchiesi et al. (Ref. 3). The fact that Storonkin et al. (Ref. 4), by employing a DTA technique, where not able to detect the intermediate compound seems rather surprising. This fact, however, might be related to the large supercooling effect found by the latter authors in the region of the ternary eutectic and difficult to prevent also in the region of the binary eutectic. Efficient stirring and slow cooling rate have likely allowed Berchiesi et al. (Ref. 3) to avoid this drawback. The presence of some impurity in Storonkin et al.'s (Ref. 4) methanoate is even possible, inasmuch as their $\mathbf{T}_{fus}(1)/K$ value (528) is some 3 K lower than those reported in Ref.s 1 (531), 2 (531), and 3 (530.65), and in Table 3 [530.46±0.04 (adiabatic calorimetry); 530.7±0.5 (DSC)].

REFERENCES:

(1) Sokolov, N.M.; **Zh. Obshch. Khim.** 1954, **24**, 1150-1156.
(2) Tsindrik, N.M.; **Zh. Obshch. Khim.** 1958, **28**, 830-834.
(3) Berchiesi, M.A.; Cingolani, A.; Berchiesi, G.; **J. Chem. Eng. Data,** 1972, **17**, 61-64.
(4) Storonkin, A.V.; Vasil'kova, I.V.; Potemin, S.S.; **Vestn. Leningr. Univ., Fiz., Khim.** 1974, (10), 84-88.
(5) Sokolov, N.M.; **Tezisy Nauch. Konf. S.M.I.** 1956[a].
(6) Sokolov, N.M.; Khaitina, M.V.; **Zh. Obshch. Khim.** 1971, **41**, 1417.
(7) Leonesi, D.; Piantoni, G.; Berchiesi, G.; Franzosini, P.; **Ric. Sci.** 1968, **38**, 702.
(8) Sinistri, C.; Franzosini, P.; **Ric. Sci.** 1963, **33(II-A)**, 419-430.
(9) Braghetti, M.; Berchiesi, G.; Franzosini, P.; **Ric. Sci.** 1969, **39**, 576.

[a] This quotation as given by Tsindrik (Ref. 2) is probaly to be completed as follows: **Tezisy Dokl. X Nauch. Konf. S.M.I.** 1956. The evaluator did not succeed in obtaining a reprint from the author, but it is highly probable that numerical data are not given in the Tezisy, since such documents usually report only summaries of the discussions held at the pertinent conferences.

COMPONENTS:	ORIGINAL MEASUREMENTS:
(1) Sodium methanoate (sodium formate); $NaCHO_2$; [141-53-7] (2) Sodium nitrate; $NaNO_3$; [7631-99-4]	Sokolov, N.M. **Zh. Obshch. Khim.** 1954, **24**, 1150-1156.
VARIABLES:	PREPARED BY:
Temperature.	D'Andrea, G.

EXPERIMENTAL VALUES:

$t/^{o}C$	T/K^{a}	$100x_2$
258	531	0
252	525	5
242	515	10
232	505	15
214	487	25
206	479	30
198	471	35
192	465	40
188	461	45
186[b]	459	49
190	463	50
206	479	55
220	493	60
235	508	65
262	535	75
284	557	85
302	575	95
308	581	100

[a] T/K values calculated by the compiler.
[b] Eutectic temperature (author).

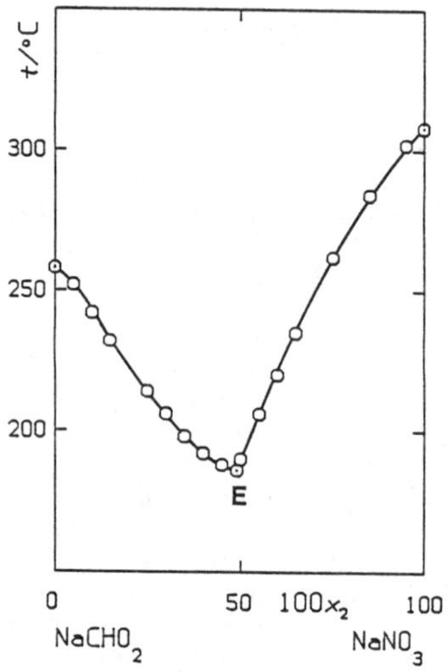

Characteristic point(s):

Eutectic, E, at 186 °C and $100x_2$= 49 (author).

AUXILIARY INFORMATION

METHOD/APPARATUS/PROCEDURE:	SOURCE AND PURITY OF MATERIALS:
Visual polythermal analysis. Salt(s) melted in a test tube. Temperature measured with a Nichrome-Constantane thermocouple and a millivoltmeter with mirror reading to 17 mV.	Component 1 synthetized from methanoic acid and $NaHCO_3$. Commercial component 2 further purified by the author according to Laiti.
	ESTIMATED ERROR:
	Temperature: accuracy probably $\pm$2 K (compiler).
	REFERENCES:

COMPONENTS:

(1) Sodium methanoate (sodium formate); $NaCHO_2$; [141-53-7]
(2) Sodium nitrate; $NaNO_3$; [7631-99-4]

ORIGINAL MEASUREMENTS:

Tsindrik, N.M.
Zh. Obshch. Khim. 1958, **28**, 830-834.

VARIABLES:

Temperature.

PREPARED BY:

D'Andrea, G.

EXPERIMENTAL VALUES:

$t/^{\circ}C$	T/K^a	$100x_2$
258	531	0
252	525	5
245	518	10
236	509	15
226	499	20
219	492	25
212	485	30
204	477	35
196	469	40
188	461	45
192	465	50
210	483	55
226	499	60
240	513	65
250	523	70
260	533	75
270	543	80
278	551	85
288	561	90
298	571	95
308	581	100

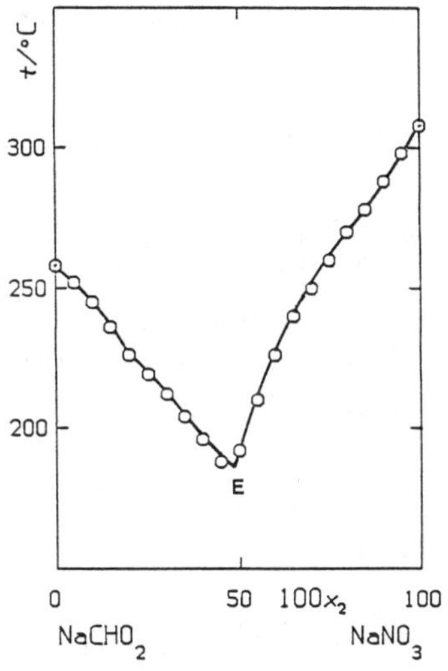

[a] T/K values calculated by the compiler.

Characteristic point(s):

Eutectic, E, at 187 °C and $100x_2$= 48 (author, Ref. 1)

AUXILIARY INFORMATION

METHOD/APPARATUS/PROCEDURE:

Visual polythermal analysis; temperatures measured with a Nichrome-Constantane thermocouple. Salt(s) melted in a test tube, hand-stirred.

SOURCE AND PURITY OF MATERIALS:

Materials of analytical purity twice recrystallized.
Component 1 undergoes a solid state transition at $t_{trs}(1)/^{\circ}C$= 242 (Ref. 1).
Component 2 undergoes a solid state transition at $t_{trs}(2)/^{\circ}C$= 275 (current literature).

ESTIMATED ERROR:

Temperature: accuracy probably ±2 K (compiler).

REFERENCES:

(1) Sokolov, N.M.
Tezisy Nauchn. Konf. S.M.I. 1956.

COMPONENTS:	ORIGINAL MEASUREMENTS:
(1) Sodium methanoate (sodium formate); $NaCHO_2$; [141-53-7] (2) Sodium nitrate; $NaNO_3$; [7631-99-4]	Berchiesi,M.A.; Cingolani,A.; Berchiesi,G. **J. Chem. Eng. Data,** 1972, **17,** 61-64.
VARIABLES:	PREPARED BY:
Temperature.	D'Andrea, G.

EXPERIMENTAL VALUES:

t/°C	T/K[a]	$100x_2$	t/°C	T/K[a]	$100x_2$
257.50	530.65	0	205.85	479.00	33.07
249.50	522.65	5.17	204.50	477.65	33.99
242.73	515.88	9.88	203.50	476.65	35.98
237.58	510.73	12.69	202.13	475.28	38.09
234.43	507.58	15.04	201.25	474.40	39.99
231.65	504.80	16.59	199.25	472.40	42.01
229.03	502.18	18.12	195.38	468.53	44.96
226.58	499.73	19.41	193.50	466.65	46.61
226.13	499.28	20.08	192.10	465.25	48.57
223.98	497.13	21.53	195.75	468.90	49.93
221.35	494.50	22.78	200.87	474.02	51.67
219.85	493.00	24.00	208.93	482.08	55.09
217.73	490.88	26.08	220.70	493.85	60.01
215.08	488.23	27.19	243.03	516.18	70.02
213.55	486.70	28.04	263.95	537.10	79.90
213.05	486.20	28.61	284.98	558.13	90.04
210.30	483.45	30.07	306.00	579.15	100.00
207.54	480.69	32.03			

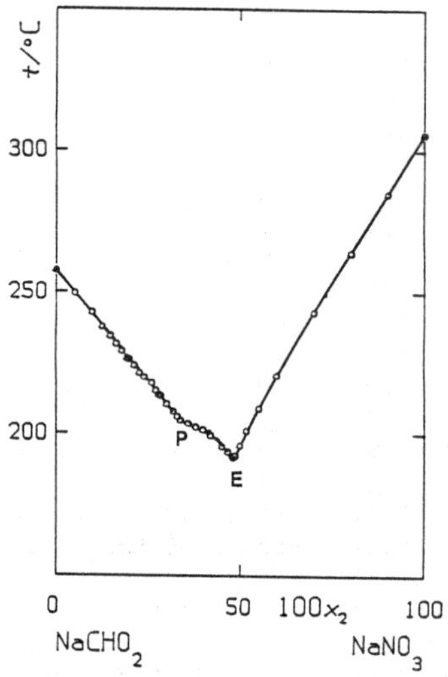

[a] T/K values calculated by the compiler.

Characteristic points: Peritectic, P, at 204 °C and $100x_2$= 34.4 (authors).
Eutectic, E, at 191 °C and $100x_2$= 48.1 (authors).

Intermediate compound: $Na_4(CHO_2)_3NO_3$, incongruently melting (authors).

AUXILIARY INFORMATION

METHOD/APPARATUS/PROCEDURE:

Visual method, supplemented by DSC analysis. Salt(s) melted in a Pyrex device (1) put into a furnace whose temperature was controlled by means of a Chromel-Alumel thermocouple connected with a L&N CAT control unit. Temperature of the melt measured with a second thermocouple checked by comparison with a certified Pt resistance thermometer, and a L&N K-5 potentiometer. Stirring by a Chemap Mod.E-1 Vibro-mixer.

SOURCE AND PURITY OF MATERIALS:

C.Erba (Milano,Italy) $NaCHO_2$ and $NaNO_3$ of stated purity not less than 99% were used after thorough dehydration.

ESTIMATED ERROR:

Temperature: accuracy ±0.03 K (authors).

REFERENCES:

(1) Braghetti,M.; Leonesi,D.; Franzosini,P. **Ric. Sci.** 1968, **38,** 116-118.

COMPONENTS:
(1) Sodium methanoate (sodium formate); $NaCHO_2$; [141-53-7]
(2) Sodium nitrate; $NaNO_3$; [7631-99-4]

ORIGINAL MEASUREMENTS:

Storonkin, A.V.; Vasil'kova, I.V.; Potemin, S.S.
Vestn. Leningr. Univ., Fiz., Khim. 1974, (10), 84-88.

VARIABLES:

Temperature.

PREPARED BY:

D'Andrea, G.

EXPERIMENTAL VALUES:

t/°C	T/K[a]	$100x_2$
255	528	0
232	505	10
218	491	20
202	475	30
182	455	40
190	463	50
215	488	60
234	507	70
252	525	80
276	549	90
306	579	100

[a] T/K values calculated by the compiler.

Note - The data tabulated were drawn by the compiler from Fig. 1 of the original paper.

Characteristic point(s):

Eutectic, E, at 176 °C and $100x_2$= 44 (authors).

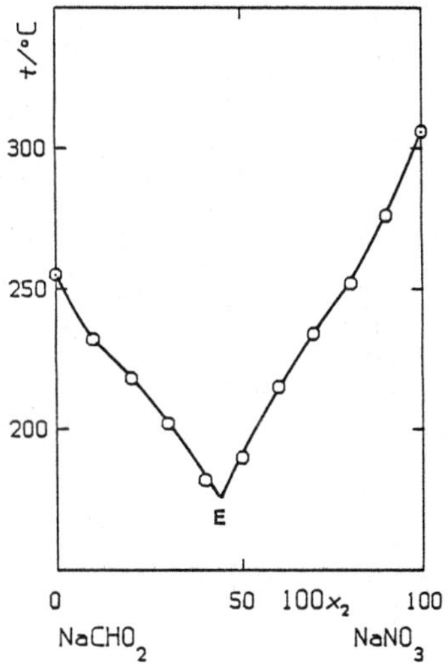

AUXILIARY INFORMATION

METHOD/APPARATUS/PROCEDURE:

DTA: Thermograph with photorecorder. Salt(s) sealed under vacuum in Pyrex ampoules. No other information given.

SOURCE AND PURITY OF MATERIALS:

$NaCHO_2$ of analytical purity and "chemically pure" $NaNO_3$, heated 10-15 h at temperatures 50-60 °C below their fusion temperatures, were employed.

ESTIMATED ERROR:

Temperature: accuracy probably ±2 K (compiler).

REFERENCES:

COMPONENTS:

(1) Sodium ethanoate (sodium acetate); $NaC_2H_3O_2$; [127-09-3]
(2) Sodium propanoate (sodium propionate); $NaC_3H_5O_2$; [137-40-6]

ORIGINAL MEASUREMENTS:

Sokolov, N.M.; Pochtakova, E.I.
Zh. Obshch. Khim. 1958, **28**, 1397-1404.

VARIABLES:

Temperature.

PREPARED BY:

D'Andrea, G.

EXPERIMENTAL VALUES:

$t/^{o}C$	T/K^{a}	$100x_2$
331	604	0
326	599	5
322	595	10
314	587	15
311	584	20
307	580	25
303	576	30
301	574	35
298	571	40
300	573	45
299	572	50
299	572	55
298	571	60
297	570	65
296	569	70
295	568	75
294	567	85
293	566	90
291	564	95
295	568	97.5
298	571	100

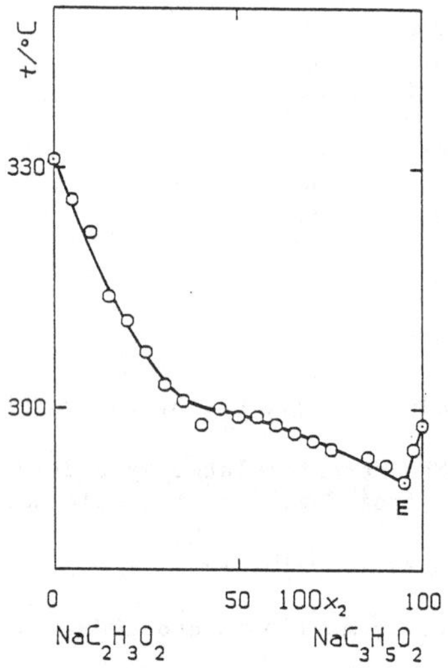

[a] T/K values calculated by the compiler.

Characteristic point(s): Eutectic, E, at 291 ^{o}C and $100x_2$= 95 (authors).

AUXILIARY INFORMATION

METHOD/APPARATUS/PROCEDURE:

Visual polythermal analysis.

NOTE:

The fusion and solid state transition temperatures reported for component 1 (604 and 527 K, respectively) agree reasonably with the $T_{fus}(1)$ and $T'_{trs}(1)$ values (601.3$\pm$0.5 and 527$\pm$15 K, respectively) listed in Table I of the Preface. Concerning component 2, the fusion temperature (571 K) looks, on the contrary, as somewhat too high; moreover, doubts are to be cast about the reliability of the lowest (350 K) and highest (560 K) transition temperatures quoted by the author from Ref. 2, inasmuch as both DSC (Table 1) and adiabatic calorimetry (Table 3) proved the occurrence of solid state transformations only at 491-494 and 467-470 K, respectively.

SOURCE AND PURITY OF MATERIALS:

Component 1: "chemically pure" material; it undergoes a phase transition at $t_{trs}(1)/^{o}C$= 254 (Ref.1).
Component 2: prepared from commercial propanoic acid (distilled before use) and "chemically pure" sodium carbonate; the solid recovered was recrystallized from **n**-butanol; it undergoes phase transitions at $t_{trs}(2)/^{o}C$= 77, 195, 217, 287 (Ref. 2).

ESTIMATED ERROR:

Temperature: accuracy probably $\pm$2 K (compiler).

REFERENCES:

(1) Bergman, A.G.; Evdokimova, K.A. **Izv. Sektora Fiz.-Khim. Anal.** 1956, **27**, 296-314.
(2) Sokolov, N.M.; **Tezisy Dokl. X Nauch. Konf. S.M.I.** 1956.

COMPONENTS:	EVALUATOR:
(1) Sodium ethanoate (sodium acetate); $NaC_2H_3O_2$; [127-09-3] (2) Sodium butanoate (sodium butyrate); $NaC_4H_7O_2$; [156-54-7]	Ferloni, P., Dipartimento di Chimica Fisica, Universita' di Pavia (ITALY).

CRITICAL EVALUATION:

The visual polythermal method was employed by Sokolov (Ref. 1) to study the lower boundary of the isotropic liquid field: the results were subsequently reviewed by Sokolov and Pochtakova (Ref. 2). According to these authors, the [congruently melting at 546 K (273 °C)] intermediate compound $Na_5(C_2H_3O_2)_3(C_4H_7O_2)_2$ is formed, and two invariants exist, i.e., a eutectic E_1 [at 539 K (266 °C), and $100\mathbf{x}_2$= 33.5], and a eutectic E_2 [at 523 K (250 °C), and $100\mathbf{x}_2$= 69].

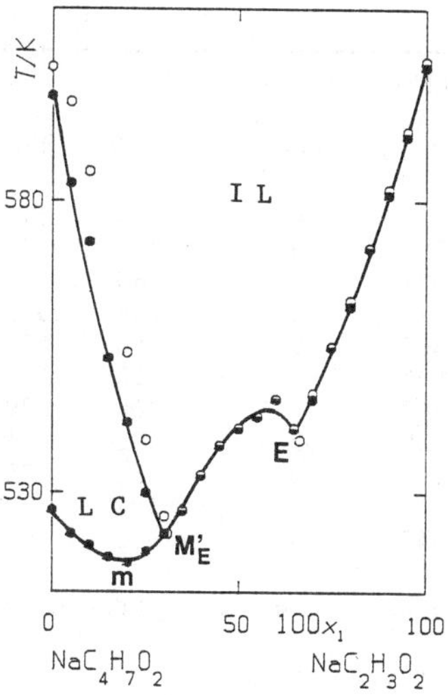

Component 2, however, forms liquid crystals. Accordingly, the fusion temperature, $\mathbf{T}_{fus}(2)$= 603 K (330 °C), reported in Refs. 1, 2 should be identified with the clearing temperature, $\mathbf{T}_{clr}(2)$, of component 2, the corresponding value from Table 1 of the Preface being 600.4±0.2 K.

For the same component, Table 1 of the Preface [besides the $\mathbf{T}_{clr}(2)$ value] provides the values 450.4±0.5, 489.8±0.2, 498.3±0.3, and 508.4±0.5 K respectively, for $\mathbf{T}^{iv}_{trs}(2)$ to $\mathbf{T}^{i}_{trs}(2)$, and $\mathbf{T}_{fus}(2)$/K= 524.5±0.5. These phase relations, first stated on the basis of DSC records, were subsequently confirmed by Schiraldi and Chiodelli's conductometric results (Ref. 3). Phase transformations are quoted in Ref. 2 from Ref. 4 as occurring at 390, 505, 525, and 589 K, respectively. A comparison of the two sets of data allows one to compare the two intermediate transition temperatures from Ref. 4 with $\mathbf{T}^{i}_{trs}(2)$ and $\mathbf{T}_{fus}(2)$ from Table 1 of the Preface. Reasonable doubts can be cast, on the contrary, about the actual existence of Ref. 4 highest and lowest transformations (the former - if present - ought to represent the transformation from one liquid crystalline phase into another).

More recently, Prisyazhnyi et al. (Ref. 5) - to whom Refs. 1, 2 seem to be unknown - carried out a derivatographical re-investigation of the system, which allowed them to draw the lower boundaries of both the isotropic liquid, and the liquid crystal field. Their clearing [$\mathbf{T}_{clr}(2)$= 598 K (325 °C)] and fusion [$\mathbf{T}_{fus}(1)$= 603 K (330 °C); $\mathbf{T}_{fus}(2)$= 527 K (254 °C)] temperatures substantially agree with those from Table 1 of the Preface; moreover, it is to be stressed that they do not mention any transition intermediate between $\mathbf{T}_{clr}(2)$ and $\mathbf{T}_{fus}(2)$.

Prisyazhnyi et al.'s, and Sokolov's results (filled and empty circles, respectively) are compared in the figure (IL: isotropic liquid; LC: liquid crystals), an inspection of which allows one to state that: (i) the invariant at about $100\mathbf{x}_2$= 70 is actually an M'_E point, and (ii) a further characteristic point exists (at about $100\mathbf{x}_2$= 80) which escaped Sokolov's attention, and is probably a minimum, m, in a series of solid solutions. Prisyazhnyi et al.'s results suggest at $0 \leq 100\mathbf{x}_1 \leq 60$ a behavior similar to that shown in Scheme A.3 of the Preface.

The two two-phase regions pertinent to the liquid crystal - isotropic liquid equilibria, and to solid solutions formation, respectively, might be so narrow as to have prevented Prisyazhnyi et al. to observe two distinct sets of points in each of these regions, whereas one cannot explain the lack of information by the same authors about eutectic fusion at $60 \leq 100\mathbf{x}_1 \leq 100$.

REFERENCES:

(1) Sokolov, N.M.; **Zh. Obshch. Khim.** 1954, 24, 1581-1593.
(2) Sokolov, N.M.; Pochtakova, E.I.; **Zh. Obshch. Khim.** 1960, **30**, 1401-1405 (*); **Russ. J. Gen. Chem.(Engl. Transl.)** 1960, **30**, 1429-1433.
(3) Schiraldi, A.; Chiodelli, G.; **J. Phys. E: Sci. Instr.** 1977, 10, 596-599.
(4) Sokolov, N.M. ; **Tezisy Dokl. X Nauch. Konf. S.M.I.** 1956.
(5) Prisyazhnyi, V.D.; Mirnyi, V.N.; Mirnaya, T.A.; **Zh. Neorg. Khim.** 1983, **28**, 253-255; **Russ. J. Inorg. Chem. (Engl. Transl.)** 1983, 28, 140-141 (*).

COMPONENTS:	ORIGINAL MEASUREMENTS:
(1) Sodium ethanoate (sodium acetate); $NaC_2H_3O_2$; [127-09-3] (2) Sodium butanoate (sodium butyrate); $NaC_4H_7O_2$; [156-54-7]	Sokolov, N.M. **Zh. Obshch. Khim.** 1954, **24**, 1581-1593.
VARIABLES:	PREPARED BY:
Temperature.	D'Andrea, G.

EXPERIMENTAL VALUES:

t/°C	T/K[a]	$100x_2$	t/°C	T/K[a]	$100x_2$
331	604	0	268	541	50
319	592	5	265	538	55
309	582	10	260	533	60
299	572	15	254	527	65
290	563	20	250	523	69
282	555	25	253	526	70
274	547	30	266	539	75
266	539	33.5	281	554	80
268	541	35	312	585	90
273	546	40	324	597	95
270	543	45	330	603	100

[a] T/K values calculated by the compiler.

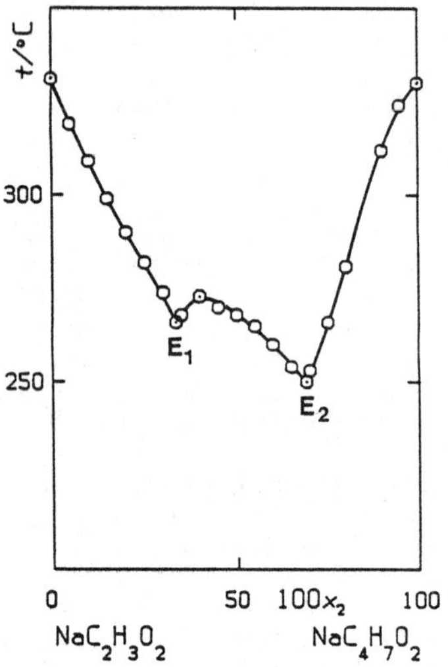

Characteristic point(s):

Eutectic, E_1, at 266 °C and $100x_2$= 33.5 (author).
Eutectic, E_2, at 250 °C and $100x_2$= 69 (author).

Intermediate compound:

$Na_5(C_2H_3O_2)_3(C_4H_7O_2)_2$ congruently melting at 273 °C.

AUXILIARY INFORMATION

METHOD/APPARATUS/PROCEDURE:

Visual polythermal analysis. Melts contained in a glass tube and stirred. Temperatures measured with a Nichrome-Constantane thermocouple and a 17 mV full scale millivoltmeter. The temperature readings refer to the disappearance of isotropicity in the melt on cooling.

SOURCE AND PURITY OF MATERIALS:

Component 1: "chemically pure" material. Component 2: prepared by reacting aqueous ("chemically pure") Na_2CO_3 with a slight excess of n-butanoic acid of analytical purity. The solvent and excess acid were removed by heating to 160 °C.

ESTIMATED ERROR:

Temperature: accuracy probably ±2 K (compiler).

REFERENCES:

COMPONENTS:

(1) Sodium ethanoate (sodium acetate); $NaC_2H_3O_2$; [127-09-3]
(2) Sodium butanoate (sodium butyrate); $NaC_4H_7O_2$; [156-54-7]

ORIGINAL MEASUREMENTS:

Sokolov, N.M.; Pochtakova, E.I.
Zh. Obshch. Khim. 1960, 30, 1401-1405 (*);
Russ. J. Gen. Chem. (Engl. Transl.) 1960, 30, 1429-1433.

VARIABLES:

Temperature.

PREPARED BY:

D'Andrea, G.

EXPERIMENTAL VALUES:

Characteristic point(s):

Eutectic, E_1, at 266 °C and $100\mathbf{x}_2$= 33.5 (authors).
Eutectic, E_2, at 250 °C and $100\mathbf{x}_2$= 69 (authors).

Intermediate compound(s):

$Na_5(C_2H_3O_2)_3(C_4H_7O_2)_2$, congruently melting at 273 °C.

AUXILIARY INFORMATION

METHOD/APPARATUS/PROCEDURE:

Visual polythermal analysis.

SOURCE AND PURITY OF MATERIALS:

Component 1: "chemically pure" material recrystallized; it undergoes a phase transition at $t_{trs}(1)/^{\circ}C$= 254 (Ref. 1), and melts at $t_{fus}(1)/^{\circ}C$= 331.
Component 2: prepared by reacting $NaHCO_3$ with n-butanoic acid, and recrystallized from n-butanol (Ref. 2, where, however, carbonate insted of hydrogen carbonate was employed; compiler); it undergoes phase transitions at $t_{trs}(2)/^{\circ}C$= 117, 232, 252, 316 (Ref. 3), and melts at $t_{fus}(2)/^{\circ}C$= 330.

ESTIMATED ERROR:

Temperature: accuracy probably ±2 K (compiler).

REFERENCES:

(1) Bergman, A.G.; Evdokimova, K.A. **Izv. Sektora Fiz.-Khim. Anal.** 1956, 27 296-314.
(2) Sokolov, N.M. **Zh. Obshch. Khim.** 1954, 24, 1581-1593.
(3) Sokolov, N.M. **Tezisy Dokl. X Nauch. Konf. S.M.I.** 1956.

COMPONENTS:

(1) Sodium ethanoate (sodium acetate); $NaC_2H_3O_2$; [127-09-3]
(2) Sodium butanoate (sodium butyrate); $NaC_4H_7O_2$; [156-54-7]

ORIGINAL MEASUREMENTS:

Prisyazhnyi, V.D.; Mirnyi, V.N.; Mirnaya, T.A. **Zh. Neorg. Khim.** 1983, **28**, 253-255; **Russ. J. Inorg. Chem. (Engl. Transl.)** 1983, **28**, 140-141 (*).

VARIABLES:

Temperature.

PREPARED BY:

D'Andrea, G.

EXPERIMENTAL VALUES:

The results are reported only in graphical form (see figure; data read with a digitizer by the compiler from Fig. 1 of the original paper; empty circles: liquid crystal - isotropic liquid equilibria; filled circles: solid - liquid crystal or solid - isotropic liquid equilibria).

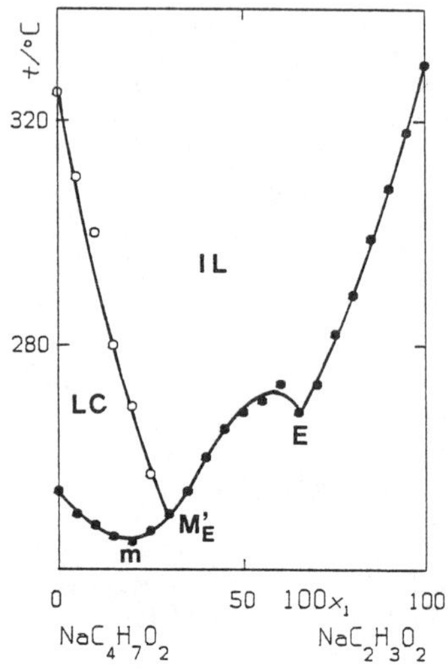

Characteristic point(s):

Eutectic, E, at about 268 °C and $100\mathbf{x}_1$ about 65 (compiler).
Minimum, m, at about 245 °C and $100\mathbf{x}_1$ about 20 (compiler).
Invariant point, M'_E, at about 250 °C and $100\mathbf{x}_1$ about 30 (compiler).

Intermediate compound(s):

$Na_5(C_2H_3O_2)_3(C_4H_7O_2)_2$, congruently melting at about 273 °C (compiler).

AUXILIARY INFORMATION

METHOD/APPARATUS/PROCEDURE:

The heating and cooling traces were recorded in an atmosphere of purified argon with an OD-102 derivatograph (MOM, Hungary) working at a rate of 6-8 K min^{-1}, and using Al_2O_3 as the reference material. Temperatures were measured with a Pt/Pt-Rh thermocouple. A hot-stage Amplival polarizing microscope was employed to detect the transformation points from the liquid crystalline into the isotropic liquid phase.

SOURCE AND PURITY OF MATERIALS:

Not stated.
Component 1: $t_{fus}(1)/°C$ about 329 (compiler).
Component 2: $t_{fus}(2)/°C$ about 254; $t_{clr}(2)/°C$ about 325 (compiler).

ESTIMATED ERROR:

Temperature: accuracy not evaluable (compiler).

REFERENCES:

COMPONENTS:	ORIGINAL MEASUREMENTS:
(1) Sodium ethanoate (sodium acetate); $NaC_2H_3O_2$; [127-09-3] (2) Sodium **iso**.butanoate (sodium **iso**.butyrate); $Na\textbf{i}.C_4H_7O_2$; [996-30-5]	Sokolov, N.M. **Zh. Obshch. Khim.** 1954, **24**, 1581-1593.
VARIABLES:	PREPARED BY:
Temperature.	D'Andrea, G.

EXPERIMENTAL VALUES:

$t/^{o}C$	T/K^{a}	$100x_2$
331	604	0
323	596	5
314	587	10
305	578	15
297	570	20
288	561	25
277	550	30
265	538	35
254	527	40
242	515	45
230	503	50
218	491	55
208	481	58
215	488	60
226	499	65
236	509	70
242	515	75
246	519	80
250	523	85
254	527	90
257	530	95
260	533	100

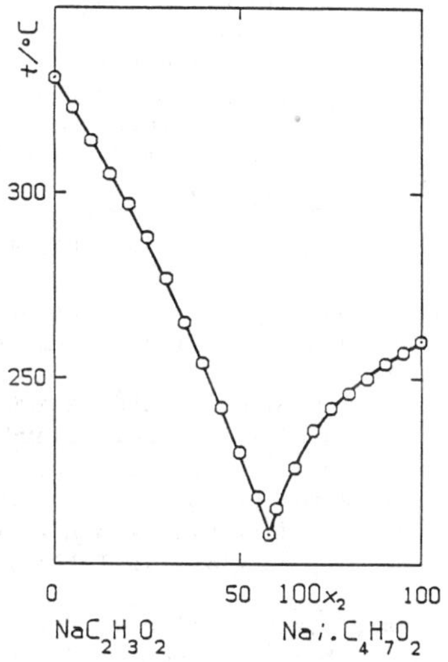

[a] **T**/K values calculated by the compiler.

Characteristic point: Eutectic, E, at 208 °C and $100x_2$= 58 (author).

AUXILIARY INFORMATION

METHOD/APPARATUS/PROCEDURE:

Visual polythermal analysis. Melts contained in a glass tube and stirred. Temperatures measured with a Nichrome-Constantane thermocouple and a 17 mV full scale millivoltmeter. The temperature readings refer to the disappearance of isotropicity in the melt on cooling.

SOURCE AND PURITY OF MATERIALS:

Component 1: "chemically pure" material. Component 2: prepared by reacting aqueous ("chemically pure") Na_2CO_3 with a slight excess of **iso**.butanoic acid of analytical purity. The solvent and excess acid were removed by heating to 160 °C.

ESTIMATED ERROR:

Temperature: accuracy probably ±2 K (compiler).

REFERENCES:

COMPONENTS:

(1) Sodium ethanoate (sodium acetate); $NaC_2H_3O_2$; [127-09-3]
(2) Sodium pentanoate (sodium valerate); $NaC_5H_9O_2$; [6106-41-8]

EVALUATOR:

Ferloni, P.,
Dipartimento di Chimica Fisica,
Universita´ di Pavia (ITALY).

CRITICAL EVALUATION:

This system was studied only by Pochtakova (Ref. 1), who claimed the existence of: (i) a eutectic, E_1, at 537 K (264 ^oC) and $100x_2$= 31.5; (ii) a eutectic, E_2, at 526 K (253 ^oC) and $100x_2$= 54; and (iii) an intermediate compound, $Na_3(C_2H_3O_2)_2C_5H_9O_2$, congruently melting at 541 K (268 ^oC).

Component 2, however, forms liquid crystals. Therefore, the fusion temperature reported in Ref. 1, $T_{fus}(2)$= 630 K (357 ^oC), has to be identified with the clearing temperature, the corresponding value from Table 1 of the Preface being $T_{clr}(2)$= 631±4 K. This Table provides also $T_{fus}(2)$= 498±2 K, a figure which can be identified (even if not fully satisfactorily) with that (489 K) corresponding to the highest phase transformation temperature quoted by Pochtakova from Ref. 2. For the same component, Table 1 of the Preface reports no solid state transition, whereas Pochtakova quotes (from Ref. 2) $T_{trs}(2)$/K= 482 and 453. It is, however, to be stressed that the single transition observed (at 479±1 K) with DTA in sodium **n**-pentanoate by Duruz et al. (Ref. 3) was no more mentioned in a subsequent DSC investigation by the same group (Ref. 5).

Concerning component 1, the fusion temperature, $T_{fus}(1)$= 604 K (331 ^oC; Ref. 1), is reasonably identified with the corresponding value from Table 1 of the Preface, viz., 601.3±0.5 K.
Allowance being made for the remarkable discrepancy, one might also connect the phase transition quoted from Ref. 2 and occurring at 511 K (238 ^oC) with that at 527±15 K reported in Table 1 of the Preface.
No reasonable correspondence, however, can be hazarded between the other T_{trs} values quoted from Ref. 2 [viz., 403 K (130 ^oC), 391 K (118 ^oC), and 331 K (58 ^oC)] and the superambient T_{trs}´s given in Table 1.

On the basis of the available data, the phase diagram of this system could be supposed to be similar to that shown in Scheme D.1 of the Preface, Pochtakova´s eutectic E_2 being intended as an M'_E point.

REFERENCES:

(1) Pochtakova, E.I.
Zh. Obshch. Khim. 1966, **36**, 3-8.

(2) Sokolov, N.M.
Tezisy Dokl. X Nauch. Konf. S.M.I. 1956.

(3) Duruz, J.J.; Michels, H.J.; Ubbelohde, A.R.
Proc. R. Soc. London 1971, **A322**, 281-299.

(4) Michels, H.J.; Ubbelohde, A.R.
JCS Perkin II 1972, 1879-1881.

COMPONENTS:

(1) Sodium ethanoate (sodium acetate); $NaC_2H_3O_2$; [127-09-3]
(2) Sodium pentanoate (sodium valerate); $NaC_5H_9O_2$; [6106-41-8]

ORIGINAL MEASUREMENTS:

Pochtakova, E.I.
Zh. Obshch. Khim. 1966, **36**, 3-8.

VARIABLES:

Temperature.

PREPARED BY:

D'Andrea, G.

EXPERIMENTAL VALUES:

$t/^{\circ}C$	T/K^a	$100x_2$	$t/^{\circ}C$	T/K^a	$100x_2$
331	604	0	262	535	45
321	594	5	258	531	50
309	582	15	254	527	52.5
298	571	20	253	526	54
280	553	27.5	254	527	55
272	545	30	266	539	57.5
264	537	31.5	277	550	60
266	539	32.5	315	588	70
268	541	33	339	612	80
267	540	35	351	624	90
265	538	37.5	357	630	100
264	537	40			

[a] **T**/K values calculated by the compiler.

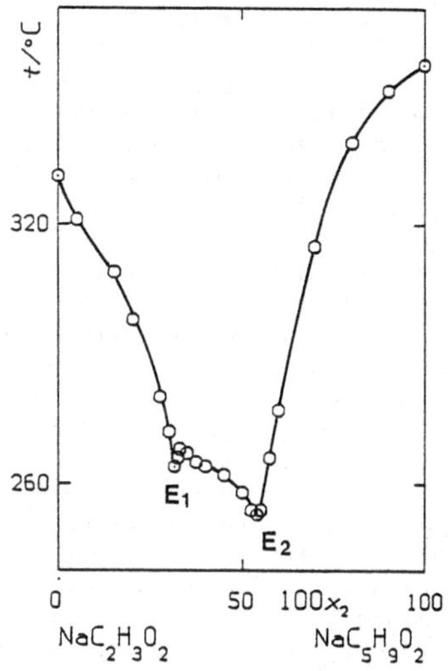

Characteristic points:

Eutectic, E_1 at 264 °C and $100x_2$= 31.5 (author).
Eutectic, E_2 at 253 °C and $100x_2$= 54 (author).

Intermediate compound:

$Na_3(C_2H_3O_2)_2C_5H_9O_2$, congruently melting at 268 °C (author).

AUXILIARY INFORMATION

METHOD/APPARATUS/PROCEDURE:

Visual polythermal analysis.

SOURCE AND PURITY OF MATERIALS:

Component 1: "chemically pure" material. Component 2: prepared from n-pentanoic acid and the hydrogen carbonate (Ref. 1, where, however, carbonate instead of hydrogen carbonate was employed; compiler). Component 1 undergoes phase transitions at $t_{trs}(1)/^{\circ}C$= 58, 118, 130, 238 (Ref. 2). Component 2 undergoes phase transitions at $t_{trs}(2)/^{\circ}C$= 180, 209, 216 (Ref. 2).

ESTIMATED ERROR:

Temperature: accuracy probably ±2 K (compiler).

REFERENCES:

(1) Sokolov, N.M.
Zh. Obshch. Khim. 1954, **24**, 1581-1593.
(2) Sokolov, N.M.
Tezisy Dokl. X Nauch. Konf. S.M.I. 1956

COMPONENTS:

(1) Sodium ethanoate (sodium acetate); $NaC_2H_3O_2$; [127-09-3]
(2) Sodium **iso**.pentanoate (sodium **iso**.valerate); $Na\mathbf{i}.C_5H_9O_2$; [539-66-2]

EVALUATOR:

Ferloni, P.,
Dipartimento di Chimica Fisica,
Universita´ di Pavia (ITALY).

CRITICAL EVALUATION:

This system was studied by Sokolov (Ref. 1), and by Pochtakova (Ref. 2) who reviewed Sokolov´s results. Both of them suggested the phase diagram to be of the eutectic type, with the invariant point at either 429 K (156 °C) and $100\mathbf{x}_2$= 73 (Ref. 1), or 433 K (160 °C) and $100\mathbf{x}_2$= 80.0 (Ref. 2).

Component 2, however, forms liquid crystals. Therefore, the fusion temperature, $\mathbf{T}_{fus}(2)$= 535 K (262 °C; Ref. 1) or 533 K (260 °C; Ref. 2), should be identified with the clearing temperature, the corresponding value from Table 2 of the Preface being $\mathbf{T}_{clr}(2)$= 559$\pm$1 K. The remarkable discrepancy between the latter value and the former ones might be attributed to some impurity in the samples of the Russian authors, inasmuch as the value from Table 2 meets rather satisfactorily those reported by Ubbelohde et al. (556 K; Ref. 3) and by Duruz et al. (553 K; Ref. 4).

For the same component, Pochtakova quotes from Ref. 5 two phase transition temperatures, viz., 451 K (178 °C), and 425 K (152 °C). The higher one can be reasonably identified with the actual fusion temperature, and compared with the value $\mathbf{T}_{fus}(2)$= 461.5$\pm$0.6 K reported in Table 2 of the Preface, whereas the lower one has no correspondence in the same Table.

Both authors report $\mathbf{T}_{fus}(1)$= 604 K (331 °C; Ref.s 1, 2), which may be satisfactorily identified with the value from Table 1 of the Preface, viz., 601.3$\pm$0.5 K. Allowance being made for the discrepancy, one might also connect the phase transition quoted (by Pochtakova) from Ref. 5 as occurring at 511 K (238 °C), with that at 527$\pm$15 K reported in Table 1. No reasonable correspondence, however, can be hazarded between the other $\mathbf{T}_{trs}$ values quoted by Pochtakova from Ref. 5 [viz., 403 K (130 °C), 391 K (118 °C), and 331 K (58 °C)] and the superambient $\mathbf{T}_{trs}$´s given in Table 1.

Taking into account the available experimental data, one may suggest that the phase diagram of this system should not be far from those shown either in Scheme A.1, or in Scheme A.3 of the Preface, the eutectic being actually intended as an M'_E point.

REFERENCES:

(1) Sokolov, N.M.
Zh. Obshch. Khim. 1954, **24**, 1581-1593.

(2) Pochtakova, E.I.
Zh. Obshch. Khim. 1963, **33**, 342-347.

(3) Ubbelohde, A.R.; Michels, H.J.; Duruz, J.J.
Nature 1970, **228**, 50-52.

(4) Duruz,J.J.; Michels,H.J.; Ubbelohde,A.R.
Proc. R. Soc. London 1971, **A 322**, 281-299.

(5) Sokolov, N.M.
Tezisy Dokl. X Nauch. Konf. S.M.I. 1956.

COMPONENTS:	ORIGINAL MEASUREMENTS:
(1) Sodium ethanoate (sodium acetate); $NaC_2H_3O_2$; [127-09-3] (2) Sodium **iso**.pentanoate (sodium **iso**.valerate); $Na\mathbf{i}.C_5H_9O_2$; [539-66-2]	Sokolov, N.M. **Zh. Obshch. Khim.** 1954, 24, 1581-1593.
VARIABLES:	PREPARED BY:
Temperature.	D'Andrea, G.

EXPERIMENTAL VALUES:

t/°C	T/K[a]	$100x_2$
331	604	0
320	593	5
311	584	10
304	577	15
295	568	20
287	560	25
280	553	30
269	542	35
260	533	40
248	521	45
232	505	50
215	488	55
199	472	60
184	457	65
166	439	70
156	429	73
163	436	75
185	458	80
207	480	85
228	501	90
247	520	95
262	535	100

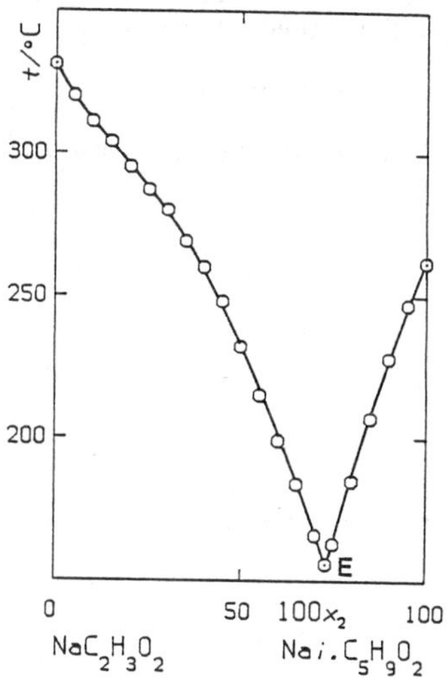

[a] T/K values calculated by the compiler.

Characteristic point: Eutectic, E, at 156 °C and $100x_2$= 73 (author).

AUXILIARY INFORMATION

METHOD/APPARATUS/PROCEDURE:

Visual polythermal analysis. Melts contained in a glass tube and stirred. Temperatures measured with a Nichrome-Constantane thermocouple and a 17 mV full scale millivoltmeter. The temperature readings refer to the disappearance of isotropicity in the melt on cooling.

SOURCE AND PURITY OF MATERIALS:

Component 1: "chemically pure" material.
Component 2: prepared by reacting aqueous ("chemically pure") Na_2CO_3 with a slight excess of **iso**.pentanoic acid of analytical purity. The solvent and excess acid were removed by heating to 160 °C.

ESTIMATED ERROR:

Temperature: accuracy probably ±2 K (compiler).

REFERENCES:

COMPONENTS:

(1) Sodium ethanoate (sodium acetate); $NaC_2H_3O_2$; [127-09-3]
(2) Sodium **iso**.pentanoate (sodium **iso**.valerate); $Na\mathbf{i}.C_5H_9O_2$; [539-66-2]

ORIGINAL MEASUREMENTS:

Pochtakova, E.I.
Zh. Obshch. Khim. 1963, **33**, 342-347.

VARIABLES:

Temperature.

PREPARED BY:

Baldini, P.

EXPERIMENTAL VALUES:

The results are reported only in graphical form (see figure).

Characteristic point(s):

Eutectic, E, at 160 °C and $100\mathbf{x}_2$= 80.0.

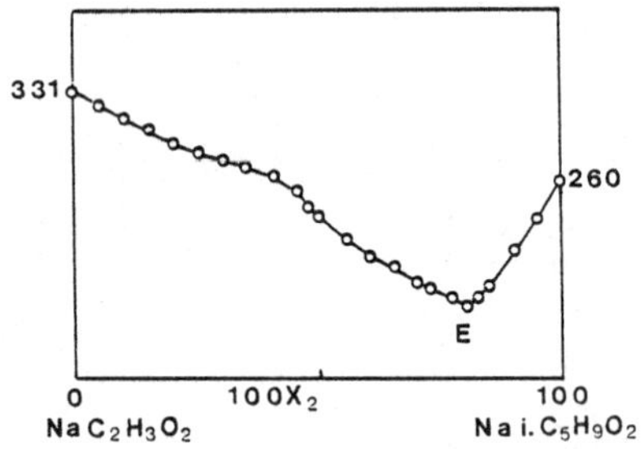

AUXILIARY INFORMATION

METHOD/APPARATUS/PROCEDURE:

Visual polythermal analysis.

SOURCE AND PURITY OF MATERIALS:

Component 1: "chemically pure" material. Component 2: prepared from commercial **iso**.pentanoic acid (distilled twice before use) and the "chemically pure" hydrogen carbonate (Ref. 1).
Component 1 undergoes phase transitions at $\mathbf{t}_{trs}(1)/^{o}C$= 58, 118, 130, 238 (Ref. 2) and melts at $\mathbf{t}_{fus}(1)/^{o}C$= 331.
Component 2 undergoes phase transitions at $\mathbf{t}_{trs}(2)/^{o}C$= 152, 178 (Ref. 2) and melts at $\mathbf{t}_{fus}(2)/^{o}C$= 260.

ESTIMATED ERROR:

Temperature: precision probably ±2 K (compiler).

REFERENCES:

(1) Sokolov, N.M.
Zh. Obshch. Khim. 1954, **24**, 1581-1593.
(2) Sokolov, N.M.
Tezisy Dokl. X Nauch. Konf. S.M.I. 1956.

COMPONENTS:	EVALUATOR:
(1) Sodium ethanoate (sodium acetate); $NaC_2H_3O_2$; [127-09-3] (2) Sodium hexanoate (sodium caproate); $NaC_6H_{11}O_2$; [10051-44-2]	Ferloni, P., Dipartimento di Chimica Fisica, Universita´ di Pavia (ITALY).

CRITICAL EVALUATION:

This system was studied by Sokolov (Ref. 1), and by Pochtakova (Ref. 2). The former author claims the existence of two eutectics [E_1, at 541 K (268 °C) and $100\mathbf{x}_2$= 34.5; E_2, at 533 K (260 °C) and $100\mathbf{x}_2$= 49.5], and of the intermediate compound $Na_8(C_2H_3O_2)_5(C_6H_{11}O_2)_3$, which congruently melts at 543 K (270 °C). The latter author claims in turn the existence of a eutectic [at 546 K (273 °C) and $100\mathbf{x}_2$= 48.5], the incongruently melting compound $Na_5(C_2H_3O_2)_4C_6H_{11}O_2$, and a "perekhodnaya tochka" [at 550 K (277 °C) and $100\mathbf{x}_2$= 34.0].

Component 2, however, forms liquid crystals. Therefore, the fusion temperature, $\mathbf{T}_{fus}(2)$= 638 K (365 °C; Ref.s 1, 2), should be identified with the clearing temperature, the corresponding value from Table 1 of the Preface being $\mathbf{T}_{clr}(2)$= 639.0±0.5 K. The transition temperature $\mathbf{T}_{trs}(2)$= 499 K (226 °C) quoted by Pochtakova from Ref. 3 has in turn to be intended as the fusion temperature, the corresponding value from Table 1 being 499.6±0.6 K.

The following point also deserves attention. Two more transitions are quoted in Ref. 2 from Ref. 3 as occurring in component 2 at 615 K (342 °C) and 476 K (203 °C), respectively. The latter agrees with that reported at 473±2 K in Table 1, whereas no evidence was obtained by subsequent investigators (Ref. 4) for a transition comparable with the former one: should it exist, it might mean that two different mesomorphic phases are present in sodium hexanoate.

As for component 1, Sokolov and Pochtakova report $\mathbf{T}_{fus}(1)$= 603 K (330 °C) and 604 K (331 °C), respectively, i.e., values which favorably meet that from Table 1 (601.3±0.5 K). For the same component, Pochtakova quotes from Ref. 3 a few other phase transition temperatures, viz., 511 K (238 °C), 403 K (130 °C), 391 K (118 °C), and 331 K (58 °C), of which only the first one finds some correspondence with one of the $\mathbf{T}_{trs}$ values from Table 1, i.e., $\mathbf{T}'_{trs}$= 527±15 K.

In conclusion, either author´s suggestions for the phase diagram require modifications. Indeed, the invariant occurring at 533 K and $100\mathbf{x}_2$= 49.5 (Ref. 1), or at 546 K and $100\mathbf{x}_2$= 48.5 (Ref. 2), should likely be identified with an M'_E point, the actual coordinates of which, however, should be verified with better accuracy. Moreover, the composition of the intermediate compound and the nature of the second invariant are not sufficiently supported by the available data, and need as well a further investigation, e.g., by DSC or DTA.

REFERENCES:

(1) Sokolov, N.M.
Zh. Obshch. Khim. 1954, **24**, 1581-1593.

(2) Pochtakova, E.I.
Zh. Obshch. Khim. 1959, **29**, 3183-3189 (*); **Russ. J. Gen. Chem. (Engl. Transl.)** 1959, **29**, 3149-3154.

(3) Sokolov, N.M.
Tezisy Dokl. X Nauch. Konf. S.M.I. 1956.

(4) Sanesi, M.; Cingolani, A.; Tonelli, P.L.; Franzosini, P.
Thermal Properties, in **Thermodynamic and Transport Properties of Organic Salts**, IUPAC Chemical Data Series No. 28 (Franzosini, P.; Sanesi, M.; Editors), Pergamon Press, Oxford, 1980, 29-115.

COMPONENTS:

(1) Sodium ethanoate (sodium acetate); $NaC_2H_3O_2$; [127-09-3]
(2) Sodium hexanoate (sodium caproate); $NaC_6H_{11}O_2$; [10051-44-2]

ORIGINAL MEASUREMENTS:

Sokolov, N.M.
Zh. Obshch. Khim. 1954, **24**, 1581-1593.

VARIABLES:

Temperature.

PREPARED BY:

D'Andrea, G.

EXPERIMENTAL VALUES:

$t/^oC$	T/K^a	$100x_2$	$t/^oC$	T/K^a	$100x_2$
331	604	0	260	533	49.5
321	594	5	265	538	50
312	585	10	300	573	55
304	577	15	321	594	60
296	569	20	332	605	65
288	561	25	342	615	70
279	552	30	349	622	75
268	541	34.5	353	626	80
269	542	35	360	633	90
269	542	40	363	636	95
265	538	45	365	638	100

[a] T/K values calculated by the compiler.

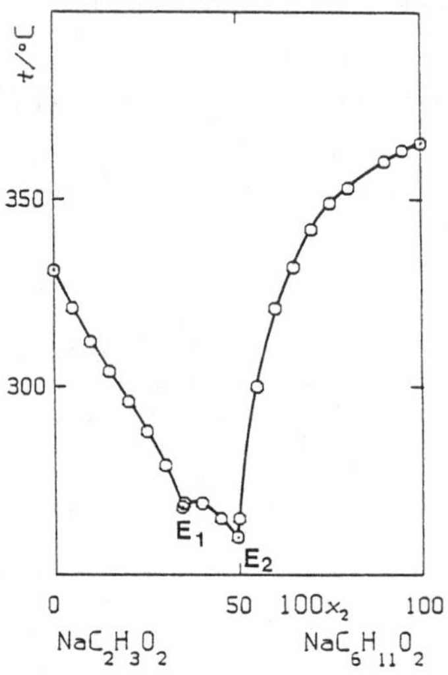

Characteristic point(s):

Eutectic, E_1, at 268 ^oC and $100x_2$= 34.5 (author).
Eutectic, E_2, at 260 ^oC and $100x_2$= 49.5 (author).

Intermediate compound(s):

$Na_8(C_2H_3O_2)_5(C_6H_{11}O_2)_3$ (author), congruently melting at 270 ^oC (compiler).

AUXILIARY INFORMATION

METHOD/APPARATUS/PROCEDURE:

Visual polythermal analysis.
Melts contained in a glass tube and stirred. Temperatures measured with a Nichrome-Constantane thermocouple and a 17 mV full scale millivoltmeter. The temperature readings refer to the disappearance of isotropicity in the melt on cooling.

SOURCE AND PURITY OF MATERIALS:

Component 1: "chemically pure" material.
Component 2: prepared by reacting aqueous ("chemically pure") Na_2CO_3 with a slight excess of n-hexanoic acid of analytical purity. The solvent and excess acid were removed by heating to 160 ^oC.

ESTIMATED ERROR:

Temperature: accuracy probably ± 2 K (compiler).

REFERENCES:

COMPONENTS:

(1) Sodium ethanoate (sodium acetate); $NaC_2H_3O_2$; [127-09-3]
(2) Sodium hexanoate (sodium caproate); $NaC_6H_{11}O_2$; [10051-44-2]

ORIGINAL MEASUREMENTS:

Pochtakova, E.I.
Zh. Obshch. Khim. 1959, **29**, 3183-3189 (*); **Russ. J. Gen. Chem. (Engl. Transl.)** 1959, **29**, 3149-3154.

VARIABLES:

Temperature.

PREPARED BY:

D'Andrea, G.

EXPERIMENTAL VALUES:

The results are reported only in graphical form (see figure).

Characteristic point(s):

Eutectic, E at 273 ^oC and $100\mathbf{x}_2$= 48.5 (author).
Characteristic point, P (**perekhodnaya tochka** in the original text; see the Introduction), at 277 ^oC and $100\mathbf{x}_2$= 34.0.

Intermediate compound:

$Na_5(C_2H_3O_2)_4C_6H_{11}O_2$ incongruently melting. (the composition is approximate).

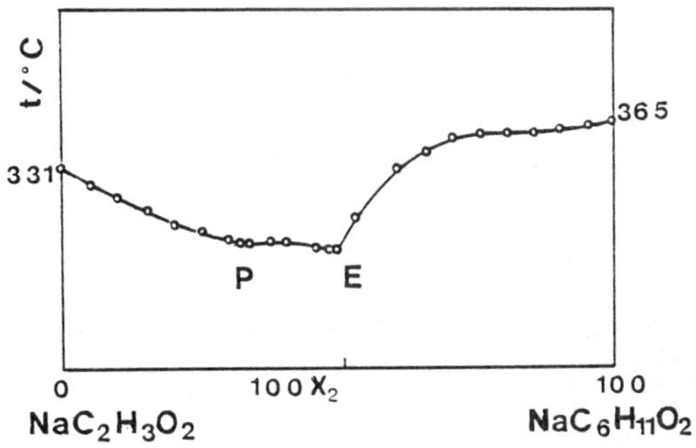

AUXILIARY INFORMATION

METHOD/APPARATUS/PROCEDURE:

Visual polythermal analysis.

SOURCE AND PURITY OF MATERIALS:

"Chemically pure" $NaC_2H_3O_2$ and $NaC_6H_{11}O_2$ prepared by reacting Na_2CO_3 with n-hexanoic acid (Ref. 1).
Component 1 undergoes phase transitions at $\mathbf{t}_{trs}(1)/^oC$= 58, 118, 130, 238 (Ref. 2).
Component 2 undergoes phase transitions at $\mathbf{t}_{trs}(2)/^oC$= 203, 226, 342 (Ref. 2).

ESTIMATED ERROR:

Temperature: accuracy probably ±2 K (compiler).

REFERENCES:

(1) Sokolov, N.M.
Zh. Obshch. Khim. 1954, **24**, 1581-1593.
(2) Sokolov, N.M.
Tezisy Dokl. X Nauch. Konf. S.M.I. 1956

COMPONENTS:

(1) Sodium ethanoate (sodium acetate); $NaC_2H_3O_2$; [127-09-3]
(2) Sodium benzoate; $NaC_7H_5O_2$; [532-32-1]

ORIGINAL MEASUREMENTS:

Sokolov, N.M.
Zh. Obshch. Khim. 1954, **24**, 1581-1593.

VARIABLES:

Temperature.

PREPARED BY:

D'Andrea, G.

EXPERIMENTAL VALUES:

$t/^{\circ}C$	T/K^a	$100x_2$
331	604	0
315	588	2.6
350	623	5
380	653	10
400	673	15
411	684	20
421	694	25
428	701	30
431	704	33
465	738	100

[a] T/K values calculated by the compiler.

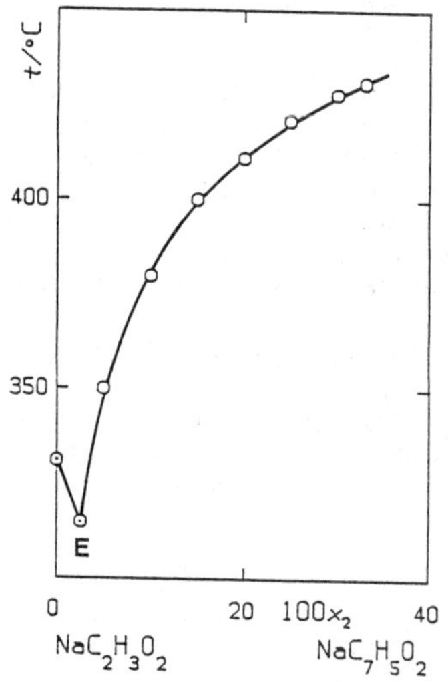

Characteristic point(s):

Eutectic, E, at 315 °C and $100x_2$= 2.6 (author).

Note - The system was investigated at $0 \leq 100x_2 \leq 33$ due to thermal instability of the benzoate.

AUXILIARY INFORMATION

METHOD/APPARATUS/PROCEDURE:

Visual polythermal analysis.
Melts contained in a glass tube and stirred. Temperatures measured with a Nichrome-Constantane thermocouple and a 17 mV full scale millivoltmeter . The temperature readings refer to the disappearance of isotropicity in the melt on cooling.

SOURCE AND PURITY OF MATERIALS:

"Chemically pure" materials.

ESTIMATED ERROR:

Temperature: accuracy probably ±2 K (compiler).

REFERENCES:

COMPONENTS:

(1) Sodium ethanoate (sodium acetate); $NaC_2H_3O_2$; [127-09-3]
(2) Sodium chloride; NaCl; [7647-14-5]

ORIGINAL MEASUREMENTS:

Il'yasov, I.I.; Bergman, A.G.
Zh. Obshch. Khim. 1960, **30**, 355-358.

VARIABLES:

Temperature.

PREPARED BY:

D'Andrea, G.

EXPERIMENTAL VALUES:

$t/^{o}C$	T/K^{a}	$100x_2$
328	601	0
328	601	2.5
328	601	5.0
328	601	7.0
328	601	10.0
368	641	12.5
398	671	15.0
427	700	17.5

[a] T/K values calculated by the compiler.

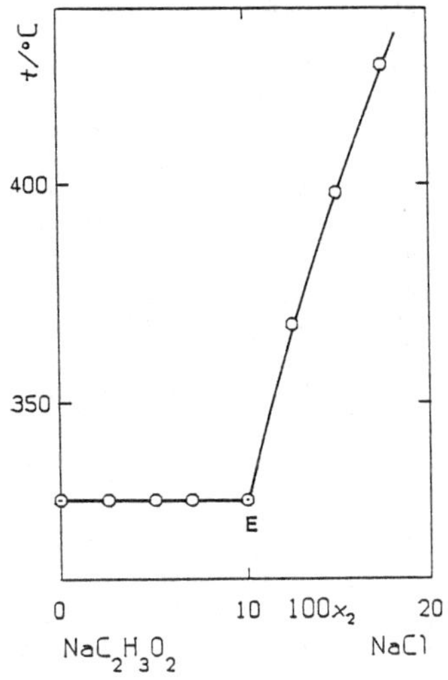

Characteristic point(s):

Eutectic, E, at 328 °C and $100x_2$= 10 (authors).

Note - The system was investigated at $0 \leq 100x_2 \leq 17.5$ due to thermal instability of component 1.

AUXILIARY INFORMATION

METHOD/APPARATUS/PROCEDURE:

Visual polythermal analysis; temperatures measured with a Nichrome-Constantane thermocouple and a millivoltmeter.

SOURCE AND PURITY OF MATERIALS:

Not stated.

NOTE:

See the NOTE relevant to the investigation by Piantoni et al. (Ref. 1) on the same system (next Table).

ESTIMATED ERROR:

Temperature: accuracy probably ±2 K (compiler).

REFERENCES:

(1) Piantoni, G.; Leonesi, D.; Braghetti, M.; Franzosini, P. **Ric. Sci.**, 1968, **38**, 127-132.

COMPONENTS:

(1) Sodium ethanoate (sodium acetate); $NaC_2H_3O_2$; [127-09-3]
(2) Sodium chloride; NaCl; [7647-14-5]

ORIGINAL MEASUREMENTS:

Piantoni, G.; Leonesi, D.; Braghetti, M.; Franzosini, P.
Ric. Sci., 1968, **38**, 127-132.

VARIABLES:

Temperature.

PREPARED BY:

D'Andrea, G.

EXPERIMENTAL VALUES:

t/°C	T/K[a]	$100x_1$	t/°C	T/K[a]	$100x_1$
328.1	601.3	100	323.9	597.1	97.2
327.4	600.6	99.5	323.4	596.6	97.1
327.0	600.2	99.3	323.7	596.9	97.0
326.5	599.7	98.9	323.8	597.0	96.9
326.1	599.3	98.7	322.0	595.2	96.0
326.0	599.2	98.5	322.1	595.3	95.7
325.9	599.1	98.5	321.4	594.6	95.2
325.4	598.6	98.3	321.2	594.4	95.0
325.3	598.5	98.1	327.4	600.6	94.0
325.0	598.2	97.9	332.5	605.7	93.9
324.8	598.0	97.8	338.9	612.1	93.6

[a] T/K values calculated by the compiler.

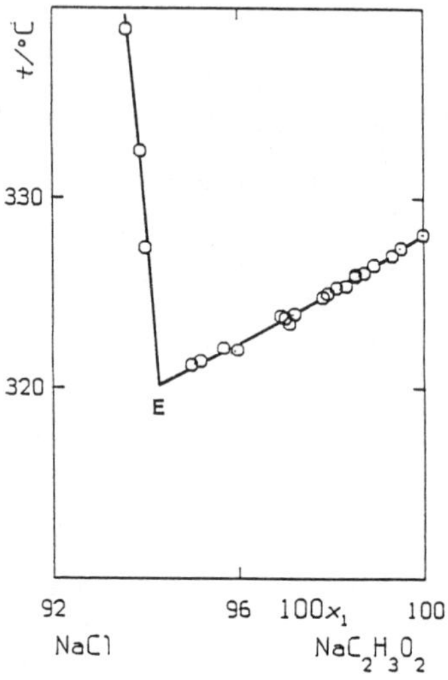

The system was investigated at $0 \leq 100x_2 \leq 6.5$.

Characteristic point(s):

Eutectic, E, at 320.1 °C and $100x_1 = 94.3$ (authors).

Note - In the original paper the results were shown in graphical form. The above listed numerical values represent a private communication by one of the authors (F., P.) to the compiler.

AUXILIARY INFORMATION

METHOD/APPARATUS/PROCEDURE:

A Pyrex device, suitable for work under an inert atmosphere, and allowing one to observe the system visually, was employed (for details, see Ref. 1). The initial crystallization temperatures were measured with a Chromel-Alumel thermocouple checked by comparison with a certified Pt resistance thermometer, and connected with a L&N Type K-3 potentiometer.

NOTE:

The authors discuss their own results in comparison with both the expected ideal behaviour of the molten mixtures and the previous data from Ref. 1. They observed that the liquidus branch richer in sodium chloride is not far from ideality.

SOURCE AND PURITY OF MATERIALS:

C. Erba RP materials, dried by heating under vacuum.

ESTIMATED ERROR:

Temperature: accuracy probably ±0.1 K.

REFERENCES:

(1) Il'yasov. I.I.; Bergman, A.G.
Zh. Obshch. Khim. 1960, **30**, 355-358.

COMPONENTS:

(1) Sodium ethanoate (sodium acetate); $NaC_2H_3O_2$; [127-09-3]
(2) Sodium thiocyanate; NaCNS; [540-72-7]

ORIGINAL MEASUREMENTS:

Sokolov, N.M.
Zh. Obshch. Khim. 1954, **24**, 1150-1156.

VARIABLES:

Temperature.

PREPARED BY:

D'Andrea, G.

EXPERIMENTAL VALUES:

$t/^{o}C$	T/K^{a}	$100x_2$
331	604	0
326	599	5
320	593	10
313	586	15
302	575	25
295	568	30
287	560	35
278	551	40
268	541	45
256	529	50
244	517	54.5
245	518	55
258	531	60
266	539	65
282	555	75
302	575	90
311	584	100

[a] T/K values calculated by the compiler.

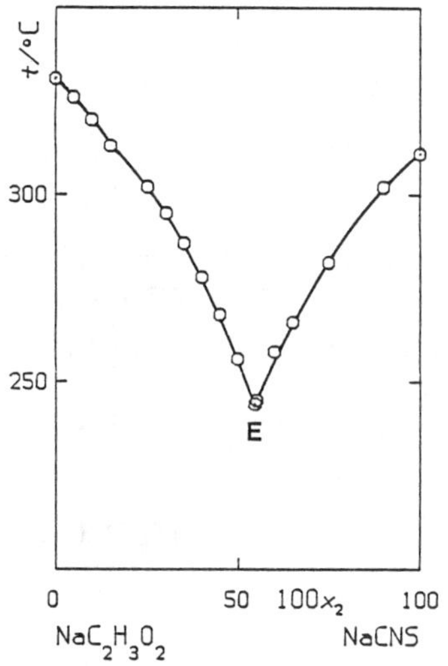

Characteristic point(s):

Eutectic, E, at 244 °C and $100x_2$= 54.5 (author).

AUXILIARY INFORMATION

METHOD/APPARATUS/PROCEDURE:

Visual polythermal analysis.
Salt(s) melted in a test tube. Temperature measured with a Nichrome-Constantane thermocouple and a millivoltmeter (17 mV full scale) with mirror reading.

SOURCE AND PURITY OF MATERIALS:

Component 1 synthetized from ethanoic acid and $NaHCO_3$. Component 2 of analytical purity recrystallized once from water and once from ethanol.

NOTE:

See the NOTE attached to the investigation by Storonkin et al. (Ref.1) on the same system.

ESTIMATED ERROR:

Temperature: accuracy probably ±2 K (compiler).

REFERENCES:

(1) Storonkin, A.V.; Vasil'kova, I.V.; Potemin, S.S.; **Vestn. Leningr. Univ., Fiz., Khim.** 1974, (16), 73-76.

COMPONENTS:

(1) Sodium ethanoate (sodium acetate); $NaC_2H_3O_2$; [127-09-3]
(2) Sodium thiocyanate; NaCNS; [540-72-7]

ORIGINAL MEASUREMENTS:

Golubeva, M.S.; Aleshkina, N.N.; Bergman, A.G.
Zh. Neorg. Khim. 1959, 4, 2606-2610; **Russ. J. Inorg. Chem., Engl. Transl.,** 1959, 4, 1201-1203 (*).

VARIABLES:

Temperature.

PREPARED BY:

D'Andrea, G.

EXPERIMENTAL VALUES:

The results are reported only in graphical form (see figure).

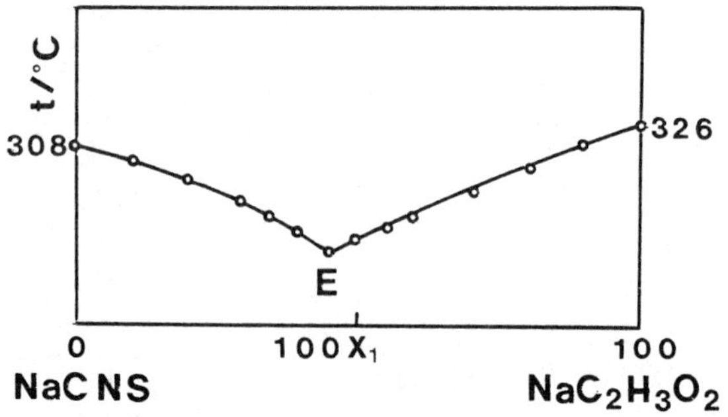

Characteristic point(s):

Eutectic, E, at 236 °C and $100x_1$= 44.5 (authors).

AUXILIARY INFORMATION

METHOD/APPARATUS/PROCEDURE:

Visual observation of fusion of the salt mixtures contained in a glass tube surrounded by a wider tube to secure a more uniform heating. Temperatures measured with a Chromel-Alumel thermocouple.

SOURCE AND PURITY OF MATERIALS:

Materials of analytical purity twice recrystallized.

NOTE:

See the NOTE attached to the investigation by Storonkin et al. (Ref.1) on the same system (see following Table).

ESTIMATED ERROR:

Temperature: accuracy probably ±2 K (compiler).

REFERENCES:

(1) Storonkin, A.V.; Vasil'kova, I.V.; Potemin, S.S.; **Vestn. Leningr. Univ., Fiz., Khim.** 1974, (16), 73-76.

COMPONENTS:

(1) Sodium ethanoate (sodium acetate); $NaC_2H_3O_2$; [127-09-3]
(2) Sodium thiocyanate; NaCNS; [540-72-7]

ORIGINAL MEASUREMENTS:

Storonkin, A.V.; Vasil´kova, I.V.; Potemin, S.S.
Vestn. Leningr. Univ., Fiz., Khim. 1974, (16), 73-76.

VARIABLES:

Temperature.

PREPARED BY:

D´Andrea, G.

EXPERIMENTAL VALUES:

$t/^{o}C$	T/K^{a}	$100\mathbf{x}_2$
328	601	0
314	587	10
298	571	20
284	557	30
264	537	40
245	518	50
240	513	60
260	533	70
278	551	80
293	566	90
308	581	100

[a] T/K values calculated by the compiler.

Note - The tabulated data were drawn by the compiler from Fig. 1 of the original paper.

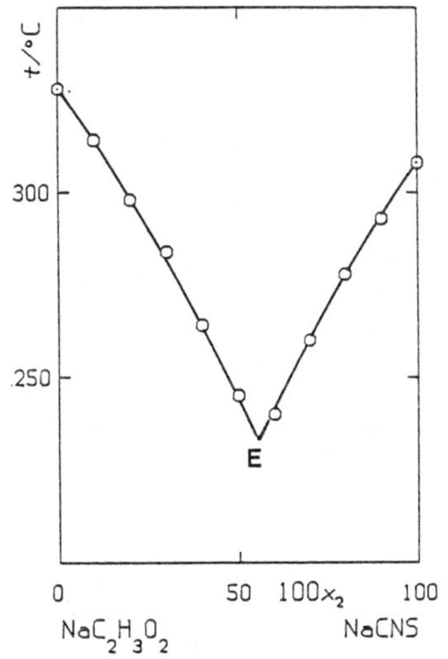

Characteristic point(s):

Eutectic, E, at 234 ^{o}C and $100\mathbf{x}_2$= 55 (authors).

AUXILIARY INFORMATION

METHOD/APPARATUS/PROCEDURE:

DTA. Thermograph with photorecorder. Salt(s) sealed under vacuum in Pyrex ampoules.
No other information given.

SOURCE AND PURITY OF MATERIALS:

$NaC_2H_3O_2$ of analytical purity and "chemically pure" NaCNS, heated 10-15 h at temperatures 50-60 ^{o}C below their fusion temperatures, were employed.

NOTE:

This binary was also submitted to visual polythermal analysis by Sokolov (Ref. 1), and Golubeva et al. (Ref. 2). The eutectic composition detected by Storonkin et al. ($100\mathbf{x}_2$= 55) fairly agrees with those reported both in Ref. 1 (54.5) and Ref. 2 (55.5). Sokolov´s eutectic temperature (517 K), on the contrary, is significantly higher than those given both by Storonkin et al. (507 K) and Golubeva et al. (509 K; Ref. 2).

ESTIMATED ERROR:

Temperature: accuracy probably ± 2 K (compiler).

REFERENCES:

(1) Sokolov, N.M.
Zh. Obshch. Khim. 1954, **24**, 1150-1156.
(2) Golubeva, M.S.; Aleshkina, N.N.; Bergman, A.G.;
Zh. Neorg. Khim. 1959, **4**, 2606-2610; **Russ. J. Inorg. Chem. (Engl. Transl.)** 1959, **4**, 1201-1203 (*).

COMPONENTS:	ORIGINAL MEASUREMENTS:
(1) Sodium ethanoate (sodium acetate); $NaC_2H_3O_2$; [127-09-3] (2) Sodium iodide; NaI; [7681-82-5]	Diogenov, G.G.; Erlykov, A.M. **Nauch. Dokl. Vysshei Shkoly, Khim. i Khim. Tekhnol.** 1958, No. 3, 413-416.
VARIABLES:	PREPARED BY:
Temperature.	D'Andrea, G.

EXPERIMENTAL VALUES:

$t/^{o}C$	T/K^{a}	$100\mathbf{x}_1$
337	610	100
336	609	97.9
332	605	94.0
326	599	92.1
324	597	87.5
320	593	84.5
314	587	80.0
311	584	77.8
312	585	76.3
326	599	74.2
346	619	70.7
360	633	68.3

[a] T/K values calculated by the compiler.

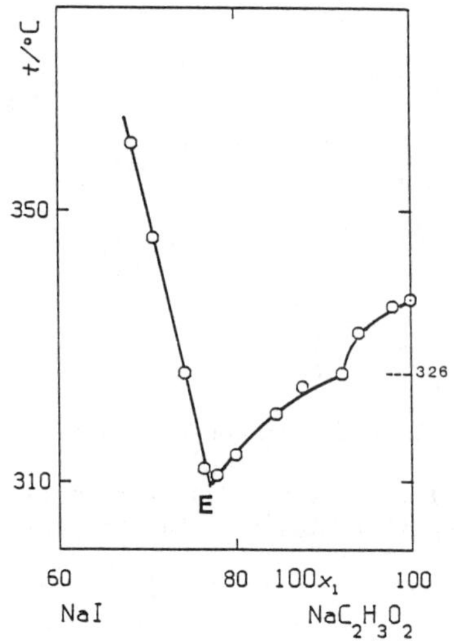

Note - The system was investigated at $100 \geq 100\mathbf{x}_1 \geq 68.3$.

Characteristic point(s):

Eutectic, E, at 310 ^{o}C and $100\mathbf{x}_2 = 23$.

AUXILIARY INFORMATION

METHOD/APPARATUS/PROCEDURE:

Visual polythermal analysis.

SOURCE AND PURITY OF MATERIALS:

Not stated.
Component 1 undergoes a phase transition at $t_{trs}(1)/^{o}C = 326$.
Component 2 melts at $t_{fus}(1)/^{o}C = 670$.

NOTE:

See the NOTE relevant to the investigation by Piantoni et al. (Ref. 1) on the same system.

ESTIMATED ERROR:

Temperature: accuracy probably ± 2 K (compiler).

REFERENCES:

(1) Piantoni, G.; Leonesi, D.; Braghetti, M.; Franzosini, P.
Ric. Sci., 1968, **38**, 127-132.

COMPONENTS:	ORIGINAL MEASUREMENTS:
(1) Sodium ethanoate (sodium acetate); $NaC_2H_3O_2$; [127-09-3] (2) Sodium iodide; NaI; [7681-82-5]	Piantoni, G.; Leonesi, D.; Braghetti, M.; Franzosini, P. **Ric. Sci.**, 1968, **38**, 127-132.
VARIABLES:	PREPARED BY:
Temperature.	D'Andrea, G.

EXPERIMENTAL VALUES:

$t/^{o}C$	T/K^{a}	$100\mathbf{x}_1$	$t/^{o}C$	T/K^{a}	$100\mathbf{x}_1$
328.1	601.3	100	315.3	588.5	87.0
327.7	600.9	99.6	313.6	586.8	85.2
327.0	600.2	98.9	312.6	585.8	84.2
326.4	599.6	98.4	311.1	584.3	82.6
325.8	599.0	97.8	310.2	583.4	81.8
325.2	598.4	97.3	309.8	583.0	81.4
324.9	598.1	97.0	308.2	581.4	79.7
325.0	598.2	96.9	306.5	579.7	78.5
324.8	598.0	96.9	306.1	579.3	77.5
323.6	596.8	95.5	307.4	580.6	76.1
322.1	595.3	93.8	311.8	585.0	75.7
318.9	592.1	90.4	318.4	591.6	75.0

[a] T/K values calculated by the compiler.

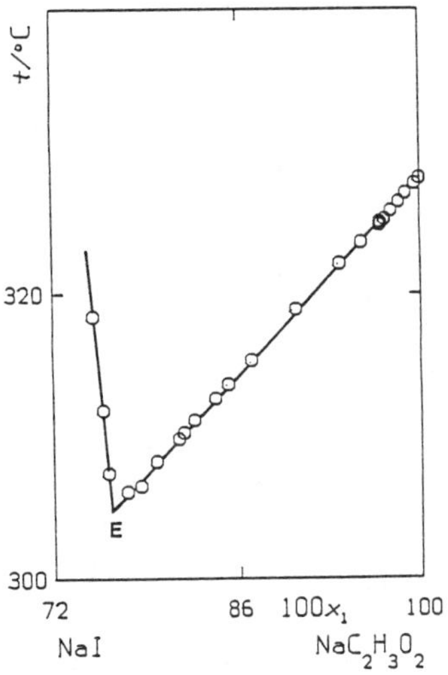

Note 1 - In the original paper the results were shown in graphical form. The above listed numerical values represent a private communication by one of the authors (F., P.) to the compiler.

Note 2 - The system was investigated at $0 \leq 100\mathbf{x}_2 \leq 25$.

Characteristic point(s): Eutectic, E, at 304.8 °C and $100\mathbf{x}_1 = 76.3$ (authors).

AUXILIARY INFORMATION

METHOD/APPARATUS/PROCEDURE:

A Pyrex device, suitable for work under an inert atmosphere, and allowing one to observe the system visually, was employed (for details, see Ref. 1). The initial crystallization temperatures were measured with a Chromel-Alumel thermocouple checked by comparison with a certified Pt resistance thermometer, and connected with a L&N Type K-3 potentiometer.

NOTE:

The authors discuss their own results in comparison with both the expected ideal behaviour of the molten mixtures and the previous data from Ref.s 1 and 2. They observed that the liquidus branch richer in sodium iodide is not far from ideality.

SOURCE AND PURITY OF MATERIALS:

C. Erba RP materials, dried by heating under vacuum.

ESTIMATED ERROR:

Temperature: accuracy probably ±0.1 K.

REFERENCES:

(1) Il'yasov. I.I.; Bergman, A.G.
Zh. Obshch. Khim. 1961, **31**, 368-370.
(2) Diogenov, G.G.; Erlykov, A.M.
Nauch. Dokl. Vysshei Shkoly, Khim. i Khim. Tekhnol. 1958, **No. 3**, 413-416.

COMPONENTS:

(1) Sodium ethanoate (sodium acetate); $NaC_2H_3O_2$; [127-09-3]
(2) Sodium nitrite; $NaNO_2$; [7632-00-0]

ORIGINAL MEASUREMENTS:

Bergman, A.G.; Evdokimova, K.A.
Izv. Sektora Fiz.-Khim. Anal., Inst. Obshchei i Neorg. Khim. Akad. Nauk SSSR 1956, **27**, 296-314.

VARIABLES:

Temperature.

PREPARED BY:

D'Andrea, G.

EXPERIMENTAL VALUES:

$t/^{\circ}C$	T/K^{a}	$100x_1$
278	551	0
275	548	4.6
265	538	11.8
259	532	15.5
247	520	23.0
237	510	28.1
228	501	33.0
228	501	34.9
236	509	37.2
240	513	39.3
248	521	41.8
251	524	44.2
258	531	46.9
265	538	49.4
276	549	55.0
287	560	61.3
294	567	66.0
297	570	68.0

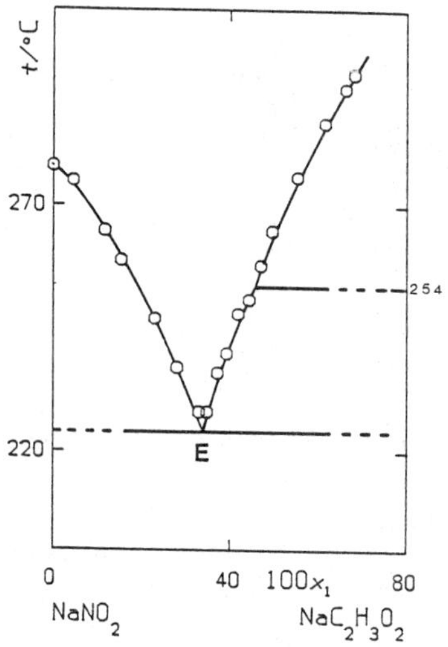

[a] T/K values calculated by the compiler.

Characteristic point(s):

Eutectic, E, at 224 °C and $100x_1 = 34$ (authors).

Note - The system was investigated at $0 \leq 100x_1 \leq 68$.

AUXILIARY INFORMATION

METHOD/APPARATUS/PROCEDURE:

Visual polythermal analysis: the temperatures of initial crystallization were measured with a Nichrome-Constantane thermocouple and a 17 mV full-scale millivoltmeter.

SOURCE AND PURITY OF MATERIALS:

Component 1: "chemically pure" $NaC_2H_3O_2 \cdot 3H_2O$ dried to constant mass; it undergoes a phase transition at $t_{trs}(1)/^{\circ}C = 254$ and fusion at $t_{fus}(1)/^{\circ}C = 326$.
Component 2: source not stated.

ESTIMATED ERROR:

Temperature: accuracy probably ± 2 K (compiler).

NOTE:

Concerning component 1, the fusion (599 K) and solid state transition (527 K) temperatures can be identified respectively with the $T_{fus}(1)$ (601.3$\pm$0.5 K) and $T'_{trs}(1)$ (527$\pm$15 K) values listed in Preface, Table 1. The coordinates of the eutectic (497 K and $100x_2 = 66$) are in reasonable agreement with those reported by both Sokolov (500-501 K) and $100x_2 = 65$; Ref. 1), and Sokolov et al. (499 K) and $100x_2 = 65$; Ref. 2).

REFERENCES:

(1) Sokolov, N.M.
Zh. Obshch. Khim. 1957, **27**, 840-844(*); **Russ. J. Gen. Chem. (Engl. Transl.)** 1957, **27**, 917-920.
(2) Sokolov, N.M.; Tsindrik, N.M.; Khaitina, M.V.
Zh. Neorg. Khim. 1970, **15**, 852-855; **Russ. J. Inorg. Chem. (Engl. Transl.)** 1970, **15**, 433-435 (*).

COMPONENTS:

(1) Sodium ethanoate (sodium acetate); $NaC_2H_3O_2$; [127-09-3]
(2) Sodium nitrite; $NaNO_2$; [7632-00-0]

ORIGINAL MEASUREMENTS:

Sokolov, N.M.
Zh. Obshch. Khim. 1957, 27, 840-844 (*);
Russ. J. Gen. Chem. (Engl. Transl.) 1957, 27, 917-920.

VARIABLES:

Temperature.

PREPARED BY:

D'Andrea, G.

EXPERIMENTAL VALUES:

$t/^oC$	T/K^a	$100\mathbf{x}_2$
331	604	0
327	600	5
322	595	10
315	588	15
306	579	20
299	572	25
292	565	30
285	558	35
279	552	40
269	542	45
259	532	50
248	521	55
237	510	60
228	501	65
237	510	70
243	516	75
249	522	80
258	531	85
266	539	90
275	548	95
284	557	100

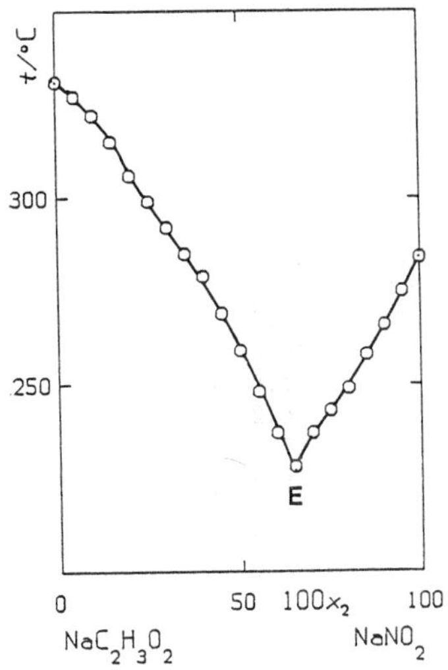

[a] T/K values calculated by the compiler.

Characteristic point(s): Eutectic, E, at 227 °C (from table 2 of the original paper) or 228 °C (according to the above tabulated data; compiler) and $100\mathbf{x}_2$= 65 (author).

AUXILIARY INFORMATION

METHOD/APPARATUS/PROCEDURE:

Visual polythermal analysis; salt mixtures melted in a glass tube (surrounded by a wider tube) and stirred with a glass thread. The temperatures of initial crystallization were measured with a Nichrome-Constantane thermocouple checked at the fusion points of water, benzoic acid, mannitol, $AgNO_3$, Cd, KNO_3, and $K_2Cr_2O_7$.

NOTE:

The fusion temperature (604 K) of component 1 can be identified with the $\mathbf{T}_{fus}(1)$ value (601.3±0.5 K) listed in Preface, Table 1. The coordinates of the eutectic (500-501 K and $100\mathbf{x}_2$= 65) are in reasonable agreement with those reported by both Bergman and Evdokimova (497 K and $100\mathbf{x}_2$= 66; Ref. 1), and by Sokolov et al. (499 K and $100\mathbf{x}_2$= 65; Ref. 2).

SOURCE AND PURITY OF MATERIALS:

"Chemically pure" materials recrystallized from water.

ESTIMATED ERROR:

Temperature: accuracy probably ±2 K (compiler).

REFERENCES:

(1) Bergman, A.G. Evdokimova, K.A. **Izv. Sektora Fiz.-Khim. Anal., Inst. Obshchei i Neorg. Khim. Akad. Nauk SSSR** 1956, **27**, 296-314.
(2) Sokolov, N.M.; Tsindrik, N.M.; Khaitina, M.V.; **Zh. Neorg. Khim.** 1970, 15, 852-855; **Russ. J. Inorg. Chem. (Engl. Transl.)** 1970, 15, 433-435 (*).

COMPONENTS:

(1) Sodium ethanoate (sodium acetate); $NaC_2H_3O_2$; [127-09-3]
(2) Sodium nitrite; $NaNO_2$; [7632-00-0]

ORIGINAL MEASUREMENTS:

Sokolov, N.M.; Tsindrik, N.M.; Khaitina, M.V. **Zh. Neorg. Khim.** 1970, **15**, 852-855; **Russ. J. Inorg. Chem. (Engl. Transl.)** 1970, **15**, 433-435 (*).

VARIABLES:

Temperature.

PREPARED BY:

D'Andrea, G.

EXPERIMENTAL VALUES:

The results are given only in graphical form (see figure).

Characteristic point(s):

Eutectic, E, at 226 °C and $100x_2$= 65 (authors).

Note - Restricted solid solutions are formed as far as $100x_2$= 15.

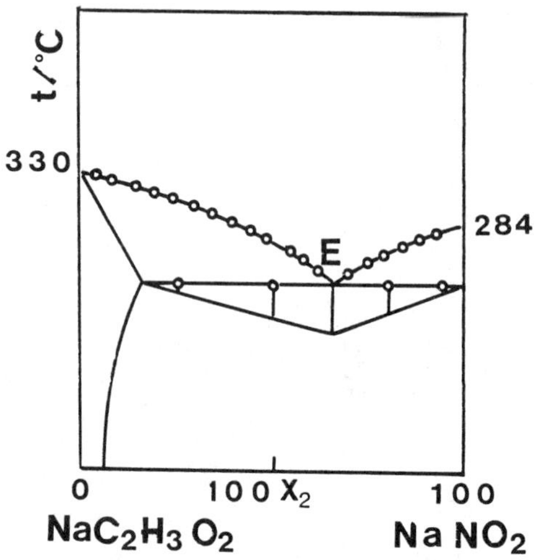

AUXILIARY INFORMATION

METHOD/APPARATUS/PROCEDURE:

Visual polythermal analysis supplemented with differential thermal analysis.

ESTIMATED ERROR:

Temperature: accuracy probably ± 2 K (compiler).

NOTE:

Concerning component 1: (i) the fusion temperature (603 K) can be identified with the $T_{fus}(1)$ value (601.3$\pm$0.5 K) listed in Preface Table 1; and (ii) among the solid state transition temperatures (331, 391, 403, and 511 K) quoted by the authors from Ref. 1, only the third and fourth ones find some correspondence in the T_{trs} values listed in Table 1. The coordinates of the eutectic (499 K and $100x_2$= 65) are in reasonable agreement with those previously reported by both Bergman and Evdokimova (497 K and $100x_2$= 66; Ref. 3), and Sokolov (500-501 K and $100x_2$= 65; Ref. 4).

SOURCE AND PURITY OF MATERIALS:

Not stated.
Component 1 undergoes phase transitions at $t_{trs}(1)/°C$= 58, 118, 180, 288 (Ref. 1; the figures 180, 288 are most probably misprints, inasmuch as the same authors quoting the same source report 130, 238 in several other papers; compiler).
Component 2 undergoes a phase transition at $t_{trs}(2)/°C$= 170 (Ref. 2).

REFERENCES:

(1) Sokolov, N.M.; **Tezisy Dokl. X Nauch. Konf. S.M.I.** 1956.
(2) Bergman, A.G.; Berul', S.I.; **Izv. Sektora Fiz.-Khim.Anal.** 1958, 21, 178-183.
(3) Bergman, A.G. Evdokimova, K.A.; **Izv. Sektora Fiz.-Khim. Anal., Inst. Obshchei i Neorg. Khim. Akad. Nauk SSSR** 1956, **27**, 296-314.
(4) Sokolov, N.M.; **Zh. Obshch. Khim.** 1957, **27**, 840-844 (*); **Russ. J. Gen. Chem. (Engl. Transl.)** 1957, **27**, 917-920.

COMPONENTS:	EVALUATOR:
(1) Sodium ethanoate (sodium acetate); $NaC_2H_3O_2$; [127-09-3] (2) Sodium nitrate; $NaNO_3$; [7631-99-4]	Ferloni, P., Dipartimento di Chimica Fisica, Universita´ di Pavia (ITALY).

CRITICAL EVALUATION:

The system $Na/C_2H_3O_2$, NO_3 was studied by Sokolov (Ref. 1), Bergman and Evdokimova (as a side of the reciprocal ternary K, $Na/C_2H_3O_2$, NO_3; Ref. 2), Diogenov (as a side of the reciprocal ternary Li, $Na/C_2H_3O_2$, NO_3; Ref. 2), Gimel´shtein and Diogenov (as a side of the reciprocal ternary Cs, $Na/C_2H_3O_2$, NO_3; Ref. 4), Storonkin et al. (as a side of the ternary $Na/C_2H_3O_2$, CNS, NO_3; Ref. 5), and Diogenov and Chumakova (as a side of the reciprocal ternary K, $Na/C_2H_3O_2$, NO_3; Ref. 6). The visual polythermal analysis, and DTA were employed in Ref.s 1-4 and 6, and in Ref. 5, respectively; moreover, in Ref. 4, X-ray diffraction patterns were taken on some compositions.

The fusion temperature of component 1 should be 604, 599, 610, 600, 601, and 599 K according to Ref.s 1,2,3,4,5, and 6, respectively, the corresponding value listed in Preface, Table 1 being 601.3 ± 0.5 K. For the same component, a solid state transition is reported by Ref.s 2, 3, and 4. The transition temperatures given by Ref.s 2 and 4 (527 and 543 K, respectively) can be identified with the $T'_{trs}(1)$ value (527 ± 15 K) listed in Table 1 of the Preface, whereas no reliability is to be attached to Diogenov´s figures (596 K; Ref.3) which has no correspondence in Table 1, and, moreover, was not confirmed in subsequent investigations by the same group (Ref. 4).

Diogenov (Ref. 3) claimed the existence of two intermediate compounds, i.e.: (i) $Na_3(C_2H_3O_2)_2NO_3$, incongruently melting, with a peritectic at 539 K and $100\mathbf{x}_2 = 38.5$; and (ii) $Na_5C_2H_3O_2(NO_3)_4$, congruently melting, with a distectic at 545 K. In the evaluator´s opinion, however, the discontinuities Diogenov (Ref. 3) found on either branch of his liquidus are relevant rather to the occurrence of solid state transitions in either component, than to the formation of any intermediate compound. In fact, in their re-investigations of the binary $Na/C_2H_3O_2$, NO_3 neither Gimel´shtein and Diogenov (who supplemented their visual observations with some X-ray diffraction patterns; Ref. 4), nor Diogenov and Chumakova (Ref. 6) could confirm Diogenov´s former point.

Therefore, the system can be safely classified as of the eutectic type, with the invariant at 494 ± 4 K and $100\mathbf{x}_2$ at about 58.

REFERENCES:

(1) Sokolov, N.M.
Zh. Obshch. Khim. 1954, **24**, 1150-1156.

(2) Bergman, A.G.; Evdokimova, K.A.
Izv. Sektora Fiz.-Khim. Anal., Inst. Obshchei i Neorg. Khim. Akad. Nauk SSSR 1956, **27**, 296-314.

(3) Diogenov, G.G.
Zh. Neorg. Khim. 1956, 1, 799-805 (*); **Russ. J. Inorg. Chem. (Engl. Transl.)** 1956, 1 (4), 199-205.

(4) Gimel´shtein, V.G.; Diogenov, G.G.
Tr. Irkutsk. Politekh. Inst., Ser. Khim., 1966, **27**, 69-75.

(5) Storonkin, A.V.; Vasil´kova, I.V.; Potemin, S.S.
Vestn. Leningr. Univ., Fiz., Khim. 1974, (16), 73-76.

(6) Diogenov, G.G.; Chumakova, V.P.
Fiz.-Khim. Issled. Rasplavov Solei, Irkutsk, 1975, 7-12.

COMPONENTS:

(1) Sodium ethanoate (sodium acetate); $NaC_2H_3O_2$; [127-09-3]
(2) Sodium nitrate; $NaNO_3$; [7631-99-4]

ORIGINAL MEASUREMENTS:

Sokolov, N.M.
Zh. Obshch. Khim. 1954, **24**, 1150-1156.

VARIABLES:

Temperature.

PREPARED BY:

D'Andrea, G.

EXPERIMENTAL VALUES:

$t/^{\circ}C$	T/K^{a}	$100\mathbf{x}_2$
331	604	0
328	601	5
324	597	10
318	591	15
304	577	25
296	569	30
286	559	35
276	549	40
263	536	45
247	520	50
233	506	55
224	497	58
231	504	60
242	515	65
264	537	75
284	557	85
304	577	95
308	581	100

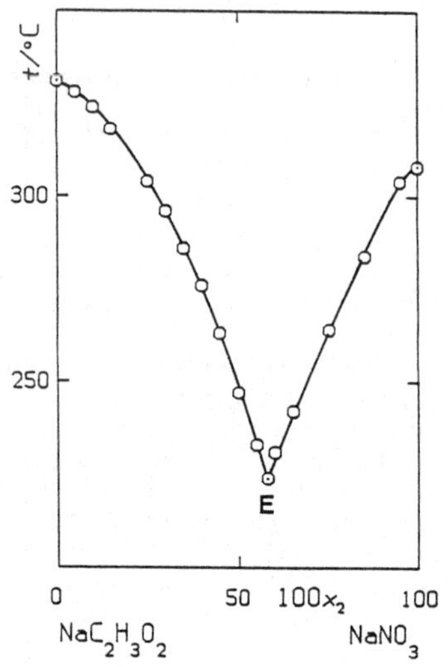

[a] T/K values calculated by the compiler.

Characteristic point(s):

Eutectic, E, at 224 °C and $100\mathbf{x}_2$= 58 (author).

AUXILIARY INFORMATION

METHOD/APPARATUS/PROCEDURE:

Visual polythermal analysis.
Salt(s) melted in a test tube. Temperature measured with a Nichrome-Constantane thermocouple and a millivoltmeter with mirror reading to 17 mV.

SOURCE AND PURITY OF MATERIALS:

Component 1 synthetized from ethanoic acid and $NaHCO_3$. Commercial component 2 further purified by the author according to Laiti.

ESTIMATED ERROR:

Temperature: accuracy probably ± 2 K (compiler).

REFERENCES:

COMPONENTS:

(1) Sodium ethanoate (sodium acetate); $NaC_2H_3O_2$; [127-09-3]
(2) Sodium nitrate; $NaNO_3$; [7631-99-4]

ORIGINAL MEASUREMENTS:

Bergman, A.G.; Evdokimova, K.A.
Izv. Sektora Fiz.-Khim. Anal., Inst. Obshchei i Neorg. Khim. Akad. Nauk SSSR 1956, **27**, 296-314.

VARIABLES:

Temperature.

PREPARED BY:

D'Andrea, G.

EXPERIMENTAL VALUES:

t/°C	T/K[a]	$100x_2$	t/°C	T/K[a]	$100x_2$
326	599	0	235	508	62.3
323	596	3.3	248	521	67.1
317	590	8.0	258	531	71.2
312	585	11.9	264	537	74.7
308	581	16.2	271	544	76.8
300	573	20.7	276	549	79.6
293	566	25.2	281	554	82.4
285	558	29.8	287	560	85.2
279	552	34.2	293	566	88.0
268	541	39.1	298	571	90.4
258	531	44.0	303	576	94.3
241	514	50.7	304	577	97.0
226	499	56.8	308	581	100

[a] T/K values calculated by the compiler.

Characteristic point(s):

Eutectic, E, at 222 °C and $100x_2$= 58 (authors).

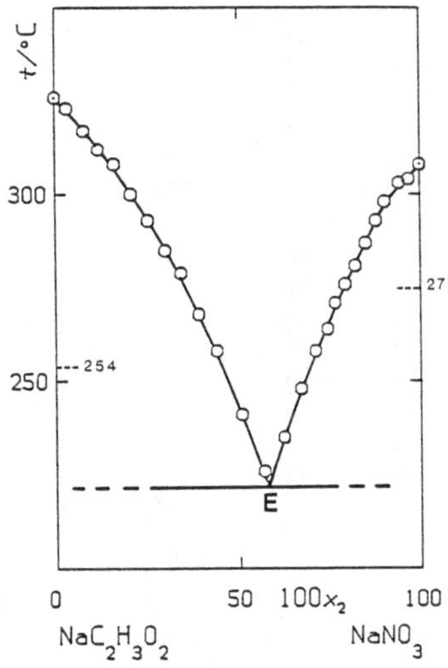

AUXILIARY INFORMATION

METHOD/APPARATUS/PROCEDURE:

Visual polythermal analysis: the temperatures of initial crystallization were measured with a Nichrome-Constantane thermocouple and a 17 mV full-scale millivoltmeter.

SOURCE AND PURITY OF MATERIALS:

Component 1: "chemically pure" $NaC_2H_3O_2.3H_2O$ dried to constant mass; it undergoes a phase transition at $t_{trs}(1)$/°C= 254.
Component 2: source not stated; it undergoes a phase transition at $t_{trs}(2)$/°C= 275 (Ref. 1).

ESTIMATED ERROR:

Temperature: accuracy probably ±2 K (compiler).

REFERENCES:

(1) Bergman, A.G.; Berul', S.I.
Izv. Sektora Fiz.-Khim. Anal. 1952, **21**, 178-183.

COMPONENTS:	ORIGINAL MEASUREMENTS:
(1) Sodium ethanoate (sodium acetate); $NaC_2H_3O_2$; [127-09-3] (2) Sodium nitrate; $NaNO_3$; [7631-99-4]	Diogenov, G.G. **Zh. Neorg. Khim.** 1956, 1, 799-805 (*); **Russ. J. Inorg. Chem. (Engl. Transl.)** 1956, 1 (4), 199-205.
VARIABLES:	PREPARED BY:
Temperature.	D'Andrea, G.

EXPERIMENTAL VALUES:

t/°C	T/K[a]	$100x_2$	t/°C	T/K[a]	$100x_2$
337	610	0	235	508	62
323	596	3	243	516	65.2
321	594	7	245	518	66
319	592	8.5	255	528	69.7
312	585	14.5	257	530	70.5
306	579	19	266	539	75.5
297	570	25	270	543	78.4
292	565	27	272	545	80
287	560	29.5	270	543	81.5
278	551	33.3	270	543	83.5
270	543	36.5	278	551	85.5
263	536	44.5	290	563	90
257	530	48	294	567	91.5
253	526	50	299	572	94.5
240	513	55	304	577	96.7
225	498	57.3	308	581	100
230	503	59.5			

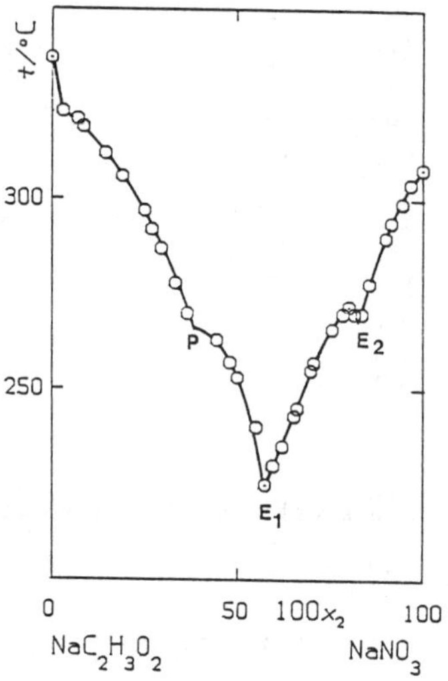

[a] T/K values calculated by the compiler.

Characteristic point(s):

Peritectic, P, at 266 °C (author) and $100x_2$= 38.5 (compiler).
Eutectic, E_1, at 225 °C and $100x_2$= 57.5 (author).
Eutectic, E_2, at about 268 °C and $100x_2$ about 82.5 (compiler).

Intermediate compound(s):

$Na_3(C_2H_3O_2)_2NO_3$, incongruently melting (author).
$Na_5C_2H_3O_2(NO_3)_4$, congruently melting at 272 °C (author).

Note - On the branch rich in component 1 an inflexion at 323 °C corresponds to a phase transition of $NaC_2H_3O_2$.

AUXILIARY INFORMATION

METHOD/APPARATUS/PROCEDURE:	SOURCE AND PURITY OF MATERIALS:
Visual polythermal analysis.	Not stated.
	ESTIMATED ERROR:
	Temperature: accuracy probably ±2 K (compiler).

COMPONENTS:

(1) Sodium ethanoate (sodium acetate); $NaC_2H_3O_2$; [127-09-3]
(2) Sodium nitrate; $NaNO_3$; [7631-99-4]

ORIGINAL MEASUREMENTS:

Gimel´shtein, V.G.; Diogenov, G.G.
Tr. Irkutsk. Politekh. Inst., Ser. Khim., 1966, **27**, 69-75.

VARIABLES:

Temperature.

PREPARED BY:

D´Andrea, G.

EXPERIMENTAL VALUES:

$t/^{\circ}C$	T/K^{a}	$100x_2$
327	600	0
320	593	6.5
315	588	11.5
307	580	18.3
299	572	23.5
287	560	30.0
277	550	34.3
266	539	39.5
257	530	46.2
247	520	51.5
235	508	56.4
230	503	60.2
241	514	64.7
253	526	70.6
270	543	80.5
284	557	87.0
293	566	90.7
303	576	96.4
308	581	100

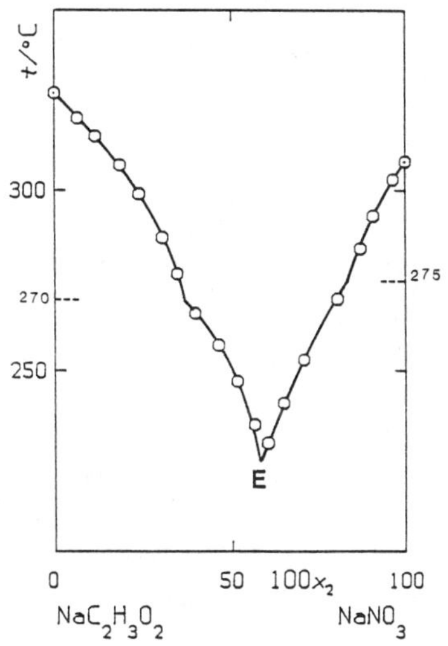

[a] T/K values calculated by the compiler.

Characteristic point(s):

Eutectic, E, at 225 $^{\circ}C$ and $100x_2 = 58$ (authors).

AUXILIARY INFORMATION

METHOD/APPARATUS/PROCEDURE:

Visual polythermal analysis supplemented with X-ray investigations. Temperatures measured with a Chromel-Alumel thermocouple and a 17 mV millivoltmeter.

SOURCE AND PURITY OF MATERIALS:

Not stated.
Component 1 undergoes a phase transition at $t_{trs}(1)/^{\circ}C = 270$.
Component 2 undergoes a phase transition at $t_{trs}(2)/^{\circ}C = 275$.

ESTIMATED ERROR:

Temperature: accuracy probably ± 2 K (compiler).

REFERENCES:

COMPONENTS:

(1) Sodium ethanoate (sodium acetate); $NaC_2H_3O_2$; [127-09-3]
(2) Sodium nitrate; $NaNO_3$; [7631-99-4]

ORIGINAL MEASUREMENTS:

Storonkin, A.V.; Vasil´kova, I.V.; Potemin, S.S.; **Vestn. Leningr. Univ., Fiz., Khim.** 1974, (16), 73-76.

VARIABLES:

Temperature.

PREPARED BY:

D´Andrea, G.

EXPERIMENTAL VALUES:

$t/^oC$	T/K^a	$100x_2$	$t/^oC$	T/K^a	$100x_2$
328	601	0	224	497	60
314	587	10	242	515	70
300	573	20	261	534	80
284	557	30	281	554	90
259	532	40	306	579	100
235	508	50			

[a] T/K values calculated by the compiler.

Note - The tabulated data were drawn by the compiler from Fig. 1 of the original paper.

Characteristic point(s):

Eutectic, E, at 218 ^oC and $100x_2$= 56 (authors).

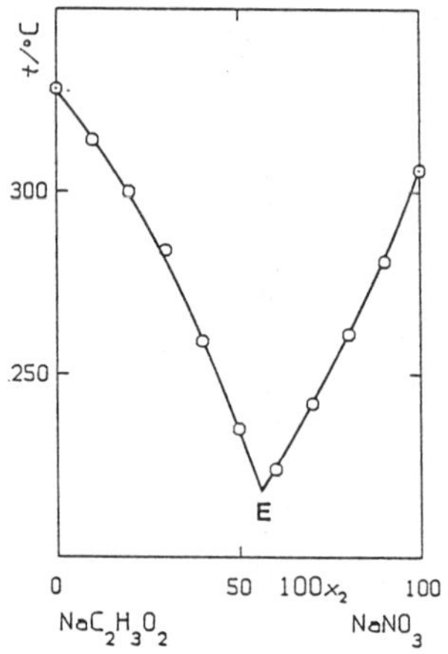

AUXILIARY INFORMATION

METHOD/APPARATUS/PROCEDURE:

DTA.
Thermograph with photorecorder.
Salt(s) sealed under vacuum in Pyrex ampoules. No other information given.

SOURCE AND PURITY OF MATERIALS:

$NaC_2H_3O_2$ of analytical purity and "chemically pure" $NaNO_3$, heated 10-15 h at temperatures 50-60 ^oC below their fusion temperatures, were employed.

ESTIMATED ERROR:

Temperature: accuracy probably ± 2 K (compiler).

REFERENCES:

COMPONENTS:

(1) Sodium ethanoate (sodium acetate); $NaC_2H_3O_2$; [127-09-3]
(2) Sodium nitrate; $NaNO_3$; [7631-99-4]

ORIGINAL MEASUREMENTS:

Diogenov, G.G.; Chumakova, V.P. **Fiz.-Khim. Issled. Rasplavov Solei, Irkutsk,** 1975, 7-12.

VARIABLES:

Temperature.

PREPARED BY:

D´Andrea, G.

EXPERIMENTAL VALUES:

Eutectic, E, at 222 ^oC (Fig. 1 of the original paper); composition not stated ($100x_1$ about 43 in compiler´s graphical estimation).

AUXILIARY INFORMATION

METHOD/APPARATUS/PROCEDURE:

Visual polythermal analysis.

SOURCE AND PURITY OF MATERIALS:

Not stated. Component 1: $t_{fus}(1)/^oC$= 326; component 2: $t_{fus}(2)/^oC$= 308 (Fig. 1 of the original paper).

ESTIMATED ERROR:

Temperature: accuracy probably ± 2 K (compiler).

REFERENCES:

COMPONENTS:

(1) Sodium propanoate (sodium propionate); $NaC_3H_5O_2$; [137-40-6]
(2) Sodium thiocyanate; NaCNS; [540-72-7]

ORIGINAL MEASUREMENTS:

Sokolov, N.M.
Zh. Obshch. Khim. 1954, **24**, 1150-1156.

VARIABLES:

Temperature.

PREPARED BY:

D'Andrea, G.

EXPERIMENTAL VALUES:

$t/^{\circ}C$	T/K^{a}	$100x_2$
298	571	0
292	565	5
289	562	10
287	560	15
282	555	25
280	553	30
276	549	35
273	546	40
270	543	45
265	538	50
258	531	54
261	534	55
269	542	60
276	549	65
285	558	75
298	571	90
311	584	100

[a] T/K values calculated by the compiler.

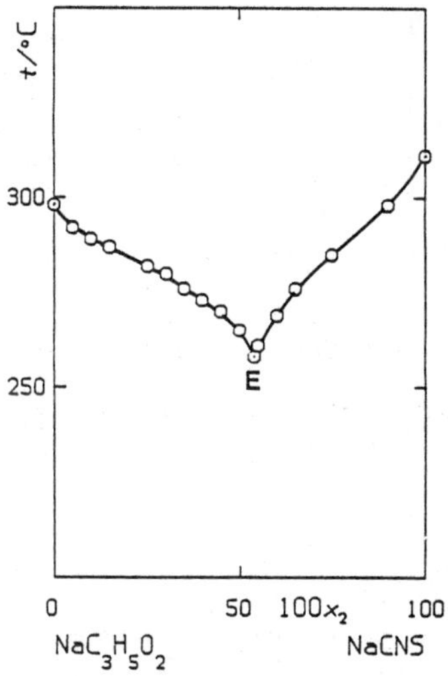

Characteristic point(s):

Eutectic, E, at 258 °C and $100x_2$= 54 (author).

AUXILIARY INFORMATION

METHOD/APPARATUS/PROCEDURE:

Visual polythermal analysis.
Salt(s) melted in a test tube. Temperature measured with a Nichrome-Constantane thermocouple and a millivoltmeter with mirror reading to 17 mV.

SOURCE AND PURITY OF MATERIALS:

Component 1 synthetized from propanoic acid and $NaHCO_3$. Component 2 of analytical purity recrystallized once from water and once from ethanol.

ESTIMATED ERROR:

Temperature: accuracy probably ±2 K (compiler).

NOTE:

See the NOTE relevant to the investigation by Storonkin et al. (Ref. 1) on the same system.

REFERENCES:

(1) Storonkin, A.V.; Vasil'kova, I.V.; Potemin, S.S.
Vestn. Leningr. Univ., Fiz., Khim. 1974, (10), 84-88.

COMPONENTS:

(1) Sodium propanoate (sodium propionate); $NaC_3H_5O_2$; [137-40-6]
(2) Sodium thiocyanate; NaCNS; [540-72-7]

ORIGINAL MEASUREMENTS:

Storonkin, A.V.; Vasil'kova, I.V.; Potemin, S.S.;
Vestn. Leningr. Univ., Fiz., Khim. 1974, (10), 84-88.

VARIABLES:

Temperature.

PREPARED BY:

D'Andrea, G.

EXPERIMENTAL VALUES:

$t/^oC$	T/K^a	$100x_2$
290	563	0
284	557	10
278	551	20
272	545	30
264	537	40
252	525	50
259	532	60
273	546	70
285	558	80
295	568	90
308	581	100

[a] T/K values calculated by the compiler.

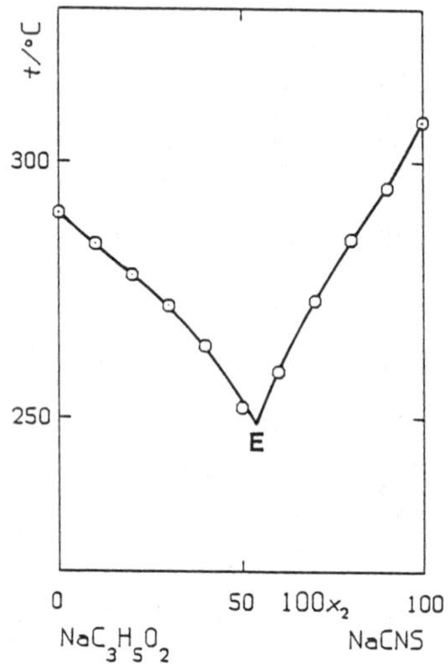

Note - The tabulated data were drawn by the compiler from Fig. 3 of the original paper.

Characteristic point(s):

Eutectic, E, at 249 ^oC and $100x_2$= 54 (authors).

AUXILIARY INFORMATION

METHOD/APPARATUS/PROCEDURE:

DTA.
Thermograph with photorecorder.
Salt(s) sealed under vacuum in Pyrex ampoules.
No other information given.

NOTE:

Concerning component 1, the fusion temperature (563 K) fairly agrees with the values listed in Preface, Tables 1 and 3 [562.4±0.5 K (DSC) and 561.88±0.03 K (adiabatic calorimetry) respectively], whereas the figure by Sokolov (571; Ref. 1) seems somewhat too high. An approximately equal difference exists also between Storonkin et al.'s and Sokolov's eutectic temperatures (522 and 531 K, respectively). The temperature values measured by Storonkin et al. are likely more reliable.

SOURCE AND PURITY OF MATERIALS:

$NaC_3H_5O_2$ prepared from propanoic acid and NaOH, and "chemically pure" NaCNS, heated 10-15 h at temperatures 50-60 ^oC below their fusion temperatures, were employed.

ESTIMATED ERROR:

Temperature: accuracy probably ±2 K (compiler).

REFERENCES:

(1) Sokolov, N.M.
Zh. Obshch. Khim. 1954, **24**, 1150-1156.

COMPONENTS:	ORIGINAL MEASUREMENTS:
(1) Sodium propanoate (sodium propionate); $NaC_3H_5O_2$; [137-40-6] (2) Sodium nitrite; $NaNO_2$; [7632-00-0]	Sokolov, N.M. **Zh. Obshch. Khim.** 1957, **27**, 840-844 (*); **Russ. J. Gen. Chem. (Engl. Transl.)** 1957, **27**, 917-920.
VARIABLES:	PREPARED BY:
Temperature.	D´Andrea, G.

EXPERIMENTAL VALUES:

$t/^{o}C$	T/K^{a}	$100x_2$	$t/^{o}C$	T/K^{a}	$100x_2$
298	571	0	296	569	55
306	579	5	286	559	60
311	584	10	284	557	65
312	585	15	276	549	70
314	587	20	269	542	75
315	588	25	256	529	80
313	586	30	262	535	85
311	584	35	267	540	90
308	581	40	272	545	95
306	579	45	284	557	100
303	576	50			

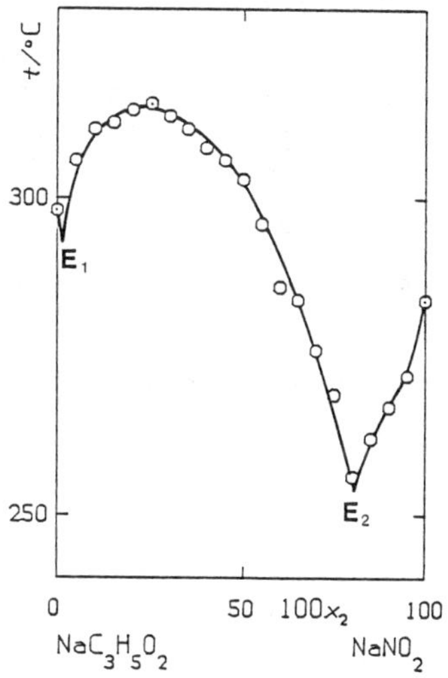

[a] **T**/K values calculated by the compiler.

Characteristic point(s):

Eutectic, E_1, at 293 °C and $100x_2$= 1.4 (author).
Eutectic, E_2, at 254 °C and $100x_2$= 80.5 (author).

Note - The coordinates of the first eutectic are given in table 2 of the original paper; they cannot, however, be drawn from the tabulated data.

Intermediate compound(s):

$Na_4(C_3H_5O_2)_3NO_2$ congruently melting at 315 °C.

AUXILIARY INFORMATION

METHOD/APPARATUS/PROCEDURE:

Visual polythermal analysis; salt mixtures melted in a glass tube (surrounded by a wider tube) and stirred with a glass thread. The temperatures of initial crystallization were measured with a Nichrome-Constantane thermocouple checked at the fusion points of water, benzoic acid, mannitol, $AgNO_3$, Cd, KNO_3, and $K_2Cr_2O_7$.

SOURCE AND PURITY OF MATERIALS:

Component 1: prepared from "chemically pure" sodium hydrogen carbonate (carbonate in the reference quoted; compiler) and commercial propanoic acid distilled before use (Ref. 1); the recovered salt was recrystallized from **n**-butanol.
Component 2: "chemically pure" material recrystallized from water.

NOTE:

The fusion temperature of component 1 (571 K) is somewhat too high: both DSC and adiabatic calorimetry provide a value close to 562 K (see Preface, Table 3).

ESTIMATED ERROR:

Temperature: accuracy probably ±2 K (compiler).

REFERENCES:

(1) Sokolov, N.M.
Zh. Obshch. Khim. 1954, **24**, 1581-1593.

COMPONENTS:

(1) Sodium propanoate (sodium propionate); $NaC_3H_5O_2$; [137-40-6]
(2) Sodium nitrate; $NaNO_3$; [7631-99-4]

ORIGINAL MEASUREMENTS:

Sokolov, N.M.
Zh. Obshch. Khim. 1954, **24**, 1150-1156.

VARIABLES:

Temperature.

PREPARED BY:

D'Andrea, G.

EXPERIMENTAL VALUES:

$t/^{\circ}C$	T/K^a	$100x_2$	$t/^{\circ}C$	T/K^a	$100x_2$
298	571	0	264	537	50
294	567	5	258	531	55
291	564	10	255	528	56.5
287	560	15	261	534	60
282	555	25	270	543	65
280	553	30	280	553	75
276	549	35	290	563	85
273	546	40	301	574	95
269	542	45	308	581	100

[a] T/K values calculated by the compiler.

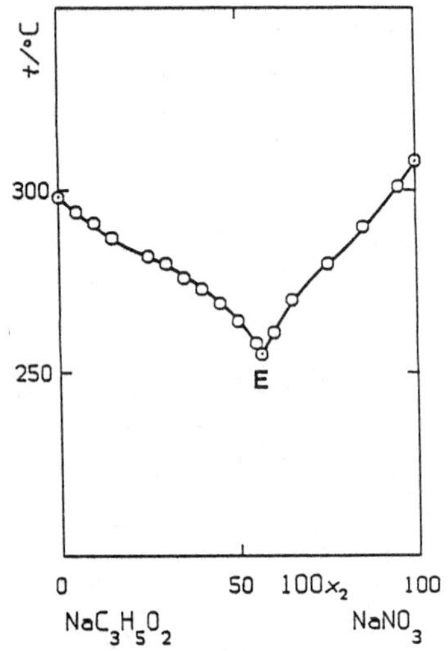

Characteristic point(s):

Eutectic, E, at 255 °C and $100x_2$= 56.5 (author).

AUXILIARY INFORMATION

METHOD/APPARATUS/PROCEDURE:

Visual polythermal analysis.
Salt(s) melted in a test tube. Temperature measured with a Nichrome-Constantane thermocouple and a millivoltmeter with mirror reading to 17 mV.

SOURCE AND PURITY OF MATERIALS:

Component 1 synthetized from propanoic acid and $NaHCO_3$. Commercial component 2 further purified by the author according to Laiti.

NOTE:

The fusion temperature of component 1 (571 K) is somewhat too high: both DSC and adiabatic calorimetry provide a value close to 562 K (see Preface, Table 3).

ESTIMATED ERROR:

Temperature: accuracy probably ±2 K (compiler).

REFERENCES:

(1) Sokolov, N.M.
Zh. Obshch. Khim. 1954, **24**, 1581-1593.

COMPONENTS:	EVALUATOR:
(1) Sodium butanoate (sodium butyrate); $NaC_4H_7O_2$; [156-54-7] (2) Sodium **iso**.butanoate (sodium **iso**.butyrate); $Na\mathbf{i}.C_4H_7O_2$; [996-30-5]	Ferloni, P., Dipartimento di Chimica Fisica, Universita' di Pavia (ITALY).

CRITICAL EVALUATION:

The system was studied only by Sokolov (Ref. 1), who claimed the existence of a continuous series of solid solutions, with a minimum at 494 K and $100\mathbf{x}_2$=72.5 .

Component 1, however, forms liquid crystals in a stability field ranging between $T_{clr}(1)/K=600.4\pm0.2$ and $T_{fus}(1)/K=524.5\pm0.5$ (according to Preface, Table 1). Consequently: (i) Sokolov's fusion temperature of component 1 (603 K) should be identified with the clearing temperature; (ii) at low values of $100\mathbf{x}_2$, Sokolov's points should refer to the formation of liquid crystals (pseudo-liquidus), and not of solid solutions (true liquidus). Besides the minimum, m, an M point should exist (although its coordinates are hard to detect on the basis of the available data, and the phase diagram should be not too different from that shown in Scheme B.3 of the Preface.

REFERENCES

(1) Sokolov, N.M.; **Zh. Obshch. Khim.** 1954, **24**, 1581-1593.

COMPONENTS:	ORIGINAL MEASUREMENTS:
(1) Sodium butanoate (sodium butyrate); $NaC_4H_7O_2$; [156-54-7] (2) Sodium **iso**.butanoate (sodium **iso**.butyrate); $Na\mathbf{i}.C_4H_7O_2$; [996-30-5]	Sokolov, N.M. **Zh. Obshch. Khim.** 1954, **24**, 1581-1593.
VARIABLES:	**PREPARED BY:**
Temperature.	D'Andrea, G.

EXPERIMENTAL VALUES:

t/°C	T/K[a]	$100\mathbf{x}_2$	t/°C	T/K[a]	$100\mathbf{x}_2$
330	603	0	235	508	55
317	590	5	229	502	60
306	579	10	224	497	65
297	570	15	222	495	70
287	560	20	221	494	72.5
279	552	25	222	495	75
270	543	30	225	498	80
264	537	35	228	501	85
257	530	40	235	508	90
250	523	45	248	521	95
242	515	50	260	533	100

[a] **T**/K values calculated by the compiler.

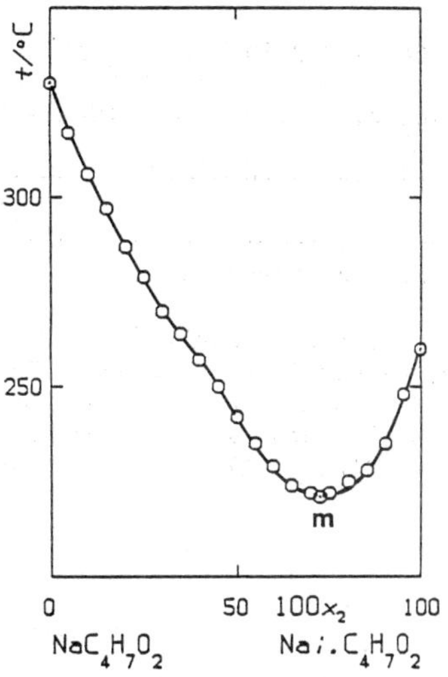

Characteristic point(s):

Continuous series of solid solutions with a minimum, m, at 221 °C and $100\mathbf{x}_2$= 72.5 (author).

METHOD/APPARATUS/PROCEDURE:	SOURCE AND PURITY OF MATERIALS:
Visual polythermal analysis. Melts contained in a glass tube and stirred. Temperatures measured with a Nichrome-Constantane thermocouple and a 17 mV full scale millivoltmeter. The temperature readings refer to the disappearance of isotropicity in the melt on cooling.	Materials prepared by reacting aqueous ("chemically pure") Na_2CO_3 with a slight excess of the proper acid of analytical purity. The solvent and excess acid were removed by heating to 160 °C. **ESTIMATED ERROR:** Temperature: accuracy probably ±2 K (compiler).

COMPONENTS:	EVALUATOR:
(1) Sodium butanoate (sodium butyrate); $NaC_4H_7O_2$; [156-54-7] (2) Sodium **iso**.pentanoate (sodium **iso**.valerate); $Na\,i.C_5H_9O_2$; [539-66-2]	Ferloni, P., Dipartimento di Chimica Fisica, Universita´ di Pavia (ITALY).

CRITICAL EVALUATION:

This system was studied only by Sokolov (Ref. 1), who suggests a eutectic phase diagram, the invariant point being at 530 K (257 °C) and $100\mathbf{x}_2$ =90.5. Both components, however, form liquid crystals.

Therefore, the fusion temperatures, $T_{fus}(1)$=603 K (330 °C) and $T_{fus}(2)$=535 K (262 °C), should be identified with the clearing temperatures, the corresponding values from Tables 1, 2 of the Preface being $T_{clr}(1)=600.4\pm0.2$ K, and $T_{clr}(2)=559\pm1$ K, respectively. The discrepancy between the values concerning component 2 might be attributed to some impurity of Sokolov´s samples, inasmuch as the value from Preface (Table 2) meets rather satisfactorily those reported by Ubbelohde et al. (556 K; Ref. 2) and by Duruz et al. (553 K; Ref. 3). No mention is made by the author of other phase transitions occurring in either component, including those corresponding to the actual fusion, which should be $T_{fus}(1)=524\pm0.5$ K (Preface, Table 1) and $T_{fus}(2)= 461.5\pm0.5$ K (Table 2).

Accordingly, the phase diagram of the system should be modified. The available data do not allow one to rule out neither of the following possibilities: (i) the eutectic point should be identified with a minimum point in a continuous series of liquid crystal solutions; (ii) the eutectic point should be identified with an M''_E point, at which the isotropic liquid should be in equilibrium with two liquid crystal solutions of different composition (Preface, Scheme C.3, Fig. 3.3).

REFERENCES:

(1) Sokolov, N.M.
Zh. Obshch. Khim. 1954, **24**, 1581-1593.

(2) Ubbelohde, A.R.; Michels, H.J.; Duruz, J.J.
Nature 1970, **228**, 50-52.

(3) Duruz, J.J.; Michels, H.J.; Ubbelohde, A.R.
Proc. R. Soc. London 1971, A322, 281-299.

COMPONENTS:

(1) Sodium butanoate (sodium butyrate); $NaC_4H_7O_2$; [156-54-7]
(2) Sodium **iso**.pentanoate (sodium **iso**.valerate); $Na\textbf{i}.C_5H_9O_2$; [539-66-2]

ORIGINAL MEASUREMENTS:

Sokolov, N.M.
Zh. Obshch. Khim. 1954, **24**, 1581-1593.

VARIABLES:

Temperature.

PREPARED BY:

D'Andrea, G.

EXPERIMENTAL VALUES:

$t/^{\circ}C$	T/K[a]	$100x_2$	$t/^{\circ}C$	T/K[a]	$100x_2$
330	603	0	287	560	55
326	599	5	284	557	60
323	596	10	281	554	65
320	593	15	277	550	70
316	589	20	273	546	75
312	585	25	269	542	80
308	581	30	263	536	85
305	578	35	258	531	90
300	573	40	257	530	90.5
295	568	45	263	536	95
292	565	50	262	535	100

[a] T/K values calculated by the compiler.

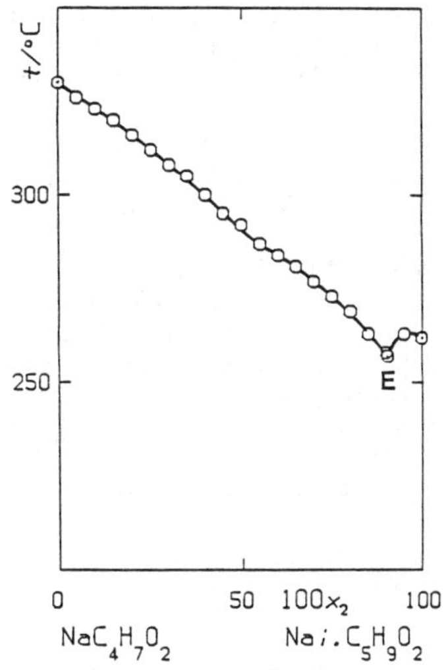

Characteristic point(s): Eutectic, E, at 257 °C and $100x_2$= 90.5 (author).

AUXILIARY INFORMATION

METHOD/APPARATUS/PROCEDURE:

Visual polythermal analysis. Melts contained in a glass tube and stirred. Temperatures measured with a Nichrome-Constantane thermocouple and a 17 mV full scale millivoltmeter. The temperature readings refer to the disappearance of isotropicity in the melt on cooling.

SOURCE AND PURITY OF MATERIALS:

Materials prepared by reacting aqueous ("chemically pure") Na_2CO_3 with a slight excess of the proper acid of analytical purity. The solvent and excess acid were removed by heating to 160 °C.

ESTIMATED ERROR:

Temperature: accuracy probably ±2 K (compiler).

COMPONENTS:

(1) Sodium butanoate (sodium butyrate); $NaC_4H_7O_2$; [156-54-7]
(2) Sodium hexanoate (sodium caproate); $NaC_6H_{11}O_2$; [10051-44-2]

EVALUATOR:

Ferloni, P.,
Dipartimento di Chimica Fisica,
Universita´ di Pavia (ITALY).

CRITICAL EVALUATION:

This system was studied only by Sokolov (Ref. 1), who claimed the existence of two eutectics [E_1, at 590 K (317 ^oC) and 100$\mathbf{x}_2$= 22.5; E_2, at 590 K (317 ^oC) and 100$\mathbf{x}_2$= 27.5], and of the intermediate compound $Na_4(C_4H_7O_2)_3C_6H_{11}O_2$, congruently melting at 594 K (321 ^oC).

Both components, however, form liquid crystals. Therefore, Sokolov´s fusion temperatures, $T_{fus}(1)$= 603 K (330 ^oC) and $\mathbf{T}_{fus}(2)$= 638 K (365 ^oC), should be identified with clearing temperatures, the corresponding values from Preface, Table 1 being $\mathbf{T}_{clr}(1)$= 600.4$\pm$0.2 K and $\mathbf{T}_{clr}(2)$= 639.0$\pm$0.5 K, respectively.

No mention is made by the author of other phase transitions of either component, including those corresponding to their actual fusions, which ought to occur at $\mathbf{T}_{fus}(1)$= 524.5$\pm$0.5 K and $\mathbf{T}_{fus}(2)$= 499.6$\pm$0.6 K, respectively (see Table 1).

Concerning the phase diagram, the available data suggest the following interpretations as possible. If the maximum at 594 K (321 ^oC) and 100$\mathbf{x}_2$= 25 does exist, Sokolov´s eutectics could be identified with either M'_E points at the opposite sides of the distectic pertinent to a congruently melting intermediate compound (Preface, Scheme D.2), or m points in a situation similar to that shown in Scheme C.3. Conversely, if the occurrence of the maximum is considered as insufficiently proved, one might think of the existence of either an M''_E point (with limited series of liquid crystal solutions on both sides; Scheme C.2), or a (single) minimum in a continuous series of liquid crystal solutions.

REFERENCES:

(1) Sokolov, N.M.
Zh. Obshch. Khim. 1954, **24**, 1581-1593.

COMPONENTS:

(1) Sodium butanoate (sodium butyrate); $NaC_4H_7O_2$; [156-54-7]
(2) Sodium hexanoate (sodium caproate); $NaC_6H_{11}O_2$; [10051-44-2]

ORIGINAL MEASUREMENTS:

Sokolov, N.M.
Zh. Obshch. Khim. 1954, **24**, 1581-1593.

VARIABLES:

Temperature.

PREPARED BY:

D'Andrea, G.

EXPERIMENTAL VALUES:

t/°C	T/K[a]	$100x_2$	t/°C	T/K[a]	$100x_2$
330	603	0	334	607	50
328	601	5	340	613	55
324	597	10	343	616	60
320	593	15	344	617	65
318	591	20	349	622	70
317	590	22.5	353	626	75
321	594	25	356	629	80
317	590	27.5	359	632	85
319	592	30	360	633	90
323	596	35	363	636	95
326	599	40	365	638	100
331	604	45			

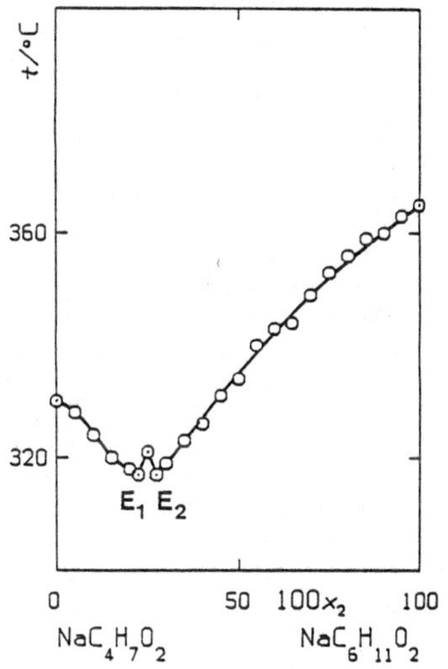

[a]T/K values calculated by the compiler.

Characteristic point(s):

Eutectic, E_1, at 317 °C and $100x_2$= 22.5 (author).
Eutectic, E_2, at 317 °C and $100x_2$= 27.5 (author).

Intermediate compound(s):

$Na_4(C_4H_7O_2)_3C_6H_{11}O_2$ [erroneously indicated as $Na_5(C_4H_7O_2)_4C_6H_{11}O_2$ in the text, compiler], congruently melting at 321 °C.

AUXILIARY INFORMATION

METHOD/APPARATUS/PROCEDURE:

Visual polythermal analysis.
Melts contained in a glass tube and stirred.
Temperatures measured with a Nichrome-Constantane thermocouple and a 17 mV full scale millivoltmeter. The temperature readings refer to the disappearance of isotropicity in the melt on cooling.

SOURCE AND PURITY OF MATERIALS:

Materials prepared by reacting aqueous ("chemically pure") Na_2CO_3 with a slight excess of the proper acid of analytical purity. The solvent and excess acid were removed by heating to 160 °C.

ESTIMATED ERROR:

Temperature: precision probably ±2 K (compiler).

REFERENCES:

COMPONENTS:	ORIGINAL MEASUREMENTS:
(1) Sodium butanoate (sodium butyrate); $NaC_4H_7O_2$; [156-54-7] (2) Sodium benzoate; $NaC_7H_5O_2$; [532-32-1]	Sokolov, N.M. **Zh. Obshch. Khim.** 1954, 24, 1581-1593.
VARIABLES:	PREPARED BY:
Temperature.	D'Andrea, G.

EXPERIMENTAL VALUES:

$t/^{o}C$	T/K^{a}	$100\mathbf{x}_2$
330	603	0
330	603	0.13
349	622	5
361	634	10
370	643	15
378	651	20
386	659	25
394	667	30
401	674	35
408	681	40
415	688	45
421	694	50
427	700	55
434	707	60
463	736	100

[a] T/K values calculated by the compiler.

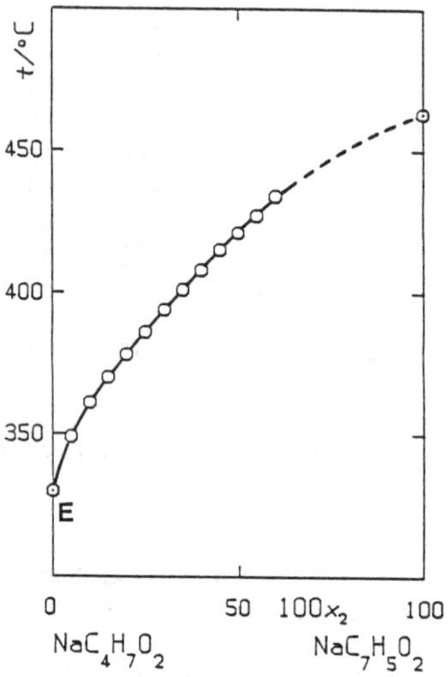

Characteristic point(s):

Eutectic, E, at 330 °C and $100\mathbf{x}_2$= 0.13 (author).

Note - The system was investigated at $0 \leq 100\mathbf{x}_2 \leq 60$ due to thermal instability of the butanoate.

AUXILIARY INFORMATION

METHOD/APPARATUS/PROCEDURE:

Visual polythermal analysis.
Melts contained in a glass tube and stirred.
Temperatures measured with a Nichrome-Constantane thermocouple and a 17 mV full scale millivoltmeter. The temperature readings refer to the disappearance of isotropicity in the melt on cooling.

SOURCE AND PURITY OF MATERIALS:

Component 1: prepared by reacting aqueous ("chemically pure") Na_2CO_3 with a slight excess of **n**-butanoic acid of analytical purity. The solvent and excess acid were removed by heating to 160 °C. Component 2: "chemically pure" material.

NOTE:

Component 1 forms liquid crystals. Therefore Sokolov's fusion temperature, $\mathbf{T}_{fus}(1)$= 603 K, should be identified with the clearing temperature, the corresponding value in Table 1 of the Preface being 600.4±0.2 K. It is hard to infer the topology of the system from the available data: indeed, the phase diagram might be similar to that shown in Preface, Scheme A.1, but other possibilities remain open.

ESTIMATED ERROR:

Temperature: accuracy probably ±2 K (compiler).

REFERENCES:

COMPONENTS:	EVALUATOR:
(1) Sodium butanoate (sodium butyrate); $NaC_4H_7O_2$; [156-54-7] (2) Sodium octadecanoate (sodium stearate); $NaC_{18}H_{35}O_2$; [822-16-2]	Spinolo, G., Dipartimento di Chimica Fisica, Universita´ di Pavia (ITALY).

CRITICAL EVALUATION:

This system was studied only by Sokolov (Ref. 1) who employed the visual polythermal analysis to draw the lower boundary of the isotropic liquid field. From the shape of this boundary, he concluded that the intermediate compound $Na_5(C_4H_7O_2)_3(C_{18}H_{35}O_2)_2$ [congruently melting at 663 K (390 ^oC)] was formed, and that the limits of the stability field of this compound were a eutectic at 521 K (248 ^oC) and $100\mathbf{x}_2$= 15, and a "perekhodnaya tochka" at 582 K (309 ^oC) and $100\mathbf{x}_2$= 96.5.

Actually, both components form liquid crystals, the liquid crystalline phases being one for component 1 (see Preface, Table 1), and two for component 2 (see Table 4 of the Preface). Sokolov´s fusion temperatures, $\mathbf{T}_{fus}(1)$= 603 K (330 ^oC), and $\mathbf{T}_{fus}(2)$= 581 K (308 ^oC), consequently should be identified with the clearing temperatures, the corresponding values from Tables 1 and 4 being 600.4$\pm$0.2 and 552.7 K, respectively.

Since the complete topology of the binary can hardly be interpreted from the available data, it is more realistic to list here the few points which, in the evaluator´s opinion, seem to be sufficiently reliable.

(i) At intermediate compositions it seems reasonable to assume that a continuous series of liquid crystal solutions is formed, with an azeotrope at 663 K and $100\mathbf{x}_2$= 40.

(ii) Accordingly, the left hand section ($0 \leq 100\mathbf{x}_2 \leq 40$) of the phase diagram might be interpreted with reference to Preface, Scheme C.2: in this case, Sokolov´s eutectic should be identified with an $M"_E$ point.

Conversely, no definite interpretation of the phase diagram at high $100\mathbf{x}_2$ values seems possible. Indeed, it is not clear how Sokolov could argue the occurrence of an invariant (the "perekhodnaya tochka" at $100\mathbf{x}_2$= 96.5) from the trend of his experimental data which does not support unambiguously any significant slope change of the curve in this region. Moreover, Sokolov´s "fusion" temperature of component 2 (581 K) looks as fully unreliable, being 18 K higher than the second highest $\mathbf{T}_{clr}$ value determined during the last 30 years (Ref. 2), and 28 K higher than the clearing temperature listed in Table 4 of the Preface.

REFERENCES:

(1) Sokolov, N.M.
Zh. Obshch. Khim. 1954, **24**, 1581-1593.

(2) Sanesi, M.; Cingolani, A.; Tonelli, P.L.; Franzosini, P.
Thermal Properties, in **Thermodynamic and Transport Properties of Organic Salts**, IUPAC Chemical Data Series No. 28 (Franzosini, P.; Sanesi, M.; Editors), Pergamon Press, Oxford, 1980, 29-115.

COMPONENTS:	ORIGINAL MEASUREMENTS:
(1) Sodium butanoate (sodium butyrate); $NaC_4H_7O_2$; [156-54-7] (2) Sodium octadecanoate (sodium stearate); $NaC_{18}H_{35}O_2$; [822-16-2]	Sokolov, N.M. **Zh. Obshch. Khim.** 1954, **24**, 1581-1593.
VARIABLES:	PREPARED BY:
Temperature.	D'Andrea, G.

EXPERIMENTAL VALUES:

$t/^{\circ}C$	T/K^a	$100x_2$	$t/^{\circ}C$	T/K^a	$100x_2$
330	603	0	376	649	60
289	562	5	370	643	65
261	534	10	364	637	70
248	521	15	358	631	75
277	550	20	350	623	80
317	590	25	340	613	85
351	624	30	330	603	90
379	652	35	314	587	95
390	663	40	309	582	96.5
389	662	45	312	585	98.5
386	659	50	308	581	100
380	653	55			

[a] T/K values calculated by the compiler.

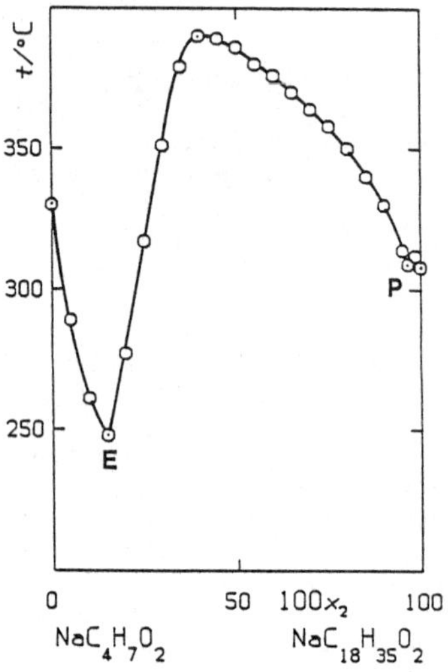

Characteristic point(s):

Eutectic, E, at 248 °C and $100x_2$= 15 (author).
Characteristic point, P ("perekhodnaya tochka" in the original text; see the Introduction), at 309 °C and $100x_2$= 96.5 (author).

Intermediate compound(s):

$Na_5(C_4H_7O_2)_3(C_{18}H_{35}O_2)_2$, congruently melting at 390 °C.

AUXILIARY INFORMATION

METHOD/APPARATUS/PROCEDURE:	SOURCE AND PURITY OF MATERIALS:
Visual polythermal analysis. Melts contained in a glass tube and stirred. Temperatures measured with a Nichrome-Constantane thermocouple and a 17 mV full scale millivoltmeter. The temperature readings refer to the disappearance of isotropicity in the melt on cooling.	Component 1: prepared by reacting aqueous ("chemically pure") Na_2CO_3 with a slight excess of n-butanoic acid of analytical purity. The solvent and excess acid were removed by heating to 160 °C. Component 2: "chemically pure" material.
	ESTIMATED ERROR:
	Temperature: precision probably ±2 K (compiler).
	REFERENCES:

COMPONENTS:	EVALUATOR:
(1) Sodium butanoate (sodium butyrate); $NaC_4H_7O_2$; [156-54-7] (2) Sodium thiocyanate; NaCNS; [540-72-7]	Spinolo, G., Dipartimento di Chimica Fisica, Universita' di Pavia (ITALY).

CRITICAL EVALUATION:

This system was studied only by Sokolov (Ref. 1), who restricted his visual polythermal investigation to the lower boundary of the isotropic liquid field. He asserted the existence of the intermediate compound $Na_4(C_4H_7O_2)_3CNS$, which melts incongruently at 541 K (268 ^{o}C), and of a eutectic at 535 K (262 ^{o}C) and $100\mathbf{x}_2$= 48.5.

Component 1, however, forms liquid crystals, which are stable between $\mathbf{T}_{fus}(1)$= 524.5±0.5 K and $\mathbf{T}_{clr}(1)$= 600.4±0.2 (see Preface, Table 1). Sokolov's fusion temperature (603 K) consequently should be identified with the clearing temperature, whereas the T_{trs} value (525 K), reported by the same author in a subsequent paper (Ref. 2), is in close agreement with the fusion temperature given in Table 1.

In the evaluator's opinion, Sokolov's findings are not sufficient to prove unambiguously the existence of the intermediate compound. Consequently, more than one interpretation can be given for the topology of this binary.

Indeed, if the compound does exist:

(i) the phase diagram could be similar to that shown in Preface, Scheme D.3,
(ii) Sokolov's "Perekhodnaya tochka" should to be identified with an M'_P point; and
(iii) the occurrence of a (so far undetected) M_E point is required.

If, on the contrary, one assumes that the intermediate coumpound does not exist, Sokolov's invariant at 541 K and $100\mathbf{x}_2$= 31.5 might be connected with the fusion of component 1 in the way shown in Scheme B.2 of the Preface.

REFERENCES:

(1) Sokolov, N.M.
Zh. Obshch. Khim. 1954, **24**, 1150-1156.

(2) Sokolov, N.M.
Tezisy Dokl. X Nauch. Konf. S.M.I. 1956.

COMPONENTS:

(1) Sodium butanoate (sodium butyrate); $NaC_4H_7O_2$; [156-54-7]
(2) Sodium thiocyanate; NaCNS; [540-72-7]

ORIGINAL MEASUREMENTS:

Sokolov, N.M.
Zh. Obshch. Khim. 1954, **24**, 1150-1156.

VARIABLES:

Temperature.

PREPARED BY:

D'Andrea, G.

EXPERIMENTAL VALUES:

$t/^{o}C$	T/K^{a}	$100x_2$	$t/^{o}C$	T/K^{a}	$100x_2$
330	603	0	263	536	45
328	601	5	262	535	48.5
324	597	10	269	542	50
316	589	15	280	553	55
291	564	25	287	560	60
275	548	30	290	563	65
268	541	31.5	298	571	75
266	539	35	304	577	90
264	537	40	311	584	100

[a] T/K values calculated by the compiler.

Characteristic point(s):

Eutectic, E, at 262 ^{o}C and $100x_2$= 48.5 (author).
Invariant point, P ("perekhodnaya tochka" in the original text, see the Introduction), at 268 ^{o}C and $100x_2$= 31.5 (author).

Intermediate compound(s):

$Na_4(C_4H_7O_2)_3CNS$, incongruently melting (author).

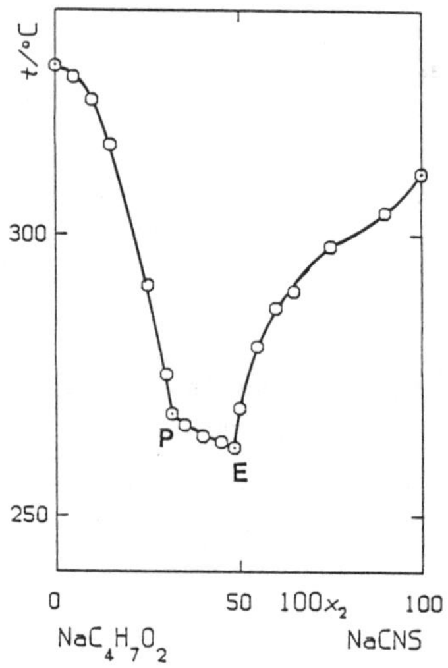

AUXILIARY INFORMATION

METHOD/APPARATUS/PROCEDURE:

Visual polythermal analysis.
Salt(s) melted in a test tube. Temperature measured with a Nichrome-Constantane thermocouple and a millivoltmeter with mirror reading to 17 mV.

SOURCE AND PURITY OF MATERIALS:

Component 1 synthetized from n-butanoic acid and $NaHCO_3$.
Component 2 of analytical purity recrystallized once from water and once from ethanol.

ESTIMATED ERROR:

Temperature: precision probably ± 2 K (compiler).

REFERENCES:

COMPONENTS:	EVALUATOR:
(1) Sodium butanoate (sodium butyrate); $NaC_4H_7O_2$; [156-54-7] (2) Sodium nitrite; $NaNO_2$; [7632-00-0]	Spinolo, G., Dipartimento di Chimica Fisica, Universita´ di Pavia (ITALY).

CRITICAL EVALUATION:

This system was studied only by Sokolov (Ref. 1) who restricted his polythermal investigation to the lower boundary of the isotropic liquid field. He claimed that an intermediate compound, i.e., $Na_4(C_4H_7O_2)_3NO_2$, exists which forms eutectics with either pure component at 590 K (317 °C) and $100\mathbf{x}_2$= 17.5, and at 347 K (274 °C) and $100\mathbf{x}_2$= 96, respectively.

No data on the solidus are available, and consequently the existence of the intermediate compound is not fully proved. Nevertheless, the evaluator is inclined to accept - at least in part - Sokolov´s interpretation of the topology of the system.

It must, however, be specified that, due to the fact that component 1 forms liquid crystals stable between 524.5±0.5 K and 600.4±0.2 K (see Preface, Table 1),
(i) the first eutectic at 590 K ought to be identified with an M'_E point; and
(ii) a further (so far undetected) invariant, presumably an M_E point, should exist.

In conclusion, the phase diagram ought to be similar to that shown in Scheme D.1 of the Preface.

REFERENCES:

(1) Sokolov, N.M.
Zh. Obshch. Khim. 1957, **27**, 840-844 (*); **Russ. J. Gen. Chem. (Engl. Transl.)** 1957, **27**, 917-920.

COMPONENTS:	ORIGINAL MEASUREMENTS:
(1) Sodium butanoate (sodium butyrate); $NaC_4H_7O_2$; [156-54-7] (2) Sodium nitrite; $NaNO_2$; [7632-00-0]	Sokolov, N.M. **Zh. Obshch. Khim.** 1957, 27, 840-844 (*); **Russ. J. Gen. Chem., Engl. Transl.,** 1957, 27, 917-920.
VARIABLES:	PREPARED BY:
Temperature.	D'Andrea, G.

EXPERIMENTAL VALUES:

t/°C	T/K[a]	$100x_2$	t/°C	T/K[a]	$100x_2$
330	603	0	312	585	55
328	601	5	309	582	60
325	598	10	306	579	65
320	593	15	304	577	70
320	593	20	300	573	75
324	597	25	296	569	80
323	596	30	291	564	85
322	595	35	285	558	90
320	593	40	276	549	95
316	589	45	284	557	100
314	587	50			

[a] T/K values calculated by the compiler.

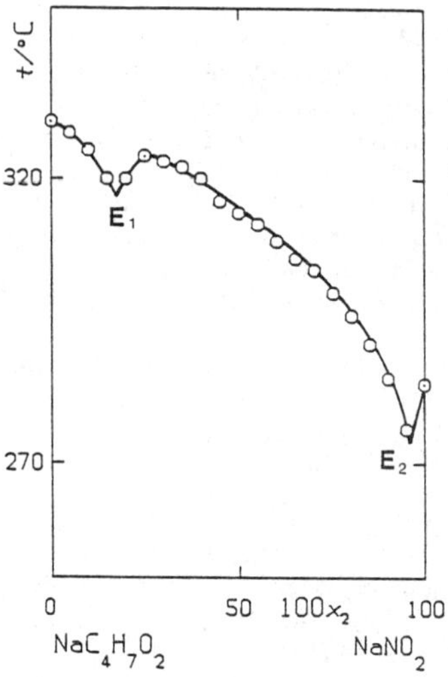

Characteristic point(s):

Eutectic, E_1, at 317 °C and $100x_2$= 17.5 (author).
Eutectic, E_2, at 274 °C and $100x_2$= 96 (author).

Note - The coordinates of the second eutectic are given in Table 2 of the original paper; they cannot, however, be drawn from the tabulated data; compiler).

Intermediate compound(s):

$Na_4(C_4H_7O_2)_3NO_2$ congruently melting at 324 °C.

AUXILIARY INFORMATION

METHOD/APPARATUS/PROCEDURE:

Visual polythermal analysis; salt mixtures melted in a glass tube (surrounded by a wider tube) and stirred with a glass thread. The temperatures of initial crystallization were measured with a Nichrome-Constantane thermocouple checked at the fusion points of water, benzoic acid, mannitol, $AgNO_3$, Cd, KNO_3, and $K_2Cr_2O_7$.

SOURCE AND PURITY OF MATERIALS:

Component 1: prepared from "chemically pure" sodium hydrogen carbonate (carbonate in the reference quoted; compiler) and commercial n-butanoic acid distilled before use (Ref. 1); the salt recovered was recrystallized from n-butanol.
Component 2: "chemically pure" material recrystallized from water.

ESTIMATED ERROR:

Temperature: accuracy probably ±2 K (compiler).

REFERENCES:

(1) Sokolov, N.M.
Zh. Obshch. Khim. 1954, **24**, 1581-1593.

COMPONENTS:	EVALUATOR:
(1) Sodium butanoate (sodium butyrate); $NaC_4H_7O_2$; [156-54-7] (2) Sodium nitrate; $NaNO_3$; [7631-99-4]	Ferloni, P., Dipartimento di Chimica Fisica, Universita´ di Pavia (ITALY).

CRITICAL EVALUATION:

The visual polythermal method was employed by Dmitrevskaya (Ref. 1) [see also Sokolov, (Ref. 2)] to study the lower boundary of the isotropic liquid field: according to this author, an incongruently melting intermediate compound of probable composition $Na_4(C_4H_7O_2)_3NO_3$ is formed, and two invariants exist, i.e., a eutectic, E [at 540 K (267 °C), and $100\mathbf{x}_2$= 50], and a "perekhodnaya tochka", P [at 549 K (276 °C), and $100\mathbf{x}_2$= 27].

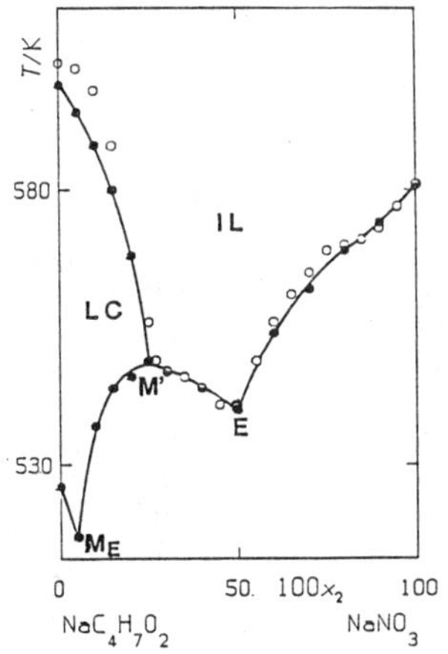

Component 1, however, forms liquid crystals. Accordingly, the fusion temperature, $T_{fus}(1)$= 603 K (330 °C), reported in Ref. 1 should be identified with the clearing temperature, $T_{clr}(1)$, of component 1, the corresponding value from Preface, Table 1 being 600.4±0.2 K.

For the same component, Table 1 of the Preface [besides the $T_{clr}(1)$ value] provides four solid state transitions (at 450.4±0.5, 489.8±0.2, 498.3±0.3, and 508.4±0.5) and $T_{fus}(1)$/K= 524.5±0.5.These phase relations, first stated on the basis of DSC records, were subsequently confirmed by Schiraldi and Chiodelli´s conductometric results (Ref. 3). Phase transformations are quoted in Ref. 1 from Ref. 4 as occurring at 390, 505, 525, and 589 K, respectively. A comparison of the two sets of data allows one to identify the two intermediate transition temperatures from Ref. 4 with the first $T_{trs}(2)$ and $T_{fus}(2)$ from Table 1. Reasonable doubts can be raised, on the contrary, about the actual existence of Ref. 4 highest transition (which - if present - should represent the transformation from a liquid crystalline phase into another one) and of the lowest transformations.

More recently, Prisyazhnyi et al. (Ref. 5) - to whom Refs. 1, 2 seem to be unknown - carried out a derivatographical re-investigation of the system, which allowed them to draw the lower boundaries of both the isotropic liquid, and the liquid crystal field. Concerning component 1, their clearing [$T_{clr}(1)$= 599 K (326 °C)] and fusion [$T_{fus}(1)$= 526 K (253 °C)] temperatures substantially agree with those from Table 1 of the Preface; it is moreover to be stressed that they do not mention any transition intermediate between $T_{clr}(1)$ and $T_{fus}(1)$.

Prisyazhnyi et al.´s, and Dmitrevskaya´s results (filled and empty circles, respectively) are compared in the figure (IL: isotropic liquid; LC: liquid crystals), an inspection of which allows one to make the following remarks. An invariant exists, which escaped Dmitrevskaya´s attention, and is reasonably to be classified as an M_E point. Moreover, the invariant at about $100\mathbf{x}_2$= 25 is actually an M´ point: its abscissa being known only approximately, it can hardly be decided if it is of the M'_E or of the M'_P type: in the former case, the complete phase diagram should be similar to Scheme D.1 of the Preface; in the latter one, to Scheme D.3.

The two-phase region pertinent to the liquid crystal - isotropic liquid equilibria might be so narrow as to have prevented Prisyazhnyi et al. to observe two distinct sets of points in this region, whereas the lack of information by the same authors about eutectic fusion in the differrent samples studied by derivatographical analysis remains rather surprising.

REFERENCES:

(1) Dmitrevskaya, O.I.; **Zh. Obshch. Khim.** 1958, **28**, 2007-2013 (*); **Russ. J. Gen. Chem. (Engl. Transl.)** 1958, **28**, 2046-2051.
(2) Sokolov, N.M.; **Zh. Obshch. Khim.** 1954, **24**, 1150-1156.
(3) Schiraldi, A.; Chiodelli, G.; **J. Phys. E: Sci. Instr.** 1977, **10**, 596-599.
(4) Sokolov, N.M. ; **Tezisy Dokl. X Nauch. Konf. S.M.I.** 1956.
(5) Prisyazhnyi, V.D.; Mirnyi, V.N.; Mirnaya, T.A.; **Zh. Neorg. Khim.** 1983, **28**, 253-255.

COMPONENTS:

(1) Sodium butanoate (sodium butyrate); $NaC_4H_7O_2$; [156-54-7]
(2) Sodium nitrate; $NaNO_3$; [7631-99-4]

ORIGINAL MEASUREMENTS:

Dmitrevskaya, O.I.
Zh. Obshch. Khim. 1958, **28**, 2007-2013 (*);
Russ. J. Gen. Chem. (Engl. Transl.) 1958, **28**, 2046-2051.

VARIABLES:

Temperature.

PREPARED BY:

D'Andrea, G.

EXPERIMENTAL VALUES:

$t/^oC$	T/K^a	$100x_2$	$t/^oC$	T/K^a	$100x_2$
330[b]	603	0	276[b]	549	55
329[b]	602	5	283[b]	556	60
325[b]	598	10	288[b]	561	65
315[b]	588	15	292[b]	565	70
283[b]	556	25	296[b]	569	75
276[b]	549	27	297	570	80
274[b]	547	30	298[b]	571	85
273[b]	546	35	300	573	90
271[b]	544	40	304[c]	577	95
268[b]	541	45	308[b]	581	100
267[b]	540	50			

[a] T/K values calculated by the compiler.
[b] Value already reported in a previous paper by Sokolov (Ref. 1); the compiler preferred to employ the values tabulated by Dmitrevskaya which are more complete.
[c] 302 in Sokolov's paper (Ref. 1).

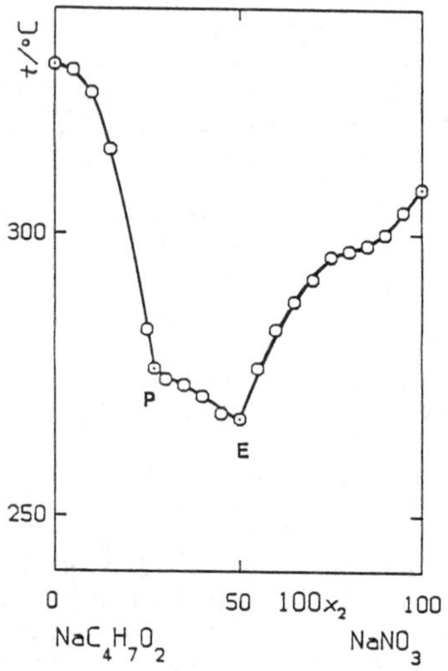

Characteristic point(s):
Eutectic, E, at 267 ^oC and $100x_2$= 50 (author).
Characteristic point, P ("perekhodnaya tochka" in the original text; see the Introduction), at 276 ^oC and $100x_2$= 27 (author).

Intermediate compound(s):
Probably $Na_4(C_4H_7O_2)_3NO_3$, incongruently melting (author).

AUXILIARY INFORMATION

METHOD/APPARATUS/PROCEDURE:

Visual polythermal analysis.
Temperatures measured with a Nichrome-Constantane thermocouple.

SOURCE AND PURITY OF MATERIALS:

Component 1 synthetized from "chemically pure" sodium hydrogen carbonate and **n**-butanoic acid that first had been distilled twice.
"Chemically pure" component 2 recrystallized and dried to constant mass.
Component 1 undergoes phase transitions at $t_{trs}(1)/^oC$= 117, 232, 252, 316 (Ref. 2).
Component 2 undergoes a phase transition at $t_{trs}(2)/^oC$= 270 (current literature).

ESTIMATED ERROR:

Temperature: accuracy probably ±2 K (compiler).

REFERENCES:

(1) Sokolov, N.M.
Zh. Obshch. Khim. 1954, **24**, 1150-1156.
(2) Sokolov, N.M.
Tezisy Dokl. X Nauch. Konf. S.M.I. 1956.

COMPONENTS:

(1) Sodium butanoate (sodium butyrate); $NaC_4H_7O_2$; [156-54-7]
(2) Sodium nitrate; $NaNO_3$; [7631-99-4]

ORIGINAL MEASUREMENTS:

Prisyazhnyi, V.D.; Mirnyi, V.N.; Mirnaya, T.A.
Zh. Neorg. Khim. 1983, **28**, 253-255; **Russ. J. Inorg. Chem. (Engl. Transl.)** 1983, **28**, 140-141 (*).

VARIABLES:

Temperature.

PREPARED BY:

D'Andrea, G.

EXPERIMENTAL VALUES:

The results are reported only in graphical form (see figure; data read with a digitizer by the compiler on Fig. 1 of the original paper; empty circles: liquid crystal - isotropic liquid equilibria; filled circles: solid - liquid crystal or solid - isotropic liquid equilibria).

Characteristic point(s):

Invariant point, M_E, at about 244 °C and $100x_2$ about 5 (compiler).
Eutectic, E, at about 267 °C and $100x_2$ about 50 (compiler).
Invariant point, M', at about 276 °C and $100x_2$ about 25 (compiler).

Intermediate compound(s):

$Na_4(C_4H_7O_2)_3NO_3$, melting at about 276 °C (compiler).

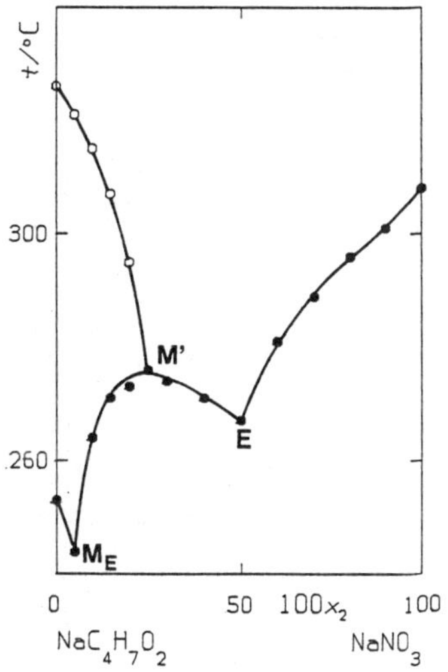

AUXILIARY INFORMATION

METHOD/APPARATUS/PROCEDURE:

The heating and cooling traces were recorded in an atmosphere of purified argon with an OD-102 derivatograph (MOM, Hungary) working at a rate of 6-8 K min^{-1}, and using Al_2O_3 as the reference material. Temperatures were measured with a Pt/Pt-Rh thermocouple. A hot-stage Amplival polarizing microscope was employed to detect the transformation points from the liquid crystalline into the isotropic liquid phase.

SOURCE AND PURITY OF MATERIALS:

Not stated.
Component 1: $t_{fus}(1)/°C$ about 253; $t_{clr}(1)/°C$ about 326 (compiler).
Component 2: $t_{fus}(2)/°C$ about 308 (compiler).

ESTIMATED ERROR:

Temperature: accuracy is not evaluable (compiler).

REFERENCES:

COMPONENTS:	EVALUATOR:
(1) Sodium **iso**.butanoate (sodium **iso**.butyrate); Na**i**.$C_4H_7O_2$; [996-30-5] (2) Sodium **iso**.pentanoate (sodium **iso**.valerate); Na**i**.$C_5H_9O_2$; [539-66-2]	Ferloni, P., Dipartimento di Chimica Fisica, Universita´ di Pavia (ITALY).

CRITICAL EVALUATION:

This system was studied only by Sokolov (Ref. 1), who claimed the existence of a continuous series of solid solutions, with a minimum at 461-462 K and $100\mathbf{x}_2$= 50.

The fusion temperature of component 1 (533 K) is not far from that reported in Preface, Table 2 (526.9±0.7 K).

Component 2, however, forms liquid crystals in a stability field ranging between $\mathbf{T}_{clr}(2)$/K= 559±1 and $\mathbf{T}_{fus}(2)$/K= 461.5±0.6 (according to Table 2).

Consequently, Sokolov´s fusion temperature of component 2 should reasonably be identified as the clearing temperature of this component. Its value, i.e., 535 K, is remarkably lower than that listed in Table 2, i.e., 559±1 K: the latter figure, however, meets rather satisfactorily those reported by Ubbelohde et al. (556 K; Ref. 2), and by Duruz et al. (553 K; Ref. 3), so that the discrepancy might be attributed to insufficient purity of Sokolov´s sample (indeed, due to the - usually small - value of the enthalpy change associated with clearing, a small amount of impurities is often sufficient to cause a dramatic drop of the clearing temperature).

Many of Sokolov´s points should represent isotropic liquid - liquid crystal, rather than isotropic liquid - solid equilibria.

Details of the phase diagram, however, are hard to be inferred from the available data.

REFERENCES

(1) Sokolov, N.M.
Zh. Obshch. Khim. 1954, **24**, 1581-1593.

(2) Ubbelohde, A.R.; Michels, H.J.; Duruz, J.J.
Nature 1970, **228**, 50-52.

(3) Duruz, J.J.; Michels, H.J.; Ubbelohde, A.R.
Proc. R. Soc. London 1971, **A322**, 281-299.

COMPONENTS:

(1) Sodium **iso**.butanoate (sodium **iso**.butyrate); $NaiC_4H_7O_2$; [996-30-5]
(2) Sodium **iso**.pentanoate (sodium **iso**.valerate); $NaiC_5H_9O_2$; [539-66-2]

ORIGINAL MEASUREMENTS:

Sokolov, N.M.
Zh. Obshch. Khim. 1954, **24**, 1581-1593.

VARIABLES:

Temperature.

PREPARED BY:

D'Andrea, G.

EXPERIMENTAL VALUES:

$t/^{\circ}C$	T/K^a	$100x_2$	$t/^{\circ}C$	T/K^a	$100x_2$
260	533	0	189	462	55
248	521	5	191	464	60
238	511	10	194	467	65
229	502	15	199	472	70
220	493	20	207	480	75
213	486	25	215	488	80
207	480	30	225	498	85
201	474	35	237	510	90
197	470	40	248	521	95
193	466	45	262	535	100
188	461	50			

[a] T/K values calculated by the compiler.

Characteristic point(s):

Minimum, m, at 189 $^{\circ}C$ (188 $^{\circ}C$, according to the table, compiler) and $100x_2$= 50 (author).

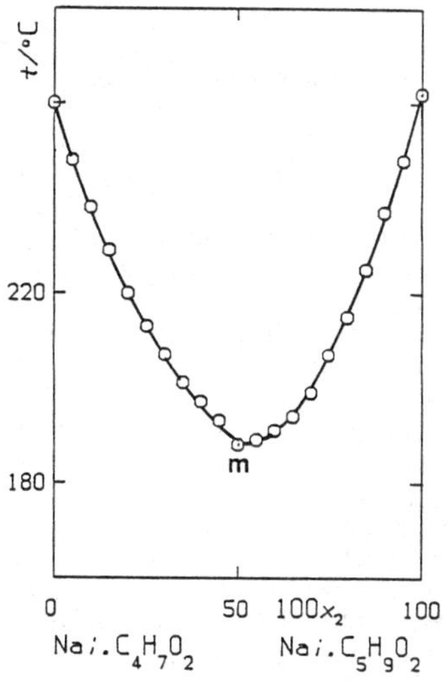

AUXILIARY INFORMATION

METHOD/APPARATUS/PROCEDURE:

Visual polythermal analysis.
Melts contained in a glass tube and stirred.
Temperatures measured with a Nichrome-Constantane thermocouple and a 17 mV full scale millivoltmeter. The temperature readings refer to the disappearance of isotropicity in the melt on cooling.

SOURCE AND PURITY OF MATERIALS:

Materials prepared by reacting aqueous ("chemically pure") Na_2CO_3 with a slight excess of the proper acid of analytical purity. The solvent and excess acid were removed by heating to 160 $^{\circ}C$.

ESTIMATED ERROR:

Temperature: accuracy is probably ± 2 K (compiler).

REFERENCES:

COMPONENTS:

(1) Sodium **iso**.butanoate (sodium **iso**.butyrate); $NaiC_4H_7O_2$; [996-30-5]
(2) Sodium hexanoate (sodium caproate); $NaC_6H_{11}O_2$; [10051-44-2]

EVALUATOR:

Spinolo, G.,
Dipartimento di Chimica Fisica,
Universita' di Pavia (ITALY).

CRITICAL EVALUATION:

This system was studied only by Sokolov (Ref. 1) who restricted his visual polythermal investigations to the lower boundary of the isotropic liquid field; and claimed the existence of a single eutectic at 433 K (160 ^{o}C) and $100x_2$= 23.5.

Component 2, however, forms liquid crystals which are stable between 639.0$\pm$0.5 K and 499.6$\pm$0.6 K (see Preface, Table 1). Consequently, the fusion temperature 638 K (365°C; Ref. 1) should be identified with the clearing temperature, and Sokolov's outline of the phase diagram is incomplete. In particular, at least two invariants should exist, although the available data do not allow one to state with certainty their nature.

The following hypotheses can be tentatively suggested.

(i) Sokolov's invariant should be considered as an M'_E point; a second one (an M_E point so far undetected) should exist at a lower temperature and at a higher x_2 value.

(ii) Sokolov's invariant is actually a eutectic, E, and a second invariant (an M'_P point so far undetected) should exist at higher temperature and at a higher x_2 value.

If hypothesis (i) is the correct one, the phase diagram ought to be similar to that shown in Scheme A.2 of the Preface.

However, taking into account that T_{fus}(2) (499.6$\pm$0.6 K; Table 1 of the Preface) is significantly higher than the fusion temperature of Sokolov's invariant, and that the enthalpy change pertinent to fusion is usually much larger than that pertinent to clearing, the evaluator is inclined to prefer hypothesis (ii). Reference should be therefore made to Preface, Scheme B.1 or B.2.

REFERENCES:

(1) Sokolov, N.M.
Zh. Obshch. Khim. 1954, **24**, 1581-1593.

COMPONENTS:

(1) Sodium **iso.**butanoate (sodium **iso.**butyrate); $Nai.C_4H_7O_2$; [996-30-5]
(2) Sodium hexanoate (sodium caproate); $NaC_6H_{11}O_2$; [10051-44-2]

ORIGINAL MEASUREMENTS:

Sokolov, N.M.
Zh. Obshch. Khim. 1954, **24**, 1581-1593.

VARIABLES:

Temperature.

PREPARED BY:

D'Andrea, G.

EXPERIMENTAL VALUES:

t/°C	T/K[a]	$100x_2$	t/°C	T/K[a]	$100x_2$
260	533	0	205	478	50
237	510	5	220	493	55
218	491	10	235	508	60
195	468	15	252	525	65
175	448	20	270	543	70
160	433	23.5	290	563	75
161	434	25	329	602	85
168	441	30	345	618	90
175	448	35	356	629	95
182	455	40	365	638	100
191	464	45			

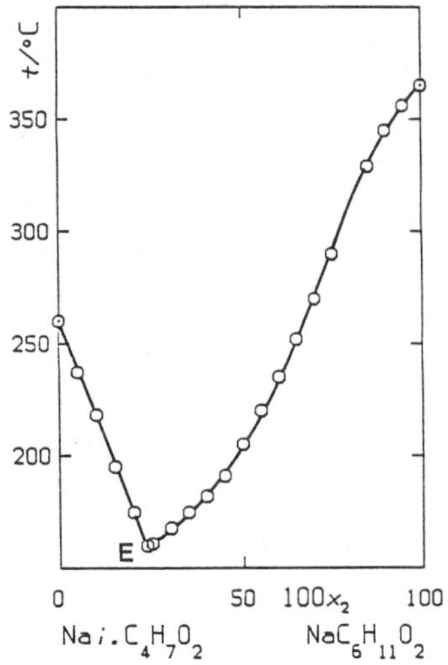

[a] T/K values calculated by the compiler.

Characteristic point(s):

Eutectic, E, at 160 °C and $100x_2$= 23.5 (author).

AUXILIARY INFORMATION

METHOD/APPARATUS/PROCEDURE:

Visual polythermal analysis.
Melts contained in a glass tube and stirred.
Temperatures measured with a Nichrome-Constantane thermocouple and a 17 mV full scale millivoltmeter. The temperature readings refer to the disappearance of isotropicity in the melt on cooling.

SOURCE AND PURITY OF MATERIALS:

Materials prepared by reacting aqueous ("chemically pure") Na_2CO_3 with a slight excess of the proper acid of analytical purity. The solvent and excess acid were removed by heating to 160 °C.

ESTIMATED ERROR:

Temperature: accuracy is probably ±2 K (compiler).

REFERENCES:

COMPONENTS:

(1) Sodium **iso**.butanoate (sodium **iso**.butyrate); $Nai.C_4H_7O_2$; [996-30-5]
(2) Sodium benzoate; $NaC_7H_5O_2$; [532-32-1]

ORIGINAL MEASUREMENTS:

Sokolov, N.M.
Zh. Obshch. Khim. 1954, **24**, 1581-1593.

VARIABLES:

Temperature.

PREPARED BY:

D'Andrea, G.

EXPERIMENTAL VALUES:

$t/^{o}C$	T/K^{a}	$100x_2$	$t/^{o}C$	T/K^{a}	$100x_2$
260	533	0	335	608	40
228	501	3.5	344	617	45
235	508	5	355	628	50
256	529	10	367	640	55
272	545	15	379	652	60
288	561	20	389	662	65
301	574	25	399	672	70
312	585	30	408	681	75
322	595	35	463	736	100

[a] T/K values calculated by the compiler.

Characteristic point(s):

Eutectic, E, at 228 ^{o}C and $100x_2$= 3.5 (author).

Note - The system was investigated at $0 \leq 100x_2 \leq 80$ due to thermal instability of the **iso**.butanoate.

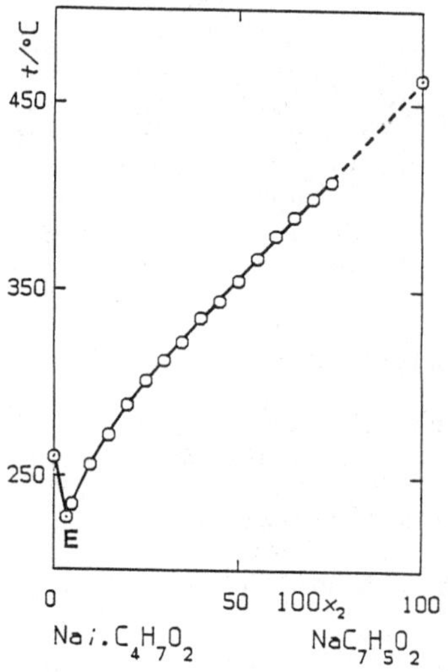

AUXILIARY INFORMATION

METHOD/APPARATUS/PROCEDURE:

Visual polythermal analysis.
Melts contained in a glass tube and stirred.
Temperatures measured with a Nichrome-Constantane thermocouple and a 17 mV full scale millivoltmeter. The temperature readings refer to the disappearance of isotropicity in the melt on cooling.

SOURCE AND PURITY OF MATERIALS:

Component 1: prepared by reacting aqueous ("chemically pure") Na_2CO_3 with a slight excess of **iso**.butanoic acid of analytical purity. The solvent and excess acid were removed by heating to 160 ^{o}C.
Component 2: "chemically pure" material.

ESTIMATED ERROR:

Temperature: accuracy is probably ± 2 K (compiler).

REFERENCES:

COMPONENTS:

(1) Sodium **iso.**butanoate (sodium **iso.**butyrate); $NaiC_4H_7O_2$; [996-30-5]
(2) Sodium octadecanoate (sodium stearate); $NaC_{18}H_{35}O_2$; [822-16-2]

EVALUATOR:

Ferloni, P.,
Dipartimento di Chimica Fisica,
Universita´ di Pavia (ITALY).

CRITICAL EVALUATION:

This system was studied only by Sokolov (Ref. 1) who employed the visual polythermal analysis to draw the lower boundary of the isotropic liquid field. From the shape of this boundary, he concluded that the intermediate compound $Na_5(i.C_4H_7O_2)_2(C_{18}H_{35}O_2)_3$ [congruently melting at 596 K (323 °C)] was formed, and that the limits of the stability field of this compound were a eutectic at 435 K (162 °C) and $100\mathbf{x}_2$= 25.5, and a "perekhodnaya tochka" at 584 K (311 °C) and $100\mathbf{x}_2$= 94.5.

Component 2, however, forms liquid crystals. Thence, the fusion temperature by Sokolov, viz., $\mathbf{T}_{fus}(2)$= 581 K (308 °C), should be identified with the clearing temperature and compared with the $\mathbf{T}_{clr}(2)$ value reported in Preface, Table 4 (552.7 K). Conversely, Sokolov´s $T_{fus}(1)$ [533 K (260 °C)] seems sufficiently reliable, being not far from the value (526.9$\pm$0.7 K) reported in Table 2 of the Preface.

In the evaluator´s opinion, the phase diagram at $0 \leq 100\mathbf{x}_2 \leq 60$ is to be reconsidered, e.g., with reference to Preface, Scheme A.2: Sokolov´s eutectic could be an M'_E point, whereas the maximum at $100\mathbf{x}_2$= 60 could represent an azeotrope.

On the contrary, no definite interpretation of the phase diagram at high $100\mathbf{x}_2$ values seems possible. Indeed, it is not clear how Sokolov could argue the occurrence of an invariant (the "perekhodnaya tochka" at $100\mathbf{x}_2$= 94.5) from the trend of his experimental data which does not unambiguously support any significant slope change of the curve in this region. Moreover, Sokolov´s "fusion" temperature of component 2 (581 K) looks as fully unreliable, being 18 K higher than the second highest $\mathbf{T}_{clr}$ value determined during the last 30 years (Ref. 2), and 28 K higher than the clearing temperature listed in Table 4 of the Preface.

REFERENCES:

(1) Sokolov, N.M.
Zh. Obshch. Khim. 1954, **24**, 1581-1593.

(2) Sanesi, M.; Cingolani, A.; Tonelli, P.L.; Franzosini, P.
Thermal Properties, in **Thermodynamic and Transport Properties of Organic Salts**, IUPAC Chemical Data Series No. 28 (Franzosini, P.; Sanesi, M.; Editors), Pergamon Press, Oxford, 1980, 29-115.

COMPONENTS:

(1) Sodium **iso**.butanoate (sodium **iso**.butyrate); $Na\mathit{i}.C_4H_7O_2$; [996-30-5]
(2) Sodium octadecanoate (sodium stearate); $NaC_{18}H_{35}O_2$; [822-16-2]

ORIGINAL MEASUREMENTS:

Sokolov, N.M.
Zh. Obshch. Khim. 1954, **24**, 1581-1593.

VARIABLES:

Temperature.

PREPARED BY:

D'Andrea, G.

EXPERIMENTAL VALUES:

t/°C	T/K[a]	$100x_2$	t/°C	T/K[a]	$100x_2$
260	533	0	319	592	50
240	513	5	321	594	55
215	488	10	323	596	60
196	469	15	322	595	65
177	450	20	321	594	70
163	436	25	320	593	75
162	435	25.5	317	590	85
217	490	30	314	587	90
260	533	35	311	584	94.5
291	564	40	312	585	97.5
309	582	45	308	581	100

[a] T/K values calculated by the compiler.

Characteristic point(s):
Eutectic, E, at 162 °C and $100x_2$= 25.5 (author).
Characteristic point, P ("perekhodnaya tochka" in the original text; see the Introduction), at 311 °C (author) and $100x_2$= 94.5 (erroneously reported as 312 °C and $100x_2$= 97.5 in the text, compiler).

Intermediate compound(s):
$Na_5(i.C_4H_7O_2)_2(C_{18}H_{35}O_2)_3$, congruently melting at 323 °C.

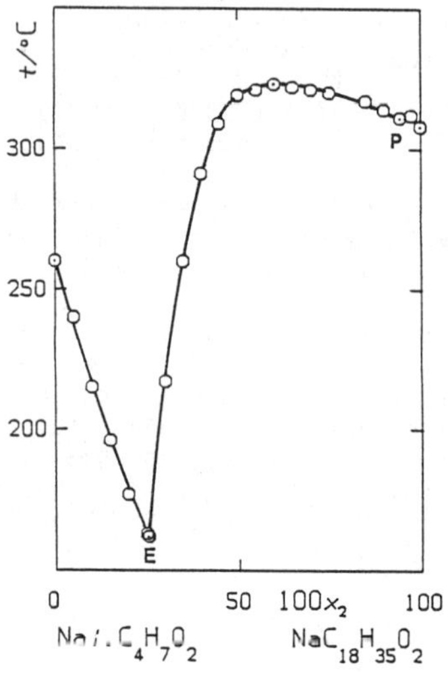

AUXILIARY INFORMATION

METHOD/APPARATUS/PROCEDURE:

Visual polythermal analysis.
Melts contained in a glass tube and stirred.
Temperatures measured with a Nichrome-Constantane thermocouple and a 17 mV full scale millivoltmeter. The temperature readings refer to the disappearance of isotropicity in the melt on cooling.

SOURCE AND PURITY OF MATERIALS:

Component 1: prepared by reacting aqueous ("chemically pure") Na_2CO_3 with a slight excess of **iso**.butanoic acid of analytical purity. The solvent and excess acid were removed by heating to 160 °C. Component 2: "chemically pure" material.

ESTIMATED ERROR:

Temperature: accuracy probably ±2 K (compiler).

REFERENCES:

COMPONENTS:	ORIGINAL MEASUREMENTS:
(1) Sodium **iso**.butanoate (sodium **iso**.butyrate); **Nai**.$C_4H_7O_2$; [996-30-5] (2) Sodium thiocyanate; NaCNS; [540-72-7]	Sokolov, N.M. **Zh. Obshch. Khim.** 1954, **24**, 1150-1156.
VARIABLES:	PREPARED BY:
Temperature.	D'Andrea, G.

EXPERIMENTAL VALUES:

$t/^{o}C$	T/K^{a}	$100x_2$
260	533	0
247	520	5
237	510	10
231	504	15
221	494	25
214	487	27.4
221	494	30
240	513	35
255	528	40
266	539	45
274	547	50
280	553	55
284	557	60
288	561	65
295	568	75
300	573	90
311	584	100

[a] T/K values calculated by the compiler.

Characteristic point(s):

Eutectic, E, at 214 ^{o}C (compiler; erroneously reported as 240 ^{o}C in table 3 of the original paper) and $100x_2$= 27.4 (author).

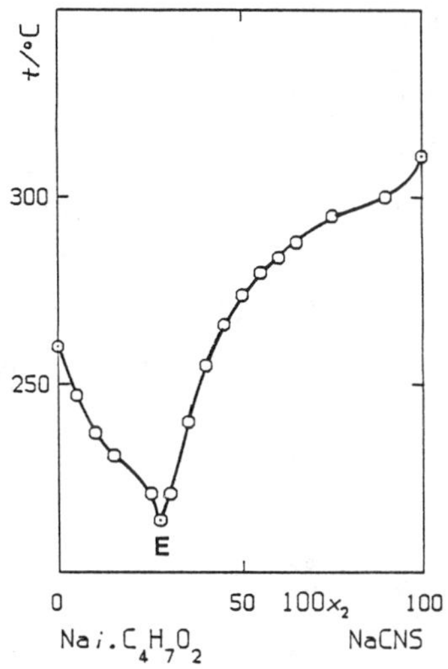

AUXILIARY INFORMATION

METHOD/APPARATUS/PROCEDURE:	SOURCE AND PURITY OF MATERIALS:
Visual polythermal analysis. Salt(s) melted in a test tube. Temperature measured with a Nichrome-Constantane thermocouple and a millivoltmeter with mirror full scale 17 mV.	Component 1 synthetized from **iso**.butanoic acid and $NaHCO_3$. Component 2 of analytical purity recrystallized once from water and once from ethanol.
	ESTIMATED ERROR:
	Temperature: accuracy probably ± 2 K (compiler).
	REFERENCES:

COMPONENTS:	ORIGINAL MEASUREMENTS:
(1) Sodium **iso.**butanoate (sodium **iso.**butyrate); $Na\mathbf{i}.C_4H_7O_2$; [996-30-5] (2) Sodium nitrite; $NaNO_2$; [7632-00-0]	Sokolov, N.M. **Zh. Obshch. Khim.** 1957, 27, 840-844 (*); **Russ. J. Gen. Chem. (Engl. Transl.)** 1957, 27, 917-920.
VARIABLES:	PREPARED BY:
Temperature.	D'Andrea, G.

EXPERIMENTAL VALUES:

$t/^{o}C$	T/K^{a}	$100\mathbf{x}_2$
260	533	0
253	526	5
253	526	10
265	538	15
271	544	20
275	548	25
277	550	30
279	552	35
279	552	40
279	552	45
278	551	50
278	551	55
278	551	60
277	550	65
276	549	70
275	548	75
276	549	80
277	550	85
279	552	90
280	553	95
284	557	100

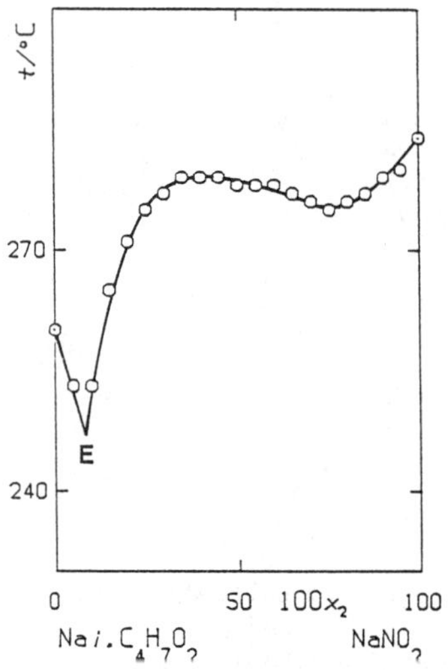

[a] T/K values calculated by the compiler.

Characteristic point(s):

Eutectic, E, at 247 ^{o}C and $100\mathbf{x}_2$= 8 (author).

AUXILIARY INFORMATION

METHOD/APPARATUS/PROCEDURE:	SOURCE AND PURITY OF MATERIALS:
Visual polythermal analysis; salt mixtures melted in a glass tube (surrounded by a wider tube) and stirred with a glass thread. The temperatures of initial crystallization were measured with a Nichrome-Constantane thermocouple checked at the fusion points of water, benzoic acid, mannitol, $AgNO_3$, Cd, KNO_3, and $K_2Cr_2O_7$.	Component 1: prepared from "chemically pure" sodium hydrogen carbonate (carbonate in the reference quoted; compiler) and commercial **iso.**butanoic acid distilled before use (Ref. 1); the salt recovered was recrystallized from **n**-butanol. Component 2: "chemically pure" material recrystallized from water.
NOTE:	ESTIMATED ERROR:
The author does not comment on the minimum at 548 K and $100\mathbf{x}_2$= 75. A possible explanation might be that liquid layering occurs: in this case, the points at $25 \leq 100\mathbf{x}_2 \leq 75$ should represent liquid-liquid instead of solid-liquid equilibria, the monotectic temperature being 548 K. It is worth mentioning that stratification was reported by the same author in the same paper for the binary $Na/\mathbf{i}.C_5H_9O_2$, NO_2.	Temperature: accuracy probably ± 2 K (compiler). REFERENCES: (1) Sokolov, N.M. **Zh. Obshch. Khim.** 1954, **24**, 1581-1593.

COMPONENTS:

(1) Sodium **iso.**butanoate (sodium **iso.**butyrate); $Nai.C_4H_7O_2$; [996-30-5]
(2) Sodium nitrate; $NaNO_3$; [7631-99-4]

ORIGINAL MEASUREMENTS:

Sokolov, N.M.
Zh. Obshch. Khim. 1954, 24, 1150-1156.

VARIABLES:

Temperature.

PREPARED BY:

D'Andrea, G.

EXPERIMENTAL VALUES:

$t/^{\circ}C$	T/K^a	$100x_2$
260	533	0
248	521	5
242	515	10
238	511	15
219	492	25
233	506	30
244	517	35
258	531	40
267	540	45
274	547	50
276	549	55
280	553	60
284	557	65
288	561	75
292	565	85
300	573	95
308	581	100

[a] T/K values calculated by the compiler.

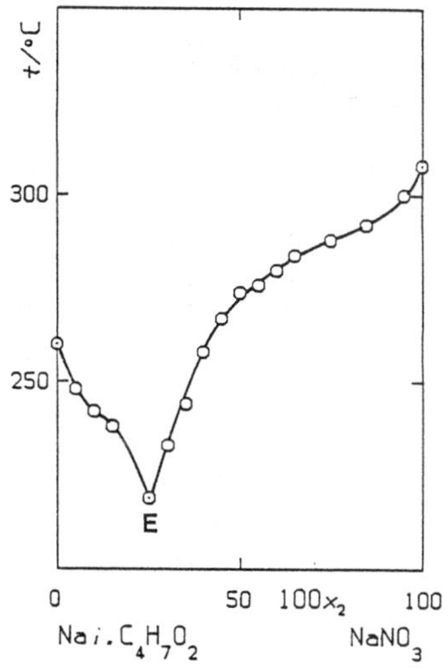

Characteristic point(s):

Eutectic, E, at 219 °C and $100x_2$= 25 (author).

AUXILIARY INFORMATION

METHOD/APPARATUS/PROCEDURE:

Visual polythermal analysis.
Salt(s) melted in a test tube. Temperature measured with a Nichrome-Constantane thermocouple and a millivoltmeter with mirror reading to 17 mV.

SOURCE AND PURITY OF MATERIALS:

Component 1 synthetized from **iso.**butanoic acid and $NaHCO_3$. Commercial component 2 further purified by the author according to Laiti.

ESTIMATED ERROR:

Temperature: accuracy probably ±2 K (compiler).

REFERENCES:

COMPONENTS:

(1) Sodium **iso.**butanoate (sodium **iso.**butyrate); $NaI.C_4H_7O_2$; [996-30-5]
(2) Sodium nitrate; $NaNO_3$; [7631-99-4]

ORIGINAL MEASUREMENTS:

Dmitrevskaya, O.I.; Sokolov, N.M.
Zh. Obshch. Khim. 1960, **30**, 20-25 (*);
Russ. J. Gen. Chem. (Engl. Transl.) 1960, **30**, 19-24.

VARIABLES:

Temperature.

PREPARED BY:

D'Andrea, G.

EXPERIMENTAL VALUES:

Characteristic point(s):

The paper reports - **inter alia** - on a refinement of the title binary, previously studied by one of the authors (Ref. 1). According to the present investigation, the coordinates of the eutectic are:

Eutectic, E, at 220 ^{o}C and $100\mathbf{x}_2$ = 25 (authors).

AUXILIARY INFORMATION

METHOD/APPARATUS/PROCEDURE:

Visual polythermal analysis.

NOTE:

Concerning component 1, no mention is made in Table 2 of solid state phase transformations, although three transitions are quoted by the authors (from Ref. 3), at 493, 364, and 340 K (220, 91, and 67 ^{o}C), respectively. Duruz et al. (Ref. 4) report in turn $\mathbf{T}'_{trs}(1)$= 493 K (in agreement with the highest transition temperature from Ref. 3), and $\mathbf{T}''_{trs}(1)$= 468 K (a figure which has no correspondence in Ref. 3). Finally, Ferloni et al. (Ref. 5) are inclined to think that Sokolov's transformation at 340 K (Ref. 3) actually represents a transition of a hydrated form of the salt.

SOURCE AND PURITY OF MATERIALS:

Component 1 synthetized from **iso.**butanoic acid and Na_2CO_3 (Ref. 2). "Chemically pure" component 2 recrystallized.
Component 1 undergoes phase transitions at $\mathbf{t}_{trs}(1)/^{o}C$= 67, 91, 220 (Ref. 3).
Component 2 undergoes a phase transition at $\mathbf{t}_{trs}(2)/^{o}C$= 270 (current literature).

ESTIMATED ERROR:

Temperature: accuracy probably ± 2 K (compiler).

REFERENCES:

(1) Sokolov, N.M. **Zh. Obshch. Khim.** 1954, **24**, 1150-1156.
(2) Sokolov, N.M. **Zh. Obshch. Khim.** 1954, **24**, 1581-1593.
(3) Sokolov, N.M. **Tezisy Dokl. X Nauch. Konf. S.M.I.** 1956.
(4) Duruz, J.J.; Michels, H.J.; Ubbelohde, A.R. **Proc. Roy. Soc. London** 1971, **A 322**, 281-299.
(5) Ferloni, P.; Sanesi, M.; Tonelli, P.L.; Franzosini, P. **Z. Naturforsch.** 1978, **A 33**, 240-242.

COMPONENTS:

(1) Sodium pentanoate (sodium valerate); $NaC_5H_9O_2$; [6106-41-8]
(2) Sodium thiocyanate; NaCNS; [540-72-7]

EVALUATOR:

Spinolo, G.,
Dipartimento di Chimica Fisica,
Universita´ di Pavia (ITALY).

CRITICAL EVALUATION:

This system was studied by Sokolov (Ref. 1) and by Sokolov and Khaitina (Ref. 2): in both papers the visual polythermal investigation was restricted to the lower boundary of the isotropic liquid field. The authors claimed the existence of a 1:1 intermediate compound which melts congruently at 564 K (291 °C; Ref. 1), and forms eutectics with either pure component, at eutectics at 562 K (289 °C) and $100\mathbf{x}_2$= 46, and at 560 K (287 °C) and $100\mathbf{x}_2$= 56.5 or 55, respectively.

Component 1, however, forms liquid crystals, which are stable between 498±2 K and 631±4 K (Preface, Table 1). The latter value fairly agrees with the fusion temperature (630 K) given in Ref. 1 and 2; the former can be identified (even if not fully satisfactorily) with that (489 K) corresponding to the highest phase transformation temperature quoted by Ref. 2 from Ref. 3. Once more for component 1, Table 1 reports no solid state transition, whereas Sokolov and Khaitina quote (from Ref. 3) $\mathbf{T}_{trs}(2)$/K= 482 and 453. It is, however, to be stressed that the single transition observed (at 479±1 K) with DTA in sodium **n**-pentanoate by Duruz et al. (Ref. 4) was not more mentioned in a subsequent DSC investigation by the same group (Ref. 5).

In the evaluator´s opinion, therefore,
i) the invariant at 562 K (289 °C) and $100\mathbf{x}_2$= 46 should be identified with an M'_E point,
ii) a (so far undetected) M_E invariant should exist within the composition range between M'_E and pure component 1, and
iii) the phase diagram ought to be similar to that shown in Scheme D.1 of the Preface, but for the fact that the liquid crystal-isotropic liquid diphasic field exhibits a maximum.

REFERENCES:

(1) Sokolov, N.M.
Zh. Obshch. Khim. 1954, **24**, 1150-1156.

(2) Sokolov, N.M.; Khaitina, M.V.
Zh. Obshch. Khim. 1972, **42**, 2121-2123.

(3) Sokolov, N.M.
Tezisy Dokl. X Nauch. Konf. S.M.I. 1956.

(4) Duruz, J.J.; Michels, H.J.; Ubbelohde, A.R.
Proc. Roy. Soc. London 1971, **A322**, 281-299.

(5) Michels, H.J.; Ubbelohde, A.R.
JCS Perkin II 1972, 1879-1881.

COMPONENTS:

(1) Sodium pentanoate (sodium valerate); $NaC_5H_9O_2$; [6106-41-8]
(2) Sodium thiocyanate; NaCNS; [540-72-7]

ORIGINAL MEASUREMENTS:

Sokolov, N.M.
Zh. Obshch. Khim. 1954, **24**, 1150-1156.

VARIABLES:

Temperature.

PREPARED BY:

D´Andrea, G.

EXPERIMENTAL VALUES:

$t/^{o}C$	T/K^{a}	$100x_2$	$t/^{o}C$	T/K^{a}	$100x_2$
357	630	0	289	562	46
370	643	5	291	564	50
378	651	10	288	561	55
375	648	15	287	560	56.5
356	629	25	290	563	60
344	617	30	293	566	65
331	604	35	297	570	75
316	589	40	302	575	90
296	569	45	311	584	100

[a] T/K values calculated by the compiler.

Characteristic point(s):

Eutectic, E_1, at 289 °C and $100x_2$= 46 (author).
Eutectic, E_2, at 287 °C and $100x_2$= 56.5 (author).

Intermediate compound(s):

$Na_2C_5H_9O_2CNS$, congruently melting at 291 °C (author).

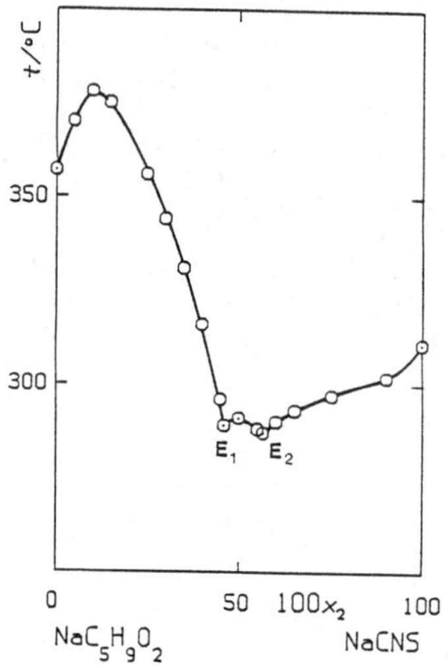

AUXILIARY INFORMATION

METHOD/APPARATUS/PROCEDURE:

Visual polythermal analysis.
Salt(s) melted in a test tube. Temperature measured with a Nichrome-Constantane thermocouple and a millivoltmeter with mirror reading to 17 mV.

SOURCE AND PURITY OF MATERIALS:

Component 1 synthetized from n-pentanoic acid and $NaHCO_3$. Component 2 of analytical purity recrystallized once from water and once from ethanol.

ESTIMATED ERROR:

Temperature: accuracy probably ± 2 K (compiler).

REFERENCES:

COMPONENTS:	ORIGINAL MEASUREMENTS:
(1) Sodium pentanoate (sodium valerate); $NaC_5H_9O_2$; [6106-41-8] (2) Sodium thiocyanate; NaCNS; [540-72-7]	Sokolov, N.M.; Khaitina, M.V. **Zh. Obshch. Khim.** 1972, **42**, 2121-2123.
VARIABLES:	PREPARED BY:
Temperature.	D'Andrea, G.

EXPERIMENTAL VALUES:

Characteristic point(s):

Eutectic, E_1, at 289 ^{o}C and $100\mathbf{x}_2$ about 46 (estimated by the compiler from Fig. 1 of the original paper).
Eutectic, E_2, at 287 ^{o}C and $100\mathbf{x}_2$ about 55 (estimated by the compiler from Fig. 1 of the original paper).

Intermediate compound(s):

$Na_2C_5H_9O_2CNS$, congruently melting.

AUXILIARY INFORMATION

METHOD/APPARATUS/PROCEDURE:	SOURCE AND PURITY OF MATERIALS:
Visual polythermal analysis.	Not stated. Component 1 undergoes phase transitions at $t_{trs}(1)/^{o}C$= 180, 209, 216 (Ref. 1) and melts at $t_{fus}(1)/^{o}C$= 356. Component 2 melts at $t_{fus}(2)/^{o}C$= 311.
	ESTIMATED ERROR:
	Temperature: accuracy probably ± 2 K (compiler).
	REFERENCES:
	(1) Sokolov, N.M. **Tezisy Dokl. X Nauch. Konf. S.M.I.** 1956.

COMPONENTS:	EVALUATOR:
(1) Sodium pentanoate (sodium valerate); $NaC_5H_9O_2$; [6106-41-8] (2) Sodium nitrite; $NaNO_2$; [7632-00-0]	Ferloni, P., Dipartimento di Chimica Fisica, Universita´ di Pavia (ITALY).

CRITICAL EVALUATION:

This binary was studied only by Sokolov (Ref. 1) who, on the basis of his visual polythermal observations, claimed the phase diagram to be of the eutectic type, the invariant occuring at 555 K (282 °C) and $100\mathbf{x}_1 = 0.04$. This investigation was restricted to the range $0 \leq 100\mathbf{x}_1 \leq 55$, because of decomposition of mixtures richer in component 1.

Component 1, however, forms liquid crystals. Thence, Sokolov´s $\mathbf{T}_{fus}(1)$ [i.e., 610 K (357°C)] should be identified with a clearing temperature, and compared with the value $\mathbf{T}_{clr}(1) = 631 \pm 4$ K reported in Preface, Table 1.

The topology of the phase diagram has therefore to be reconsidered with reference to Preface, Schemes A, among which, however, the available data, unfortunately, do not allow one to make a definite choice.

Anyway, Sokolov´s invariant should be an M'_E point and not a conventional eutectic.

REFERENCES:

(1) Sokolov, N.M.
Zh. Obshch. Khim. 1957, 27, 840-844 (*); **Russ. J. Gen. Chem. (Engl. Transl.)** 1957, 27, 917-920.

COMPONENTS:	ORIGINAL MEASUREMENTS:
(1) Sodium pentanoate (sodium valerate); $NaC_5H_9O_2$; [6106-41-8] (2) Sodium nitrite; $NaNO_2$; [7632-00-0]	Sokolov, N.M. **Zh. Obshch. Khim.** 1957, **27**, 840-844 (*); **Russ. J. Gen. Chem. (Engl. Transl.)** 1957, **27**, 917-920.
VARIABLES:	PREPARED BY:
Temperature.	D'Andrea, G.

EXPERIMENTAL VALUES:

$t/^oC$	T/K^a	$100x_2$
323	596	45
323	596	50
323	596	55
323	596	60
323	596	65
323	596	70
323	596	75
323	596	80
323	596	85
321	594	90
319	592	95
284	557	100

[a] T/K values calculated by the compiler.

Note - The system was investigated at $0 \leq 100x_1 \leq 55$ since further increase in component 1 content causes decomposition of the mixtures.

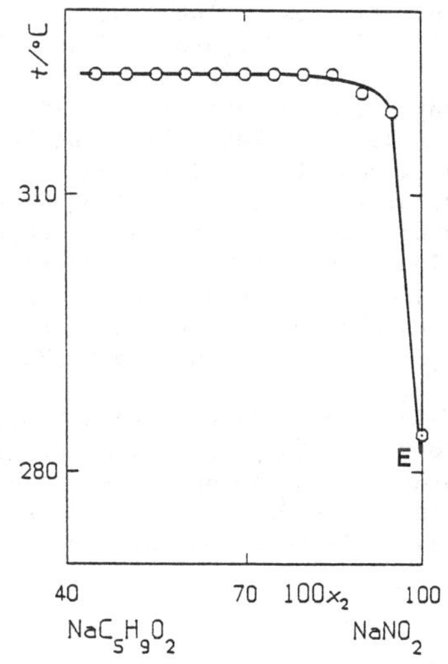

Characteristic point(s):

Eutectic, E, at 282 °C and $100x_1$= 0.04 (both figures, listed in table 2 of the original paper, cannot be drawn from the tabulated data; moreover, in the same table the eutectic composition is erroneously reported as $100x_2$= 0.04; compiler).

AUXILIARY INFORMATION

METHOD/APPARATUS/PROCEDURE:

Visual polythermal analysis; salt mixtures melted in a glass tube (surrounded by a wider tube) and stirred with a glass thread. The temperatures of initial crystallization were measured with a Nichrome-Constantane thermocouple checked at the fusion points of water, benzoic acid, mannitol, $AgNO_3$, Cd, KNO_3, and $K_2Cr_2O_7$.

SOURCE AND PURITY OF MATERIALS:

Component 1: prepared from "chemically pure" sodium hydrogen carbonate (carbonate in the reference quoted; compiler) and commercial pentanoic acid distilled before use (Ref. 1); the salt recovered was re-crystallized from n-butanol;
$t_{fus}(1)/^oC$= 357.

Component 2: "chemically pure" material recrystallized from water.

ESTIMATED ERROR:

Temperature: accuracy probably ±2 K (compiler).

REFERENCES:

(1) Sokolov, N.M.
Zh. Obshch. Khim. 1954, **24**, 1581-1593.

COMPONENTS:

(1) Sodium pentanoate (sodium valerate); $NaC_5H_9O_2$; [6106-41-8]
(2) Sodium nitrate; $NaNO_3$; [7631-99-4]

EVALUATOR:

Spinolo, G.,
Dipartimento di Chimica Fisica,
Universita' di Pavia (ITALY).

CRITICAL EVALUATION:

This system was studied by Sokolov (Ref. 1), and by Sokolov and Khaitina (Ref. 2). In both cases, the visual polythermal analysis was employed to detect the lower boundary of the isotropic liquid field. Accordingly, the authors claimed that a 1:1 intermediate compound forms, which melts congruently at 568 K (295 °C), and gives eutectics with either component. Concerning the precise location of these invariants, some values given in the text of the original papers should be corrected with a closer inspection of the pertinent figures. The correct values seem therefore to be **T**= 564 K (291 °C) and $100\mathbf{x}_2$= 40.5 (Ref. 2), and **T**= 554 K (281 °C) and $100\mathbf{x}_2$= 58.5, respectively.

Component 1, however, forms liquid crystals, which are stable between 498±2 K and 631±4 K (Preface, Table 1). The latter value fairly agrees with the fusion temperature (630 K) given in Ref. 1 and 2; the former can be identified (even if not fully satisfactorily) with that (489 K) corresponding to the highest phase transformation temperature quoted by Ref. 3. Once more for component 1, Table 1 reports no solid state transition, whereas Sokolov quotes (Ref. 3) $\mathbf{T}_{trs}(1)/K$= 482 and 453. It is, however, to be stressed that the single transition observed (at 479±1 K) with DTA in sodium **n**-pentanoate by Duruz et al. (Ref. 4) was no more mentioned in a subsequent DSC investigation by the same group (Ref. 5).

Taking into account the above remaks, the eutectic at 564 K (291°C) and $100\mathbf{x}_2$= 40.4 ought to be an M''_E point, and the occurrence of a further invariant (so far undetected and probably an M_E point) is to be expected. The phase diagram could be similar to that shown in Scheme D.1 of the Preface, but for the fact that the liquid crystal-isotropic liquid field is splitted into two parts by a maximum.

REFERENCES:

(1) Sokolov, N.M.
Zh. Obshch. Khim. 1954, **24**, 1150-1156
(2) Sokolov, N.M.; Khaitina, M.V.
Zh. Obshch. Khim. 1972, **42**, 2121-2123
(3) Sokolov, N.M.
Tezisy Dokl. X Nauch. Konf. S.M.I. 1956.
(4) Duruz, J.J.; Michels, H.J.; Ubbelohde, A.R.
Proc. Roy. Soc. London 1971, **A322**, 281-299.
(5) Michels, H.J.; Ubbelohde, A.R.
JCS Perkin II 1972, 1879-1881.

COMPONENTS:

(1) Sodium pentanoate (sodium valerate); $NaC_5H_9O_2$; [6106-41-8]
(2) Sodium nitrate; $NaNO_3$; [7631-99-4]

ORIGINAL MEASUREMENTS:

Sokolov, N.M.
Zh. Obshch. Khim. 1954, **24**, 1150-1156.

VARIABLES:

Temperature.

PREPARED BY:

D'Andrea, G.

EXPERIMENTAL VALUES:

$t/^{\circ}C$	T/K^a	$100x_2$	$t/^{\circ}C$	T/K^a	$100x_2$
357	630	0	295	568	50
366	639	5	288	561	55
372	645	10	281[c]	554	58.5
369	642	15	285	558	60
350	623	25	293	566	65
336	609	30	298	571	75
320	593	35	300	573	85
296	569	40	305	578	95
291[b]	564	40.5	308	581	100
294	567	45			

[a] T/K values calculated by the compiler.
[b] 295 in the original table, corrected by the compiler on the basis of Fig. 2 of the original paper.
[c] 291 in the original table, corrected by the compiler on the basis of Fig. 2 of the original paper.

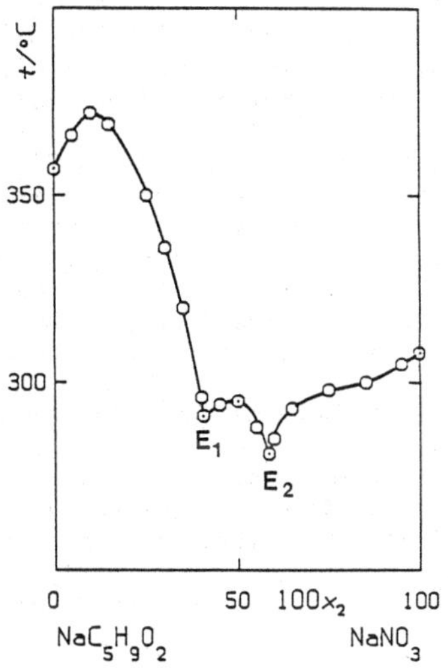

Characteristic point(s):

Eutectic, E_1, at 291 °C and $100x_2$= 40.5 (author).
Eutectic, E_2, at 281 °C and $100x_2$= 58.5 (author).

Intermediate compound(s):

$Na_2C_5H_9O_2NO_3$, congruently melting at 295 °C (compiler).

AUXILIARY INFORMATION

METHOD/APPARATUS/PROCEDURE:

Visual polythermal analysis.
Salt(s) melted in a test tube. Temperature measured with a Nichrome-Constantane thermocouple and a millivoltmeter with mirror reading to 17 mV.

SOURCE AND PURITY OF MATERIALS:

Component 1 synthetized from n-pentanoic acid and $NaHCO_3$. Commercial component 2 further purified by the author according to Laiti.

ESTIMATED ERROR:

Temperature: accuracy probably ±2 K (compiler).

REFERENCES:

COMPONENTS:

(1) Sodium pentanoate (sodium valerate); $NaC_5H_9O_2$; [6106-41-8]
(2) Sodium nitrate; $NaNO_3$; [7631-99-4]

ORIGINAL MEASUREMENTS:

Sokolov, N.M.; Khaitina, M.V.
Zh. Obshch. Khim. 1972, **42**, 2121-2123.

VARIABLES:

Temperature.

PREPARED BY:

D'Andrea, G.

EXPERIMENTAL VALUES:

Characteristic point(s):

Eutectic, E_1, at 291 ^oC and $100\mathbf{x}_2$ about 40.5 (estimated by the compiler from Fig. 1 of the original paper).
Eutectic, E_2, at 281 ^oC and $100\mathbf{x}_2$ about 58.5 (estimated by the compiler from Fig. 1 of the original paper).

Intermediate compound(s):

$Na_2C_5H_9O_2NO_3$, congruently melting.

AUXILIARY INFORMATION

METHOD/APPARATUS/PROCEDURE:

Visual polythermal analysis.

SOURCE AND PURITY OF MATERIALS:

Not stated.
Component 1 undergoes phase transitions at $t_{trs}(1)/^oC$= 180, 209, 216 (Ref. 1) and melts at $t_{fus}(1)/^oC$= 356.
Component 2 undergoes a phase transition at $t_{trs}(2)/^oC$= 275 (current literature value), and melts at $t_{fus}(2)/^oC$= 308.

ESTIMATED ERROR:

Temperature: accuracy probably ± 2 K (compiler).

REFERENCES:

(1) Sokolov, N.M.
Tezisy Dokl. X Nauch. Konf. S.M.I. 1956.

COMPONENTS:	EVALUATOR:
(1) Sodium **iso**.pentanoate (sodium **iso**.valerate); $NaiC_5H_9O_2$; [539-66-2] (2) Sodium hexanoate (sodium caproate); $NaC_6H_{11}O_2$; [10051-44-2]	Ferloni, P., Dipartimento di Chimica Fisica, Universita' di Pavia (ITALY).

CRITICAL EVALUATION:

This system was studied only by Sokolov (Ref. 1), who claimed that a continuous series of solid solutions is formed, with a minimum, m, at 512 K (239 °C), and $100\mathbf{x}_2$= 20.

Both components, however, form liquid crystals (see Preface, Tables 2, 1). Therefore, Sokolov's fusion temperatures, $\mathbf{T}_{fus}(1)/K$= 535 (262 °C) and $\mathbf{T}_{fus}(2)/K$= 638 (365 °C), should be identified with clearing temperatures, the corresponding data from Tables 2 and 1 being 559±1 K and 639.0±0.5 K, respectively.

Concerning component 1, the remarkable discrepancy might be attributed to insufficient purity of Sokolov's sample, inasmuch as the value from Table 2 (559±1) meets rather satisfactorily those reported by Ubbelohde et al. (556 K; Ref. 2), and by Duruz et al. (553 K; Ref. 3). Indeed, due to the - usually small - value of the enthalpy change associated with clearing, very small amounts of impurities are often sufficient to cause a dramatic drop of the clearing temperature.

A continuous series of liquid crystal (instead of solid) solutions should form, and the complete phase diagram should be similar to that shown in Scheme C.1 of the Preface, with a common minimum of the curves limiting the isotropic liquid - liquid crystal diphasic field.

REFERENCES

(1) Sokolov, N.M.
Zh. Obshch. Khim. 1954, **24**, 1581-1593.
(2) Ubbelohde, A.R.; Michels, H.J.; Duruz, J.J.
Nature 1970, **228**, 50-52.
(3) Duruz, J.J.; Michels, H.J.; Ubbelohde, A.R.
Proc. R. Soc. London 1971, **A322**, 281-299.

COMPONENTS:

(1) Sodium **iso**.pentanoate (sodium **iso**.valerate); Na*i*.$C_5H_9O_2$; [539-66-2]
(2) Sodium hexanoate (sodium caproate); $NaC_6H_{11}O_2$; [10051-44-2]

ORIGINAL MEASUREMENTS:

Sokolov, N.M.
Zh. Obshch. Khim. 1954, **24**, 1581-1593.

VARIABLES:

Temperature.

PREPARED BY:

D'Andrea, G.

EXPERIMENTAL VALUES:

t/°C	T/K[a]	$100x_2$	t/°C	T/K[a]	$100x_2$
262	535	0	283	556	55
255	528	5	291	564	60
247	520	10	296	569	65
243	516	15	304	577	70
239	512	20	313	586	75
242	515	25	322	595	80
248	521	30	331	604	85
256	529	35	341	614	90
263	536	40	354	627	95
271	544	45	365	638	100
277	550	50			

[a] T/K values calculated by the compiler.

Characteristic point(s):

Continuous series of solid solutions with a minimum, m, at 239 °C and $100x_2$ = 20 (author).

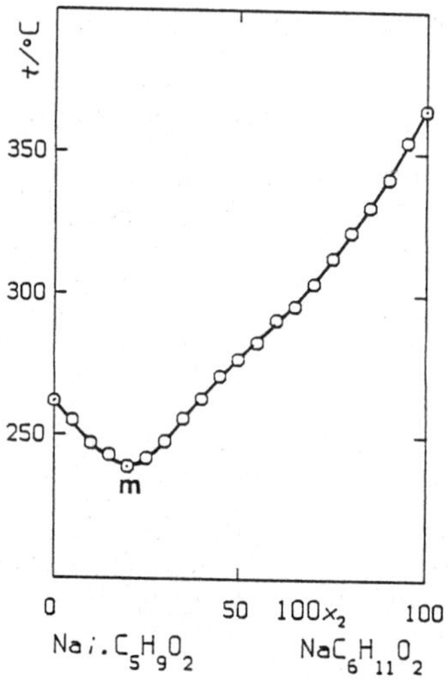

AUXILIARY INFORMATION

METHOD/APPARATUS/PROCEDURE:

Visual polythermal analysis.
Melts contained in a glass tube and stirred.
Temperatures measured with a Nichrome-Constantane thermocouple and a 17 mV full scale millivoltmeter. The temperature readings refer to the disappearance of isotropicity in the melt on cooling.

SOURCE AND PURITY OF MATERIALS:

Materials prepared by reacting aqueous ("chemically pure") Na_2CO_3 with a slight excess of either **iso**.pentanoic or **n**-hexanoic acid of analytical purity. The solvent and excess acid were removed by heating to 160 °C.

ESTIMATED ERROR:

Temperature: accuracy probably ±2 K (compiler).

REFERENCES:

COMPONENTS:

(1) Sodium **iso**.pentanoate (sodium **iso**.valerate); $NaiC_5H_9O_2$; [539-66-2]
(2) Sodium benzoate; $NaC_7H_5O_2$; [532-32-1]

EVALUATOR:

Spinolo, G.,
Dipartimento di Chimica Fisica,
Universita´ di Pavia (ITALY) .

CRITICAL EVALUATION:

This binary was studied only by Sokolov (Ref. 1), who restricted his polythermal analysis to the lower boundary of the isotropic liquid field, and claimed the existence of a eutectic at 534 K (261 °C) and $100x_2$= 3.

Component 1, however, forms liquid crystals [at $T_{fus}(1)$= 461.5±0.6 K; Preface, Table 2] before being transformed in a clear melt. Therefore, Sokolov´s fusion temperature, (535 K) should be identified with the clearing temperature, the corresponding value from Table 2 being 559±1 K. The latter figure is remarkably higher than that given by Ref. 1, and it agrees rather satisfactorily with those reported by Ubbelohde et al. (556 K, Ref. 2) and by Duruz et al. (553 K, Ref. 3).

Thus, in the evaluator´s opinion, the phase diagram could be more correctly interpreted with reference to Scheme A.1 of the Preface, and Sokolov´s eutectic should be identified with an M'_E point.

REFERENCES:

(1) Sokolov, N.M.
Zh. Obshch. Khim. 1954, **24**, 1581-1593.
(2) Ubbelohde, A.R., Michels, H.J., and Duruz, J.J.
Nature 1970, **228**, 50-52.
(3) Duruz, J.J., Michels, H.J., and Ubbelohde, A.R.
Proc. R. Soc. London 1971,**A322**, 281-299.

COMPONENTS:

(1) Sodium **iso**.pentanoate (sodium **iso**.valerate); $Nai.C_5H_9O_2$; [539-66-2]
(2) Sodium benzoate; $NaC_7H_5O_2$; [532-32-1]

ORIGINAL MEASUREMENTS:

Sokolov, N.M.
Zh. Obshch. Khim. 1954, **24**, 1581-1593.

VARIABLES:

Temperature.

PREPARED BY:

D'Andrea, G.

EXPERIMENTAL VALUES:

$t/^{\circ}C$	T/K^a	$100x_2$
262	535	0
261	534	3
275	548	5
298	571	10
317	590	15
337	610	20
349	622	25
365	638	30
379	652	35
389	662	40
396	669	45
401	674	50
407	680	55
415	688	60
421	694	65
426	699	70
463	736	100

[a] T/K values calculated by the compiler.

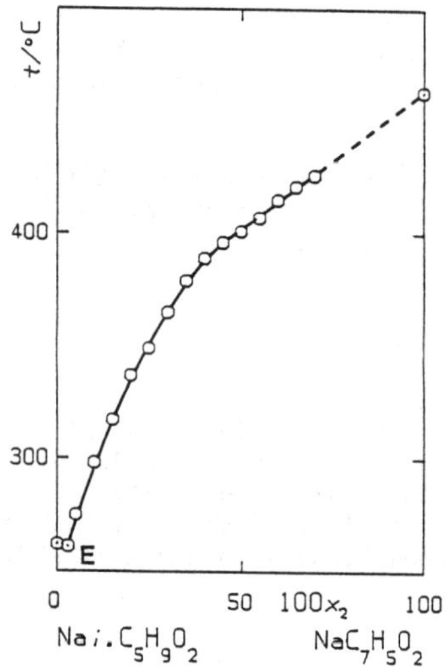

Characteristic point(s):

Eutectic, E, at 261 °C and $100x_2 = 3$ (author).

Note - The system was investigated at $0 \leq 100x_2 \leq 70$ due to thermal instability of the **iso**.pentanoate.

AUXILIARY INFORMATION

METHOD/APPARATUS/PROCEDURE:

Visual polythermal analysis.
Melts contained in a glass tube and stirred.
Temperatures measured with a Nichrome-Constantane thermocouple and a 17 mV full scale millivoltmeter. The temperature readings refer to the disappearance of isotropicity in the melt on cooling.

SOURCE AND PURITY OF MATERIALS:

Component 1: prepared by reacting aqueous ("chemically pure") Na_2CO_3 with a slight excess of **iso**.pentanoic acid of analytical purity. The solvent and excess acid were removed by heating to 160 °C.
Component 2: "chemically pure" material.

ESTIMATED ERROR:

Temperature: accuracy probably ±2 K (compiler).

REFERENCES:

COMPONENTS:	EVALUATOR:
(1) Sodium **iso**.pentanoate (sodium **iso**.valerate); $Na\mathbf{i}.C_5H_9O_2$; [539-66-2] (2) Sodium octadecanoate (sodium stearate); $NaC_{18}H_{35}O_2$; [822-16-2]	Ferloni, P., Dipartimento di Chimica Fisica, Universita´ di Pavia (ITALY).

CRITICAL EVALUATION:

This system was studied only by Sokolov (Ref. 1) who employed the visual polythermal analysis to draw the lower boundary of the isotropic liquid field. From the shape of this boundary, he concluded that the intermediate compound $Na_3(\mathbf{i}.C_5H_9O_2)(C_{18}H_{35}O_2)_2$ [congruently melting at 596 K (323 °C)] was formed, and that the limits of the stability field of this compound were a eutectic at 413 K (140 °C) and $100\mathbf{x}_2$= 17.3, and a "perekhodnaya tochka" at 582 K (309 °C) and $100\mathbf{x}_2$= 93.5.

Actually, both components form liquid crystals, the liquid crystalline phases being one for component 1 (see Preface, Table 2), and two for component 2 (see Preface, Table 4). Therefore, Sokolov´s fusion temperatures, $\mathbf{T}_{fus}(1)$= 535 K (262 °C), and $\mathbf{T}_{fus}(2)$= 581 K (308 °C), should be identified with clearing temperatures, the corresponding values from Tables 2 and 4 being 559±1 and 552.7 K, respectively.

At intermediate compositions it seems reasonable to assume that a continuous series of liquid crystal solutions is formed, with an azeotrope at 596 K and $100\mathbf{x}_2$= 70. Accordingly, the left hand section ($0 \leq 100\mathbf{x}_2 \leq 70$) of the phase diagram might be interpreted with reference to Scheme C.2 of the Preface: in this case, Sokolov´s eutectic should be intended as an M''_E point, allowance being made for the fact that Sokolov´s "fusion" temperature of component 1 is 24 K lower than the relevant $\mathbf{T}_{clr}$ value listed in Table 2, i.e., 559±1 K. It is, however, to be stressed that the latter figure agrees rather satisfactorily with those reported by Ubbelohde et al. (556 K; Ref. 3) and by Duruz et al. (553 K; Ref. 4).

Conversely, no definite interpretation of the phase diagram at high $100\mathbf{x}_2$ values seems possible. Indeed, it is not clear how Sokolov could argue the occurrence of an invariant (the "perekhodnaya tochka" at $100\mathbf{x}_2$= 93.5) from the trend of his experimental data which does not unambiguously support any significant slope change of the curve in this region. Moreover, Sokolov´s "fusion" temperature of component 2 (581 K) looks as fully unreliable, being 18 K higher than the second highest $\mathbf{T}_{clr}$ value determined during the last 30 years (Ref. 2), and 28 K higher than the clearing temperature listed in Table 4.

REFERENCES:

(1) Sokolov, N.M.
Zh. Obshch. Khim. 1954, **24**, 1581-1593.

(2) Sanesi, M.; Cingolani, A.; Tonelli, P.L.; Franzosini, P.
Thermal Properties, in **Thermodynamic and Transport Properties of Organic Salts,** IUPAC Chemical Data Series No. 28 (Franzosini, P.; Sanesi, M.; Editors), Pergamon Press, Oxford, 1980, 29-115.

(3) Ubbelohde, A.R.; Michels, H.J.; Duruz, J.J.
Nature 1970, **228**, 50-52.

(4) Duruz, J.J.; Michels, H.J.; Ubbelohde, A.R.
Proc. R. Soc. London 1971, **A 322**, 281-299.

COMPONENTS:	ORIGINAL MEASUREMENTS:
(1) Sodium **iso**.pentanoate (sodium **iso**.valerate); $Na\textbf{i}.C_5H_9O_2$; [539-66-2] (2) Sodium octadecanoate (sodium stearate); $NaC_{18}H_{35}O_2$; [822-16-2]	Sokolov, N.M. **Zh. Obshch. Khim.** 1954, **24**, 1581-1593.
VARIABLES:	PREPARED BY:
Temperature.	D'Andrea, G.

EXPERIMENTAL VALUES:

t/°C	T/K[a]	$100x_2$	t/°C	T/K[a]	$100x_2$
262	535	0	305	578	55
201	474	5	315	588	60
167	440	10	322	595	65
147	420	15	323	596	70
140	413	17.3	321	594	75
162	435	20	318	591	85
225	498	30	314	587	90
247	520	35	309	582	93.5
266	539	40	312	585	95
282	555	45	308	581	100
295	568	50			

[a] T/K values calculated by the compiler.

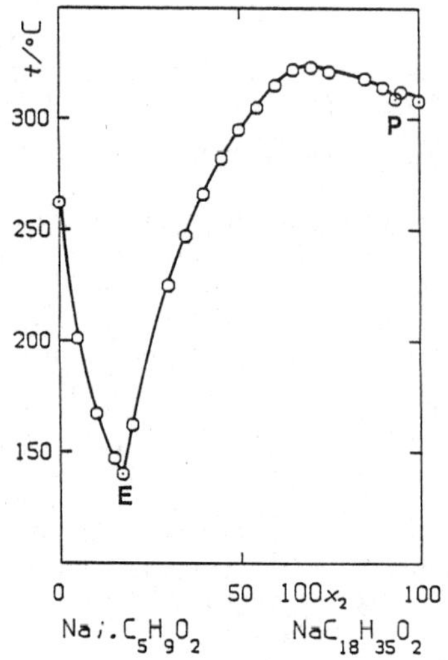

Characteristic point(s):

Eutectic, E, at 140 °C and $100x_2$= 17.3 (author).
Characteristic point, P ("perekhodnaya tochka" in the original text; see the Introduction), at 309 °C (author) and $100x_2$= 93.5 (erroneously reported as 92 in the text, compiler).

Intermediate compound(s):

$Na_3\textbf{i}.C_5H_9O_2(C_{18}H_{35}O_2)_2$, congruently melting at 323 °C (compiler).

AUXILIARY INFORMATION

METHOD/APPARATUS/PROCEDURE:	SOURCE AND PURITY OF MATERIALS:
Visual polythermal analysis. Melts contained in a glass tube and stirred. Temperatures measured with a Nichrome-Constantane thermocouple and a 17 mV full scale millivoltmeter. The temperature readings refer to the disappearance of isotropicity in the melt on cooling.	Component 1: prepared by reacting aqueous ("chemically pure") Na_2CO_3 with a slight excess of **iso**.pentanoic acid of analytical purity. The solvent and excess acid were removed by heating to 160 °C. Component 2: "chemically pure" material. ESTIMATED ERROR: Temperature: accuracy probably ±2 K (compiler). REFERENCES:

COMPONENTS:

(1) Sodium **iso**.pentanoate (sodium **iso**.valerate); $NaI.C_5H_9O_2$; [539-66-2]
(2) Sodium thiocyanate; NaCNS; [540-72-7]

EVALUATOR:

Spinolo, G.,
Dipartimento di Chimica Fisica,
Universita´ di Pavia (ITALY).

CRITICAL EVALUATION:

This binary was studied only by Sokolov (Ref. 1), who restricted his polythermal investigation to the lower boundary of the isotropic liquid field, and claimed the existence of a eutectic at 523 K (250 ^oC) and $100\mathbf{x}_2$= 32.

Component 1, however, forms liquid crystals [at $\mathbf{T}_{fus}(1)$= 461.5$\pm$0.6 K; Preface, Table 2] before turning into a clear melt. Sokolov´s fusion temperature (535 K) consequently should be identified with the clearing temperature, the corresponding value from Table 2 being 559$\pm$1 K. The latter figure is remarkably higher that that given by Ref. 1, altough meeting rather satisfactorily those reported by Ubbelohde et al. (556 K, Ref. 2) and by Duruz et al. (553 K, Ref. 3).

Therefore, in the evaluator´s opinion, the phase diagram could be more correctly interpreted with reference to Scheme A.2. of the Preface. Accordingly, Sokolov´s eutectic should be identified with an M'_E point.

REFERENCES:

(1) Sokolov, N.M.
Zh. Obshch. Khim. 1954, **24**, 1150-1156.
(2) Ubbelohde, A.R., Michels, H.J., and Duruz, J.J.
Nature 1970, **228**, 50-52.
(3) Duruz, J.J., Michels, H.J., and Ubbelohde, A.R.
Proc. R. Soc. London 1971,**A322**, 281-299.

COMPONENTS:

(1) Sodium **iso**.pentanoate (sodium **iso**.valerate); Na**i**.$C_5H_9O_2$; [539-66-2]
(2) Sodium thiocyanate; NaCNS; [540-72-7]

ORIGINAL MEASUREMENTS:

Sokolov, N.M.
Zh. Obshch. Khim. 1954, **24**, 1150-1156.

VARIABLES:

Temperature.

PREPARED BY:

D'Andrea, G.

EXPERIMENTAL VALUES:

t/°C	T/K[a]	$100x_2$
262	535	0
277	550	5
287	560	10
288	561	15
270	543	25
256	529	30
250	523	32
260	533	35
272	545	40
285	558	45
291	564	50
297	570	55
300	573	60
303	576	65
305	578	75
306	579	90
311	584	100

[a] T/K values calculated by the compiler.

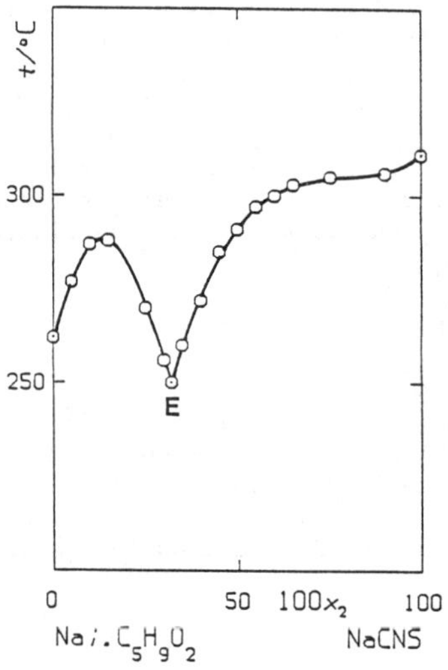

Characteristic point(s):

Eutectic, E, at 250 °C and $100x_2$= 32 (author).

AUXILIARY INFORMATION

METHOD/APPARATUS/PROCEDURE:

Visual polythermal analysis.
Salt(s) melted in a test tube. Temperature measured with a Nichrome-Constantane thermocouple and a millivoltmeter with mirror reading to 17 mV.

SOURCE AND PURITY OF MATERIALS:

Component 1 synthetized from **iso**.pentanoic acid and $NaHCO_3$.
Component 2 of analytical purity recrystallized once from water and once from ethanol.

ESTIMATED ERROR:

Temperature: accuracy probably ±2 K (compiler).

REFERENCES:

COMPONENTS:	EVALUATOR:
(1) Sodium **iso**.pentanoate (sodium **iso**.valerate); $NaI.C_5H_9O_2$; [539-66-2] (2) Sodium nitrite; $NaNO_2$; [7632-00-0]	Spinolo, G., Dipartimento di Chimica Fisica, Universita´ di Pavia (ITALY).

CRITICAL EVALUATION:

This binary was studied only by Sokolov (Ref. 1), who restricted his polythermal investigation to the lower boundary of the isotropic liquid field, and claimed the existence of a eutectic at 542 K (269 °C) and $100\mathbf{x}_2$= 21.

Component 1, however, forms liquid crystals [at $\mathbf{T}_{fus}(1)$= 461.5±0.6 K; Preface, Table 2] before turning into a clear melt. Sokolov´s fusion temperature (535 K) consequently should be identified with the clearing temperature, the corresponding value from Table 2 being 559±1 K. The latter figure is remarkably higher that that given by Ref. 1, although meeting rather satisfactorily those reported by Ubbelohde et al. (556 K, Ref. 2) and by Duruz et al. (553 K, Ref. 3).

Allowance being made for the fact that a liquid-liquid miscibility gap impinges on the liquidus branch richer in the higher melting component ($NaNO_2$), the phase diagram could be more correctly interpreted with reference to Scheme A.2 of the Preface, and Sokolov´s eutectic ought to be identified with an M'_E point.

REFERENCES:

(1) Sokolov, N.M.
Zh. Obshch. Khim. 1957, **27**, 840-844 (*); **Russ. J. Gen. Chem. (Engl. Transl.)** 1957, **27**, 917-920.
(2) Ubbelohde, A.R., Michels, H.J., and Duruz, J.J.
Nature 1970, **228**, 50-52.
(3) Duruz, J.J., Michels, H.J., and Ubbelohde, A.R.
Proc. R. Soc. London 1971, **A322**, 281-299.

COMPONENTS:	ORIGINAL MEASUREMENTS:
(1) Sodium **iso**.pentanoate (sodium **iso**.valerate); Na*i*.$C_5H_9O_2$; [539-66-2] (2) Sodium nitrite; $NaNO_2$; [7632-00-0]	Sokolov, N.M. **Zh. Obshch. Khim.** 1957, 27, 840-844 (*); **Russ. J. Gen. Chem. (Engl. Transl.)** 1957, 27, 917-920.
VARIABLES:	PREPARED BY:
Temperature.	D'Andrea, G.

EXPERIMENTAL VALUES:

t/°C	T/K[a]	$100x_2$
262	535	0
273	546	5
278	551	10
276	549	15
270	543	20
272	545	25
273	546	30
273	546	35
274	547	40
275	548	45
277	550	50
278	551	55
279	552	60
282	555	65

[a] T/K values calculated by the compiler.

Note - Liquid layering occurs at $66 \leq 100x_2 \leq 98.4$ (author).

Characteristic point(s):

Eutectic, E, at 269 °C and $100x_2$= 21 (author).

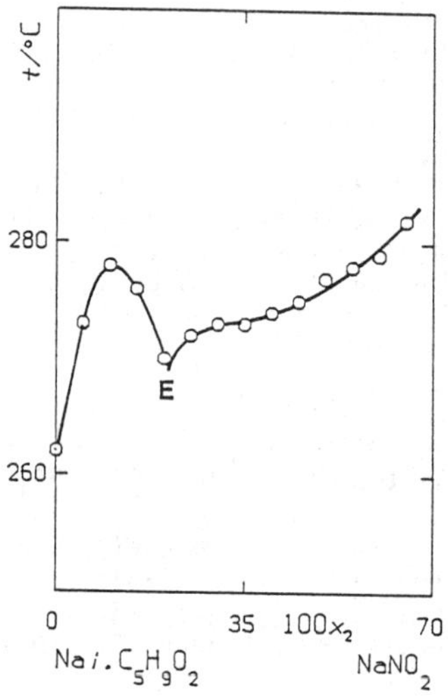

AUXILIARY INFORMATION

METHOD/APPARATUS/PROCEDURE:	SOURCE AND PURITY OF MATERIALS:
Visual polythermal analysis; salt mixtures melted in a glass tube (surrounded by a wider tube) and stirred with a glass stirrer. The temperatures of initial crystallization were measured with a Nichrome-Constantane thermocouple checked at the fusion points of water, benzoic acid, mannitol, $AgNO_3$, Cd, KNO_3, and $K_2Cr_2O_7$.	Component 1: prepared from "chemically pure" sodium hydrogen carbonate (carbonate in the reference quoted by the author; compiler), and commercial **iso**.pentanoic acid distilled before use (Ref. 1); the recovered salt was recrystallized from **n**-butanol. Component 2: "chemically pure" material recrystallized from water; $t_{fus}(2)/°C$= 284.
	ESTIMATED ERROR:
	Temperature: accuracy probably ±2 K (compiler).
	REFERENCES:
	(1) Sokolov, N.M. **Zh. Obshch. Khim.** 1954, 24, 1581-1593.

COMPONENTS:

(1) Sodium **iso**.pentanoate (sodium **iso**.valerate)
$Na\mathbf{i}.C_5H_9O_2$; [539-66-2]
(2) Sodium nitrate;
$NaNO_3$; [7631-99-4]

EVALUATOR:

Ferloni, P.,
Dipartimento di Chimica Fisica,
Universita´ di Pavia (ITALY).

CRITICAL EVALUATION:

This binary was studied by visual polythermal and thermographical analysis by Sokolov (Ref. 1), and Dmitrevkaya and Sokolov (Ref. 2), respectively, with substantially analogous results. The phase diagram was claimed by these authors to be of the eutectic type with the invariant at either 527 K (254 °C) and $100\mathbf{x}_1$= 31 (Ref. 1), or 526 K (253 °C) and $100\mathbf{x}_1$= 31.5 (Ref. 2).

Component 1, however, forms liquid crystals. Consequently, the fusion temperature, $\mathbf{T}_{fus}(1)$= 535 K (262°C; Ref.s 1, 2) should be identified with the clearing temperature, the corresponding value from Table 2 of the Preface being 559±1 K. The latter figure is remarkably higher than that by the above mentioned investigators, and agrees rather satisfactorily with those reported by Ubbelohde et al. (556 K; Ref. 3) and by Duruz et al. (553 K; Ref. 4).

For the same component: (i) the transition at 451 K (178 °C) quoted in Ref. 2 from Ref. 5 should be identified with the actual fusion temperature, the corresponding value from Table 2 being 461.5±0.6 K, whereas (ii) the transition at 425 K (152 °C) also quoted in Ref. 2 from Ref. 5 has no correspondence in Table 2.

Thus the whole phase diagram should be re-interpreted, e.g., with reference to Scheme A.2 of the Preface. In particular, the invariant at 526 K and $100\mathbf{x}_1$= 31.5 should be an M'_E point and not a conventional eutectic.

REFERENCES:

(1) Sokolov, N.M.
Zh. Obshch. Khim. 1954, **24**, 1150-1156.

(2) Dmitrevskaya, O.I.; Sokolov, N.M.
Zh. Obshch. Khim. 1967, **37**, 2160-2166 (*); **Russ. J. Gen. Chem. (Engl. Transl.)** 1967, **37**, 2050-2054.

(3) Ubbelohde, A.R., Michels, H.J., and Duruz, J.J.
Nature 1970, **228**, 50-52.

(4) Duruz, J.J., Michels, H.J., and Ubbelohde, A.R.
Proc. R. Soc. London 1971,**A322**, 281-299.

(5) Sokolov, N.M.
Tezisy Dokl. X Nauch. Konf. S.M.I. 1956.

COMPONENTS:

(1) Sodium **iso**.pentanoate (sodium **iso**.valerate); Na**i**.$C_5H_9O_2$; [539-66-2]
(2) Sodium nitrate; $NaNO_3$; [7631-99-4]

ORIGINAL MEASUREMENTS:

Sokolov, N.M.
Zh. Obshch. Khim. 1954, **24**, 1150-1156.

VARIABLES:

Temperature.

PREPARED BY:

D'Andrea, G.

EXPERIMENTAL VALUES:

$t/^{o}C$	T/K^{a}	$100x_2$
262	535	0
273	546	5
280	553	10
280	553	15
265	538	25
257	530	30
254	527	31.2
260	533	35
272	545	40
282	555	45
288	561	50
294	567	55
299	572	60
302	575	65
304	577	75
306	579	85
307	580	95
308	581	100

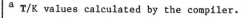

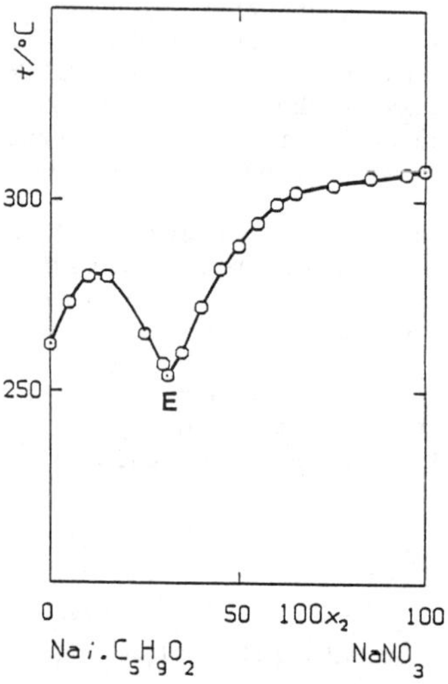

[a] T/K values calculated by the compiler.

Characteristic point(s):

Eutectic, E, at 254 °C and $100x_2$= 31 (author).

AUXILIARY INFORMATION

METHOD/APPARATUS/PROCEDURE:

Visual polythermal analysis.
Salt(s) melted in a test tube. Temperature measured with a Nichrome-Constantane thermocouple and a millivoltmeter with mirror reading to 17 mV.

SOURCE AND PURITY OF MATERIALS:

Component 1 synthetized from **iso**.pentanoic acid and $NaHCO_3$.
Commercial component 2 further purified by the author according to Laiti.

ESTIMATED ERROR:

Temperature: accuracy probably ± 2 K (compiler).

REFERENCES:

COMPONENTS:

(1) Sodium **iso**.pentanoate (sodium **iso**.valerate); $NaiC_5H_9O_2$; [539-66-2]
(2) Sodium nitrate; $NaNO_3$; [7631-99-4]

ORIGINAL MEASUREMENTS:

Dmitrevskaya, O.I.; Sokolov, N.M.
Zh. Obshch. Khim. 1967, 37, 2160-2166 (*);
Russ. J. Gen. Chem. (Engl. Transl.) 1967, 37, 2050-2054.

VARIABLES:

Temperature.

PREPARED BY:

D'Andrea, G.

EXPERIMENTAL VALUES:

$t/^oC$	T/K^a	$100x_2$	$t/^oC$	T/K^a	$100x_2$
262	535	0	120^b	393	31.5
178^b	451	0	282	555	50
152^b	425	0	250^c	523	50
272	545	5	140^b	413	50
250^c	523	5	120^b	393	50
120^b	393	5	300	573	75
90^b	363	5	246^c	519	75
270	543	15	280^b	553	75
256^c	529	15	114^b	387	75
64^b	337	15	305	578	95
116^b	389	15	276^b	549	95
92^b	365	15	110^b	383	95
253	526	31.5	308	581	100
253^c	526	31.5	275^b	548	100

[a] T/K values calculated by the compiler.
[b] Transformation in the solid state.
[c] Eutectic temperature.

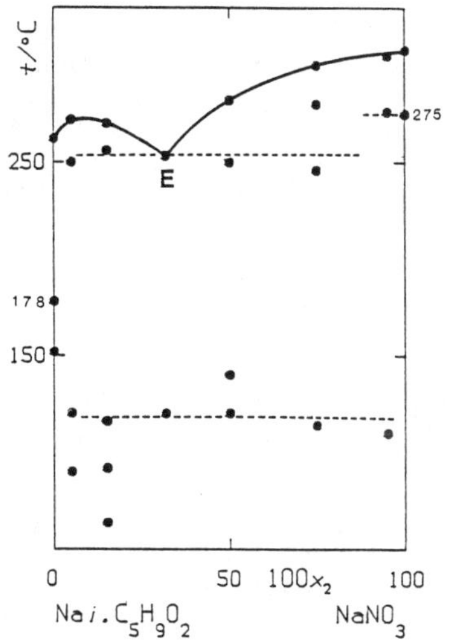

Characteristic point(s):

Note - The present thermographical data supplement the previous visual polythermal investigation by Sokolov (Ref. 1).

Eutectic, E, at 253 °C and $100x_2$= 31.5.

AUXILIARY INFORMATION

METHOD/APPARATUS/PROCEDURE:

Thermographical analysis (heating curves recorded automatically).

SOURCE AND PURITY OF MATERIALS:

Component 1: synthetized from **iso**.pentanoic acid and the carbonate (Ref. 2).
Component 2: "chemically pure" material recrystallized.
Component 1 undergoes phase transitions at $t_{trs}(1)/^oC$= 152, 178 (Ref. 3).
Component 2 undergoes a phase transition at $t_{trs}(2)/^oC$= 275 (current literature).

ESTIMATED ERROR:

Temperature: accuracy probably ±2 K (compiler).

REFERENCES:
(1) Sokolov, N.M.
Zh. Obshch. Khim. 1954, 24, 1150-1156.
(2) Sokolov, N.M.
Zh. Obshch. Khim. 1954, 24, 1581-1593.
(3) Sokolov, N.M.
Tezisy Dokl. X Nauch. Konf. S.M.I. 1956.

COMPONENTS:	EVALUATOR:
(1) Sodium hexanoate (sodium caproate); $NaC_6H_{11}O_2$; [10051-44-2] (2) Sodium benzoate; $NaC_7H_5O_2$; [532-32-1]	Spinolo, G., Dipartimento di Chimica Fisica, Universita´ di Pavia (ITALY).

CRITICAL EVALUATION:

This binary was studied only by Sokolov (Ref. 1), who restricted his polythermal investigation to the lower boundary of the isotropic liquid field, and claimed the existence of a "perekhodnaya tochka" (P) at 644 K (371 °C) and $100\mathbf{x}_2$= 13.

Component 1, however, forms liquid crystals [above $\mathbf{T}_{fus}(1)$= 499.6±0.6 K; Table 1 of the Preface] before turning into a clear melt. Sokolov´s fusion temperature (638 K) should be consequently identified with the clearing temperature, the corresponding value from Table 1 being 639.0±0.5 K.

Sokolov´s P point at $100\mathbf{x}_2$= 13 corresponds to a slightly marked minimum of the data listed in Ref. 1: the experimental temperature values at $5 \leq 100\mathbf{x}_2 \leq 15$ actually range between 642 and 647 K, i.e. approximately within the accuracy limits estimated by the compiler.

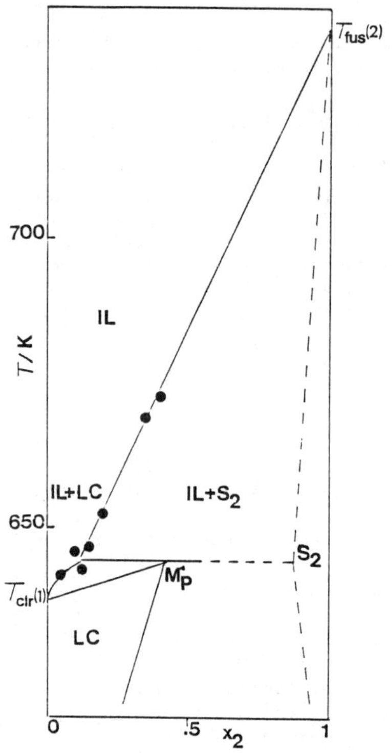

If the temperature differences between the maximum at 646 K (and $100\mathbf{x}_2$= 10) and the P point at 644 K is thought to be meaningful, the phase diagram could be interpreted with reference to Scheme A.2 of the Preface: accordingly, Sokolov´s invariant should be identified with an M'_E point).

If, on the contrary, the above mentioned temperature difference is thought to be meaningless, reference can be made to the front figure, where Sokolov´s data are reported. In this case a peritectic equilibrium should exist (at about 644 K) among a liquid crystal, an isotropic liquid and a solid crystal. Accordingly, Sokolov´s P point should be identified with an M'_P point, and a further invariant should exist, e.g. an M_E at $\mathbf{T} \leq \mathbf{T}_{fus}(1)$.

REFERENCES:

(1) Sokolov, N.M.
Zh. Obshch. Khim. 1954, **24**, 1581-1593.

COMPONENTS:

(1) Sodium hexanoate (sodium caproate); $NaC_6H_{11}O_2$; [10051-44-2]
(2) Sodium benzoate; $NaC_7H_5O_2$; [532-32-1]

ORIGINAL MEASUREMENTS:

Sokolov, N.M.
Zh. Obshch. Khim. 1954, **24**, 1581-1593.

VARIABLES:

Temperature.

PREPARED BY:

D'Andrea, G.

EXPERIMENTAL VALUES:

$t/^oC$	T/K^a	$100x_2$
365	638	0
369	642	5
373	646	10
371	644	13
374	647	15
380	653	20
396	669	35
400	673	40
463	736	100

[a] T/K values calculated by the compiler.

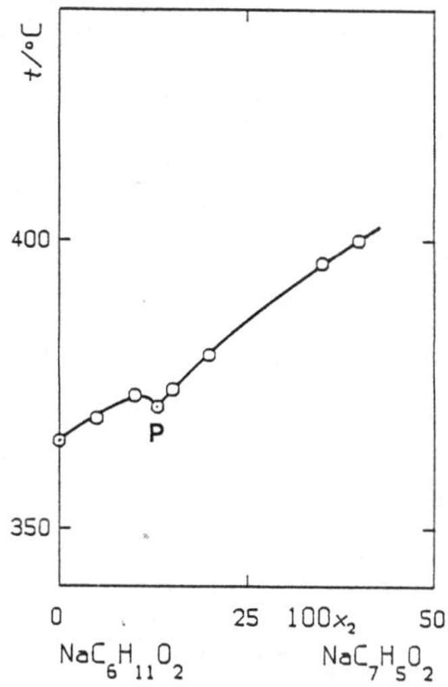

Characteristic point(s):

Characteristic point, P ("perekhodnaya tochka" in the original text; see the Introduction), at 371 ^oC and $100x_2$= 13 (author).

Note - The system was investigated between $0 \leq 100x_2 \leq 40$ due to thermal instability of the hexanoate.

AUXILIARY INFORMATION

METHOD/APPARATUS/PROCEDURE:

Visual polythermal analysis. Melts contained in a glass tube and stirred. Temperatures measured with a Nichrome-Constantane thermocouple and a 17 mV full scale millivoltmeter. The temperature readings refer to the disappearance of isotropicity in the melt on cooling.

SOURCE AND PURITY OF MATERIALS:

Component 1: prepared by reacting aqueous ("chemically pure") Na_2CO_3 with a slight excess of hexanoic acid of analytical purity. The solvent and excess acid were removed by heating to 160 ^oC. Component 2: "chemically pure" material.

ESTIMATED ERROR:

Temperature: accuracy probably ± 2 K (compiler).

REFERENCES:

COMPONENTS:

(1) Sodium hexanoate (sodium caproate); Na $C_6H_{11}O_2$; [10051-44-2]
(2) Sodium octadecanoate (sodium stearate); Na $C_{18}H_{35}O_2$; [822-16-2]

EVALUATOR:

Ferloni, P.,
Dipartimento di Chimica Fisica,
Universita´ di Pavia (ITALY).

CRITICAL EVALUATION:

This system was studied only by Sokolov (Ref. 1) who employed the visual polythermal analysis to draw the lower boundary of the isotropic liquid field. From the shape of this boundary, he concluded that the intermediate compound $Na_5(C_6H_{11}O_2)_2(C_{18}H_{35}O_2)_3$ [congruently melting at 602 K (329 °C)] was formed, and that the limits of the stability field of this compound were a eutectic at 512 K (239 °C) and $100\mathbf{x}_2$= 17.5, and a "perekhodnaya tochka" at 587 K (314 °C) and $100\mathbf{x}_2$= 94.5.

Actually, both components form liquid crystals, the liquid crystalline phases being one for component 1 (see Table 1 of the Preface), and two for component 2 (see Table 4 of the Preface). Sokolov´s fusion temperatures, $\mathbf{T}_{fus}(1)$= 638 K (365 °C), and $\mathbf{T}_{fus}(2)$= 581 K (308 °C), are consequently to be identified with the clearing temperatures, the corresponding values from Tables 1 and 4 being 639.0±0.5 and 552.7 K, respectively.

Since the complete topology of the binary can hardly be interpreted from the data available, it is more realistic to list here the few points which, in the evaluator´s opinion, seem to be sufficiently reliable.

(i) At intermediate compositions it seems reasonable to assume that a continuous series of liquid crystal solutions is formed, with an azeotrope at 602 K and $100\mathbf{x}_2$= 60.

(ii) Accordingly, the left hand section ($0 \le 100\mathbf{x}_2 \le 60$) of the phase diagram might be interpreted with reference to Scheme C.2 of the Preface: in this case, Sokolov´s eutectic should be intended as an M''_E point.

Conversely, no definite interpretation of the phase diagram at high $100\mathbf{x}_2$ values seems possible. Indeed, it is not clear how Sokolov could argue the occurrence of an invariant (the "perekhodnaya tochka" at $100\mathbf{x}_2$= 94.5) from the trend of his experimental data which does not unambiguously support any significant slope change of the curve in this region. Moreover, Sokolov´s "fusion" temperature of component 2 (581 K) looks as fully unreliable, being 18 K higher than the second highest $\mathbf{T}_{clr}$ value determined during the last 30 years (Ref. 2), and 28 K higher than the clearing temperature listed in Table 4.

REFERENCES:

(1) Sokolov, N.M.
Zh. Obshch. Khim. 1954, **24**, 1581-1593.
(2) Sanesi, M.; Cingolani, A.; Tonelli, P.L.; Franzosini, P.
Thermal Properties, in **Thermodynamic and Transport Properties of Organic Salts**, IUPAC Chemical Data Series No. 28 (Franzosini, P.; Sanesi, M.; Editors) Pergamon Press, Oxford, 1980, 29-115.

COMPONENTS:

(1) Sodium hexanoate (sodium caproate); $NaC_6H_{11}O_2$; [10051-44-2]
(2) Sodium octadecanoate (sodium stearate); $NaC_{18}H_{35}O_2$; [822-16-2]

ORIGINAL MEASUREMENTS:

Sokolov, N.M.
Zh. Obshch. Khim. 1954, **24**, 1581-1593.

VARIABLES:

Temperature.

PREPARED BY:

D'Andrea, G.

EXPERIMENTAL VALUES:

t/°C	T/K[a]	$100x_2$	t/°C	T/K[a]	$100x_2$
365	638	0	326	599	55
320	593	5	329	602	60
272	545	10	328	601	65
242	515	15	327	600	70
239	512	17.5	326	599	75
248	521	20	324	597	80
264	537	25	321	594	85
280	553	30	319	592	90
293	566	35	314	587	94.5
305	578	40	316	589	95
313	586	45	308	581	100
320	593	50			

[a] T/K values calculated by the compiler.

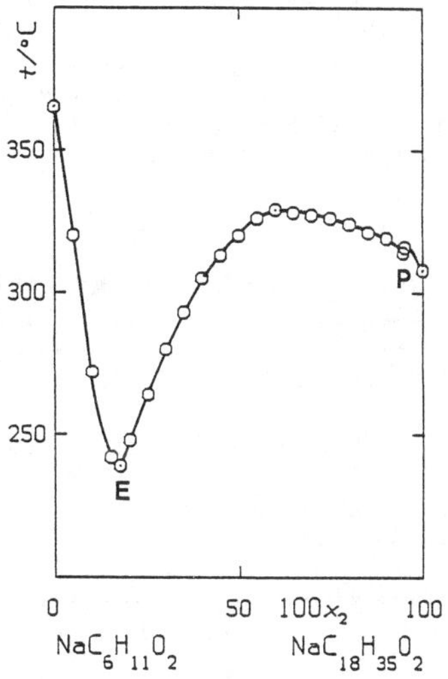

Characteristic point(s):

Characteristic point, P ("perekhodnaya tochka" in the original text; see the Introduction), at 314 °C and $100x_2$= 94.5 (author).
Eutectic, E, at 239 °C and $100x_2$= 17.5 (author).

Intermediate compound(s):

$Na_5(C_6H_{11}O_2)_2(C_{18}H_{35}O_2)_3$ (author), congruently melting at 329 °C (compiler).

AUXILIARY INFORMATION

METHOD/APPARATUS/PROCEDURE:

Visual polythermal analysis. Melts contained in a glass tube and stirred. Temperatures measured with a Nichrome-Constantane thermocouple and a 17 mV full scale millivoltmeter. The temperature readings refer to the disappearance of isotropicity in the melt on cooling.

SOURCE AND PURITY OF MATERIALS:

Component 1: prepared by reacting aqueous ("chemically pure") Na_2CO_3 with a slight excess of **n**-hexanoic acid of analytical purity. The solvent and excess acid were removed by heating to 160 °C. Component 2: "chemically pure" material.

ESTIMATED ERROR:

Temperature: accuracy probably ±2 K (compiler).

REFERENCES:

COMPONENTS:

(1) Sodium hexanoate (sodium caproate); $NaC_6H_{11}O_2$; [10051-44-2]
(2) Sodium thiocyanate; NaCNS; [540-72-7]

EVALUATOR:

Spinolo, G.,
Dipartimento di Chimica Fisica,
Universita´ di Pavia (ITALY).

CRITICAL EVALUATION:

This binary was studied only by Sokolov (Ref. 1), who restricted his polythermal investigation to the lower boundary of the isotropic liquid field, and claimed the existence of a eutectic at 568 K (295 °C) and $100x_2$= 63.

Component 1, however, forms liquid crystals [at $T_{fus}(1)$= 499.6±0.6 K; Preface, Table 1] before turning into a clear melt. Sokolov's fusion temperature (638 K) is consequently to be identified with the clearing temperature, the corresponding value from Table 1 being 639.0±0.5 K.

Therefore, in the evaluator's opinion, the phase diagram could be more correctly interpreted with reference to Scheme A.2 of the Preface, and Sokolov's eutectic ought to be intended as an M'_E point.

REFERENCES:

(1) Sokolov, N.M.; **Zh. Obshch. Khim.** 1954, **24**, 1150-1156.

COMPONENTS:

(1) Sodium hexanoate (sodium caproate); $NaC_6H_{11}O_2$; [10051-44-2]
(2) Sodium thiocyanate; NaCNS; [540-72-7]

ORIGINAL MEASUREMENTS:

Sokolov, N.M.
Zh. Obshch. Khim. 1954, **24**, 1150-1156.

VARIABLES:

Temperature.

PREPARED BY:

D'Andrea, G.

EXPERIMENTAL VALUES:

t/°C	T/K[a]	$100x_2$	t/°C	T/K[a]	$100x_2$
365	638	0	330	603	50
383	656	5	316	589	55
397	670	10	305	578	60
399	672	15	295	568	63
386	659	25	297	570	65
377	650	30	304	577	75
366	639	35	307	580	90
352	625	40	311	584	100
342	615	45			

[a] T/K values calculated by the compiler.

Characteristic point(s):

Eutectic, E, at 295 °C and $100x_2$= 63 (author).

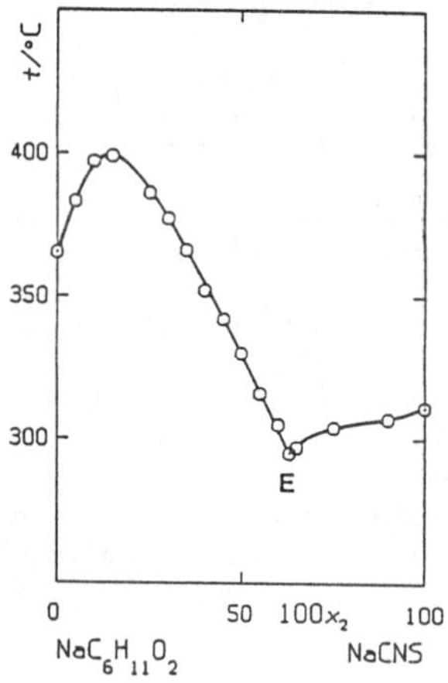

AUXILIARY INFORMATION

METHOD/APPARATUS/PROCEDURE:

Visual polythermal analysis. Salt(s) melted in a test tube. Temperature measured with a Nichrome-Constantane thermocouple and a millivoltmeter with mirror reading to 17 mV.

SOURCE AND PURITY OF MATERIALS:

Component 1 synthetized from n-hexanoic acid and $NaHCO_3$. Component 2 of analytical purity recrystallized once from water and once from ethanol.

ESTIMATED ERROR:

Temperature: accuracy probably ±2 K (compiler).

COMPONENTS:

(1) Sodium hexanoate (sodium caproate); $NaC_6H_{11}O_2$; [10051-44-2]
(2) Sodium nitrate; $NaNO_3$; [7631-99-4]

EVALUATOR:

Ferloni, P.,
Dipartimento di Chimica Fisica,
Universita' di Pavia (ITALY).

CRITICAL EVALUATION:

This binary was studied only by Sokolov (Ref. 1), who restricted his polythermal investigation to the lower boundary of the isotropic liquid field, and claimed the existence of a eutectic at 560 K (287 °C) and $100x_2$= 56.5, and the occurrence of liquid layering at 576 K (302 °C) and $100x_2$> 60.

Component 1, however, forms liquid crystals [at $T_{fus}(1)$= 499.6±0.6 K; Preface, Table 1] before turning into a clear melt. Sokolov's fusion temperature (638 K) is consequently to be identified with the clearing temperature, the corresponding value from Table 1 being 639.0±0.5 K. Therefore, in the evaluator's opinion, the phase diagram could be more correctly interpreted with reference to Scheme A.2 of the Preface, allowance being made for the fact that a liquid-liquid miscibility gap impinges on the liquidus branch richer in the higher melting component ($NaNO_3$). Consequently, Sokolov's eutectic should be an M'_E point.

REFERENCES:

(1) Sokolov, N.M.; **Zh. Obshch. Khim.** 1954, 24, 1150-1156.

COMPONENTS:

(1) Sodium hexanoate (sodium caproate); $NaC_6H_{11}O_2$; [10051-44-2]
(2) Sodium nitrate; $NaNO_3$; [7631-99-4]

ORIGINAL MEASUREMENTS:

Sokolov, N.M.
Zh. Obshch. Khim. 1954, 24, 1150-1156.

VARIABLES:

Temperature.

PREPARED BY:

D'Andrea, G.

EXPERIMENTAL VALUES:

t/°C	T/K[a]	$100x_2$
365	638	0
376	649	5
383	656	10
385	658	15
375	648	25
367	640	30
357	630	35
342	615	40
326	599	45
313	586	50
296	569	55
287	560	56.5
300	573	59
302	575	60

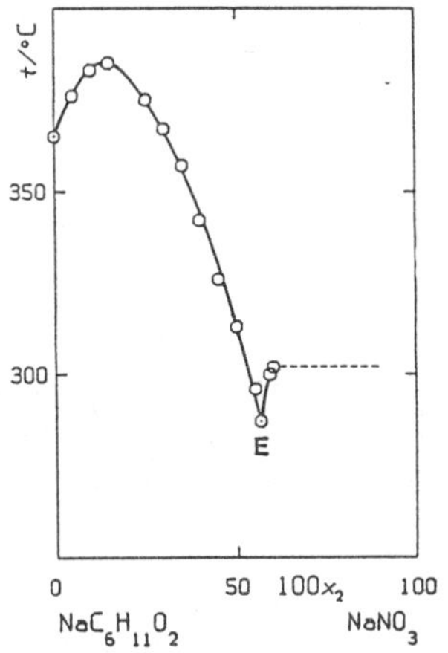

[a] T/K values calculated by the compiler.
Characteristic point(s): Eutectic, E, at 287 °C and $100x_2$= 56.5 (author).

AUXILIARY INFORMATION

METHOD/APPARATUS/PROCEDURE:

Visual polythermal analysis. Salt(s) melted in a test tube. Temperature measured with a Nichrome-Constantane thermocouple and a millivoltmeter with mirror reading to 17 mV.

NOTE:

At $100x_2$ > 60, liquid layering occurs.

SOURCE AND PURITY OF MATERIALS:

Component 1 synthetized from n-hexanoic acid and $NaHCO_3$. Commercial component 2 further purified by the author according to Laiti.

ESTIMATED ERROR:

Temperature: accuracy probably ±2 K (compiler).

COMPONENTS:	EVALUATOR:
(1) Sodium benzoate; $NaC_7H_5O_2$; [532-32-1] (2) Sodium octadecanoate (sodium stearate); $NaC_{18}H_{35}O_2$; [822-16-2]	Ferloni, P. Dipartimento di Chimica Fisica. Universita´ di Pavia (ITALY).

CRITICAL EVALUATION:

This binary was studied only by Sokolov (Ref. 1) who reported a phase diagram of the eutectic type with the invariant at 574 K (301 °C) and $100\mathbf{x}_1$= 1.3.

Component 2, however, forms liquid crystals. Thence, the fusion temperature by Sokolov, viz., $\mathbf{T}_{fus}(2)$= 581 K (308 °C), should be intended as a clearing temperature and compared with the $T_{clr}(2)$ value reported in Table 4 (552.7 K). It is to be stressed that Sokolov´s "fusion" temperature looks as fully unreliable, being 18 K higher than the second highest $\mathbf{T}_{clr}$ value determined during the last 30 years (Ref. 2), and 28 K higher than the clearing temperature listed in Table 4.

The whole phase diagram is therefore to be reconsidered.

REFERENCES:

(1) Sokolov, N.M.
Zh. Obshch. Khim. 1954, **24**, 1581-1593.

(2) Sanesi, M.; Cingolani, A.; Tonelli, P.L.; Franzosini, P.
Thermal Properties, in **Thermodynamic and Transport Properties of Organic Salts**, IUPAC Chemical Data Series No. 28 (Franzosini, P.; Sanesi, M.; Editors), Pergamon Press, Oxford, 1980, 29-115.

COMPONENTS:	ORIGINAL MEASUREMENTS:
(1) Sodium benzoate; $NaC_7H_5O_2$; [532-32-1] (2) Sodium octadecanoate (sodium stearate); $NaC_{18}H_{35}O_2$; [822-16-2]	Sokolov, N.M. **Zh. Obshch. Khim.** 1954, **24**, 1581-1593.
VARIABLES:	PREPARED BY:
Temperature.	D'Andrea, G.

EXPERIMENTAL VALUES:

t/°C	T/K[a]	100x_1
308	581	0
301	574	1.3
310	583	5
321	594	10
332	605	15
344	617	20
353	626	24
362	635	30
369	642	35
376	649	40
384	657	45
390	663	50
396	669	55
463	736	100

[a] T/K values calculated by the compiler.

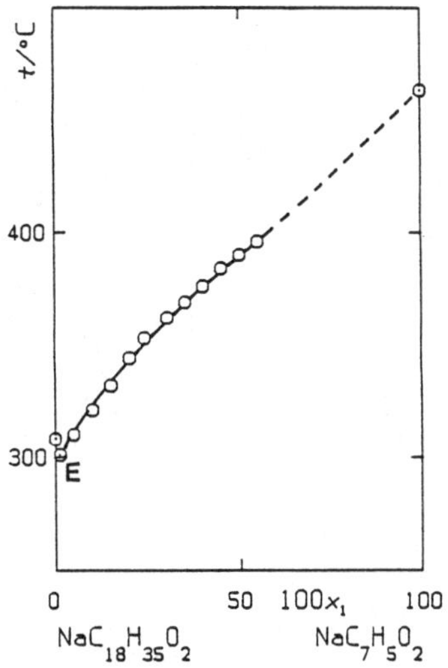

Characteristic point(s):

Eutectic, E, at 301 °C and 100x_1= 1.3 (author).

Note - The system was investigated at $0 \leq 100x_1 \leq 55$ due to thermal instability of the octadecanoate.

AUXILIARY INFORMATION

METHOD/APPARATUS/PROCEDURE:

Visual polythermal analysis. Melts contained in a glass tube and stirred. Temperatures measured with a Nichrome-Constantane thermocouple and a 17 mV full scale millivoltmeter. The temperature readings refer to the disappearance of isotropicity in the melt on cooling.

NOTE:

Component 2 forms liquid crystals. Thence, the fusion temperature by Sokolov, viz., $T_{fus}(2)$= 581 K (308 °C), should be intended as a clearing temperature and compared with the $T_{clr}(2)$ value (552.7 K) reported on Preface, Table 4. It is to be stressed that Sokolov's "fusion" temperature looks as fully unreliable, being 18 K higher than the second highest T_{clr} value determined during the last 30 years (Ref. 1), and 28 K higher than the clearing temperature listed in Table 4. The whole phase diagram is therefore to be reconsidered.

SOURCE AND PURITY OF MATERIALS:

"Chemically pure" materials.

ESTIMATED ERROR:

Temperature: accuracy probably ±2 K (compiler).

REFERENCES:

(1) Sanesi, M.; Cingolani, A.; Tonelli, P.L.; Franzosini, P. **Thermal Properties**, in **Thermodynamic and Transport Properties of Organic Salts**, IUPAC Chemical Data Series No. 28 (Franzosini, P.; Sanesi, M.; Editors), Pergamon Press, Oxford, 1980, 29-115.

COMPONENTS:

(1) Rubidium ethanoate (rubidium acetate); $RbC_2H_3O_2$; [563-67-7]
(2) Rubidium nitrate; $RbNO_3$; [13126-12-0]

EVALUATOR:

Ferloni, P.,
Dipartimento di Chimica Fisica,
Universita' di Pavia (ITALY).

CRITICAL EVALUATION:

This binary was studied for the first time by Gimel'shtein and Diogenov (Ref. 1) who reported the lower boundary of the isotropic liquid region in the reciprocal ternary $Na,Rb/C_2H_3O_2,NO_3$ on the basis of visual polythermal observations. They claimed the existence of the intermediate compound $Rb_3(C_2H_3O_2)_2NO_3$ (congruently melting at 475 K [202 °C]) and of two eutectics, E_1, at 471 K (198 °C) and $100\mathbf{x}_1 = 81.5$, and E_2, at 454 K (181 °C) and $100\mathbf{x}_1 = 35.5$.

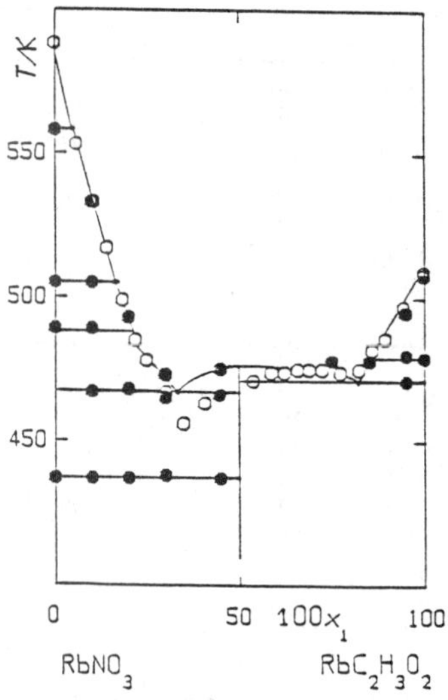

In a subsequent paper on the reciprocal ternary $Cs,Rb/C_2H_3O_2,NO_3$ (Ref. 2), the same Authors reported refined values of the coordinates of the second eutectic (467 K [194 °C] and $100\mathbf{x}_1 = 33$), obtained with the same experimental approach. The new data suggest that those reported in Ref. 1 should be affected by a systematic error (as high as 13 K) in the composition range $20 \leq 100\mathbf{x}_1 \leq 50$.

The system was once more investigated by Gimel'shtein (Ref. 3) who directly measured with D.T.A. the temperatures of the characteristic points previously inferred by the shape of the liquidus. Besides a substantial agreement with the findings of Ref. 2, he reported two more solid state transitions of pure $RbNO_3$, at 437 and 505 K. Moreover, the existence of an intermediate compound was more reliably proved by a comparison of the X-ray powder pattern of an intermediate mixture, which showed diffraction lines not pertinent to either pure component.

Finally, the composition of the compound was adjusted by Diogenov, Erlykov and Gimel'shtein (Ref. 4) during an investigation of the reciprocal ternary $Li,Rb/C_2H_3O_2,NO_3$ with coupled visual polythermal and thermographical analysis. According to this paper, the compound has a 1:1 composition and congruently melts at 476 K (203 °C) and the eutectics are at 467 K (194 °C) and $100\mathbf{x}_1 = 33.5$ and at 471 K (198 °C) and $100\mathbf{x}_1 = 82$, respectively.

In spite of the lack of any comment by the Authors on their previous results, the evaluator is inclined to recommend the last interpretation as the more reliable. The figure reports phase boundaries drawn accordingly. On the same figure, the two sets of experimental data, available in numerical form (the visual polythermal data from Ref. 1 and the thermographical data from Ref. 3), have also been plotted as open and full circles, respectively. As a further remark, the melting and transition points of component 1 reported on Ref.s 1-4 reasonably match the most recent literature data: $\mathbf{T}_{fus}(1) = 510$, 515, 514, 514, 509; $\mathbf{T}_{trs}(1) = 498$ (Ref. 5). For what concerns the phase transitions of component 2, reference can be made to the recent work by Kennedy et al. (Ref. 6 and the bibliography therein quoted).

REFERENCES:

(1) Gimel'shtein, V.G.; Diogenov, G.G.; **Zh. Neorg. Khim.** 1958, **3**, 1644-1649 (*); **Russ. J. Inorg. Chem. (Engl. Transl.)** 1958, **3** (7), 230-237.
(2) Diogenov, G.G.; Gimel'shtein, V.G.; **Zh. Neorg. Khim.** 1966, **11**, 207-209 (*); **Russ. J. Inorg. Chem. (Engl. Transl.)** 1966, **11**, 113-114.
(3) Gimel'shtein, V.G.; **Tr. Irkutsk. Politekh. Inst.** 1971, No. 66, 80-100.
(4) Diogenov, G.G.; Erlykov, A.M.; Gimel'shtein, V.G.; **Zh. Neorg. Khim.** 1974, **19**, 1955-1960; **Russ. J. Inorg. Chem. (Engl. Transl.)** 1974, **19**, 1069-1073 (*).
(5) Sanesi, M.; Cingolani, A.; Tonelli, P.L.; Franzosini, P.; **Thermal Properties**, in **Thermodynamic and Transport Properties of Organic Salts**, IUPAC Chemical Data Series No. 28 (Franzosini, P.; Sanesi, M.; Editors), Pergamon Press, Oxford, 1980, 29-115.
(6) Kennedy, S.W.; Kriven, W.M.; **J. Mater. Sci.** 1976, **11**, 1767-1769.

COMPONENTS:

(1) Rubidium ethanoate (rubidium acetate); $RbC_2H_3O_2$; [563-67-7]
(2) Rubidium nitrate; $RbNO_3$; [13126-12-0]

ORIGINAL MEASUREMENTS:

Gimel´shtein, V.G.; Diogenov, G.G.
Zh. Neorg. Khim. 1958, 3, 1644-1649 (*);
Russ. J. Inorg. Chem. (Engl. Transl.) 1958, 3 (7), 230-237.

VARIABLES:

Temperature

PREPARED BY:

D´Andrea, G.

EXPERIMENTAL VALUES:

t/°C	T/K[a]	$100x_1$	t/°C	T/K[a]	$100x_1$	t/°C	T/K[a]	$100x_1$	t/°C	T/K[a]	$100x_1$
317	590	0	212	485	22	201	474	59	202	475	82.5
300	573	3	205	478	25	201	474	62.5	209	482	86
280	553	6	194	467	30.2	202	475	66	213	486	89.5
260	533	10.5	183	456	35	202	475	69	224	497	94.5
244	517	14.3	190	463	41	202	475	72.5	236	509	100
226	499	18.5	198	471	54	201	474	77.5			

[a] T/K values calculated by the compiler.

Characteristic point(s): Eutectic, E_1, at 198 °C and $100x_1$ = 81.5 (authors).
Eutectic, E_2, at 181 °C and $100x_1$ = 35.5 (authors).

Intermediate compound: $Rb_3(C_2H_3O_2)_2NO_3$, congruently melting at 202 °C (authors).

METHOD/APPARATUS/PROCEDURE:

Visual polythermal method. Temperatures measured with a Chromel-Alumel thermocouple and a 17-mV-range millivoltmeter. Mixtures being hygroscopic, the method of additions with determination of the sample mass by difference was employed in order to avoid hydration.

SOURCE AND PURITY OF MATERIALS:

Not stated. Component 1 undergoes a phase transition at t/°C= 216. Component 2 undergoes phase transitions at t/°C= 210, 290.

ESTIMATED ERROR:

Temperature: accuracy probably ±2 K (compiler).

COMPONENTS:

(1) Rubidium ethanoate (rubidium acetate); $RbC_2H_3O_2$; [563-67-7]
(2) Rubidium nitrate; $RbNO_3$; [13126-12-0]

ORIGINAL MEASUREMENTS:

Diogenov, G.G.; Gimel´shtein, V.G.
Zh. Neorg. Khim. 1966, 11, 207-209 (*);
Russ. J. Inorg. Chem. (Engl. Transl.) 1966, **11**, 113-114.

VARIABLES:

Temperature

PREPARED BY:

D´Andrea, G.

EXPERIMENTAL VALUES:

The paper reports - **inter alia** - on a refinement of the title binary, previously studied by the same authors (Ref. 1). According to the present investigation, the coordinates of the second eutectic are:

Eutectic, E_2, at 194 °C and $100x_1$ = 33 (authors).

METHOD/APPARATUS/PROCEDURE:

Visual polythermal method. Temperatures measured with a Chromel-Alumel thermocouple.

SOURCE AND PURITY OF MATERIALS:

Not stated.

ESTIMATED ERROR:

Temperature: accuracy probably ±2 K (compiler).

REFERENCES:

(1) Gimel´shtein, V.G.; Diogenov, G.G. **Zh. Neorg. Khim.** 1958, 3, 1644-1649 (*); **Russ. J. Inorg. Chem. (Engl. Transl.)**, 1958, 3 (7), 230-237.

COMPONENTS:

(1) Rubidium ethanoate (rubidium acetate); $RbC_2H_3O_2$; [563-67-7]
(2) Rubidium nitrate; $RbNO_3$; [13126-12-0]

ORIGINAL MEASUREMENTS:

Gimel'shtein, V.G.
Tr. Irkutsk. Politekh. Inst. 1971, No. 66, 80-100.

VARIABLES:

Temperature.

PREPARED BY:

D'Andrea, G.

EXPERIMENTAL VALUES:

$t/^{o}C$	T/K^{a}	$100x_2$	$t/^{o}C$	T/K^{a}	$100x_2$	$t/^{o}C$	T/K^{a}	$100x_2$	$t/^{o}C$	T/K^{a}	$100x_2$
235	508	0	202	475	55.0	195	468	80.0	164	437	90.0
206	479	0	193	466	55.0	164	437	80.0	315	588	100
222	495	5.0	164	437	55.0	260	533	90.0	285	558	100
207	480	5.0	200	473	70.0	232	505	90.0	232	505	100
198	471	5.0	192	465	70.0	216	489	90.0	216	489	100
205	478	15.0	165	438	70.0	194	467	90.0	164	437	100
205	478	25.0	220	493	80.0						

[a] T/K values calculated by the compiler.

The meaning of the data listed in the table becomes apparent by observing the figure reported in the critical evaluation.

Characteristic point(s):

Eutectic, E_1, at 198 ^{o}C (composition not reported) (author).
Eutectic, E_2, at 194 ^{o}C (composition not reported) (author).

Intermediate compound(s):

$Rb_3(C_2H_3O_2)_2NO_3$, congruently melting at 202 ^{o}C (author), and undergoing a phase transition at 65 ^{o}C (author).

AUXILIARY INFORMATION

METHOD/APPARATUS/PROCEDURE:

Differential thermal analysis (using a derivatograph with automatic recording of the heating curves) and room temperature X-ray diffractometry (using a URS-501M apparatus) were employed. X-ray patterns were taken at $100x_2$= 40.

SOURCE AND PURITY OF MATERIALS:

Not stated.
Component 1 melts at $t_{fus}/^{o}C$= 235, and undergoes a phase transition at $t/^{o}C$= 206.
Component 2 melts at $t_{fus}/^{o}C$= 315, and undergoes phase transitions at $t/^{o}C$= 164, 216, 232, 285.

ESTIMATED ERROR:

Temperature: accuracy probably ±2 K (compiler).

REFERENCES:

COMPONENTS:

(1) Rubidium ethanoate (rubidium acetate); $RbC_2H_3O_2$; [563-67-7]
(2) Rubidium nitrate; $RbNO_3$; [13126-12-0]

ORIGINAL MEASUREMENTS:

Diogenov, G.G.; Erlykov, A.M.; Gimel´shtein, V.G.
Zh. Neorg. Khim. 1974, **19**, 1955-1960; **Russ. J. Inorg. Chem. (Engl. Transl.)** 1974, **19**, 1069-1073 (*).

VARIABLES:

Temperature.

PREPARED BY:

D´Andrea, G.

EXPERIMENTAL VALUES:

Characteristic point(s):

Eutectic, E_1, at 198 ^oC and $100\mathbf{x}_2$= 18 (authors).
Eutectic, E_2, at 194 ^oC and $100\mathbf{x}_2$= 66.5 (authors).

Intermediate compound(s):

$Rb_2C_2H_3O_2NO_3$, congruently melting at 203 ^oC (authors).

AUXILIARY INFORMATION

METHOD/APPARATUS/PROCEDURE:

The data were obtained by visual polythermal and thermographic methods, supplemented with a few X-ray diffraction patterns.

SOURCE AND PURITY OF MATERIALS:

Not stated. Component 1 undergoes a phase transition at $\mathbf{t}/^oC$= 206 and melts at $\mathbf{t}_{fus}/^oC$= 236.
Component 2 melts at $\mathbf{t}_{fus}/^oC$= 317.

ESTIMATED ERROR:

Temperature: accuracy probably $\pm$2 K (compiler).

REFERENCES:

SYSTEM INDEX

Page numbers preceeded by E refer to evaluation texts whereas those not preceeded by E refer to compiled tables.

REGISTRY NUMBER INDEX

Page numbers preceeded by E refer to evaluation texts whereas those not preceeded by E refer to compiled tables.

AUTHOR INDEX

Page numbers preceded by E refer to evaluation texts whereas page numbers not preceded by E refer to compiled tables.

SOLUBILITY DATA SERIES